W9-ANK-236

Algebraic Varieties

Canadian Mathematical Society
Conference Proceedings
Volume 2, Part 2 (1982)

FIXED POINT FORMULA FOR SINGULAR VARIETIES

Paul Baum

Dedicated to my algebraic topology teacher John C. Moore

This expository note reviews the fixed point formulas of Lefschetz [17] and Atiyah-Bott [3]. Then an extension of the Atiyah-Bott formula to singular varieties [13] is stated. This extension is joint work with W. Fulton and G. Quart. The main problem in formulating the extension is giving an explicit description for the local contribution of a fixed point. Two descriptions of the local contribution will be given: one based on evaluating a power series $Q(t)$ at $t = 1$, and another based on the notion of equivariant intersection number. The local numbers so defined are equal. Each of the two points of view leads to a quite different proof of the fixed point formula.

Thanks go to my co-workers W. Fulton and G. Quart, and also to D. Kazhdan, A. Landman, and Y. L. L. Tong for stimulating and enlightening discussions.

1980 Mathematics Subject Classification. 55N15, 55M20, 58G10, 14C17, 14F12.

Key Words and Phrases. Lefschetz fixed point formula, Atiyah-Bott fixed point formula, Holomorphic Lefschetz number, Bott periodicity, equivariant intersection number.

Research partially supported by a National Science Foundation grant.

© 1982 American Mathematical Society
0731-1036/82/0000-0457/$06.00

§1. Lefschetz fixed point formula

Let M be a compact oriented C^∞ manifold. Suppose that $f : M \longrightarrow M$ is a C^∞ map. Recall that the Lefschetz number $L(f)$ is:

(1.1) $$L(f) = \sum_{i=0}^{n} (-1)^i \mathrm{Trace}[f^* : H^i(M,\mathbb{Q}) \longrightarrow H^i(M;\mathbb{Q})]$$

In (1.1) $n = \dim_{\mathbb{R}} M$ and $H^i(M;\mathbb{Q})$ is the i-th singular cohomology group of M with rational coefficients.

A fixed point p of f is non-degenerate if the derivative map $f_{*p} : T_pM \longrightarrow T_pM$ does not have 1 as an eigen-value. Here T_pM is the tangent space of M at p in the sense of C^∞ manifolds. Thus T_pM is a vector-space over the real numbers $\mathbb{R}$. If p is a non-degenerate fixed-point of f, then there exists an open subset U of M with $p \in U$ and p the only fixed point of f in U.

I denotes the identity mapping of T_pM onto itself. For a non-degenerate fixed-point p, $I - f_{*p}$ is then a non-singular $\mathbb{R}$-linear transformation of the $\mathbb{R}$-vector-space T_pM onto itself. Define $\varepsilon(p)$ by

(1.2) $$\varepsilon(p) = +1 \quad \text{if} \quad \det(I-f_{*p}) > 0$$

(1.3) $$\varepsilon(p) = -1 \quad \text{if} \quad \det(I-f_{*p}) < 0$$

In the space of all C^∞ maps from M to M the maps having only non-degenerate fixed points are dense.

Assume now that f has only non-degenerate fixed points. M^f denotes the fixed-point set of f.

(1.4) $$M^f = \{p \in M \mid f(p) = p\}$$

The Lefschetz fixed-point formula [17] is:

(1.5) $$L(f) = \sum_{p \in M^f} \varepsilon(p)$$

§2. Holomorphic fixed-point formula

Atiyah-Bott [3] and also Atiyah-Segal-Singer [7], [8] have proved a refinement of the Lefschetz formula. To state their theorem, let M be a compact complex-analytic manifold. Let $f : M \longrightarrow M$ be a holomorphic map. Assume that as a map of the C^∞ manifold underlying M, f has only non-degenerate fixed points. With $\varepsilon(p)$ as above, each $\varepsilon(p)$ is then $+1$. So the classical Lefschetz formula becomes

(2.1) $$L(f) = \sum_{p \in M^f} 1$$

or

(2.2) $L(f) =$ The number of fixed points of f

Set $n = \dim_{\mathbb{C}} M$. The holomorphic structure on M and f determines numbers $L_0(f), L_1(f), \ldots, L_n(f)$ with

(2.3) $$L(f) = \sum_{j=0}^{n} (-1)^j L_j(f)$$

To define $L_j(f)$ let Ω^j be the sheaf of germs of holomorphic j-forms on M. For $i = 0,1,\ldots,n$ each cohomology group $H^i(M,\Omega^j)$ is a finite-dimensional vector-space over $\mathbb{C}$. f induces a map $f^* : H^i(M,\Omega^j) \longrightarrow H^i(M,\Omega^j)$. Then:

(2.4) $$L_j(f) = \sum_{i=0}^{n} (-1)^i \mathrm{Trace}[f^* : H^i(M,\Omega^j) \longrightarrow H^i(M,\Omega^j)]$$

From the elliptic-operator point of view, $L_j(f)$ can be defined by using the $\bar{\partial}$-complex of $\Lambda^j T^*$. Here T denotes the holomorphic tangent bundle of M, T^* is the dual vector-bundle of T, and $\Lambda^j T^*$ is the j-th exterior power of T^*. Denote this $\bar{\partial}$-complex by

(2.5) $$0 \to C^\infty(\Lambda^j T^*) \xrightarrow{\bar{\partial}} C^\infty(\Lambda^j T^* \otimes \bar{T}^*) \to C^\infty(\Lambda^j T^* \otimes \Lambda^2 \bar{T}^*) \to \ldots \to C^\infty(\Lambda^j T^* \otimes \Lambda^n \bar{T}^*) \to 0$$

f induces an endomorphism of this elliptic complex and hence gives rise to a map $f^* : H^{j,i} \longrightarrow H^{j,i}$, where $H^{j,i}$ is the i-th homology of the complex, (2.5). Then:

$$(2.6) \qquad L_j(f) = \sum_{i=0}^{n} (-1)^i \mathrm{Trace}[f^* : H^{j,i} \longrightarrow H^{j,i}]$$

According to the Dolbeault theorem [15], $H^{j,i}$ and $H^i(M,\Omega^j)$ are canonically isomorphic. So (2.4) and (2.6) give the same number $L_j(f)$.

At a fixed point p of f consider the holomorphic tangent space T_pM. T_pM is a $\mathbb{C}$-vector-space with

$$(2.7) \qquad \dim_{\mathbb{C}} T_pM = n = \dim_{\mathbb{C}} M$$

The holomorphic derivative map $f_{*p} : T_pM \longrightarrow T_pM$ is $\mathbb{C}$-linear. Let $\lambda_1, \lambda_2, \ldots, \lambda_n$ be the eigen-values of this $\mathbb{C}$-linear transformation. The non-degeneracy of p implies each $\lambda_i \neq 1$. As usual $\sigma_j(X_1,X_2,\ldots,X_n)$ denotes the j-th elementary symmetric function of the indeterminates $X_1,X_2,\ldots,X_n$. Define $\varepsilon_j(p)$ by

$$(2.8) \qquad \varepsilon_j(p) = \frac{\sigma_j(\lambda_1,\lambda_2,\ldots,\lambda_n)}{\prod_{i=1}^{n}(1-\lambda_i)} \qquad j = 1,2,\ldots,n$$

$$(2.9) \qquad \varepsilon_0(p) = \frac{1}{\prod_{i=1}^{n}(1-\lambda_i)}$$

The formula of Atiyah-Bott [3] and of Atiyah-Segal-Singer [7], [8] is

$$(2.10) \qquad L_j(f) = \sum_{p \in M^f} \varepsilon_j(p) \qquad j = 0,1,\ldots,n$$

As noted above, $L(f) = \sum_{j=0}^{n} (-1)^j L_j(f)$. It is immediate that $\sum_{j=0}^{n} (-1)^j \varepsilon_j(p) = 1$. So (2.10) implies $L(f) = \sum_{p \in M^f} 1$. Thus (2.10) is a refinement of the classical Lefschetz formula.

§3. Fixed-point formula for singular varieties

Let X be a compact complex-analytic variety. X may be singular. Assume that X admits a complex-analytic embedding into a complex projective space.

Let $\alpha : X \longrightarrow X$ be a holomorphic automorphism of X such that

(3.1) α is of finite order

(3.2) α has only a finite number of fixed points

Set $n = \dim_{\mathbb{C}} X$ and define $L_0(\alpha)$ by

$$(3.3) \qquad L_0(\alpha) = \sum_{i=0}^{n} (-1)^i \mathrm{Trace}[\alpha^* : H^i(X,\mathcal{O}_X) \longrightarrow H^i(X,\mathcal{O}_X)]$$

$\mathcal{O}_X$ is the structure sheaf of X. If X is non-singular, then $L_0(\alpha)$ is equal to $L_0(\alpha)$ as defined by (2.4) and (2.6).

Set $X^\alpha = \{p \in X \mid \alpha(p) = p\}$. For $p \in X^\alpha$, $\varepsilon_0(p)$ can be defined by considering two cases.

Case 1. p is a non-singular point of X.

Case 2. p is a possibly singular point of X.

For Case 1, let T_pX be the holomorphic tangent space to X at p. α induces a C-linear transformation $\alpha_{*p} : T_pX \longrightarrow T_pX$. Let $\lambda_1, \lambda_2, \ldots, \lambda_n$ be the eigen-values of α_{*p}. (3.1) and (3.2) imply that each $\lambda_i \neq 1$. $\varepsilon_0(p)$ is then defined by

$$(3.4) \qquad \varepsilon_0(p) = \frac{1}{\prod_{i=1}^{n} (1-\lambda_i)}$$

So at a non-singular point p, $\varepsilon_0(p)$ is equal to $\varepsilon_0(p)$ as in (2.9).

For Case 2, let $\mathcal{O}_p$ be the local ring of X at p. Thus $\mathcal{O}_p$ is the ring of germs of holomorphic functions at p. α induces an automorphism.

$$(3.5) \qquad \hat{\alpha} : \mathcal{O}_p \longrightarrow \mathcal{O}_p$$

of $\mathcal{O}_p$. If $u \in \mathcal{O}_p$, then

$$(3.6) \qquad \hat{\alpha}(u) = u \circ \alpha$$

In $\mathcal{O}_p$ let I_p be the ideal generated by all elements of the form $u - \hat{\alpha}(u)$. Then for r a non-negative integer I_p^r / I_p^{r+1} is a finite dimensional vector space over $\mathbb{C}$. $\hat{\alpha} : \mathcal{O}_p \longrightarrow \mathcal{O}_p$ maps each I_p^r to itself and so gives a linear transformation

$$\hat{\alpha}_r : I_p^r/I_p^{r+1} \longrightarrow I_p^r/I_p^{r+1} \tag{3.7}$$

With t an indeterminate consider the formal power series

$$Q_p(t) = \sum_{r=0}^{\infty} \text{Trace}(\hat{\alpha}_r)t^r \tag{3.8}$$

This power series $Q_p(t)$ is a rational function which does not have a pole at $t = 1$. Hence $Q_p(1)$ makes sense and $\varepsilon_0(p)$ is defined by

$$\varepsilon_0(p) = Q_p(1) \tag{3.9}$$

If p is a non-singular point of X , then I_p is the maximal ideal m_p of the local ring $\mathcal{O}_p$. With $\lambda_1,\lambda_2,\ldots,\lambda_n$ as in (3.4), $Q_p(t)$ is:

$$Q_p(t) = \frac{1}{\prod_{i=1}^{n}(1-\lambda_i t)} \tag{3.10}$$

so that for a non-singular point p

$$Q_p(1) = \frac{1}{\prod_{i=1}^{n}(1-\lambda_i)} \tag{3.11}$$

The holomorphic fixed point formula [13] for α satisfying (3.1) and (3.2) is:

$$L_0(\alpha) = \sum_{p\in X^\alpha} \varepsilon_0(p) \tag{3.12}$$

§4. Equivariant Bott periodicity

An alternate description for the local contribution $\varepsilon_0(p)$ will now be developed. This alternate description is based on the notion of equivariant intersection number, and is somewhat more intuitive and geometric than (3.9).

K^* denotes the Atiyah-Hirzebruch cohomology theory, K-theory [1], [5], [14]. Let A be a compact Hausdorff topological space. $K^0(A)$ is the Grothendieck group of complex vector-bundles on A. Let B be a closed subspace of A. A complex $0 \longrightarrow E_r \longrightarrow E_{r-1} \longrightarrow \cdots \longrightarrow E_0 \longrightarrow 0$ of vector-bundles on A which is exact on B determines an element of $K^0(A,B)$. See [6], [4].

G is a finite group. A G-space is a compact Hausdorff topological space A on which there is given a continuous left G-action

$$G \times A \longrightarrow A \tag{4.1}$$

$K^0_G(A)$ is the Grothendieck group of complex G-vector-bundles on A. See [20]. Let B be a closed subset of A such that for all $g \in G$

$$gB = B \tag{4.2}$$

Form $A \cup cB$ where cB is the cone on B. The G-action on B extends in the evident way to cB so $A \cup cB$ is a G-space. Let v be the vertex of cB. Restriction gives a map $K^0_G(A \cup cB) \longrightarrow K^0_G(v)$. $K^0_G(A,B)$ is the kernel of this map.

As in [18], a complex $0 \longrightarrow E_r \longrightarrow E_{r-1} \longrightarrow \cdots \longrightarrow E_0 \longrightarrow 0$ of G-vector-bundles on A which is exact on B determines an element of $K^0_G(A,B)$.

$R(G)$ denotes the representation ring of G. $K^0_G(A)$ and $K^0_G(A,B)$ are modules over $R(G)$.

Let V be a G-module. V is a finite-dimensional vector-space over $\mathbb{C}$ on which there is given a $\mathbb{C}$-linear left G-action

$$G \times V \longrightarrow V \tag{4.3}$$

As an abelian group, $R(G)$ is the Grothendieck group of the category of G-modules. Tensor product of G-modules gives the multiplication in $R(G)$.

If V is a G-module, set $V^* = \mathrm{Hom}_{\mathbb{C}}(V,\mathbb{C})$. Make V^* into a G-module by

$$(g\phi)(u) = \phi(g^{-1}u) \qquad g \in G\ ,\ \phi \in V^*\ ,\ u \in V \tag{4.4}$$

Set $n = \dim_{\mathbb{C}} V$. For $i = 1,2,\ldots,n$ make the i-th exterior power of V^* , $\Lambda^i V^*$, into a G-module by

(4.5) $g(\phi_1 \wedge \phi_2 \wedge \ldots \wedge \phi_i) = g\phi_1 \wedge g\phi_2 \wedge \ldots \wedge g\phi_i \qquad \phi_j \in V^*$

$\Lambda^0 V^* = \mathbb{C}$, and G acts on $\Lambda^0 V^* = \mathbb{C}$ by

(4.6) $g\lambda = \lambda \qquad g \in G , \lambda \in \mathbb{C}$

Choose a Hermitian metric $< , >$ for V which is preserved by G . Let D and S be the unit disc and unit sphere.

(4.7) $D = \{u \in V \mid <u,u> \leq 1\}$

(4.8) $S = \{u \in V \mid <u,u> = 1\}$

With its usual topology D is a G-space. On D consider the G-vector-bundle $D \times \Lambda^i V^*$. The projection of this G-vector-bundle onto its base D is

(4.9) $(u, \phi_1 \wedge \phi_2 \wedge \ldots \wedge \phi_i) \longrightarrow u \qquad u \in D , \phi_j \in V^*$

The G-action on this vector-bundle is

(4.10) $g(u, \phi_1 \wedge \phi_2 \wedge \ldots \wedge \phi_i) = (gu, g\phi_1 \wedge g\phi_2 \wedge \ldots \wedge g\phi_i)$

Form the complex of G-vector bundles

(4.11) $0 \longrightarrow D \times \Lambda^n V^* \longrightarrow D \times \Lambda^{n-1} V^* \longrightarrow \ldots \longrightarrow D \times \Lambda^0 V^* \longrightarrow 0$

Here the map $D \times \Lambda^i V^* \longrightarrow D \times \Lambda^{i-1} V^*$ is

(4.12) $(u, \phi_1 \wedge \phi_2 \wedge \ldots \wedge \phi_i) \longrightarrow (u, \sum_{j=1}^{i} (-1)^{j+1} \phi_j(u) \phi_1 \wedge \ldots \wedge \hat{\phi}_j \wedge \ldots \wedge \phi_i)$

Denote this complex by $\Lambda(V^*)$. $\Lambda(V^*)$ is exact on the complement of the origin. So $\Lambda(V^*)$ determines an element of $K_G^0(D,S)$. Equivariant Bott periodicity [2], asserts

(4.13) $K_G^0(D,S)$ is a free $R(G)$-module on one generator, with generator given by $\Lambda(V^*)$.

§5. Equivariant Intersection Number

Let M be a complex-analytic manifold. A complex-analytic sub-variety of M is a subset X of M which in some open neighborhood of each point of X is the set of common zeroes of a finite number of functions defined and holomorphic in that neighborhood.

Suppose that X, Y are two complex-analytic sub-varieties of M. Assume that one connected component of $X \cap Y$ is a point p. The intersection number of X, Y at p, denoted $\#_p(X,Y)$ has a well-known definition [21] and is a non-negative integer

(5.1) $$\#_p(X,Y) \in \mathbb{Z}$$

G is a finite group. Let there be given a left action of G on M.

(5.2) $$G \times M \longrightarrow M$$

with

(5.3) For each $g \in G$, the map $M \longrightarrow M$ given by $x \longrightarrow gx$ is complex-analytic

As above X, Y are complex-analytic sub-varieties of M and p is a connected component of $X \cap Y$. Assume that for each $g \in G$, $gX = X$, $gY = Y$, $gp = p$. Then the equivariant intersection number of X, Y at p, denoted $\#_p(X,Y,G)$ is an element of $R(G)$.

(5.4) $$\#_p(X,Y,G) \in R(G)$$

$\#_p(X,Y,G)$ is defined by applying equivariant Bott periodicity. To do this, first recall that a holomorphic G-vector-bundle on M is a G-vector-bundle E on M such that E is a holomorphic vector-bundle and

(5.5) For each $g \in G$, the map $E \longrightarrow E$ given by $u \longrightarrow gu$ is complex-analytic.

If E is a holomorphic G-vector-bundle on M, $\underline{E}$ denotes the sheaf of germs of holomorphic sections of E. For an open subset

U of M , $\Gamma(E|U)$ is the set of holomorphic sections of $E|U$. If $g \in G$, and $s \in \Gamma(E|U)$, define $gs \in \Gamma(E|gU)$ by

(5.6) $$(gs)(x) = g(s(g^{-1}x)) \qquad x \in gU$$

Then $\underline{E}$ is a locally free complex-analytic G-sheaf on M . For the definition of coherent complex-analytic G-sheaf see [16]. If $\mathcal{F}$ is any coherent complex-analytic G-sheaf on M , then locally $\mathcal{F}$ can be resolved by locally free complex-analytic G-sheaves. Thus given $\mathcal{F}$ and given $p \in M$, there exists an open neighborhood U of p in M with

(5.7) For each $g \in G$, $gU = U$

(5.8) On U there are holomorphic G-vector-bundles $E_r, E_{r-1},\ldots,E_0$ and an exact sequence of coherent complex-analytic G-sheaves
$0 \longrightarrow \underline{E}_r \longrightarrow \underline{E}_{r-1} \longrightarrow \cdots \longrightarrow \underline{E}_0 \longrightarrow \mathcal{F}|U \longrightarrow 0$

Now p is a connected component of $X \cap Y$. For all $g \in G$, $gX = X$, $gY = Y$, $gp = p$. Denote the inclusions of X, Y in M by $i : X \longrightarrow M$, $j : Y \longrightarrow M$. The structure sheaves $\mathcal{O}_X$, $\mathcal{O}_Y$ of X, Y have direct images $i_*(\mathcal{O}_X)$, $j_*(\mathcal{O}_Y)$. Let U be an open subset of M which satisfies

(5.9) For all $g \in G$, $gU = U$

(5.10) $X \cap U$ and $Y \cap U$ are closed subsets of U .

Then $i_*(\mathcal{O}_X)|U$ and $j_*(\mathcal{O}_Y)|U$ are coherent complex-analytic G-sheaves on U .

Choose such a U with $p = X \cap Y \cap U$ and

(5.11) On U , $i_*(\mathcal{O}_X)|U$ and $j_*(\mathcal{O}_Y)|U$ can be resolved by locally free complex-analytic G-sheaves.

Let $0 \longrightarrow \underline{E}_r \longrightarrow \underline{E}_{r-1} \longrightarrow \cdots \longrightarrow \underline{E}_0 \longrightarrow i_*(\mathcal{O}_X)|U \longrightarrow 0$ and $0 \longrightarrow \underline{F}_s \longrightarrow \underline{F}_{s-1} \longrightarrow \cdots \longrightarrow \underline{F}_0 \longrightarrow j_*(\mathcal{O}_Y)|U \longrightarrow 0$ be resolutions on U of $i_*(\mathcal{O}_X)|U$ and $j_*(\mathcal{O}_Y)|U$.

On U consider the two complexes ξ, η of G-vector-bundles given by

(5.12) $$\xi = \{0 \to E_r \to E_{r-1} \to \cdots \to E_0 \to 0\}$$

(5.13) $$\eta = \{0 \to F_s \to F_{s-1} \to \cdots \to F_0 \to 0\}$$

Form the tensor product complex $\xi \otimes \eta$. $p = X \cap Y \cap U$ so

(5.14) $$\xi \otimes \eta \text{ is exact on } U - \{p\}$$

Let TM be the holomorphic tangent bundle of M . Choose a Hermitian metric $< , >$ for TM which is preserved by G . The G-action on M makes T_pM into a G-module. In T_pM let D,S be the unit disc and unit sphere. The Hermitian metric $< , >$ gives the C^∞ manifold underlying M the structure of a Riemannian manifold. At p there is the exponential map

(5.15) $$\exp : T_pM \longrightarrow M$$

It may be assumed that $< , >$ has been chosen so that exp maps D diffeomorphically onto a disc in U .

(5.16) $$\exp : D \longrightarrow U$$

Since $< \ >$ is preserved by G , $\exp : D \longrightarrow U$ is a G-equivariant map

(5.17) $$\exp(gu) = g(\exp(u)) \qquad u \in T_pM\ ,\ g \in G$$

On D , consider the pull-back $\exp^*(\xi \otimes \eta)$ of $\xi \otimes \eta$ by exp .

Set $T^*_p = \mathrm{Hom}_{\mathbb{C}}(T_pM, \mathbb{C})$.

According to equivariant Bott periodicity $K_G^0(D,S)$ is a free R(G)-module on one generator, with generator given by $\Lambda(T^*_p)$. Define $\#_p(X,Y,G)$ by requiring that in $K_G^0(D,S)$

(5.18) $$\exp^*(\xi \otimes \eta) = \#_p(X,Y,G) \cdot \Lambda(T^*_p)$$

The ordinary (i.e. non-equivariant) intersection number of X,Y at p is equal to the dimension of the virtual G-module $\#_p(X,Y,G)$

(5.19) $$\#_p(X,Y) = \dim_{\mathbb{C}} \#_p(X,Y,G)$$

$\#_p(X,Y,G)$ is the obstruction to equivariantly pulling apart X,Y at p. Three examples will be given to clarify this remark.

§6. Examples of $\#_p(X,Y,G)$

1 denotes the trivial one-dimensional G-module. 1 is the unit in $R(G)$ since for any G-module V

(6.1) $$1 \otimes V = V$$

For any G-module V, $\lambda_{-1}(V)$ is the alternating sum of the exterior powers of V.

(6.2) $$\lambda_{-1}(V) = 1 + \sum_{i=1}^{n} (-1)^i \Lambda^i(V)\,, \quad n = \dim_{\mathbb{C}} V \qquad \lambda_{-1}(V) \in R(G)\,.$$

Example 1. p is a point such that at p X and Y are both non-singular and such that the intersection at p is transversal in M. Then:

(6.3) $$\#_p(X,Y,G) = 1$$

Example 2. $X = Y = p$. T_pM denotes the holomorphic tangent space of M at p. The action of G on M makes T_pM into a G-module. Let T^*_p be the G-module dual to T_pM. Then:

(6.4) $$\#_p(p,p,G) = \lambda_{-1}(T^*_p)$$

Example 3. For M take $\mathbb{C}^n$ with a given $\mathbb{C}$-linear G-action

(6.5) $$G \times \mathbb{C}^n \longrightarrow \mathbb{C}^n$$

$\mathbb{C}[X_1,X_2,\ldots,X_n]$ denotes the ring of polynomials in the indeterminates $X_1,X_2,\ldots,X_n$. In $\mathbb{C}[X_1,X_2,\ldots,X_n]$ let $P_1,P_2,\ldots,P_r$ be a sequence with

(6.6) For $i = 1,2,\ldots,r-1$ P_{i+1} is not a zero divisor in $\mathbb{C}[X_1,X_2,\ldots,X_n]/(P_1,P_2,\ldots,P_i)$

and

(6.7) In $\mathbb{C}[X_1,X_2,\ldots,X_n]$ the ideal generated by $P_1,P_2,\ldots,P_r$ is equal to its own radical.

Set $X = \{z \in \mathbb{C}^n \mid 0 = P_1(z) = P_2(z) = \ldots = P_r(z)\}$. X is a complete intersection in $\mathbb{C}^n$.

Let W be the vector-subspace of $\mathbb{C}^n$ on which G acts as the identity.

(6.8) $$W = \{z \in \mathbb{C}^n \mid gz = z \quad \text{for all} \quad g \in G\}$$

Make the following two assumptions:

(6.9) The origin of $\mathbb{C}^n$ is a connected component of $X \cap W$

(6.10) For each $i = 1,2,\ldots,r$ there is a group homomorphism $\chi_i : G \longrightarrow \mathbb{C}^*$ with $P_i(g^{-1}z) = \chi_i(g)P_i(z)$ whenever $g \in G$ and $z \in \mathbb{C}^n$. $\mathbb{C}^*$ denotes the multiplicative group of non-zero complex numbers.

From (6.10) it is evident that $gX = X$ for all $g \in G$.

0 denotes the origin of $\mathbb{C}^n$. To compute $\#_p(X,W,G)$ first suppose that $\chi_1,\chi_2,\ldots,\chi_\ell$ are non-trivial and $\chi_{\ell+1},\chi_{\ell+2},\ldots,\chi_r$ are trivial. Hence if (χ_i) denotes the G-module

(6.11) $$G \times \mathbb{C} \longrightarrow \mathbb{C}$$

given by

(6.12) $$g\lambda = \chi_i(g)\lambda \qquad g \in G \ , \ \lambda \in \mathbb{C}$$

then

(6.13) For $i = 1,2,\ldots,\ell$ $(\chi_i) \neq 1$, and for $i = \ell+1, \ell+2, \ldots, r$ $(\chi_i) = 1$.

Let $\tilde{X}$ be the sub-variety of $\mathbb{C}^n$ defined by $P_{\ell+1}, P_{\ell+2},\ldots,P_r$.

(6.14) $$\tilde{X} = \{z \in \mathbb{C}^n \mid 0 = P_{\ell+1}(z) = P_{\ell+2}(z) = \ldots = P_r \ .$$

Now (3.10) implies

(6.15) $$P_1, P_2, \ldots, P_\ell \quad \text{vanish on} \quad W$$

Therefore

(6.16) $$X \cap W = \tilde{X} \cap W$$

Let $\#_0(\tilde{X}, W)$ be the ordinary (i.e. non-equivariant) intersection number of $\tilde{X}$ and W at 0. Then:

(6.17) $$\#_0(X,W,G) = \#_0(\tilde{X},W) \cdot \prod_{i=1}^{\ell} (1-(\chi_i))$$

This concludes example 3.

Note that examples 2 and 3 show that $\#_p(X,Y,G)$ can be non-zero in $R(G)$ even though $\dim_{\mathbb{C}} X + \dim_{\mathbb{C}} Y < \dim_{\mathbb{C}} M$.

§7. Zariski tangent space

Let X be a complex-analytic variety. X may be singular. For $x \in X$, $\mathcal{O}_x$ denotes the local ring at x . $\mathcal{O}_x$ is the ring of germs of holomorphic functions at x . m_x is the maximal ideal of $\mathcal{O}_x$. Thus m_x consists of those germs which take the value zero at x . The Zariski tangent space $T_x X$ is the vector-space which is dual to m_x/m_x^2 .

(7.1) $$T_x X = \mathrm{Hom}_{\mathbb{C}}(m_x/m_x^2, \mathbb{C})$$

$T_x X$ is a finite dimensional vector-space over $\mathbb{C}$. If x is a non-singular point of X , then $T_x X$ is just the usual holomorphic tangent space of X at x . If Y is another complex-analytic variety and $f : X \longrightarrow Y$ is a holomorphic map, then there is an induced $\mathbb{C}$-linear map

(7.2) $$f_{*x} : T_x X \longrightarrow T_{f(x)} Y$$

Suppose now that the finite group G acts on X with

(7.3) For each $g \in G$, the map $X \longrightarrow X$ given by $x \longrightarrow gx$ is complex-analytic

Let $p \in X$ satisfy

(7.4) For all $g \in G$, $gp = p$.

The action of G on X makes T_pX into a G-module.

<u>LEMMA</u>. In X there exists an open neighborhood U of p and a complex-analytic embedding $\phi : U \longrightarrow T_pX$ such that:

(7.5) $gU = U$ for all $g \in G$

(7.6) $\phi(gx) = g\phi(x)$ for $g \in G$, $x \in U$

(7.7) ϕ takes p to the origin of T_pX

With ϕ, U as in the lemma, set $X_p = \phi(U)$. Set $W_p = \{v \in T_pX \mid gv = v \text{ for all } g \in G\}$. 0 denotes the origin of T_pX. Assume that 0 is a connected component of $X_p \cap W_p$. Then $\#_0(X_p, W_p, G)$ is relevant to the holomorphic fixed-point formula for X.

§8. The fixed point formula restated

With $\alpha : X \longrightarrow X$ as in (3.1) and (3.2), let I denote the identity mapping of X, and let s be the smallest positive integer with $\alpha^s = I$. Let G be the finite cyclic group of order s whose elements are $I, \alpha, \alpha^2, \ldots, \alpha^{s-1}$. Note that $\alpha^{-1} = \alpha^{s-1}$. G acts on X in the evident way.

(8.1) $$G \times X \longrightarrow X$$

T_pX is the Zariski tangent space to X at p. α induces a $\mathbb{C}$-linear map

(8.2) $$\alpha_{*p} : T_pX \longrightarrow T_pX$$

Hence T_pX is a G-module. According to (4.5)-(4.7), p has an open neighborhood U in X with $\alpha U = U$ and such that

(8.3) There exists a G-equivariant complex-analytic embedding

$$\phi : U \longrightarrow T_pX \text{ mapping } p \text{ to the origin of } T_pX .$$

Set $X_p = \phi(U_p)$. 0 denotes the origin of T_pX . W_p is the linear subspace of T_pX on which α_{*p} is the identity

(8.4) $$W_p = \{v \in T_pX \mid \alpha_{*p}(v) = v\}$$

Then $\#_0(X_p,W_p,G) \in R(G)$.

Given any $\psi \in R(G)$ and any $g \in G$, let $Tr\langle g,\psi\rangle$ be the trace of ψ at g . Form $Tr\langle \alpha^{-1},\#_0(X_p,W_p,G)\rangle$.

Set $\nu_p = T_pX/W_p$. α_{*p} gives a $\mathbb{C}$-linear map $\nu_p \longrightarrow \nu_p$. Let $\lambda_1,\lambda_2,\ldots,\lambda_k$ be the eigen-values of this map. Define $\varepsilon_0(p)$ by

(8.5) $$\varepsilon_0(p) = \frac{Tr\langle \alpha^{-1},\#_0(X_p,W_p,G)\rangle}{\prod_{i=1}^{k}(1-\lambda_i)}$$

$\varepsilon_0(p)$ as defined by (8.5) is equal to $\varepsilon_0(p)$ as defined by (3.9). The holomorphic fixed-point formula [13] for $\alpha : X \longrightarrow X$ is:

(8.6) $$L_0(\alpha) = \sum_{p\in X^\alpha} \varepsilon_0(p)$$

<u>Remark</u> 1. Suppose p is a point at which X is a complete intersection. For this case, Example 3 of §6 shows how to explicitly compute $\#_0(X_p,W_p,G)$.

<u>Remark</u> 2. $G = \{I,\alpha,\alpha^2,\ldots,\alpha^{s-1}\}$. Let E be a holomorphic G-vector-bundle on X . $\underline{E}$ denotes the sheaf of germs of holomorphic sections of E . $H^i(X,\underline{E})$ is a G-module. In $R(G)$ form the alternating sum of these G-modules

(8.7) $$\sum_{i=0}^{n} (-1)^i H^i(X,\underline{E}) \in R(G)$$

For each $p \in X^\alpha$, the action of G on E makes E_p into a G-module. Denote this G-module by (E_p) . Let X_p,W_p,ν_p be as above. ν_p is a G-module. ν_p^* denotes the dual G-module. $\lambda_{-1}(\nu_p^*)$ is invertible in $R(G)$. And in $R(G)$ there is the equality

(8.8) $$\sum_{i=0}^{n} (-1)^i H^i(X,\underline{E}) = \sum_{p\in X^\alpha} \frac{(E_p)\#_0(X_p,W_p,G)}{\lambda_{-1}(\nu_p^*)}$$

§9. Localization and proofs

As above, X is a compact complex-analytic variety which admits a holomorphic embedding into complex projective space. X may be singular. $\alpha : X \longrightarrow X$ is a holomorphic automorphism of X such that

(9.1) α is of finite order.

It is <u>not</u> required here that α have only finitely many fixed points.

I denotes the identity map of X. s is the smallest positive integer with $\alpha^s = I$. G is the finite cyclic group of order s whose elements are $I, \alpha, \alpha^2, \ldots, \alpha^{s-1}$. There is the evident action of G on X

(9.2) $$G \times X \longrightarrow X$$

Let $K_0^{\omega G}(X)$ be the Grothendieck group of coherent complex-analytic G-sheaves on X. $X^\alpha = \{p \in X \mid \alpha(p) = p\}$. $i : X^\alpha \longrightarrow X$ is the inclusion of X^α in X. If $\mathcal{F}$ is a coherent complex-analytic G-sheaf on X^α, then the direct image sheaf $i_*\mathcal{F}$ is a coherent complex-analytic G-sheaf on X.

(9.3) $$i_* : K_0^{\omega G}(X^\alpha) \longrightarrow K_0^{\omega G}(X)$$

Set $P = \{\psi \in R(G) \mid \mathrm{Tr}\langle\alpha,\psi\rangle = 0\}$. P is a prime ideal in $R(G)$. Denote the localization of $R(G)$ at P by $R(G)_P$. Now $K_0^{\omega G}(X^\alpha)$ and $K_0^{\omega G}(X)$ are both modules over $R(G)$. Denote their localizations at P by $K_0^{\omega G}(X^\alpha)_P$ and $K_0^{\omega G}(X)_P$. In [19] G. Quart proved:

(9.4) <u>THEOREM</u>. $i_* : K_0^{\omega G}(X^\alpha)_P \longrightarrow K_0^{\omega G}(X)_P$ is an isomorphism of $R(G)_P$ modules.

The fixed point formula (8.6) is a quite direct corollary of the Quart localization theorem. Compare [18].

The fixed point formula (8.6) can also be proved by constructing a Riemann-Roch map

(9.5) $$K_0^{\omega G}(X) \longrightarrow K_0^{tG}(X)$$

and then applying the Atiyah-Segal localization theorem [7] [20] to $K_0^{tG}(X)$. In (9.5) $K_0^{tG}(X)$ is topological equivariant K homology. The Riemann-Roch map in (9.5) is the equivariant version of the Riemann-Roch map of [9] [10] [11] [12]. This Riemann-Roch approach was used in [13]. Note that the Riemann-Roch map takes the Quart localization theorem in algebraic geometry K theory to the Atiyah-Segal localization theoren in topological K theory. This is part of the very strong analogy which the Riemann-Roch map [9] [10] [11] [12] establishes between these two K theories.

From the power series point of view the fixed point formula (3.12) is proved by an elegant and magic argument given by G. Lusztig and corrected by D. Kazhdan.

References

[1] M. F. Atiyah, K-theory, Benjamin, 1967.

[2] M. F. Atiyah, "Bott periodicity and the index of elliptic operators", Quart. J. Math. Oxford (2), 19(1968), 113-140.

[3] M. F. Atiyah and R. Bott, "A Lefschetz fixed point formula for elliptic complexes: I", Ann. of Math. 86(1967), 374-407. II. Applications, Ann. of Math. 88(1968), 451-491.

[4] M. F. Atiyah, R. Bott, and A. Shapiro, "Clifford modules", Topology 3(1964) suppl. 1, 3-38.

[5] M. F. Atiyah and F. Hirzebruch, "Vector bundles and homogeneous spaces". Differential geometry. Proceedings of Symposia in Pure Mathematics, Vol. 3, 7-38, Amer. Math. Soc. 1961.

[6] M. F. Atiyah and F. Hirzebruch, "The Riemann-Roch theorem for analytic embeddings", Topology 1(1961), 151-166.

[7] M. F. Atiyah and G. B. Segal, "The index of elliptic operators, II", Ann. of Math. (2) 98(1968), 531-545.

[8] M. F. Atiyah and I. M. Singer, "The index of elliptic operators, I", Ann. of Math. (2) 87(1968), 484-530; "The index of elliptic operators III", Ann. of Math. (2) 87(1968), 546-604.

[9] P. Baum, "Riemann-Roch theorem for singular varieties", Proceedings of Symposia in Pure Mathematics, Volume 27, Part 2, Amer. Math. Soc., Providence, 1975, 3-16.

[10] P. Baum and R. G. Douglas, "K homology and index theory", Proceedings of Symposia in Pure Mathematics, Volume 38, Amer. Math. Soc., Providence, 1981.

[11] P. Baum, W. Fulton, and R. MacPherson, "Riemann-Roch for singular varieties", Publ. Math. IHES 45(1975), 101-167.

[12] P. Baum, W. Fulton, and R. MacPherson, "Riemann-Roch and topological K theory for singular varieties", Acta Math. 143(1979), 155-192.

[13] P. Baum, W. Fulton, and G. Quart, "Lefschetz-Riemann-Roch for singular varieties", Acta Math. 143(1979), 193-211.

[14] R. Bott, Lectures on K(X), Benjamin, 1969.

[15] P. Dolbeault, "Formes différentielles et cohomologie sur une variété analytique complexe. I" Ann. of Math. (2) 64(1956), 83-130; II. Ann. of Math. (2) 65(1957), 282-330.

[16] A. Grothendieck, "Sur quelques points d'algèbre homologique", Tohoku Math. J. 9(1957), 119-221.

[17] S. Lefschetz, Topology, Amer. Math. Soc. Colloq. Publ. XII, Amer. Math. Soc., New York, 1930.

[18] H. A. Nielsen, "Diagonalizably linearized coherent sheaves", Bull. Soc. Math., France 102(1974), 85-97.

[19] G. Quart, "Localization theorem in K-theory for singular varieties", Acta Math. 143(1979), 213-217.

[20] G. B. Segal, "Equivariant K-Theory", Publ. Math. Inst. Hautes Etudes Sci., Paris 34(1968), 129-151.

[21] J. P. Serre, Algèbre locale Multiplicités. Springer lecture notes in Mathematics 11(1965).

Department of Mathematics
Brown University
Providence, Rhode Island 02912
U.S.A.

September, 1981

Canadian Mathematical Society
Conference Proceedings
Volume 2, Part 2 (1982)

FREE ACTIONS OF FINITE GROUPS ON VARIETIES, I

William Browder and Nicholas M. Katz

In 1956, [J-P. Serre] showed that any finite group G is the fundamental group of a non-singular projective variety over $\mathbb{C}$, in fact, G can be made to act freely on a non-singular complete intersection.

In this paper we ask the question: if G acts freely on an irreducible projective variety V, what restrictions does this place on V? Our principal results are relations between degree V and the order of G or its elements. In particular, for such actions of G on V which preserve the hyperplane class in $H^2(V)$, when $H^1(V) = 0$, we get (1.11):

(1) $\exp G$ divides degree V

(2) $\#(G)$ divides $(\text{degree } V)^2$

($\exp G$ = smallest m such that $g^m = 1$, all $g \in G$, $\#(G)$ = order G).

An example is given where (2) cannot be improved.

In this paper we take a purely topological point of view, actions are only assumed continuous and we use methods of algebraic topology. In our paper [II], similar results are derived using algebraic methods for algebraic actions, where the results hold for varieties over more general fields.

The proof is based on the fact that the map $\pi: V \longrightarrow V/G$ has degree $\#(G)$, (which we show in a rather general context in (1.4)), and then studying the question of when the hyperplane class $\alpha \in \pi^* H^2(V/G)$, so that the calculation of degree V involves π_*, and hence introduces a factor $\#(G)$, (see (1.6)).

Under the given hypothesis we show $\alpha \in \operatorname{im} \pi^*$ if and only if a natural

Research partially supported by the National Science Foundation.

© 1982 American Mathematical Society
0731-1036/82/0000-0458/$03.75

obstruction element $\sigma(\alpha) \in H^3(B_G)$ is zero (1.7), $(\sigma(\alpha) = d_3\alpha$ in E_3 of the Cartan-Leray spectral sequence). Since for cyclic G, $H^3(B_G) = 0$, conclusion (1) follows.

The result (2) is deduced from a result on cohomology of groups (1.9) which asserts that for a p-group G, $z \in H^3(B_G)$, there is a subgroup $G' \subset G$ such that z restricts to 0 in G' and $\#(G)$ divides $(\#(G'))^2$.

More generally if $G \times V \longrightarrow V$ is a free action with $V = \prod_{i=1}^{k} V_i$, where V_i is irreducible projective with $H^1(V_i) = 0$, for each i, and $g^*\alpha_i = \alpha_i$ for each $g \in G$ and each $\alpha_i \in H^2(V)$ corresponding to the hyperplane class of V_i, then we show that

(1a) $\exp G$ divides $\prod_{i=1}^{k} \deg V_i$

(2a) $\#(G)$ divides $(\prod_{i=1}^{k} \deg V_i)^{k+1}$.

Since $\prod \deg V_i$ is in general a proper divisor of $\deg V$, this will often yield a stronger result. The proof of this is based on the obstruction $\sigma(\alpha_i)$ of (1.7) and the degree of $V \longrightarrow V/G$, together with the more general version (2.1) of (1.9):

If $z_1,\dots,z_k \in H^3(B_G)$, there exists a subgroup $G_0 \subset G$ such that z_i restricts to zero in $H^3(B_{G_0})$ for all $i = 1,\dots,k$ and $\#(G)$ divides $(\#(G_0))^{k+1}$.

The latter has the following amusing corollary for bilinear forms:

If $\omega_1,\dots,\omega_k$ are alternating bilinear forms on a vector space U over a field $\mathbb{F}$, there exists a subspace $U_0 \subset U$, isotropic for all ω_i, $i = 1,\dots,k$, and such that $\dim U_0 \geq \frac{1}{k+1}\cdot(\dim U)$.

We also note a simple relation on the degree (V/G), and degree V in some contexts (1.12).

By variety, we will mean an irreducible projective variety $V^n \subset \mathbb{C}P^N$, possibly singular. All group actions will be continuous and effective (i.e. for $g \in G$ $gx = x$ for all $x \in X$ implies $g = 1 \in G$), and will preserve orientation, where this makes sense, but no smoothness is assumed.

§1. Groups acting.

Let W^m be an oriented connected manifold, with $[W] \in \check{H}_m(W)$ the orientation class (where $\check{H}$ denotes homology with compact supports). If a group G acts freely on W, preserving orientation, then W/G is also a connected oriented manifold and $\pi_*[W] = \#(G)[W/G]$, where $\pi: W \longrightarrow W/G$ is the projection to the orbit space, i.e., π has degree $\#(G)$.

(1.1) Definition. An n near-manifold is a compact connected space X together with a decomposition $X = W \cup S$ where

(1) W is a connected n-manifold.

(2) S is a compact space of dimension $< n-1$.

If W is oriented we will say X is oriented.

The definition is a generalization of the [Eilenberg-Steenrod] definition of n-circuit, without simplicial structure.

We will use Čech cohomology and dimension will mean covering dimension.

(1.2) Lemma. Suppose S,X are compact, $S \subset X$, and suppose $\dim S < k$. Then

$$H^i(X,S) \longrightarrow H^i(X)$$

is an isomorphism for all $i \geq k+1$.

Proof. Since $\dim S < k$, $H^i(S) = 0$ for $i \geq k$, and from the exact sequence of the pair (X,S)

$$\cdots \longrightarrow H^{i-1}(S) \longrightarrow H^i(X,S) \longrightarrow H^i(X) \longrightarrow H^i(S) \longrightarrow \cdots$$

the result follows. □

(1.3) Lemma. For an oriented n near-manifold X,

$$H^n(X) \simeq H^n(X,S) \simeq \check{H}^n(W) \simeq \mathbf{Z} .$$

Proof. The first isomorphism follows from (1.2), the second becomes $(W,\emptyset) \rightarrow (X,S)$ is a relative homeomorphism (excision), and the third because W is connected. □

An orientation will mean a specific choice of generator $\mu(X) \in H^n(X)$.

Suppose X is an n near-manifold $X = W \cup S$, with W a connected n-manifold, S compact of dimension $< n-1$. We will say that a group G acts on (X,S) if we have an action $G \times X \longrightarrow X$ such that $g(S) = S$ for all $g \in G$ (so that $g(W) = W$ also, for all $g \in G$).

(1.4) Proposition. Let G be a finite group acting on the oriented n near-manifold (X,S), preserving orientation, (so that $g^*\mu(X) = \mu(X)$ for all $g \in G$). Then X/G is again an oriented n near-manifold, and if $\pi: X \longrightarrow X/G$ is the projection, $\pi^*\mu(X/G) = \#(G)\mu(X)$.

Note that we do not assume the action is free.

Proof. Since G acts effectively on X preserving orientation, it does the same on W. It follows that the singularity set Y of the action of G on W is a closed subset of W of codimension ≥ 2, so that $W' = W - Y$ is connected and $G \times W' \longrightarrow W'$ is a free action. For if it had dimension n the action would not be effective, since a group element would fix all of W. If it had dimension $n-1$, there would be a group element g whose fixed set separates W locally, and hence g would reverse orientation. Then $X = W' \cup S'$, where $S' = S \cup Y$ and $X/G = W'/G \cup S'/G$ is an oriented n-circuit (since $\dim S' < n-1$ implies $\dim S'/G < n-1$). But $\pi^*\mu(W'/G) = \#(G)\mu(W')$ since G acts free on W', and (1.4) follows from (1.3) and the commutative diagram

$$\begin{array}{ccccc} H^n(W') & \xleftarrow{\cong} & H^n(X,S') & \xrightarrow{\cong} & H(X) \\ \downarrow{\scriptstyle \pi^*} & & \downarrow{\scriptstyle \pi^*} & & \downarrow{\scriptstyle \pi^*} \\ \check{H}^n(W'/G) & \xleftarrow{\cong} & H^n(X/G,S'/G) & \xrightarrow{\cong} & H^n(X/G) \end{array} . \quad \square$$

(1.5) Definition. A polarization of an oriented $2n$ near-manifold X is an element $x \in H^2(X)$. We define the degree of (X,x) $(\deg(X,x) \in \mathbb{Z})$ by the formula $x^n = (\deg(X,x))\mu(X)$.

(1.6) Proposition. Let G be a finite group acting on the polarized $2n$ near-manifold (X,x), $x \in H^2(X)$, preserving orientation, and suppose $x = \pi^*(x')$ for some $x' \in H^2(X/G)$. Then $\#(G)$ divides $(\deg(X,x))$. Further $(X/G,x')$ is a polarized $2n$ near-manifold with $\deg(X/G,x') = \dfrac{\deg(X,x)}{\#(G)}$.

Proof. $(\deg(X,x))\mu(X) = x^n = (\pi^*(x'))^n = \pi^*((x')^n) = \pi^*(\deg(X/G,x')\mu(X/G)) = (\#(G))(\deg(X/G,x'))\mu(X)$. $\square$

(1.7) Proposition. Let G (finite) act freely on a space Y with $y \in H^2(Y)$ such that $g^*y = y$ for all $g \in G$ and suppose further that $H^1(Y) = 0$. Then there is a natural obstruction $\sigma(y) \in H^3(B_G)$ such that $y \in \pi^*H^2(Y/G)$ if and only if $\sigma(y) = 0$.

Proof. Consider the Borel construction, i.e., the bundle

$$Y \underset{G}{\times} U \xrightarrow{p} B_G = U/G$$

(where U is the universal (contractible free) G-space). Here $Y \underset{G}{\times} U \simeq Y/G$ since G acts freely on Y , and Y = fibre of p . The spectral sequence (see [Spanier]) of this bundle (the Cartan-Leray spectral sequence) has

$$E_2^{s,t} = H^s(B_G;H^t(Y))$$

and

$$E_\infty = G(H^*(Y/G)) .$$

Since $g^*(y) = y$ all $g \in G$, y defines an element of $E_2^{0,2} = H^0(B_G;H^2(Y)) = H^2(Y)^G$, the invariants of $H^2(Y)$ under G , and since $H^1(Y) = 0$, $d_2y = 0$. Define $\sigma(y) = d_3y \in H^3(B_G;H^0(Y)) = H^3(B_G)$. If $d_3y = 0$ then y persists to E_∞ so is in the image π^* , and conversely, so (1.7) follows. □

(1.8) Corollary. Suppose $G \simeq \mathbb{Z}/k$, and G acts freely and orientably on the polarized 2n near-manifold (X,x) such that $g^*x = x$, all $g \in G$, and $H^1(X) = 0$. Then k divides deg(X,x) .

Apply (1.7) and (1.6), noting that $H^3(B_G) = 0$ for cyclic G . □

(1.9) Theorem. Let G be a finite p-group, $z \in H^3(B_G)$. Then there is a subgroup $G_2 \subset G$ (depending on z) such that

(1) $i^*z = 0$ $(i:G_2 \longrightarrow G)$ and

(2) $\#(G)$ divides $(\#(G_2))^2$.

We postpone the proof of (1.9). In fact we will prove a more general result (2.1) in §2.

(1.10) Theorem. Let G be a finite group acting freely and orientably on the polarized 2n near-manifold (X,x) with (a) $g^*x = x$ for all $g \in G$ and (b) $H^1(X) = 0$. Then

(1) exp G divides deg(X,x) .

(2) $\#(G)$ divides $(\deg(X,x))^2$.

Proof. Simply combine (1.7), (1.8) and (1.9). Note that $\exp G$ is the last common multiple of the (order g), $g \in G$. Also note that it suffices to prove (2) for each p-Sylow subgroup of G separately, to get the divisibility at each prime. □

The obvious inspiration for the definition of 'polarized' $2n$ near-manifold is an irreducible projective variety $V \subset \mathbb{C}P^N$ of dimension n.

(1.11) Corollary. Let G be a finite group acting freely and orientably on an irreducible complex projective variety V of dimension n. Suppose $g^*x = x$, all $g \in G$ where $x \in H^2(V)$ is the hyperplane class $x = f^*(\alpha)$, $\alpha \in H^2(\mathbb{C}P^N)$, $f: V \subset \mathbb{C}P^N$, and suppose further that $H^1(V) = 0$. Then

(1) $\exp G$ divides $\deg V$;

(2) $\#(G)$ divides $(\deg V)^2$.

Proof. Since V is a polarized $2n$ near-manifold this follows immediately from (1.10). □

Note the following example: Let $G \simeq \mathbb{Z}/p \times \mathbb{Z}/p$ generated by $T, S \in G$ act on $\mathbb{C}P^{p-1}$ on the homogeneous coordinates: $T[z_0, \ldots, z_{p-1}] = [z_0, z_1, \lambda^2 z_2, \ldots, \lambda^{p-1} z_{p-1}]$ (where λ is a primitive p-th root of unity), $S[z_0, \ldots, z_{p-1}] = [z_1, z_2, \ldots, z_{p-1}, z_0]$ (the cyclic permutation of coordinates). Then G leaves the Fermat hypersurface of degree p, $V(p) = \{\sum z_i^p = 0\}$ invariant and acts freely. For any hypersurface $X^N \subset \mathbb{C}P^{N+1}$, $H^1(X) = 0$ if $N \geq 2$, and since the action on $V(p)$ is the restriction of an action of G on $\mathbb{C}P^{p-1}$, homotopic to the identity if $p > 2$, the hypotheses of (1.11) are satisfied in that case. Hence the conclusion (2) of (1.11) cannot be improved in this example.

For hypersurfaces $X^N \subset \mathbb{C}P^{N+1}$ with $N \geq 3$, $H^1(X) \simeq \mathbb{Z}$ also, so for groups G of odd order acting on X, the hypotheses of (1.11) are automatically satisfied, and for G of even order if N is odd and G preserves orientation.

We also have the following elementary relation on the degree of V/G in some circumstances.

(1.12) Proposition. Let G act on the projective variety V^n, and suppose $H^2(V;\mathbb{Q}) = \mathbb{Q}$. Then

$$\deg(V/G) = \frac{k^n}{\#(G)} \deg(V)$$

for some integer k.

Proof. If $\alpha' \in H^2(V/G)$ is the hyperplane section class of V/G, then $\pi^*(\alpha') = k\alpha + t$, where $\alpha \in H^2(V)$ is the hyperplane section class of V, and t is a torsion class. Then

$$(\alpha')^n = \deg(V/G)\mu(V/G) \quad \text{and}$$

$$\begin{aligned}(\pi^*\alpha')^n &= \deg(V/G)(\#(G))(\mu(V)) \\ &= (k\alpha+t)^n = (k\alpha)^n = k^n \deg(V)\mu(V)\end{aligned}$$

(since any product involving t is a torsion class and $H^{2n}(V) \simeq \mathbb{Z}$. □

We may generalize our notion of 'polarized near-manifold' as follows:

Let $a_i \in \mathbb{Z}_+$, (= the positive integers), $\bar{a} = (a_1,\ldots,a_k)$.

(1.13) Definition. An $\bar{a}$-polarized m near-manifold is an m near-manifold X such that $m = 2\sum_{i=1}^{k} a_i$, together with elements $x_1,\ldots,x_k \in H^2(X)$. We define 'degree' of $(X,x_1,\ldots,x_k;\bar{a})$ by the equation

$$\deg(X,x_1,\ldots,x_k;\bar{a})\mu(X) = x_1^{a_1},\ldots,x_k^{a_k} .$$

An example of such would be a product of k varieties $V_1^{a_1}\times\ldots\times V_k^{a_k}$, x_i the hyperplane class of V_i, and its degree in this sense would be the product of the degrees of the factors, which is in general considerably smaller than the degree $(V_1 \times\ldots\times V_k)$ as a projective variety. For example:

$$\deg(\mathbb{C}P^1\times\mathbb{C}P^1,\alpha\otimes 1,1\otimes\alpha;(1,1)) = 1$$

$$\deg(\mathbb{C}P^1\times\mathbb{C}P^1,\alpha\otimes 1 + 1\otimes\alpha) = 2$$

(1.14) Theorem. Let G act freely on an $\bar{a}$-polarized m near-manifold $(X,x_1,\dots,x_k;\bar{a})$ preserving orientation and with $g^*x_i = x_i$ each i, every $g \in G$, and suppose $H^1(X) = 0$. Then:

(1) $\exp G$ divides $\deg(X,x_1,\dots,x_n;\bar{a})$

(2) $\#(G)$ divides $(\deg(X,x_1,\dots,x_k;\bar{a}))^{k+1}$.

(1.15) Corollary. Let $V_1,\dots,V_k$ be irreducible projective varieties with $H^1(V_i) = 0$ for all i, and suppose G acts freely on $V_1 \times\dots\times V_k$ preserving orientation and with $g^*(\alpha_i) = \alpha_i$ each $i = 1,\dots,k$, all $g \in G$, where $\alpha_i \in H^2(V)$ corresponds to the hyperplane section of V_i. Then

(1) $\exp G$ divides $\prod_{i=1}^{k} \deg(V_i)$

(2) $\#(G)$ divides $(\prod_{i=1}^{k} \deg(V_i))^{k+1}$.

The proof of (1.14)(2) will follow from a general theorem (2.1) on cohomology of groups, just the way (1.10)(2) followed from (1.9).

§2. A theorem in cohomology of finite groups and a remark on bilinear forms.

(2.1) Theorem. Let G be a finite p-group, $x_1,\dots,x_k \in H^3(B_G)$. Then there is a subgroup $G_2 \subset G$ (depending on $x_1,\dots,x_k$) such that

(1) $i^*(x_\ell) = 0$ all ℓ, $(i:G_2 \longrightarrow G)$, and

(2) $\#(G)$ divides $(\#(G_2))^{k+1}$.

Proof. Since G is a p-group, G is nilpotent so that we can find a cyclic subgroup C of order p in the center of G. Let $K = G/C$ so that

$$C \rightarrowtail G \twoheadrightarrow K$$

is a central extension, $\#(G) = p(\#(K))$.

Consider the spectral sequence for this central extension with

$$E_2^{p,q} = H^p(K;H^q(C))$$

and E_∞ the associated graded group to $H^*(G)$. (We are now using the group-cohomology notation, where $H^*(G)$ means $H^*(B_G)$ in the notation of §1).

Since C is finite cyclic $H^1(C) = H^3(C) = 0$ and the only non-zero terms in E_2 are $E_2^{3,0} = H^3(K)$ and

$$E_2^{1,2} = H^1(K;H^2(C)) = H^1(K;C) = \mathrm{Hom}(K,C) .$$

Then each element $x_\ell \in H^3(G)$, has a representative λ_ℓ in $E_2^{1,2}$ which represents the class of x_ℓ in E_∞ . Let $\lambda = \prod \lambda_\ell : K \longrightarrow \prod^k C$ be the product map, and let K_1 = kernel λ , so $\#K = p^k(\#(K_1))$.

Let G_1 be the pull-back of the extension to K_1:

$$\begin{array}{ccc} C & = & C \\ \downarrow & & \downarrow \\ G_1 & \xrightarrow{i_1} & G \\ \downarrow{\scriptstyle p_1} & & \downarrow \\ K_1 & \longrightarrow & K . \end{array}$$

Then $i_1^* x_\ell = p_1^* z_\ell$ for some elements $z_\ell \in H^3(K_1)$ (using the naturality of the spectral sequences and the fact that $K_1 \subset \ker \lambda_\ell$, each ℓ) .

By induction, we have a subgroup $j_2 : K_2 \subset K_1$ such that $j_2^* z_\ell = 0$, $\ell = 1,\ldots,k$, and $\#K_1$ divides $(\#(K_2))^{k+1}$. Let G_2 = pull-back of the extension to K_2 so that $\#G_2 = p(\#(K_2))$:

$$\begin{array}{ccccc} C & = & C & = & C \\ \downarrow & & \downarrow & & \downarrow \\ G_2 & \xrightarrow{i_2} & G_1 & \xrightarrow{i_1} & G \\ \downarrow{\scriptstyle p_2} & & \downarrow{\scriptstyle p_1} & & \downarrow \\ K_2 & \xrightarrow{j_2} & K_1 & \longrightarrow & K . \end{array}$$

If $i : G_2 \subset G$, $i = i_1 i_2$ so

$$i^* x_\ell = i_2^* i_1^* x_\ell = i_2^* p_1^* x_\ell = p_2^* j_2^* z_\ell = 0 ,$$

for each $\ell = 1,\ldots,k$.

Also

$$\#(G) = p(\#(K)) = p^{k+1}(\#(K_1))$$

divides $p^{k+1}(\#(K_2))^{k+1} = (p(\#(K_2))^{k+1} = (\#(G_2))^{k+1}$. $\square$

(2.2) Theorem. Let U be a finite dimensional vector space over a field $\mathbb{F}$ and let $\omega_1,\ldots,\omega_k$ be alternating bilinear forms $U\times U \longrightarrow \mathbb{F}$. Then there is a subspace $U_0 \subset U$ such that $\omega_i|U_0 \equiv 0$ for each i (U_0 is isotropic for each ω_i) and $\dim U_0 \geq \frac{1}{k+1} \dim U$.

Proof. If $\mathbb{F} = \mathbb{Z}/p$, (2.2) may easily be deduced from (2.1), with $G = U$ and thinking of the forms ω_i as elements of $H^2(G;\mathbb{Q}/\mathbb{Z})$, which is isomorphic to $H^3(G)$ (using the exact cohomology sequence arising from the exact coefficient sequence

$$0 \longrightarrow \mathbb{Z} \longrightarrow \mathbb{Q} \longrightarrow \mathbb{Q}/\mathbb{Z} \longrightarrow 0).$$

We give a direct proof.

Choose a non-zero element $u \in U$ and let $\lambda : U \longrightarrow \prod^k \mathbb{F}$ be given by $\lambda(x) = (\omega_1(x,u),\omega_2(x,u),\ldots,\omega_k(x,u))$. Let $U_1 = \ker \lambda$, so that

$$\dim U_1 + k \geq \dim U .$$

Then $u \in U_1$, and $\omega_i(x,u) = 0$ for $x \in U_1$, $i = 1,\ldots,k$, so $\omega_1,\ldots,\omega_k$ define forms $\overline{\omega_1},\ldots,\overline{\omega_k}$ on $\overline{U} = U_1/(u)$, and $\dim \overline{U} = \dim U_1 - 1 < \dim U$, since $u \neq 0$. By induction, there is a subspace $\overline{U}_0 \subset \overline{U}$ such that $\overline{\omega}_i|\overline{U}_0 \equiv 0$, all i , and $\dim \overline{U}_0 \geq \frac{1}{k+1}(\dim \overline{U})$. If U_0 is the inverse image of $\overline{U}_0$ in U_1 , then $\omega_i|U_0 \equiv 0$ all i and $\dim U_0 = \dim \overline{U}_0 + 1 \geq$ $\geq \frac{1}{k+1}(\dim \overline{U}) + 1 = \frac{\dim \overline{U} + k+1}{k+1} = \frac{\dim U_1 + k}{k+1} \geq \frac{1}{k+1}(\dim U)$. $\square$

Bibliography

W. Browder and N. Katz, Free actions of finite groups on varieties, II, (to appear).

S. Eilenberg and N. Steenrod, Foundations of algebraic topology, Princeton University Press.

E. Spanier, Algebraic Topology, McGraw Hill, New York, 1966.

J.-P. Serre, Sur la topologie des variétiés algébriques en caracteristique p. Symp. Top. Mexico (1956), 24-53.

DEPARTMENT OF MATHEMATICS
PRINCETON UNIVERSITY
PRINCETON, NJ 08544-0037

Department of Mathematics
Box 37 -- Fine Hall
Princeton University
Princeton, NJ 08544-0037

Canadian Mathematical Society
Conference Proceedings
Volume 2, Part 2 (1982)

Global monodromy of elliptic Lefshetz fibrations

R. Mandelbaum and J.R. Harper

Abstract

Let M, N be simply connected compact smooth 4-manifolds. Let $P = CP^2$ and $Q = \overline{CP}^2 = CP^2$ with orientation opposite to the usual. Then there exists integers $k_1, k_2 \geq 0$ such that $M \# k_1 P \# k_2 Q$ is diffeomorphic to $N \# k_1 P \# k_2 Q$. If M, N are not simply connected this need not be true. In this paper we consider minimal elliptic surfaces with no multiple fibers and at least one degenerate fiber and try to understand to what extent the above theorem is true. In order to do this we first attempt to classify the global geometric monodromes of such surfaces. Our main theorem is a complete classification for such elliptic surfaces V with base $F_g = F_1 = T^2$.

Theorem

Let V be as above. The V is diffeomorphic $T^2 \times T^2 \oplus X$ with $e(X) = e(X)$ and X a simply connected elliptic surface, where $e(X)$ = Euler Class of X and $\oplus$ is fiber connect sum. Furthermore $V \# P = S^2 \times T \# (2n+1) P \# (10n+1) Q$.

© 1982 American Mathematical Society
0731-1036/82/0000-0459/$02.75

This note is an account of work in progress and is based on the lecture given by the first author.

Let V be a compact complex surface and $\pi: V \to R$ be a holomorphic map onto a non-singular Riemann surface R with generic fibre a torus. Then $\pi: V \to R$ is called an elliptic fibration and V is called an elliptic surface. If a finite set of singular values $P = \{p_1, \ldots, p_n\}$ is removed from R, then $\pi: V - \pi^{-1} P \to R - P = R^*$ is a smooth torus bundle over R^*. We formalize the underlying topology of this situation.

Definition. (See [Msh] p. 162). Let $\pi: M \to F_g$ be a smooth map of a compact oriented 4-manifold onto a real oriented surface of genus g. We call $\pi: M \to F_g$ an elliptic Lefshetz fibration if the following conditions are satisfied;

a) there is a finite set $\phi \neq P \subset F_g$ such that $\pi: M \to \pi^{-1}P \to F_g - P = F_g^*$ is a smooth torus bundle,

b) for each p in P, $\pi^{-1}(p)$ has exactly one singular point c and $\pi^{-1}(p)$ has the homotopy type of $S^1 \vee S^2$,

c) there exist neighborhoods B of c in M and $D = \pi(B)$ of p in R, with orientations induced by the global orientation of M and R, such that in local complex coordinates λ in D (x,y) in B, $\pi|B$ is given implicitly by the equation $x^2 + y^2 = \lambda$.

Note 1. The singularity described in parts (b) and (c) is the ordinary double point of an algebraic curve.

Note 2. According to Kodaira [K], the monodromy of the torus bundle over ∂D is conjugate to the translation

$$\begin{pmatrix} 1 & 1 \\ 0 & 1 \end{pmatrix}$$

under the identification of $\mathrm{Diff}^+(T^2)$/amb. isotopy with $SL(2,\mathbb{Z})$.

Our reason for concentrating attention on the special class of elliptic Lefshetz fibrations is the following theorem of Moishezon [Msh]. Recall that a complex surface is minimal if there are no analytically embedded ∂-spheres L with self-interaction number $L^2 = -1$. See [Man 1]p.30ff for details on the topology.

Theorem [Msh] Every minimal elliptic surface without multiple fibres is diffeomorphic to an elliptic Lefshetz fibration with the same base.

In the simply connected case, Moishezon also solved the classification question.

<u>Theorem</u> [Msh] Two elliptic Lefshetz fibrations M and N with base S^2 are diffeomorphic if and only if they have the same Euler number, $e(M) = e(N)$. In particular, M is diffeomorphic to the fibre connected sum $\oplus_{i=1} V_0$ where $n = e(M)/12 \in \mathbb{Z}$ and V_0 is the rational elliptic surface.

<u>Remark</u> The cardinality of P is $12n = e(M)$.

<u>Note 3.</u> The diffeomorphism type of $V_0 = P\#9Q$.

<u>Note 4.</u> An equivalent description of $\pi: M \to S^2$ is as a pull-back via the map $z \to z^n$, $z \in S^2$, $12\,n = e(M)$,

$$\begin{array}{ccc} M & \to & V_0 \\ \pi \downarrow & & \downarrow \\ S^2 & \underset{z^n}{\to} & S^2 \end{array}$$

The proof of this theorem is by an explicit global monodromy calculation. These methods, which are combinatorial in essence, can be used to study the general case $M \to F_g$. We can regard the singular set P as lying inside a small disc on F_g. We present $\pi_1(F_g{}^*)$ with generators A_i, B_i, C_j, where $1 \le i \le g$ and there is one C_j for each point in P, and the single relation

$$1 = \prod_{i=1}^{g} [A_i, B_i] \prod_j C_j .$$

Here is a picture for the case $g = 1$.

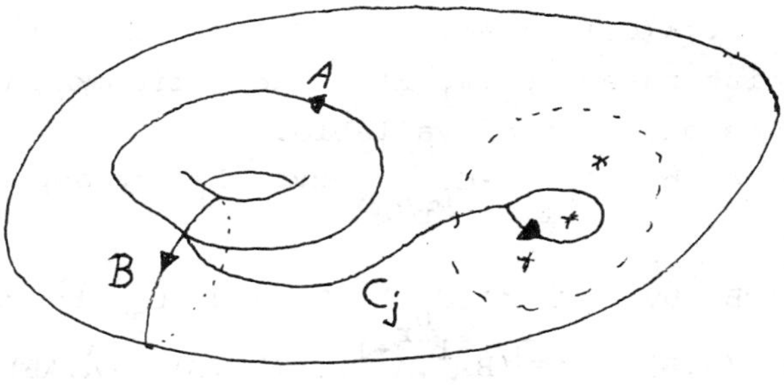

Then the <u>global</u> monodromy of $\pi: M \to F_g$ is a homomorphism

$$\phi_\pi: \pi_1(F_g{}^*) \to SL(2,Z)$$

such that a) $\phi_\pi(\prod_{i=1}^{g} [A_i, B_i] \prod_j C_j) = I$

b) $\phi_\pi(C_j)$ is conjugate to $\begin{pmatrix} 1 & 1 \\ 0 & 1 \end{pmatrix}$.

Conversely, given such data, we can use it to assemble an elliptic Lefshetz fibration.

<u>Definition</u>. Let ϕ_1, ϕ_2 be global monodromies. We say $\phi_1 \sim \phi_2$ if there is a diffeomorphism μ of $F_g{}^*$ such that $\phi_1 = \phi_2\mu_*$ where μ_* is the induced map of π_1.

If $\phi_1 \sim \phi_2$ then $M\phi_1$ and M_{ϕ_2} are diffeomorphic. A general problem is to determine the equivalence classes. In the case $g = 1$, we can prove there is only one equivalence class. We suspect this is the case in general.

<u>Theorem</u>. Let $\pi\colon M \to F_1 = T^2$ be an elliptic Lefshetz fibration. Then for some integer $n>0$; $e(M) = 12n$ and there is a presentation for $\pi_1(F_1{}^*)$

$$\langle A, B, C_j, D_j \quad 1 \le j \le 6n \mid [A,B] \prod_{j=1}^{6n} C_j D_j = 1 \rangle$$

and the global monodromy ϕ_π is equivalent to the global monodromy ψ satisfying

$$\psi(A) = \psi(B) = I$$

$$\psi(C_j) = \begin{pmatrix} 1 & 1 \\ 0 & 1 \end{pmatrix} = X$$

$$\psi(0_j) = \begin{pmatrix} 1 & 0 \\ -1 & 1 \end{pmatrix} = Y .$$

We outline the proof. We write $V = (C_1, C_2, \ldots, C_e, A, B, A^{-1}, B^{-1})$ and $V_\phi = (\phi(C_1), \ldots, \phi(C_A), \phi(A), \ldots, \phi(B^{-1})) \in SL(2,Z)^{e+4}$.

By means of changing bases in $\pi_1(F^*)$ and diffeomorphisms of F^* the following moves on V are available.

1) replace A by $C_{L_1} \cdots C_{L_r} A$ and C_j by appropriate conjugates.

2) replace B by $(C_{L_1} \cdots C_{L_r})^{-1} B$ and C_j by conjugates.

3) replace (A,B) by (B^{-1}, A^{-1}), (AB, B), (A, AB) or (A, BA^{-1}) along with appropriate conjugates of C_j.

4) Braid group moves on $(C_1, \ldots, C_e)$ leaving A, B fixed.

5) act on V by inner automorphisms.

For example, the type of move described in 1) is pictured below

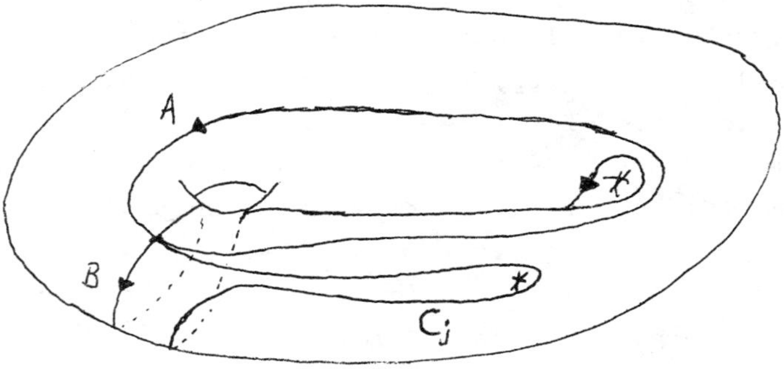

To V_ϕ we attach a triple of non-negative integers defined as follows. Consider the image $\bar{V}_\phi$ of V_ϕ in PSL(2,Z). This group is $Z_2 * Z_3$ with generators α, β such that $\alpha^3 = 1 = \beta^2$. Matrix representatives are $\alpha = XY$, $\beta = XY^2$. We also have $X = \alpha\beta\alpha$, $Y = \alpha^2\beta$. Then the conjugate of X in PSL(2,Z) are the "short words" $\alpha\beta\alpha$, $\alpha^2\beta$, $\beta\alpha^2$ or "long words" $Q\alpha\beta\alpha Q^{-1}$. Let $\ell(w)$ = syllable length of a reduced word, e.g. $\ell(\alpha^2\beta\alpha\beta) = 4$. Then $\ell(w) = 0 \iff w = 1$. For a reduced word Q we write $f(Q)$, $e(Q)$ for the front and end syllables. For a reduced word of the form $Q\alpha\beta\alpha Q^{-1}$ we can have $e(Q) = \beta$ or $Q = \alpha^i$ $i = 0,1,2$ in which case $Q\alpha\beta\alpha Q$ is short. We then define $L(Q\alpha\beta\alpha Q^{-1}) = \ell(Q), 0$ according as the first or second case holds. The triple of numbers associated with V_ϕ is

$$\ell V_\phi = (\ell\phi(A), \ \ell\phi(B), \ \sum_j L\phi(C_j))$$

Then our proof is by induction on lexicographic order to reduce to the situation considered by Moishezon [Msh]p. 180. The idea in the inductive step is that if none of the moves 1) - 4) lowers anything then all the words in V_ϕ must have either a common front or end syllable. Then there is a diffeomorphism of $M - \pi^{-1}P = k(\pi,1)$ which is the identity on the boundary and which induces an inner automorphism such that for the equivalent V_ϕ, the values of the first two entries of ℓV_ϕ are preserved while that on the third is lowered. The case $\ell V_\phi = (0,0,0)$ is handled in [Msh].

Note 5. By an indirect algebraic geometric argument [Man2] the result is known for elliptic surfaces $\pi\colon\ V \to Fg$ provided $e(M) > 3g$.

Note 6. Our proof would still work under the hypotheses

$\phi[A_i, B_i] = I$ for $1 \le i \le g-1$ and

$\phi([A_g, B_g] \prod_j C_j) = I$ for the case of genus g.

We conclude this paper with a brief account of how this program fits into a general picture of 4-manifolds.

To begin with, recall that in the simply connected case, compact 4-manifolds are homotopy equivalent if and only if they have isomorphic intersection forms [Wh] Then, by a theorem of Novikov and Wall, homotopy equivalence implies an h-cobordism. Lastly, as noted by Wall [Wa], in dimension 4, an h-cobordism implies a stable diffeomorphism, that is, there exists $k \geq 0$ such that $M \# k(S^2 \times S^2)$ and $N \# k(S^2 \times S^2)$ are diffeomorphic. One can rewrite the diffeomorphism in the form

$$M \# k_1 P \# k_2 Q = N \# k_1 P \# k_2 Q$$

for some integers k_1, k_2 where $P = CP^2$ and $Q = \overline{CP^2}$.

In general there is no effective way to determine k_1 or k_2 from the topology of M and N. If M and N are complex surfaces, upper bounds can be determined. In particular, in [MM 1,2] there are a large number of cases where $k_1 = 1$ and $k_2 = 0$. For example

Theorem. If M is a simply connected minimal elliptic surface then $M \# P = 2n P \# (10n-1)Q$ where $n = e(M)/12$.

In the non-simply connected case, very close to nothing is known. Clearly the intersection form no longer determines the homotopy type. Furthermore, the homotopy type need not determine the h-cobordism type. The remarkable examples in [CS] of fake RP^4's supply examples homotopy equivalent but not h-cobordant to RP^4.

Remark. One does get homotopy equivalence implies h-cobordant for manifolds M satisfying

$$L_5{}^h(\pi_1(M)) = H^2(\pi_1 M; Z_2) = 0$$

but although surface groups Γ satisfy $H^2(\Gamma; Z_2) = 0$, $L_5(\Gamma) \neq 0$ so the implication is not known for them. All the manifolds we consider have fundamental group a surface group.

Restricting ourselves to elliptic surfaces and elliptic Lefshetz fibration, we can ask the following questions.

Q1. Given $M \to F_g$, $N \to F_g$ Lefshetz fibration with $e(M) = e(N)$, when are they homotopy equivalent, h-cobordant, or diffeomorphic?

Q2. Does $M \# k_1 P \# k_2 Q$ decompose in any simple way for sufficiently large k_1, k_2? If so, how small can k_1 and k_2 be taken?

Q3. Is there any nice fiber-decomposition of $M \to F_g$ analogous to Moishezon's Theorem for base S^2 ? A good candidate would be $M = T^2 \times F_g \oplus X$ where X is a simply connected Lefshetz fibration. For the genus 1 case, F_g = torus, our main result can be recast to give partial answers to these questions.

Theorem. Let $V \to T$ be a minimal elliptic surface with no multiple fibers and $e(V) > 0$. Then if $g = 1$, $V = T^2 \times T^2 \oplus X$ with $e(X)=e(V)$ and X a simply connected elliptic surface. Furthermore

$$V \# P = S^2 \times T \# (2n+1) P \# (10n+1) Q$$

where $e(V) = 12n$.

We would expect that for $V \to F_g$,

$$V \# P = S^2 \times F\hat{g} \# (2g+ 2n-1) P \# (2g+10n-2) Q \ .$$

REFERENCES

CS - S.Cappell and J. Shaneson, Some New Four-manifolds, Ann. of Math (2) 104 (1976), 61-72.

K - K. Kodaira, On Compact Analytic Surfaces II, Ann. of Math (2) 77 (1963), 563-626.

Man 1 - R. Mandelbaum, Four-Dimensional Topology: An Introduction Bull. Am. Math. Soc. (New Series) #2 (1980), 1-159.

Man 2 - R. Mandelbaum, Lefshetz Fibrations of Riemann Surfaces (to appear)

MM 1 - R. Mandelbaum and B. Moishezon, On the Topological Structure of Non-singular Algebraic Surfaces in CP^3, Topology 15 (1976) 23-40.

MM 2 - R. Mandelbaum and B. Moishezon, On the Topology of Simply Connected Algebraic Surfaces, Trans. Am. Math. Soc. 280 (1980) 195-222.

Msh. - B. Moishezon, Complex Surfaces and Connected Sums of Complex Projective Planes, Lecture Notes in Math, vol. 603, Springer-Verlag.

Wa - G.T.C. Wall, On Simply Connected 4-Manifolds, J. London Math Soc. 39 (1964) 141-149.

Wh - J.H.C. Whitehead, On Simply Connected 4-Dimensional Polyhedra, Comment. Math. Helv. 22 (1949) 48-92.

Group Actions

Canadian Mathematical Society
Conference Proceedings
Volume 2, Part 2 (1982)

EXTENSIONS OF FINITE GROUP ACTIONS FROM SUBMANIFOLDS OF A DISK

Amir H. Assadi(†)

Introduction. This paper contains some indications of the proofs and supplements to the results of [A3]. The main results concern with existence and classification of smooth actions on a disk D^n having a prescribed G-manifold M^n as an invariant codimension zero submanifold (subject to the appropriate hypotheses). These results are applied to characterize stationary point sets of semifree G-actions on disks in the stable range, that is, when the dimension of the stationary point set is less than half the dimension of the disk. A complete classification of such actions on disks up to G-diffeomorphism is obtained as an application. The first result of this kind is L. Jones' converse to the P.A. Smith theorem [J1, Theorem 2.1]. Besides being of great independent interest in transformation groups, Jones' theorem has played a key role in other calculations and applications, e.g. in [A-H-V] and [J2]. The proof of Jones' theorem, however, has a gap as it is stated. Therefore, we have added a brief discussion of his theorem, and indicated the correct statement which is implied by his proof. The original statement of Jones' theorem is proved correctly in his thesis [J], as was brought to our attention by the referee. Thus, the calculations and applications in [J2] and [A-H-V] still remain valid.

The author is grateful to Professors William Browder, Ian Hambleton and Robert Stong for helpful discussions and suggestions during the course of this research. I would like to thank the School of Mathematics of The Institute for Advanced Study and the Department of Mathematics of the University of Virginia for their hospitality. I would like to take the opportunity and acknowledge the influence of L. Jones' paper [J1] on this work. Finally, the skillful and painstaking job of typing of this manuscript has been done by Marie Brown, to whom I am grateful.

The organization of the paper is as follows. Section 1 contains the statement of the results and an (intuitive) indication of the method and difficulties in extending an action. Section 2 summarizes some background

(†) This research has been partially supported by an NSF grant through The Institute for Advanced Study, Princeton, N.J., and the NSF grant MCS 80-01959.

© 1982 American Mathematical Society
0731-1036/82/0000-0460/$06.50

material and useful observations. Section 3 discusses some relationship between Stiefel-Whitney classes of a linear representation and framings. These results are applied to the situation considered by Jones in [J1], and the necessary modifications of his proof are discussed. Section 4 contains the proof of the main theorems and the applications.

Section 1. A number of problems in transformation groups are concerned with existence and classification of actions of certain kind on a given manifold. These problems can be often formulated as a special case of the following extension problem. Let $F_0 \subset F$ be two collections of subgroups of a finite group G such that they are closed under conjugation, and F contains the trivial subgroup $\{1\}$. Let W^n be a smooth manifold, and $M^m \subset W^n$ be an embedded submanifold with normal bundle ν. Furthermore, assume that $\phi: G \times M \to M$ is a smooth action on M -- not necessarily effective -- with isotropy subgroups belonging to F. We are interested in studying a very special case of the following extension problem:

(E) Does there exist a smooth action $\psi: G \times W \to W$ with isotropy groups belonging to F, such that ψ leaves M invariant, $\psi | G \times M = \phi$, and the isotropy subgroups of W-M belong to $F - F_0$? (We use the notation and terminology of [A2]).

The requirements on isotropy subgroups are imposed so that applications to more traditional problems -- such as classification and construction of actions -- can be made. For instance, if ϕ is the trivial action -- i.e., $M^G = M$, and $F_0 = \{G\}$ -- then the problem becomes to decide if it is possible to realize M as the fixed point set of an action on W with isotropy subgroups in F. (Cf. [A1] and [A2] for further applications and solutions to some problems of P.A. Smith, Bredon, and Raymond.)

Existence of an action ψ satisfying the conditions of (E) requires:

(a) existence of a compact G-CW complex X and a homotopy equivalence $f: X \to W$ such that M is an invariant G-CW complex of X and isotropy subgroups of X-M lie in $F - F_0$;

(b) existence of a G-vector bundle ξ over X which is equivalent to the pull-back of the tangent bundle of W by f, and such that $\xi | M$ splits (equivariantly) as the sum of the tangent bundle of M and a bundle equivalent to ν.

(a) has been studied in detail in [A1], and it is shown that under some (necessary) hypotheses, one may define an obstruction $\mathcal{O}(M)$ lying in a finite abelian group $\Omega(F,F_0)$ such that $\mathcal{O}(M) = 0$ if and only if (a) is satisfied. $\Omega(F,F_0)$ depends only on F and F_0, and it is obtained by algebraic constructions on appropriate categories of chain complexes over $\mathbb{Z}G$. It is possible to give criteria for $\mathcal{O}(M)$ to vanish in terms of more familiar homological invariants of M and W. For instance, in the case of $F = \{1,G\}$ and $F_0 = \{G\}$ -- the semifree actions -- $\Omega(F,F_0)$ is the image of the Swan map σ_G in the reduced projective class group of $\mathbb{Z}G$ and $\mathcal{O}(M) = \sum_{i>0} (-1)^i \sigma_G(H_i(M))$. (Cf. [A1], II.6.4 and II.6.5) If F is the family of all subgroups and $F_0 = \{G\}$, then one can demonstrate that $\Omega(F,\{G\})$ is isomorphic to the subquotient of $\tilde{K}_0(\mathbb{Z}G)$ corresponding to the set of obstructions defined by Oliver in [O]. So for $W = D^n$, the n-disk, one obtains the generalizations of the results of [J1], for example. This obstruction always vanishes when X is not required to be compact, and this simplifies the solution of the problem for the non-compact manifolds.

Only some special cases of (b) have been treated in [A1], due to complications arising in calculations involving K_G-theory.

The following theorem applies some of the results of [A1] to provide criteria for existence of extensions in (E) for the case $W = D^n$ or S^n and $F_0 = F - \{1\}$. Once some extension exists, we may classify them by applying some ideas of [A1], Chapter III reminiscent of Hirsch-Mazur type arguments. Here, we state the results for smooth actions on smooth manifolds. A piecewise linear version can also be formulated, and similar results hold for PL manifolds.

Theorem 1. Let M^m be a compact connected G-manifold with isotropy subgroups in F, and suppose $(M,\partial M) \subset (D^n,S^{n-1})$ is embedded with normal bundle ν. Assume that:

(i) ν has the structure of a G-linear bundle over M.

(ii) $H_*(M;\mathbb{Z}/|G|\mathbb{Z}) = 0$, and each $g \in G$, $g\colon M \to M$ is homotopic to the identity map of M.

(iii) $n > 2m+3 \geq 6$.

Then there is an invariant $\mathcal{O}(M) \in \Omega(F,F_0)$, $F_0 = F - \{1\}$, such that $\mathcal{O}(M) = 0$ if and only if the action on M extends to D^n with the property that $M^H \cong (D^n)^H$ for each $H \neq \{1\}$, $H \subseteq G$, and the G-normal bundle of $M \subset D^n$ is ν. (Recently, W. Browder and the author have obtained more general theorems. These results will appear in a later paper.)

The intuitive idea of the proof is the following. First, we identify the disk bundle of ν with a tubular neighborhood N^n of M in D^n. It is possible to enlarge N^n to a submanifold N_k^n by exchanging handles between closure of D^n-N and N by surgery on ∂N, so that N_k is k-connected. To extend the action to N_k, we first construct N_k abstractly, by adding equivariant handles to N along an appropriate submanifold of ∂N which is equipped with the action; then we extend the action of ∂N to these handles. At the same time, we must arrange to extend the trivialization of the tangent bundle TN of N to a trivialization of the tangent bundle of N_k, so that we can embed N_k in D^n(rel N), and obtain the desired k-connected neighborhood of N with an action extending the action of N. This process is delicate. To avoid obstructions which arise in changing a trivialization of TN to a G-trivialization (which may not exist for TN), one must appeal to special arguments and constructions using the hypotheses of the theorem. Once a contractible G-neighborhood N_n is constructed -- using the fact that $\mathcal{O}(M) = 0$ -- the extension to D^n follows routinely by observing that $\overline{D^n - N^n}$ is a collar on part of ∂N^n.

The above extension of the action to D^n, denoted by D_0^n, is obtained fairly canonically with an explicit inductive construction. This enables us to classify such extensions by constructing an equivariant embedding of D_0^n in the interior of any other extension D_1^n which satisfies the conclusion of the theorem. $\overline{D_1^n - D_0^n}$ is an equivariant h-cobordism between ∂D_1^n and ∂D_0^n -- which is a product in a neighborhood of the fixed point sets. This results in an element of Wh(G). Conversely, for each nonzero element of Wh(G) one may use the reverse process and construct a new extension from D_0^n which is not equivariantly diffeomorphic to it.

Theorem 2. The equivariant diffeomorphism classes of the extensions obtained in Theorem 1 are in one-to-one correspondence with elements of Wh(G), the Whitehead group of G.

An application to the semifree case, which is a generalization of a theorem of Lowell Jones [J1] who proved a similar result for G = cyclic group, and initiated the extension problem, follows.

Theorem 3. Let F^k be a compact submanifold of the disk, $(F, \partial F) \subset (D^{n+k}, S^{n+k-1})$, and let ν be the normal bundle with $n > k+1 \geq 2$. In order to have a semifree action on D^{n+k} with $(D^{n+k})^G = F$, it is necessary and sufficient to have:

(i) $\bar{H}_*(F;\mathbb{Z}_{|G|}) = 0$, and $\sum_{i \geq 1} (-1)^i \sigma_G(H_i(F)) = 0$.

(ii) ν is a G-linear bundle over F with a free representation at each fibre. σ_G is the Swan map mentioned above.

The above theorem together with Theorem 2 may be used to classify all semifree action on D^{n+k} subject to the dimension hypothesis $n > \dim(D^{n+k})^G \geq 2$.

The extension problem (E) for $W = \mathbb{R}^n$ or $W = S^n$ can be treated using similar results.

Section 2. In this section we summarize some background material. We refer the reader to [A1] for further details and examples. In the sequel, all G-spaces are G-CW complexes, and they are referred to as G-complexes. The results of this paper hold for smooth and PL actions, and the special case of topological G-actions on manifolds in which a reasonable theory of equivariant regular neighborhoods exist. We remark that there are many topological actions in which equivariant regular neighborhoods do not exist. For instance, let G be a finite group not of prime power order. Then G has a (smooth) fixed-point free action on some $\mathbb{R}^n$, i.e. $(\mathbb{R}^n)^G = \emptyset$.

Let S^n be the one-point compactification of $\mathbb{R}^n$ with the natural G-action on $\mathbb{R}^n$ $\{\infty\}$, leaving $\{\infty\}$ fixed. (Cf. [A2] for instance. This has been constructed first for $G = \mathbb{Z}_{pq}$, $(p,q) = 1$, $p \neq 1$, $q \neq 1$, by Conner and Floyd.) For $G = \mathbb{Z}_{pq}$, for instance, the point $\{\infty\}$ does not have an equivariant regular neighborhood. Since if N were a closed G-regular neighborhood of $\{\infty\}$, then S^n-Int(N) will be a disk D^n with a fixed point free action of $\mathbb{Z}_{pq}$, which is impossible by the Brauer fixed point theorem, for instance.

Let G be a finite group acting on a PL manifold M. M is called a PL G-manifold, (or the action is called PL), if for some simplicial subdivision of M, elements of G act by simplicial maps. The second baricentric subdivision of M gives M the structure of a G-CW complex, (cf. Bredon [B1]). By a result of Illmann, a smooth G-manifold also has the structure of a G-CW complex, for G finite [I2].

Let X be a G-manifold (smooth, PL). Then $S(X) = \bigcup_{\substack{H \subseteq G \\ H \neq 1}} X^H$ is the singular set of a smooth or PL G-action on a smooth, respectively PL, manifold has a G-invariant regular neighborhood which can be taken smooth or PL depending on the action. This G-regular neighborhood may not be unique, but this will not concern us.

Let $X \subset Y$, and $\phi: G \times X \to X$ and $\psi: G \times Y \to Y$ be actions such that $\psi(G \times X) = X$, and $\psi|G \times X =$. We call ψ a free extension of ϕ if $Y-X$ is a free G-space. If no confusion arises, we simply call Y a free extension of X. A G-complex X is called Smith acyclic, if for each prime power order subgroup $H \subsetneq G$, $|H| = p^r \neq 1$, $\tilde{H}_*(X^H;\mathbb{Z}/p\mathbb{Z}) = 0$. This definition is motivated by the celebrated theorem of P.A. Smith stating that for a G-action on a finite dimensional acyclic G-complex X, $\tilde{H}_*(X^H;\mathbb{Z}/p\mathbb{Z}) = 0$ for every prime power order subgroup H, $|H| = p^r \neq 1$. So $S(X) = \bigcup_{\substack{H \subseteq G \\ H \neq 1}} X^H$, the singular set of an action on a finite dimensional contractible complex is Smith acyclic. Naturally, one asks whether the converse is true. The first converse theorem was proved by L. Jones, who gave a constructive proof for the case of semifree actions of the cyclic group $\mathbb{Z}_n$. Namely, if F^k is a finite dimensional connected complex with $\tilde{H}_*(F^k;\mathbb{Z}/n\mathbb{Z}) = 0$, then there is a semifree action of $\mathbb{Z}_n$ on a finite dimensional contractible complex X with $X^{\mathbb{Z}_n} = F$. If F is finite, then X can be taken finite too.

Let X_0 be a Smith acyclic G-CW complex of finite dimension with $\phi: G \times X_0 \to X_0$ the action. Suppose we would like to find a free extension of ϕ to an action $\psi: G \times X \to X$, where X is contractible. To construct X, we proceed to add equivariant G-cells $G \times D^2$ via equivariant maps $\alpha: G \times S^1 \to X_0$ to obtain a 1-connected G-CW complex X_1 containing X_0 and extending the action on X_0 freely. Inductively, we can proceed in this way to obtain a k-connected G-CW complex X_k which extends X_0 freely, and such that $\tilde{H}_{k+1}(X_k)$ is the only nonvanishing homology. Then $H_{k+1}(X_k)$ is seen to be a $\mathbb{Z}G$-projective module, and as such, it represents a class $[H_{k+1}(X_k)] \in \tilde{K}_0(\mathbb{Z}G)$, since X_k is also Smith acyclic, being a free extension of the Smith acyclic complex X_0. (Cf. [A1] or [O].) Furthermore, $[H_{k+1}(X_k)]$ is independent of the free extension X_k, and depends only on $S(X_0)$. It is easily seen that the vanishing of $\mathcal{O}(X_0) = [H_{k+1}(X_k)]$ in $\tilde{K}_0(\mathbb{Z}G)$ is necessary and sufficient for the existence of a contractible finite G-CW complex which is a free extension of the finite G-complex X_0. If X_0 is not finite, or we can drop the finiteness condition on X, then this obstruction does not matter, and can be avoided by an infinite process. So the homological obstruction $\mathcal{O}(X_0)$ is represented by $[H_{k+1}(X_k)] \in \tilde{K}_0(\mathbb{Z}G)$ for any free extension X_k which is k-connected, with $\tilde{H}_{k+1}(X_k)$ as the only nontrivial reduced homology.

In general, if we have a finite G-CW complex X_0 with isotropy groups in a family F of subgroups of G, and if $F_0 \subset F$ is a subfamily, there exists a finite abelian group $\Omega(F,F_0)$, and a well-defined element $\mathcal{O}(X_0) \in (F,F_0)$ such that $\mathcal{O}(X_0) = 0$ if and only if there exists a contractible finite G-CW complex

X containing X_0 as a G-subcomplex, and $X-X_0$ has isotropy subgroups in $F - F_0$, (provided that the necessary Smith theoretic conditions imposed on X_0 are satisfied). In the case of free extensions, there is a monomorphism $\omega: \Omega(F,F-\{1\}) \to \tilde{K}_0(\mathbb{Z}G)$, such that $\omega(\mathcal{O}(X_0)) = [H_{k+1}(X_k)]$ described above. It is perhaps useful to mention that $\mathcal{O}(X_0)$ is well-defined if and only if X_0 is Smith acyclic. This fact is used frequently below without explicitly being mentioned each time. One may replace finite groups by compact Lie groups, to obtain similar results.

One further remark is helpful in the case of free extensions of actions from submanifold of disks. Let $W^n \subset D^n$ be a codimension zero submanifold, and suppose that W is an effective G-manifold. Suppose that we would like to construct a free extension of the action on W^n to D^n. The following necessary and sufficient condition proves useful. We state it under some dimension restrictions which may be relaxed slightly, or replaced by other conditions of similar nature.

2.1. Proposition. Let $W^n \subset D^n$ be as above, and let $\dim W^H < \frac{n}{2}$ for all $1 \neq H \subseteq G$, and $H_i(W^n) = 0$ for $i \geq \frac{n}{2}$. Then the action on W^n extends freely to a contractible compact G-manifold $V^n \supset W^n$ if and only if:

(i) there exists a finite contractible X which is a free extension of W^n. This is satisfied if and only if W^n is Smith acyclic and $\mathcal{O}(W) = 0$.

(ii) there exists a linear G-bundle β over X such that $\beta|W^n$ is stably equivalent to $T(W^n)$. ($T(W^n)$ = the stable tangent bundle of W^n.)

This is a special case of a more general result in [A1] and is obtained by the process of equivariant thickening. (Cf. Chapter III of [A1].)

So in the extension problem, our effort will concentrate on the construction of the G-bundle β over X. This is implicit in the proofs of the extension theorems. We have given direct arguments, however, which adapt the proof of equivariant thickening theorem of [A1] to this special case. We should remark that the G-homotopy type of the contractible G-complex X in the above proposition is uniquely determined by the G-homotopy type of W.

The proof of the following theorem is rather technical and lengthy. The details of the geometric constructions involved in the proof are included in a forthcoming paper. This theorem in fact provides us (almost) with the bundle β mentioned in 2.1 above.

2.2 Theorem. Let W^n be a compact parallelizable manifold with $\partial W \neq \emptyset$, and let $A: G \times W^n \to W^n$ be an effective G-action which is homotopic to the identity on the $(k+1)$-skeleton of W, $0 \leq k < [\frac{n}{2}]$. Assume that:

(1) $\tilde{H}_i(W^n;\mathbb{Z}_q) = 0$ for $i \leq k$, $q = |G|$.

(2) $\dim W^H < n-k$ for each subgroup $H \neq 1$.

(3) $W^G \neq \emptyset$ if $|G| =$ even.

Then there is a k-connected G-manifold V^n with the following properties:

(i) V^n is obtained from W^n by equivariant surgery (on spheres of dimension $\leq k$) in the complement of a sufficiently small equivariant regular neighborhood of the singular set of W.

(ii) The surgeries are framed (possibly non-equivariantly), so that V^n is parallelizable (and framed, if W^n is framed). (V^n is in general non-compact.)

<u>Section 3</u>. Given a linear representation $\rho: G \to O(n)$, on $\mathbb{R}^n$, we can form the universal $\mathbb{R}^n(\rho)$-bundle $\hat{\rho}$, $E_G \times_G \mathbb{R}^n \to BG$, where $E_G \to BG$ is the universal principal G-bundle. The i-th Stiefel-Whitney characteristic class of ρ, $w_i(\rho)$ is defined to be $w_i(\hat{\rho}) \in H^i(BG;\mathbb{Z}_2)$. Other characteristic classes of a linear representation may be defined similarly.

Let $S^{n-1} \subset \mathbb{R}^n$ be the unit sphere of this representation space, and $S_0(\rho)$ be the free stratum of S^{n-1}, i.e., the complement of the singular set (= union of fixed point sets of non-trivial subgroups). To avoid the trivial case, let ρ be faithful, i.e., $S_0(\rho) \neq \emptyset$. Let $p: S^{n-1} \to S^{n-1}/G$ be the orbit map.

<u>3.1. Lemma</u>. If $w_i(\rho) = 0$, then $w_i(S_0(\rho)/G) = 0$.

Proof. Since $\mathbb{TR}^n(\rho) = \mathbb{R}^n(\rho) \times \mathbb{R}^n(\rho)$ with the diagonal G-action (given via ρ), we have the following pullback diagram

$$\begin{array}{ccc} \mathbb{TR}^n(\rho) & \xrightarrow{\pi_2} & \mathbb{R}^n(\rho) \\ \downarrow{\scriptstyle \pi_1} & & \downarrow{\scriptstyle \pi} \\ \mathbb{R}^n(\rho) & \xrightarrow{\pi_0} & * \end{array}$$

To avoid excessive notation, let $\hat{X}_i$ denote $E_G \times_G X_i$ and $\hat{f}: \hat{X}_1 \to \hat{X}_2$ denote the induced map $1 \times_G f$, for any G-space X_i and any G-map $f: X_1 \to X_2$. Thus, we have the pullback diagram:

$$\begin{array}{ccc} \hat{\mathbb{TR}}^n(\rho) & \xrightarrow{\hat{\pi}_2} & \hat{\mathbb{R}}^n(\rho) \\ \downarrow{\scriptstyle \hat{\pi}_1} & & \downarrow{\scriptstyle \hat{\pi}} \\ \hat{\mathbb{R}}^n(\rho) & \xrightarrow{\hat{\pi}_0} & BG = \hat{*}. \end{array}$$

If $\hat{\xi} = (\hat{\mathbb{TR}}^n(\rho), \hat{\pi}_1, \hat{\mathbb{R}}^n(\rho))$ and $\hat{\rho} = (\hat{\mathbb{R}}^n(\rho), \hat{\pi}, BG)$, then $\hat{\pi}_0^* w_i(\hat{\rho}) = w_i(\hat{\pi}_0^*\hat{\rho}) = w_i(\hat{\xi})$. We have also the commutative diagram:

$$\begin{array}{ccc} \tau S_0(\rho) & \longrightarrow & \mathbb{TR}^n(\rho) \\ \alpha \big\downarrow & & \big\downarrow \\ S_0(\rho) & \xrightarrow{\;j\;} & \mathbb{R}^n(\rho) \end{array} \qquad (\text{where } \tau S_0(\rho) = \mathbb{TR}^n(\rho)|S_0(\rho))$$

which yields the commutative diagram:

$$\begin{array}{ccccccc} \tau S_0(\rho)/G & \rightleftarrows & \hat{\tau S_0}(\rho)/G & \rightleftarrows & \hat{\tau S_0}(\rho) & \longrightarrow & \hat{\mathbb{TR}}^n(\rho) \\ \alpha'' \big\downarrow & & \alpha' \big\downarrow & & \hat{\alpha} \big\downarrow & & \big\downarrow \\ S_0(\rho)/G & \xrightarrow[=]{\;1\;} & S_0(\rho)/G & \underset{r}{\overset{q}{\rightleftarrows}} & S_0(\hat{\rho}) & \xrightarrow{\;j\;} & \hat{\mathbb{R}}^n(\rho) \end{array}$$

where q is a homotopy equivalence with a homotopy inverse r, and the bundle $(\hat{\tau S_0}(\rho), \hat{\alpha}, \hat{S_0}(\rho))$ is equivalent to $(\tau S_0(\rho).G, \alpha'', S_0(\rho)/G)$ which denotes the Whitney sum of $TS_0(\rho)/G$ and a trivial line bundle. So $w_i(S_0(\rho)/G) = r^*\hat{j}^* w_i(\hat{\xi}) = r^*\hat{j}^*\hat{\pi}_0^* w_i(\hat{\rho})$, and $w_i(\rho) = w_i(\hat{\rho}) = 0$ implies that $w_i(S_0(\rho)/G) = 0$ as claimed above.

3.2. Remark. 1. If $|G|$ = odd, then $w_i(\rho) = 0$ for any representation ρ. See Remark 8.2.

2. The representation ρ preserves the orientation if and only if $w_1(\rho) = 0$.

3.3 Corollary. If $w_1(\rho) = w_2(\rho) = 0$, and $K \subset S_0(\rho)/G$ is any 2-complex, then $\tau(S_0(\rho)/G)|K$ is trivial.

Proof. By the above lemma $w_i(\tau(S_0(\rho)/G|K) = 0$ for $i = 1,2$, which implies the desired result, because the first and second Stiefel-Whitney classes determine a vector bundle over a 2-complex.

3.4. Proposition. Let ρ be a faithful representation of G such that $w_i(\rho) = 0$ for $i = 1,2$, and let ϕ be a framing of $\tau S^{n-1}(\rho)$. Assume further that $\dim \mathbb{R}^n(\rho)^H < n-2$ for any subgroup $H \neq 1$. Let Γ be a finite 1-dimensional free G-complex, and $f: \Gamma \to S_0(\rho)$ be a G-embedding. Then, there is a framing ϕ' of $\tau S^{n-1}(\rho)$ such that $\phi'|f(\Gamma)$ is equivariant, and ϕ' is homotopic to ϕ.

Proof. We have $\pi_1 S_0(\rho) = 0$ by van Kampen's theorem, since the fixed point set of each subgroup has codimension larger than two. We may arrange for the 2-skeleton of $S_0(\rho)/G$ to contain $f(\Gamma)/G$. So let $K \subset S_0(\rho)/G$ be a finite 2-complex such that $K^1 \supset f(\Gamma)/G$ and ${}_1(K) \xrightarrow{\cong} \pi_1(S_0(\rho)/G)$. Then $p^{-1}(K)$ contains

$f(\Gamma)$. On the other hand, $p^{-1}(K)$ is a simply-connected 2-complex, and hence homotopy equivalent to a bouquet of 2-spheres. Since $\pi_2(SO(n)) = 0$, $[p^{-1}(K), SO(n)] = 0$. By hypothesis and an application of Corollary 3.3, $\tau S_0(\rho)/G|K$ has a framing ψ, which yields an equivariant framing $p^*\psi$ on $\tau S^{n-1}(\rho)|p^{-1}(K)$. On the other hand, $p^*\psi \simeq \phi|p^{-1}(K)$, since $[p^{-1}(K), SO(n)] = 0$. Hence we can find a framing $\phi' \simeq \phi$ such that $\phi'|p^{-1}(K) = p^*\psi$ is equivariant. In particular, $\phi'|f(\Gamma)$ is equivariant.

I would like to thank the referee for providing the relevant pages of Jones' thesis [J]. His argument there differs from the published versions in [J1] and avoids the difficulty that we shall now discuss.

In his argument on pp. 62-63 [J1], Jones considers a map $p : S^1 \longmapsto S^1$ of degree n,
and calls for an appropriate choice of the framing ψ of the tangent bundle of the orbit space restricted to S^1 such that the given framing on the total space satisfies $\phi|S^1 \simeq p^*\psi$, i.e., it is claimed that ϕ can be deformed to a new framing ϕ' such that $\phi'|S^1$ is equivariant with respect to the $\mathbb{Z}_n$-action. One can show by means of examples that this is not in general possible when n is even. For instance, consider the case of S^k being the unit sphere of a linear $\mathbb{Z}_2$-representation ρ with no trivial factors, $S^k = \partial D^{k+1}(\rho)$, and p: $S^k \to \mathbb{R}P^k$ the projection. Consider a $\mathbb{Z}_2$-invariant $S^1 \subset S^k$ and search for a framing of $\tau(\mathbb{R}P^k/S^1)$ as required by Jones. One can verify that such a framing exists if and only if $w_1(\rho) = w_2(\rho) = 0$, which is not possible for the (orientation preserving) case of $k \equiv 1 \pmod 4$. (Further details and a different proof of a generalized version of Jones' Theorem will appear in a future paper of W. Browder and the author.) We will indicate below the modified hypotheses for Jones' Theorem, and an indication of the proof based on the above discussion. For the smooth case, we have a somewhat weaker condition than the PL case, and it is described in terms of the representation on a normal fibre rather than the orbit space (of the free part) of the regular neighborhood. We use his terminology and notation in [J1] Theorem 2.1.

3.5. Theorem. (L. Jones) Let $(K^k, \partial K^k) \subset (D^r, \partial D^r)$ be a submanifold with a (smooth, PL) regular neighborhood $\tau_K \subset D^r$ such that $\tau_K \cap \partial D^r$ is a regular neighborhood of ∂K in ∂D^r. Assume that

(1) $\tilde{H}_*(K;\mathbb{Z}_n) = 0$.

(2) There is a semifree periodic (PL, smooth) homeomorphism $\alpha: \tau_K \to \tau_K$ having period n and fixed point set K.

(3) In the smooth case, let ρ be the representation of $\mathbb{Z}_n$ given by α on the normal bundle of K and:

(3 Smooth) $w_i(\rho) = 0$ for $i = 1,2$

(3 PL) $w_i(\partial\tau_K - \partial K/\alpha) = 0$ for $i = 1,2$.

(4) $2k+2 < r$ and $r \geq 6$.

Then α can be extended to $\beta: D^r \to D^r$, where β is a semifree periodic (PL, smooth) homeomorphism of period n, having K as its fixed point set.

The proof of Jones is now justified by virtue of (3) and Proposition 3.4 above. His claims on pp. 62-63 regarding the framing H are seen to be true: in the first step of the induction, Proposition 3.4 in the smooth case, and a variation of its proof in the PL case, show that the desired framing on S^1 can be constructed, and the argument goes through. In the inductive step, $\pi_1(\partial W_\ell) = 0$, and the circle $S^1 \times$ (base point) can be taken as part of a simply connected $\mathbb{Z}_n$-invariant 2-complex K' in the free stratum of W_ℓ. So the argument of 3.4 shows that the framing can be deformed on $S^1 \times$ (base point) to be equivariant, as claimed by Jones.

Our generalization of Jones' theorem is not restricted by the hypothesis on vanishing of Stiefel-Whitney classes; that is, condition (3) can be eliminated altogether. Note that $\sigma_G \equiv 0$ for $G = \mathbb{Z}_n$, as it is indicated in Lemma 1.1 of [J1].

Section 4. The following theorem gives sufficient conditions for a G-manifold to be a G-submanifold of a G-disk D^n containing the singular set of the action on D^n.

4.1. Theorem. Let M^m be a compact connected submanifold of D^n ($\partial M^m \subset \partial D^n$) with normal bundle ν^k. Let $A_0: G \times M^m \to M^m$ be a G-action homotopic to identity, and $\tilde{A}_0: G \times \nu^k \to \nu^k$ an effective action which gives ν^k a G-bundle structure covering A_0. Assume

(1) $\tilde{H}_*(M;\mathbb{Z}_q) = 0$, $q = |G|$,

(2) $H_i(M) = 0$ for $i \geq r$, where $r < \min\{k, [\frac{n}{2}]\}$; $n \geq 6$.

(3) $\dim D(\nu)^H < n-r+1$, where $D(\nu)$ is the associated disk bundle of ν, and $H \neq 1$ is any subgroup.

(4) If $|G|$ = even, $M^G \neq \emptyset$.

Then there is a G-action $A: G \times D^n \to D^n$ extending $\tilde{A}: G \times D(\nu) \to D(\nu)$ such that $S(D^n) \cong S(D(\nu))$, if and only if $\mathcal{O}(M) = \sum_{i>0} (-1)^i \sigma_G(H_i(M)) \in \tilde{K}_0(\mathbb{Z}G)$ vanishes.

Proof. The necessity has been discussed in Section 2. We need to prove the sufficiency. So assume that $\mathcal{O}(M) = 0$. Then, we may add finitely many G-cells of dimension $\leq r$ to $D(\nu)$ to obtain a G-complex Y which is (r-1)-connected, and $\pi_r(Y) \cong H_r(Y)$ is a finitely generated free $\mathbb{Z}G$-module, $H_i(Y) = 0$ for $i \neq 0,r$, and Y extends $D(\nu)$ freely.

On the other hand, the action $\tilde{A}: G \times D(\nu) \to D(\nu)$ is homotopic to the identity, and $\dim D(\nu)^H < n-r+1$ by (3). Thus, we may apply Theorem 2.2 to obtain an (r-1)-connected G-manifold V which contains a $S(D(\nu)) \cup D_\varepsilon(\nu)$, where $D_\varepsilon(\nu)$ is a sufficiently small tubular neighborhood of M in $D(\nu)$. This is true because $r-1 < n-m = k$, and all surgeries in Theorem 2.2 can be done (by a general position argument) outside of a small neighborhood $D_\varepsilon(\nu)$. Choosing a G-diffeomorphism of $D(\nu)$ and $D_\varepsilon(\nu)$, we may assume that $V \supset D(\nu)$ as well.

Let $h: D(\nu) \to V$ be the inclusion map. Then, using obstruction theory (see Bredon [B2]), h can be extended to $h: Y \to V$, since $\pi_i(V) = 0$ for $i \leq r-1$, and $Y/D(\nu)$ has cells of dimension $\leq r$. Let $\xi = h^*\tau V$. Then, the G-bundle ξ over Y satisfies the properties of the equivariant thickening theorem 2.1 (see also [A1] III.2.3). Therefore, there is a G-manifold $W^n \supset D(\nu)$, and a G-map $f: W \to Y$ which is a homotopy equivalence. Further, $f^*\xi \cong \tau W^n$. So W^n is parallelizable, $H_r(W^n)$ is a free $\mathbb{Z}G$-module, and $H_i(W) = 0$ for $i \neq 0,r$.

Represent a $\mathbb{Z}G$-basis for $H_r(W^n) \cong \pi_r(W^n)$ by maps $\alpha_i: S^r \to W^n$, $1 \leq i \leq s$. By Poincaré duality, $H_j(W,\partial W) \cong H^{n-j}(W) = 0$ for $j \leq r$, since $n-j \geq n-r > r$, and $H_r(W)$ is free. It is easily checked that $\pi_1(\partial W) = \pi_1(W) = 0$, so that $\pi_j(W,\partial W) = 0$ for $j \leq r$. Thus, we may assume that $\alpha_i: S^r \to \partial W^n$ represent the basis. By general position (see condition (3)), we may assume that $\alpha_i(S^r)$ lies in the free stratum of ∂W. Let $p: W \to W/G$ be the orbit map (or any restriction thereof). The dimension condition (2) allows us to make $p\circ\alpha_i: S^r \to (\partial W - S(\partial W))/G$ into disjoint embeddings. Therefore, the induced equivariant maps $\bar{\alpha}_i: G \times S^r \to \partial W - S(\partial W)$ are disjoint embeddings as well.

Since W is parallelizable, and $r < \frac{1}{2}\dim \partial W$, the normal bundle of $\bar{\alpha}_i(e \times S^r) \subset \partial W$ is trivial. It follows easily that the G-normal bundle $\bar{\alpha}_i(G \times S^r)$ is equivariantly trivial. Hence, we may assume that we have disjoint G-embeddings $\alpha_i': G \times S^r \times D^{n-r-1} \to \partial W$ with images in $\partial W - S(\partial W)$. We may attach the (r+1)-handles $G \times D^{r+1} \times D^{n-r-1}$ to ∂W using α_i'. Clearly, this gives a contractible G-manifold U^n which is diffeomorphic to D^n by the h-cobordism theorem. It is routine to check that this action on D^n satisfies the requirements of the theorem.

<u>4.2. Theorem</u>. Let M^m be as in the above Theorem 4.1, satisfying (1)-(4). Then, there is a G-action $A: G \times \mathbb{R}^n \to \mathbb{R}^n$ extending $\tilde{A}: G \times D(\nu) \to D(\nu)$, such that $S(\mathbb{R}^n) \cong S(\mathrm{int}(D(\nu)))$. Furthermore, the obstruction to finding an equivariant boundary for $(\mathbb{R}^n;A)$ is $\mathcal{O}(M) = \sum_{i>0} (-1)^i \sigma_G(H_i(M)) \in \tilde{K}_0(\mathbb{Z}G)$.

Proof. As in the proof of Theorem 4.1, we construct the G-manifold $W^n \supset D(\nu)$, with $\bar{H}_i(W) \neq 0$ only for $i = r$. Since $H_r(W)$ is $\mathbb{Z}G$-projective, we can kill it by an infinite process as in the proof of Theorem 2.2 (the details will appear later) by adding r- and (r+1)-handles equivariantly to obtain the open contractible G-manifold U^n containing W^n. U^n is seen to be simply-connected at ∞, hence diffeomorphic (PL homeomorphic if we are dealing with PL actions), to $\mathbb{R}^n$ by Stalling's theorem [S2].

Having such an action on $\mathbb{R}^n$, we would like to find a compact G-manifold $V^n \subset \mathbb{R}^n$ such that $\mathbb{R}^n\text{-int}(V^n) \cong \partial V^n \times [0,1)$, as in Siebenmann [S4]. Since $M^m \subset \mathbb{R}^n$ is compact, we can apply the method of Siebenmann to $D(\nu)$, to obtain the compact G-manifold W with $\bar{H}_i(W') \neq 0$ only for $i = r$, (as in W above). As before, $H_r(W')$ represents $\mathcal{O}(M) = \sum_{i>0} (-1)^i \sigma_G(H_i(M)) \in \tilde{K}_0(\mathbb{Z}G)$. Let $N(\partial W')$ and $N(\mathbb{R}^n)$ be equivariant regular neighborhoods of the singular sets of actions on $\partial W'$ and $\mathbb{R}^n$ respectively. Clearly, $N(\partial W')$ can be chosen so that $N(\mathbb{R}^n)\text{-int}(W') \cong \partial W' \times [0,1)$. Let $\partial_+ W' = \partial W' - (\partial W' \cap N \subset \mathbb{R}^n)$, and $X^n = \mathbb{R}^n\text{-int}(N(\mathbb{R}^n))$. Then, the generators of $H_{r+1}(\mathbb{R}^n, W')$ $(\cong H_r(W'))$ can be represented by equivariant embeddings $G \times (D^{r+1}, S^r) \to (X^n\text{-int}(W' \cap X^n), \partial_+ W')$ by general position. Since the end of X^n/G is seen to be tame, Siebenmann's argument [S4] can be easily adapted to show that his projective obstruction is precisely $[H_r(W') \in \tilde{K}_0(\mathbb{Z}G)]$, since $\pi_1(X^n/G) \cong G$.

The proof of the following theorem is parallel to the proof of Theorem 4.1.

<u>4.3. Theorem</u>. Let M^m, ν, and A_0 be as in Theorem 4.1. Assume that they satisfy the following hypotheses:

(1)' $\bar{H}_*(M;\mathbb{Z}_q) = 0$, $q = |G|$.

(2)' $m < \min\{\frac{n+1}{2}, n\text{-dim } D(\nu)^H + 1, H \subseteq G, H \neq 1\}$, $k > 2$, $n \geq 6$.

(3)' If $|G|$ = even, then $M^G \neq \emptyset$.

Then, there is a G-action $A: G \times D^n \to D^n$ extending A_0 such that some equivariant regular neighborhoods of $S(M)$ and $S(D^n)$ are G-diffeomorphic if and only if $\mathcal{O}(M) \in \tilde{K}_0(\mathbb{Z}G)$ vanishes.

<u>4.4. Theorem</u>. Let $(F^k, \partial F^k) \subset (D^n, \partial D^n)$ be a connected submanifold of the disk with normal bundle ν, such that $k \leq [\frac{n}{2}]$, $n \geq 6$ and ν has the structure of a G-linear bundle with a free representation ρ at each fibre. Then necessary and sufficient conditions for existence of a semifree G-action on D^n with fixed point set F with the equivariant normal bundle ν are

(i) $\bar{H}_*(F;\mathbb{Z}/|G|\mathbb{Z}) = 0$,

(ii) $\mathcal{O}(F) = \sum_{i>0} (-1)^i \sigma_G(H_i(F)) = 0$, where σ_G is the Swan homomorphism.

Proof. Let $W = D(\nu)$, the equivariant disk bundle of ν. Then the G-action on W is homotopic to identity. Furthermore, $H_i(F) = 0$ for $i > k-2$, and $\widetilde{KO}^{-1}(F)$ has no $|G|$-torsion, by an Atiyah-Hirzebruch spectral sequence argument. So by Theorem 2.2 and 4.1, the theorem follows. The necessity of the conditions are discussed in Section 2.

The G-diffeomorphism types of (D^n,A) obtained in Theorems 4.1, 4.3, and 4.4 are classified under a slightly stronger dimension hypothesis to allow standard immersion-embedding techniques.

4.5. Theorem. Let M^m, ν, and A_0 be given as in Theorem 4.3, satisfying (1)'-(3)' and the following:

(4)' $\mathcal{O}(M) \in \widetilde{K}_0(\mathbb{Z}G)$ vanishes.

(5)' $m < \frac{n}{2}$.

Then, the G-diffeomorphism classes of extensions satisfying the conclusion of Theorem 4.3 are in one-to-one correspondence with Wh(G).

Proof. For a G-manifold X, let N(X) denote an equivariant regular neighborhood of the singular set $S(X)$, (see Section 2). Let (D^n,A) be the extension constructed in 4.3, and let (D_1^n,A_1) be any other extension, so that the equivariant regular neighborhoods $N_0 = N(D^n;A)$ and $N_1 = N(D_1^n;A_1)$ are G-diffeomorphic. As usual, we may assume that we have arranged so that $N_0 \cap \partial D^n = N(\partial D^n)$ and $N_1 \cap \partial D_1^n = N(\partial D_1^n)$, so that $\partial_+ N_i = \partial N_i - \mathrm{int}(N_i \cap \partial D_i^n)$ has a free action. By obstruction theory (Bredon [B2]) we can extend f to an equivariant map $F: D^n \to D_1^n$. Since D^n is obtained by adding free handles of index $< \frac{n}{2}$, F can be deformed into an isovariant map by a homotopy rel $\partial_+ N_0$, that is, $F(D^n - N_0) \subset D_1^n - N_1$. Since $\pi_1(D_i^n - N_i) = 1$, a Mayer-Vietoris argument and Whitehead's theorem show that F is indeed a G-homotopy equivalence. Add an equivariant collar to ∂D_1^n, so that $F(D^n) \subset \mathrm{int}(D_1^n)$. Immersion and embedding theory applied to the inductive stages of construction of D^n (as in the proof of Theorem 4.1, where in the k-th stage we have only added handles of index $\leq k+1$ to $\partial_+ N_0$) shows that F may be made into an equivariant embedding $h: D^n \to \mathrm{int}(D_1^n)$ extending $f: N_0 \to N_1$. The cobordism $U^n = D_1^n - \mathrm{int}\ h(D^n)$ is an equivariant h-cobordism which is a product in some equivariant regular neighborhood of the singular set, i.e., $N(U^n) \cong N(\partial D_1^n) \times [0,1] \cong N(\partial h(D^n)) \times [0,1]$. Thus, the free G-manifold $V^n = U^n - \mathrm{int}(N(U^n))$ is a relative equivariant h-cobordism. Hence V^n/G is an h-cobordism based on $(\partial_0 V^n)/G = h(\partial D^n - \mathrm{int}(\partial D^n \cap N))/G$. Let $t(D_1^n,D^n) \in Wh(G)$ be the torsion of this cobordism. It follows from the s-cobordism theorem that $t(D_1^n,D^n) = 0$, if and only if D^n and D_1^n are G-diffeomorphic. Since such Whitehead torsion elements classify V^n (based on $\partial_0 V^n$)

up to G-diffeomorphism, the mapping $D_1^n \to t(D_1^n, D^n)$ is a monomorphism.

Conversely, given an element $t_0 \in Wh(G)$, we construct a relative h-cobordism X^n with $\partial_0 X^n = (\partial D^n - \mathrm{int}(\partial D^n \cap N))/G$ and torsion $\tau(X^n, \partial_0 X^n) = t_0$. Then, the universal cover $\tilde{X}^n$ is an equivariant h-cobordism which we add to D^n along the common piece of boundary $\partial_0 X^n = \partial D^n - \mathrm{int}(\partial D^n \cap N)$. After smoothing corners equivariantly, $D_2^n = D^n \cup \tilde{X}^n$ is a new extension with $t(D_2^n, D^n) = t_0$. This establishes the desired one-to-one correspondence in which the extension D^n corresponds to $0 \in Wh(G)$.

From 4.4 and 4.5, we obtain a complete classification of semifree G-actions on disks (in the general position range).

4.6. Theorem. Semifree G-actions on D^n with fixed point sets of dimension $k < [\frac{n}{2}]$, $n \geq 6$, are classified up to G-diffeomorphism by:

(i) the diffeomorphism type of fixed point set

(ii) the G-linear bundle isomorphism type of the equivariant normal bundle to the fixed point set, and

(iii) an element of $Wh(G)$.

4.7. Remark. Unlike G-actions on disks, the Whitehead torsion element disappears when we consider the restriction of the action to interior (D^n), or equivalently, its extensions to $\mathbb{R}^n$ (obtained by adding an open collar to ∂D^n). The proof is similar to Theorem 4.5 together with an infinite process (after Stallings [5]) to show that half-open G-h-cobordisms of the type obtained in 4.5 are equivariantly a product. So all free extensions to $\mathbb{R}^n$ are G-diffeomorphic.

Section 5. One can apply the results on exention and classification on disks to obtain similar results for the case of spheres. It turns out that for the existence of extensions of G-actions from submanifolds of spheres, we encounter another obstruction, which will be shown to be a well-defined element of the Rothenberg's group of "Semilinear G-spheres". Moreover, the classification of such extensions (if any exists at all) is obtained "modulo the action of the group of semilinear spheres. Some information has been obtained about the above mentioned groups in the semifree case by Browder-Petrie [B-P] and Rothenberg [R2][R3] and Rothenberg-Sondow [R-S] among the earlier results, and the reader is referred to R. Schultz's recent paper [S] for more results, and a systematic treatment of related problems, as well as an extensive bibliography.

In order to define an obstruction for existence of an extension, we define the set of n-dimensional semilinear G-spheres to be the collection S_n of G-diffeomorphism classes of smooth G-actions on S^n such that the fixed-point sets are standard spheres of smaller dimension which are embedded with trivial normal bundle. To obtain a well-defined obstruction, we introduce the notion of semilinear concordance between such objects to be (the G-diffeomorphism class of) a G-action on $S^n \times [0,1]$ with fixed-point sets diffeomorphic to $S^k \times [0,1]$. Then the concordance classes of n-dimensional semilinear G-disks form a set C_n. In such generality, there is no obvious notion of connected sum or a well-defined base-point to induce a monoid structure. However, the obstruction of our interest lies in this set, and in in order to calculate we will discuss briefly the extra hypotheses to be imposed, as need arises.

Now suppose $W^n \subset S^n$ is a codimension zero compact smooth submanifold and $\varphi : G \times W^n \longrightarrow W^n$ is an effective smooth action. In order to apply our previous results, we make the following assumptions (some of which may be relaxed).

(S1) $H^*(W^n;\mathbb{Z}_q) = H_*(S^k;\mathbb{Z}_q)$, $q = |G|$, for some $k < \frac{n}{2}$, $n \leq 6$.

(S2) Each W^H is a closed submanifold, with dimension of each connected component less than $n-k+1$, and $W^G \neq \emptyset$.

(S3) The action φ on W^n is homotopic to the identity.

5.1. Remark. Using obstruction theory, condition (S3) can be shown to be equivalent to G acting trivially on $H_*(W)$, at least when W is simply-connected.

Let R be a G-invariant regular neighborhood of $S(W)$, and $U \subset W$ be a closed G-invariant disk neighborhood of some stationary point x_o. Smith theory together with our dimension hypotheses imply that $\dim S(W) \leq k$. Using handle-body theory, we observe that W can be given a handle-decomposition with handles of index no greater than k. R also can be given such a handle-decomposition. (In fact, using induction on the number of strata, and a straightforward generalization of non-equivariant handle-body theory one can assume that this handle-decomposition is equivariant, see also Wasserman [W]). Then, W is obtained by adding equivariant free handles of index no greater than k to ∂R. First choose $U^n \subset S^n$ to be a linear G-disk neighborhood of $x_o \in W^G$ such that $U \cap R$ is a G-invariant regular neighborhood of $S(U)$. Now we may re-embed W^n in S^n, by embedding the free G-handles of W^n to $\partial(R-U)$, since these handles are of index no greater than k and $k < \frac{n}{2}$. By abuse of notation call the image of this embedding also W, and observe that we have the following arrangement. Set $D^n_o = S^n\text{-interior}(U)$, $W_o = W\text{-interior}(U)$, $R_o = R\text{-interior}(U)$. Then $W_o \subset D^n_o$, and $\partial W_o \cap \partial D^n_o = R_o \cap \partial D^n_o$ is a G-invariant regular neighborhood of $S(\partial W_o)$, and that $U \cap R$ is a G-invariant regular neighborhood of $S(U \cap R)$. Thus $\partial D^n_o - \partial W_o \neq \emptyset$, and $W^n_o \subset D^n_o$ together with the action $\varphi_o = \varphi|W_o$ satisfies the conditions that $D(\nu)$ does in the notation and the proof of Theorem 4.1.

The proof of Theorem 4.1. applies to this situation, to give a free extension $\bar{\varphi} : G \times D^n_o \longrightarrow D^n_o$ of φ_o if and only if $\mathcal{O}'(W) = \mathcal{O}(W_o) = \sum_{i=1}^{k-1} (-1)^i \sigma_G(H_i(W))$ vanishes in $\widetilde{K}_o(\mathbb{Z}G)$. Now assume that $\mathcal{O}'(W) = 0$, and we have chosen a free extension, say $(D^n_o,\bar{\varphi})$; then the action $(\partial D^n_o,\bar{\varphi} \mid \partial D^n_o)$ is semilinear, and it represents an element of S_{n-1}. Moreover, the proof of the classification theorem shows that the concordance class of $(\partial D^n_o,\bar{\varphi} \mid \partial D^n_o)$ in C_{n-1} is well-defined, and it only depends on (W_o,φ_o) - in fact, it depends only on the "concordance class of W^n ", if we define "concordance" appropriately in this case. At this point, it is clear that a necessary condition for existence of a free smooth action to S^n is that $(\partial D^n_o,\bar{\varphi} \mid \partial D^n_o)$ bounds a semilinear action on a disk D^n, say (D^n,ψ). Conversely, if the concordance class of $(\partial D^n_o,\bar{\varphi}|\partial D^n_o)$ is represented by the unit sphere of some linear representation ρ, say $(S^{n-1},\rho \mid S^{n-1})$, then $(\partial D^n_o;\bar{\varphi}|\partial D^n_o)$ will bound a semilinear disk (D^n,ψ), and $(D^n_o,\bar{\varphi}) \cup (D^n,\psi)$ gives a smooth G-action on S^n say $(S^n,\hat{\varphi})$. However, it is not clear that $(S^n,\hat{\varphi})$ is a free extension of (W^n,φ) ; rather it is only an extension in the complement of a small linear disk - although $(S^n,\hat{\varphi})^G$ is certainly diffeomorphic to $(W^n)^G$. To obtain a somewhat better formulation, we impose the following extra hypothesis taken from [R3] (p. 293). A G-manifold M is said to satisfy <u>the codimension</u> i condition if $\dim M^H - \dim M^{H'} \neq i$, for $H \subset H' \subset G$, and $\dim M^{H'} > 0$. If M is the underlying G-space of a linear representation ρ of G, then we say ρ satisfies the codimension i condition if M does.

(S4) There is a positive dimensional component of W^G with a slice representation ρ' which satisfies the codimension 1 and 2 conditions, and $\dim W^H \geq 6$ for all $H \neq G$.

We also need the following definition :

<u>5.2. Definition</u>. A G-homeomorphism $f : X \longrightarrow Y$ between two smooth G-manifolds is called an almost G-diffeomorphism, if in the complement of some linear disk neighborhood D^n_x of a stationary point $x \in X^G$, $f : X^G - D^n_x \longrightarrow Y - f(D^n_x)$ is a G-diffeomorphism.

Write $\rho' = \rho \oplus \mathbb{R}$, where G-action on $\mathbb{R}$ is trivial. Then $(\partial D_o,\bar{\varphi} \mid \partial D^n_o)^G \neq \emptyset$, and the slice representation at each stationary point is ρ. The G-diffeomorphism classes of such semilinear spheres form an abelian group under (G,ρ)-oriented connected sum - see [R3] or [R-S] for more details. The choice of a (G,ρ)-orientation at the chosen stationary point $x_o \in W^G$ induces a (G,ρ)-orientation for $(\partial D_o,\bar{\varphi} \mid \partial D^n_o)$, and the concordance class of $(\partial D_o,\bar{\varphi} \mid \partial D^n_o)$ is a well-defined class μ in Rothenberg's group $C_{n-1}(\rho)$, which is the abelian group of concordance classes of semilinear spheres with slice representations equivalent to ρ, (if the stationary point set is connected, this is automatic, otherwise one must impose the condition at both stationary points in the disconnected case.) Then an equivariant engulfing argument, as in Illman [I3] (see also Rothenberg [R3] for the statements and indication of a proof) shows that semilinear disks with slice representation ρ' are G-homeomorphic [1].

We have the following :

5.3. Theorem. Suppose $W^n \subset S^n$ is given with an effective smooth action $\varphi : G \times W^n \longrightarrow W^n$ which satisfies conditions (S1)-(S4) above. Then there is a well-defined class $\delta(W^n) \in C_{n-1}(\rho)$ which depends only on (W^n,φ), and the choice of a stationary point $x \in W^G$ with the slice representation ρ (subject to (S4)), provided that $\mathcal{O}'(W) = \sum_{i=1}^{k-1} (-1)^i \sigma_G(H_i(W)) \in \tilde{K}_o(\mathbb{Z}G)$ vanishes.

5.4. Theorem. Let $W^n \subset S^n$, and $\varphi : G \times W^n \longrightarrow W^n$ be an effective smooth action satisfying (S1)-(S4). Then the following conditions together are necessary and sufficient in order to have a smooth G-action $\psi : G \times S^n \longrightarrow S^n$ such that (S^n,ψ) would have a G-submanifold $(W'^n,\psi \mid W'^n)$ almost G-diffeomorphic to (W^n,φ).

(1) $\mathcal{O}'(W)$ vanishes in $\tilde{K}_o(\mathbb{Z}G)$.

(2) $\delta(W^n)$ vanishes in $C_{n-1}(\rho)$, where $\rho' = \rho \oplus \mathbb{R}$.

If such an extension exists, then $(W',\psi|W')$ is G-homeomorphic to (W,φ). □

The special case of semifree actions is worthwhile to be stated separately, since it completely characterizes the stationary point sets of such actions.

5.5. Theorem. Let $F^k \subset S^n$ with normal bundle ν, $k < \frac{n}{2}$, $n \geq 6$. The necessary and sufficient conditions for existence of a semifree G-action on S^n with stationary point set F^k are the following :

(1) $H_*(F^k;\mathbb{Z}_q) = H_*(S^k;\mathbb{Z}_q)$, $q = |G|$.

(2) ν admits a G-linear bundle structure with a free G-representation ρ at each fibre.

(3) $\mathcal{O}'(F) = \sum_{i=1}^{k-1} (-1)^i \sigma_G(H_i(F)) \in \tilde{K}_o(\mathbb{Z}G)$ vanishes.

(4) $\delta(F,\nu) \in C_{n-1}(\rho)$ vanishes for some choice of a G-bundle structure on ν with slice representation $\rho' = \rho \oplus \mathbb{R}$.

---

[1] I am grateful to S. Illman for helpful discussions on his results in [I3].

5.6. Corollary. Keep the notation and assumptions of Theorem 5.5. If only hypotheses (1)-(3) are satisfied, then there is a locally smooth action on S^n (differentiable in the complement of a point) with $(S^n)^G = F^k$.

This corollary uses the theorem of Connell-Montgomery-Yang [CMY] . Similarly, using the generalization of this theorem by S. Illman [I3] (See also Rothenberg [R3]) we have :

5.7. Corollary. Let $W^n \subset S^n$, and $\varphi : G \times W^n \longrightarrow W^n$ satisfy (S1)-(S4), and that $\mathcal{O}'(W) = 0$. Then there is a locally smooth action (S^n,ψ) such that (S^n,ψ) has a G-submanifold $(W^n\psi|W')$ which is G-homeomorphic (and almost G-diffeomorphic) to (W^n,φ) . □

5.8. Remark. In Theorem 5.5. condition (4) makes sense, due to the following remark (first observed by R. Stong) that for finite groups G which admit a free representation, all such irreducible representations are of the same type, and of the same dimension. This can be verified by examination of such representations described in J. Wolf's book [W1] . Thus, if ν admits a G-bundle structure with a free representation ρ , then one can vary the G-structure of ν by varying ρ . This will give (a priori) different choices $\delta(F,\nu)$ in the corresponding $C_{n-1}(\rho)$ depending on the chosen G-linear structure on ν .

Let M be a PL manifold with G-action. The action of G on M is called a PL G-action, if for some simplicial structure on M elements of G act by simplicial maps. This notion is in general different from the notion of a G-PL action, (as defined by Illman and Rothenberg, for instance). The fixed-point sets of a PL G-action are not necessarily submanifolds, but they are only subcomplexes. Thus, if we have to deal with fixed-point sets of non-trivial subgroups, e.g. doing handle-additions, taking cobordisms etc., this type of actions do not satisfy enough regularity conditions to allow such constructions. However, for our purposes, such actions are appropriate, since we deal with the free stratum only, as for as such constructions are concerned. Thus, to decide the questions related to the existence and classifications of semifree actions, such a general definition is sufficient. In particular, this was noted long ago by L. Jones in the case of periodic homeomorphisms, and his theorems are all proved in the PL framework [J1] .

Suppose M has a PL G-action, so that elements of G act by simplicial maps for some simplicial structure on M . By taking a finer subdivision (in fact second bary-centric subdivision suffices according to Bredon [B1] chapter 3, see also Illman [I2].) M will have the structure of a G-cw complex. The above theorems and argument remain valid for PL G-actions, having in mind a few minor modifications which arise from the fact that there are no normal bundles to the fixed point sets ; rather, one must take G-invariant regular neighborhoods. Some theorems take simpler form in the case of PL G-action. For instance :

5.9. Theorem. Let $F^k \subset S^n$ be a PL submanifold, $n \geq 6$, $k < \frac{n}{2}$, with a regular neighborhood $R \subset S^n$. Then the following conditions are necessary and sufficient for the existence of a semifree PL G-action on S^n with stationary point set F :

(1PL) $H_*(F^k;\mathbb{Z}_q) = H_*(S^k;\mathbb{Z}_q)$, $q = |G|$.

(2PL) R admits a semifree PL G-action with stationary point set F^k .

(3PL) $\mathcal{O}'(F) = (-1)^i = \Sigma (-1)^i \sigma_G(H_i(F)) = 0$.

The proof uses straightforward modification of the argument in the smooth case to obtain a action on the disk D^n_o extending the action on R^n_o . Then one adds a cone $c(\partial D^n_o)$ to D^n to obtain the PL G-action $\psi: G \times S^n \longrightarrow S^n$ with $(S^n)^G = F$.

The classification of extensions of the type described in Theorem 5.4, and the semifree case in Theorem 5.5 is reduced to the case of disks, by removing a small G-invariant disk, and using our classification theorem 4.6. The argument is quite parallel to the one presented in [R3] (See also [D-R]), using theorem 4.6 instead of the classification of semilinear disks. The details are left to the reader.

Further calculations and results on classifications of actions on spheres are announced in [A1] chapter 6, where some related results of Rothenberg [R2] , Rothenberg-Sondow [R-S] , and Dovermann-Rothenberg [D-R], and their relations to our results are discussed. In particular, for $G = \mathbb{Z}_q$, where q is a prime power, Dovermann-Rothenberg have independently and simultaneously obtained classification results for semifree actions on disks, under the condition that the multiplicity of each irreducible factor in the slice representation of the stationary-point set is greater than the dimension of the stationary point set. Their classification is more intrinsic, however, using the Reidemeister torsion invariant, which excludes the need for reference to a particular fixed extension as in our case. The existence of extensions for this particular case is covered by L. Jones' theorem [J1] . The reader should consult [D-R] for a summary of their results in this direction and related topics.

I would like to express my thanks to the Institut des Hautes Etudes Scientifiques for their hospitality while this manuscript was revised.

REFERENCES

[A1] Assadi, A.H. : "Finite Group Actions on Simply-Connected Manifolds and CW Complexes," Thesis, Princeton University, 1979 (to appear as Memoir AMS, n° 269).

[A2] Assadi, A.H. : "Some Examples of Finite Group Actions," Proc. of Waterloo Conference on Topology, (June 1978), Springer-Verlag Lecture Notes.

[A3] Assadi, A.H. : "Extensions of Finite Group Actions from Submanifolds of a Disk," preprint, The Institute of Advanced Study (1979).

[A-H-V] Alexander, J.P., Hamrick, G.C. and Vick, J.W. : "Involutions on Homotopy Spheres," Inven. Math. 24, 35-50 (1974).

[B1] Bredon, G. : "Introduction to Compact Transformation Groups," Academic Press (1972).

[B2] Bredon, G. : "Equivariant Cohomology Theories," Springer-Verlag Lecture Notes.

[B-P] Browder, W. and Petrie, T. : "Diffeomorphisms of Manifolds and Semifree Actions on Homotopy Spheres", Bull. AMS 77 (1971).

[C-M-Y] Connell, E.H., Montgemery,D. , Yang, C.T. : "Compact Groups in E^n " Ann. of Math. (2) 80 (1964) 94-103 ; correction, ibid. 81 (1965) 194.

[D-R] Dovermann, K.H. , Rothenberg, M. : "An Equivariant Surgery Sequence and Equivariant Diffeomorphism and Homeomorphism Classification (A Survey)". Proc. Siegen Conference, Springer-Verlag Lecture Notes.

[I1] Illman, S. : "Equivariant Algebraic Topology", Thesis, Princeton Unversity (1972).

[I2] Illman, S. : "Smooth Equivariant Triangulations of G-manifolds for G a Finite Group", Math. Ann. 233, 99-220 (1978).

[I3] Illman, S. : "Recognition of Linear Actions on Spheres" 1979 (preprint).

[J1] Jones, L. : "The Converse to Fixed Point Theorem of P.A. Smith I," Ann. of Math. 94 (1971), 52-68.

[J2] Jones, L. : "Converse to Fixed Point Theorem of P.A. Smith II," Indiana U. Math. J. 22 (1972), 309-325.

[J] L. Jones: Ph. Thesis, Yale University 1970.

[O] Olivier, R. : "Fixed-point Sets of Groups Actions on Finite Acyclic Complexes," Commen. Math. Helv. 50 (1975), 155-177.

[R1] Rim, D.S. : "Modules Over Finite Groups", Ann. of Math. 69 (1959), 700-712

[R2] Rothenberg, M. : "Differential Group Actions on Spheres", Proc. Ad. Study Inst. of Alg. Top., Aarhus (1970).

[R] Rothenberg, M. : "Torsion Invariants and Finite Transformation Groups", Proc. Symp. Pure Math. vol. 32, Part I, AMS (1978).

[R-S] Rothenberg, M. and Sondow, J. : "Nonlinear Smooth Representations of Compact Lie Groups", Pac. J. Math. (84) 1979, 427-444.

[S1] Schultz, R. : "Differentiable Group Actions on Homotopy Spheres, II : Ultrasemi-free Actions", Trans. AMS (268) 1981, 255-297.

[S2] Swan, R. : "Periodic Resolutions for Finite Groups", Ann. of Math. 72 (1960), 267-291.

[S3] Stallings, J. : "The Piecewise Linear Structure of Euclidean Space", Proc. Camb. Phil. Soc. 58 (1962), 481-488.

[S4] Stallings, J. : "On Infinite Processes Leading to Differentiability in the Complement of a Point", Diff. and Comb. Topology, S. Cairns, Editor, Princeton University Press, 1965.

[S5] Siebenmann, L. : "Finding Boundaries for Open Manifolds", Thesis, Princeton University, 1965.

[W] Wasserman, A. : "Equivariant Differential Topology", Topology (8) 1969.

[W1] Wolf, J. : "Spaces of Constant Curvature", 4-th Ed., Publish or Perish, Berkeley Cal. 1977.

Canadian Mathematical Society
Conference Proceedings
Volume 2, Part 2 (1982)

DIHEDRAL GROUP ACTIONS ON HOMOTOPY SPHERES

Karl Heinz Dovermann[1]

1. Introduction

Suppose $G = D_q$ is the dihedral group with $2q$ elements and $q = 3, 5,$ or 7.

Main Theorem: There exist smooth actions of G on homotopy spheres Σ such that

(1.1) $$2 \dim \Sigma^{\mathbb{Z}_2} + \dim \Sigma^{\mathbb{Z}_q} < \dim \Sigma + 2 \dim \Sigma^G.$$

(We mean strict inequality.) The lowest dimensional examples constructed here ($G = D_3$) are of dimension 35. Actions of larger groups can be studied, in particular of groups D_q, $q > 7$. We omit this to avoid additional arguments treating an obstruction in $\tilde{K}_0(\mathbb{Z}[G])$.

These actions are a piece of information which fits into the following study of transformation groups:

We are given some data as

(a) the acting group G

(b) the homology of the underlying space

(c) isotropy groups

(d) regularity of the underlying space (smooth actions, G-CW complexes, etc.)

Then we want to understand how these data restrict some other invariant.

One such invariant could be a relation between dimensions of fixed point sets for subgroups of G. Here is some history to the study of such invariants.

Montgomery-Yang [M1,2,3]

(1.2) wanted to show that $\geq$ holds in (1.1) (this motivated our study)

(1.3) conjectured that = holds in (1.1), this is true for linear spheres, and they checked low dimensional examples (private communication). Conjecture (1.3) turned out to be false by

THEOREM A [DP1]: Let G be a finite group. There exists a function h_G such that for all smooth G homotopy spheres Σ

AMS(MOS) subject classification (1980): 57 S17, S25, R65.

[1]The author was supported in part by NSF grant MCS 8100751.

© 1982 American Mathematical Society
0731-1036/82/0000-0461/$06.50

$$\dim \Sigma^G = h_G\{\dim \Sigma^H | H \text{ proper subgroup of } G\}$$

if and only if G is a noncyclic group of prime power order.

The existence of this function h_G has been shown for elementary abelian p-groups by Borel [B,p.175] and for arbitrary noncyclic groups of prime power order in [DH]. For q odd we still find

$$\frac{1}{2}(\dim \Sigma - \dim \Sigma^{\mathbb{Z}_q}) \equiv \dim \Sigma^G \mod 2$$

(compare this with 1.3) even for mod q cohomology spheres in the category of G complexes, see [St]. It follows from Smith theory that $\dim \Sigma - \dim \Sigma^{\mathbb{Z}_q}$ is even. As a starting point of the study of dimension relations we can consider

THEOREM B (Artin [A]): If G is a noncyclic finite group there exists a function h_G such that for unit spheres $\Sigma = S(V)$ in orthogonal representations V

$$\dim \Sigma^G = h_G\{\dim \Sigma^H | H \text{ proper subgroup of } G\}.$$

By making rather strong assumptions on the homology of the underlying space (see (b)) and even allowing G CW actions (see (d)) tom Dieck showed

THEOREM C [tD]: Let G be nilpotent, and Σ a semilinear homotopy sphere. Then there exists a G representation V such that $\dim \Sigma^H = \dim S(V)^H$ for all $H \subseteq G$.

This general result by tom Dieck is unpublished and only the "complex" semilinear case is treated in [tD]. There are two results which relate to condition (c).

THEOREM D (Bredon, see [MY2], [M], [L]). If Σ is a smooth mod 2 homology sphere with smooth D_q action and $\mathbb{Z}_q$ is not an isotropy group then 1.3 holds.

THEOREM E (Swan, [S]). There exists a finite D_3 CW complex X which is homotopy equivalent to S^3 and such that the action is free.

THEOREM F (Pardon [Pa]). There exist smooth free D_q actions (q an odd prime) on mod p homology spheres ($p \neq 2$). (As an easy example we could even use SO_3.)

These last two theorems show how examples for (1.1) can be found if we work with a weak assumption in (b) or (d). So our Main Theorem shows how Theorem D gives the minimal assumptions to obtain a dimension relation for smooth manifolds. It was 1.2 which motivated this study. But furthermore it gives us the possibility to demonstrate the way many concepts in equivariant surgery are applied.

I want to thank Ted Petrie for many discussions which helped to solve this problem. I admire the frankness in which Deane Montgomery discussed his progress on this problem. I want to thank R. Schultz, W. Pardon, G. Carlsson and J. Milgram for helpful discussions. Ian Hambleton was very helpful in explaining the algebra of surgery obstruction groups to me.

The proof of the Main Theorem is in Section 2. From this proof we separate two parts. One of these parts is the proof that certain bundle data are appropriate for doing surgery. This together with some general remarks on surgery theory is the content of section 3. We shall also encounter a surgery obstruction whose computation is more involved. In section 4 we describe the analysis of the surgery obstruction and compute its contribution arising from semicharacteristics. The remaining part deals with the contribution from Reidemeister torsion. There we generalize (in part) the theory from [Ma2] to nonfree actions.

2. Proof of the Main Theorem

In this section we shall prove the Main Theorem. The section is organized as follows. We construct two normal maps $f_0\colon M_0 \to Y_0$ and $f\colon M \to Y$. From Theorem A we know that $\sigma(f_0)$ vanishes; $\sigma(f_0)$ is the obstruction to converting f_0 by surgery into a pseudo equivalence, i.e. an equivariant map which is a homotopy equivalence. Then we translate this information into information in terms of stepwise surgery obstructions $\sigma_L(f_0)$, where L is a subgroup of G. We conclude that all $\sigma_L(f_0)$ vanish. From the knowledge of the obstruction groups under consideration we conclude that also the stepwise obstructions $\sigma_L(f)$ for f vanish. This implies that $\sigma(f)$ vanishes. We construct $f\colon M \to Y$ such that after our surgeries M is replaced by a homotopy sphere Σ. The dimensions of fixed point sets in M are chosen so that the inequality in 1.1 holds. As dimensions of fixed point sets, for any $L \subset G$, do not change under surgery, Σ will be an example of a homotopy sphere as claimed in the Main Theorem.

We shall spell out the details in case $G = D_3$. We make some remarks why the proof can be generalized to D_q, $q = 5$ and 7 but we omit details to keep the notation simple.

Notation: $D_q = \{a,b \mid a^2 = b^q = 1,\ aba = b^{-1}\}$. The subgroups of D_q generated by a and b are called H and K respectively. D_3 will be abbreviated by G, then $K \cong \mathbb{Z}_3$.

2a.) Construction of a G normal map $\mathscr{W}_0 = (M_0, f_0, b_0, c_0)$.

Set
$$Y_0 = S(2\cdot\mathbb{C}[G])$$
$$M_0 = G \times_H Y_0 \amalg -G \times_K Y_0.$$

Here S denotes the unit sphere and - indicates a reversed orientation. The identity on Y_0 induces $f_0\colon M_0 \to Y_0$ in the obvious equivariant way, then the degree of f_0 is 1. Pick a G vector bundle isomorphism $b_0\colon TM_0 \to f^* TY_0$. This b_0 induces a stable G vector bundle isomorphism $b_0\colon TM_0 \to f^* TY_0$ and an isomorphism $c_0 = \pi(b_0)$ of bundle systems (compare [DP2, Th. B8 page 111]).

Without going into details, here is a short explaination. For equivariant surgery it does not suffice to have above b_0. It is necessary to also keep track of unstable normal bundles like $\nu(M_0^L, M_0)$ where $L \subseteq G$. The set of these normal bundles with some natural compatibility conditions give rise to a bundle system. There is a natural notion of an isomorphism between bundle systems. The condition $c_0 = \pi(b_0)$ expresses a compatibility between the stable bundle isomorphism b_0 (for the tangent bundle) and the isomorphism of bundle systems c_0 (for the set of normal bundles). See also [DP3, section 2] for a short explaination of bundle data. The data $\mathscr{U}_0 = (M_0, f_0, b_0, c_0)$ define a preambient map. The main characteristic of a preambient map is that it has all the bundle data to apply equivariant surgery to it. In fact Theorem A exploits an easy special case of the equivariant π-π Theorem [DP2, Th.B8] to show that $\mathscr{U}_0$ can be converted by equivariant surgery into a pseudo equivalence. The π-π Theorem is applied as follows. We show that $\mathscr{U}_0$ is the boundary of an equivariant normal map, this is after connecting the fixed point sets of all subgroups of G in M_0. The main characteristic of a normal map is that it is a preambient map but it also satisfies some combinatorial conditions which are always satisfied for a pseudo equivalence. So we find a normal map $\mathscr{U} = (W, F, B, C)$ where $F: W \to Z$ such that $\partial W = M_0$, $\partial Z = Y_0$ and f_0, b_0, c_0 are obtained from F, B, C by restriction. This means that $\mathscr{U}_0$ is the boundary of $\mathscr{U}$; or $\partial\mathscr{U} = \mathscr{U}_0$. The normal maps $\mathscr{U}_0$ and $\mathscr{U}$ are special cases of those constructed in [DP1].

2b.) Interpretation in terms of stepwise obstructions.

In the proof of the π-π Theorem we do surgery on $\mathscr{U}_0$ and $\mathscr{U}$ at the same time, compare [W1, section 4] or [DP2, section 5], but for our purpose it is more convenient to just consider the effect on $\mathscr{U}_0$. The map $f_0^H: M_0^H \to Y_0^H$ is the identity, and having the implications of Smith theory in mind we want to keep this map unchanged, i.e. we work relative to the $\mathbb{Z}_2$ fixed point set. We have to do surgery on f_0 such that $f_0^K: M_0^K \to Y_0^K$ becomes a mod 3 homology equivalence. By surgery we make f_0^K connected up to the middle dimension (this is done ambiently inside f_0). Then there is an obstruction

$$(2.1) \qquad \sigma_K(f_0) \in L_3^h(\mathbb{Z}_{(3)}[NK/K], 1)$$

to converting f_0^K into a mod 3 homology equivalence. As the obstruction group vanishes [Pa, 1.16] we find that $\sigma_K(f_0) = 0$. The vanishing of $\sigma_K(f_0)$ also followed from the existence of (W, F, B, C) in step 2.a).

We continue by doing surgery in the free part, i.e. leaving the H and K fixed point sets fixed. By surgery we make f_0 connected up to the middle. Now f_0 defines an element

(2.2) $\sigma_1(f_0) \in L_3^h(\mathbb{Z}[G],1)$

the obstruction for converting f_0 into a pseudo equivalence. For a precise description see [DP3, 4.5]. There we construct a formation which represents $\sigma_1(f_0)$. Formations are defined in [R1]. Following [DP3, 4.5] we construct a formation (Q, ϕ, Q_0, Q_1) where the sub Lagrangian Q_1 is a projective module over $\mathbb{Z}[G]$. But as $\tilde{K}_0(\mathbb{Z}[G]) = 0$ we in fact have a formation based on free $\mathbb{Z}[G]$ modules, representing a class in $L_3^h(\mathbb{Z}[G],1)$. The obstruction $\sigma_1(f_0)$ vanishes (by [DP3, 4.6]) as we know from 2a (and [DP1]) that f_0 can be converted into a pseudo equivalence. In general we only know that $\sigma_1(f_0)$ is invariant under surgery in the free part. In the particular situation of $G = D_3$ this last step in our surgery procedure can be done just in the free part. We would have to go back and do surgery on L fixed sets, $L \neq 1$, only if the formation was not free over $\mathbb{Z}[G]$, compare [DP2, Th.5B], and the guide through [DP2] at the end of its introduction. This argument will become clearer in section 4 where we analyse $L_3^h(\mathbb{Z}[G],1)$.

2c.) <u>Construction of a normal map</u> $\mathcal{U} = (M, f, b, c)$.

Set

$$Y = Y_0 * X * X * X.$$

Here Y_0 is as in step 2a.) and X is as constructed by Swan, see Theorem E. We could take more joins with X, but three suffice. There are G representations ${}^2U$ and ${}^3U$ such that

(2.3) $${}^2h\colon \mathrm{Res}_H\, S({}^2U) \to \mathrm{Res}_H Y$$

$${}^3h\colon \mathrm{Res}_K\, S({}^3U) \to \mathrm{Res}_K Y$$

where ${}^2h$ and ${}^3h$ are H and K homotopy equivalences respectively. Set ${}^2M = G \times_H S({}^2U)$ and ${}^3M = G \times_K S({}^3U)$. So we have a (simple) G homotopy equivalence

(2.4) $$h\colon M = {}^2M \amalg - {}^3M \to M_0 * X * X * X,$$

M_0 is as in 2a.). We set $f = (f_0 * \mathrm{Id}) \circ h$. This map is of degree 1. We will not find such nice bundle data for f as we did for f_0, so we discuss them at the place needed. Anyway we have a map

(2.5) $$f\colon M \to Y$$

and

$\dim M = 35$ $\quad\quad$ $\dim M^H = 11$

$\dim M^K = 7$ $\quad\quad$ $\dim M^G = -1$

so M satisfies the dimension relation stated in the Main Theorem.

2d.) <u>The surgery obstruction for</u> $f: M \to Y$.

First we want to do surgery on the $K = \mathbb{Z}_3$ fixed set (ambiently inside of M) to convert f^K into a mod 3 homology equivalence. We restrict our attention to ${}^3f = f|_{{}^3M}: {}^3M \to Y$. Note, $({}^3f)^K = f^K = f_0^K$. Obviously $T({}^3M) \cong ({}^3f)^*\xi$ where ξ is a G vector bundle over Y. An isomorphisms ${}^3b$ of these bundles defines bundle data $({}^3b, {}^3c)$ for the map ${}^3f$ as b_0 did in 2a.) for f_0. So we can apply surgery theory to $({}^3M, {}^3f, {}^3b, {}^3c)$. The obstruction to doing surgery on ${}^3f$ such that $({}^3f)^K$ becomes a mod 3 homology equivalence is again in $L_3^h(\mathbb{Z}_{(3)}[\mathbb{Z}_2],1)$, and hence it vanishes as in 2b.). So from now on we assume for $f: M \to Y$ that it induces already a mod 3 homology equivalence on the K fixed set and a homotopy equivalence on the H fixed set.

Now we have to continue with surgery in the free part. To do surgery we need bundle data, the existence of which we shall state below and which we shall discuss in the next section. Abbreviate the bundle data by b. Then we can do surgery on (M, f, b). There is an obstruction $\sigma_1(f)$ for converting f into a pseudo equivalence. We shall state here and prove in section 4 that this obstruction vanishes.

Use $\hat{}$ for the following functorial construction. If X is a G space then $\hat{X} = E \times_G X$ where E is a contractible space with free G action. If $f: M \to X$ is an equivariant map between G spaces then $\hat{f}: \hat{M} \to \hat{X}$ is the map induced by f.

LEMMA 2.6 Suppose $f: M \to Y$ is as above. There exists a vector bundle η over $\hat{Y}$ and a vector bundle isomorphism

$$b: \widehat{TM} \oplus \underline{\mathbb{R}} \to f^*\eta \oplus \underline{\mathbb{R}}.$$

For any L representation V we denote by $\underline{V}$ the product bundle with fibre V where the base is understood from the context. We call (b,η) as in 2.6 <u>completed bundle data</u>.

LEMMA 2.7 Completed bundle data are appropriate for surgery in the free part.

The proofs of these two lemmas and the precise interpretation of the statement of Lemma 2.7 are in the next section. We abbreviate $f: M \to Y$ with the completed bundle data (b,η) by $\mathcal{W} = (M, f, b)$ and call it a <u>weak normal map</u>. More generally a weak normal map is an equivariant degree 1 map from an oriented G manifold to a G CW complex satisfying Poincaré Duality which has completed bundle data.

By surgery on $\mathcal{W}$ in the free part below the middle dimension we can replace $\mathcal{W}$ by another weak normal map which is connected up to the middle dimension. As in 2.2 or [DP3, 4.5] $\mathcal{W}$ defines an element

$$\sigma_1(f) \in L_3^h(\mathbb{Z}[G],1).$$

LEMMA 2.8 $\sigma_1(f) = 0$.

The proof is immediate using 4.4 and 6.8.

As $\sigma_1(f)$ vanishes we can do surgery in the middle dimension of f and convert f into a pseudo equivalence. This is more or less a folklore statement. For a statement that comes close see Lemma 4.7 in [DP3]. The idea is obvious, $\sigma_1(f)$ is represented by a formation $\overline{Q} = (Q, \phi, Q_0, Q_1)$ which indicates algebraically that we could do surgery, as $\sigma_1(f) = 0$. This algebra tells us on which homotopy classes we are supposed to do surgery. We can do surgery on these classes as they can be represented by imbedded spheres with trivial normal bundle using the dimension assumptions in 2.5 and the bundle data in 2.6. This completes the proof of the Main Theorem.

2.9 A remark on finding more examples as claimed in the Main Theorem, i.e. for bigger dihedral groups. For $G = D_q$ $q = 5$ or 7 the generalization is fairly straight forward. We still have that $\tilde{K}_0(\mathbb{Z}[G]) = 0$. The analysis of the surgery obstruction stays unchanged. A space similar to X, as it was constructed by Swan and used in 2c.), still exists, see [St]. It is also easy to generalize the construction of f_0 and f for this setting. To accommodate for 3.6 and the choice we made after 3.6, we might have to take joins with more than 3 copies of X. All of this is possible. This approach does not generalize easily to bigger dihedral groups. One problem is that $\tilde{K}_0(\mathbb{Z}[G])$ does not vanish anymore. This means that the formation defining $\sigma_1(f)$ might be a projective formation (no more free). This will force us to go back and do surgery on the H and K fixed set in 2b.). Hence we might have messed up the counting principle used in the computation of the semicharacteristic in section 4.

3. Surgery and completed bundle data

We still have to prove two lemmas on surgery from the last section. Before we do this we explain a few principles from (equivariant) surgery theory. Suppose $f\colon N \to Y$ is an equivariant map between G spaces and N is a compact smooth G manifold. A diagram

$$(3.1) \qquad \mu: \quad \begin{array}{ccc} S^k & \xrightarrow{\ \iota\ } & N^L \\ \downarrow & & \downarrow f^L \\ D^{k+1} & \xrightarrow{\ \kappa\ } & Y^L \end{array}$$

represents a class α in $\pi_{k+1}(f^L) = \pi_{k+1}(M_f^L, N^L)$ where M_f denotes the mapping cylinder of f, and L is an isotropy group in N. We would like to kill α, and this should replace (N,f) by (N',f'). Let $\mathcal{U}$ denote the following extension of μ.

$$(3.2)\qquad \mathcal{U}: \begin{array}{ccc} \mathbb{D}_0 & \xrightarrow{i} & N \\ \downarrow & & \downarrow f \\ \mathbb{D} & \xrightarrow{k} & Y \end{array}$$

where $\mathbb{D}_0 = G\times_L S^k \times D(\mathbb{R}^{n-k} \times W)$

$\mathbb{D} = G\times_L D^{k+1} \times D(\mathbb{R}^{n-k} \times W)$

$n = \dim N^L$ and W the normal representation of N^L in N. (To avoid notational effort we assume that all components of N^L have the same dimension and the same slice.) If we can choose $\mathcal{U}$ such that i is an imbedding we have achieved our goal, namely $N \amalg N' = \partial Z$ with $Z = N \times I \cup_{\mathbb{D}_0} \mathbb{D}$. (For convenience assume $\partial N = \phi$.) We have a map $f \times I \cup_i k = F: Z \to Y \times I$ of triads and f' is obtained as restriction. So the point is to find $\mathcal{U}$ such that i is an immersion and $i_0 = i|_{G\times_L S^k}: G\times_L S^k \to N$ is an imbedding, i.e. we imbedded $G\times_L S^k$ with trivial normal bundle. Bundle data serve the following purpose in surgery theory. Within the homotopy class of $\imath$ (see 3.1) they determine a regular homotopy class of immersions i (see 3.2), i.e. an immersion i_0 with trivial normal bundle, and such that the bundle data extend over the handle $\mathbb{D}$ which we attach. Thus we get bundle data for F and by restriction for f'. It is then an algebraic problem whether the regular homotopy class of i_0 contains a representative which is an imbedding.

So a good bundle theory is one which provides immersions with trivial normal bundles and which is surgery invariant. The classical example [W] for free actions consists of a vector bundle η over Y/G and a stable vector bundle isomorphism $b: T(N/G) \to \bar{f}^*\eta$. Here $\bar{f}: N/G \to Y/G$ is the map induced by f on the quotients. Hirsch's Lemma is used to show that this is a good bundle theory. The bundle data (b_0, c_0) used in (2b) and the bundle data $({}^3b, {}^3c)$ used in the first part of (2d) were studied in [DP2, section 4]. There we used Hirsch's Lemma and the equivariant tubular neighborhood theorem to show that we had a good bundle theory.

<u>Proof of Lemma 2.6</u>: We choose the following notation. Set $(f|_{{}^3M}: {}^3M \to Y) = (g: N \to Y)$, compare 2.4 and 2.5. This is part of the map f we started out with. Let $g^e: N^e \to Y$ denote the map (e = eventual) constructed from g which was a mod 3 homology equivalence on the K fixed set, see first part of 2d. We shall construct a bundle ζ over $\hat{Y}$ and an isomorphism

$$(3.3)\qquad a: \hat{TM} \oplus \varepsilon \to {}^i\hat{f}^*\,\zeta,$$

ε is a trivial bundle, ${}^if$ is the initial f from 2.5. We also have the bundle ξ over Y and an isomorphism

(3.4) $b\colon T^3M \oplus \varepsilon \to g^*\xi.$

This ζ and a will have the property that (^3a denotes a restriction of a)

(3.5) $\hat{g}^*\zeta = (g^*\xi)^\wedge$ and $^3a = \hat{b}$ (although $\hat{\xi} \neq \zeta$).

From 3.5 it is clear that we can do surgery on (N,g) with the bundle data (ξ,b) from 3.4. Then we can complete b after each surgery step on (N,g) and obtain completed bundle data $(\zeta, \hat{b} = a)$ with the same compatability condition as in 3.5. (See remark after proof.) So until now we do not use these completed bundle data to do surgery but we just preserve them. Hence we shall get this pair of bundle data for (N^e,g^e). As ^2M and $f|_{^2M}$ stayed unchanged we thus have completed bundle data as required in the lemma.

To finish this proof we thus have to find (ζ,a) and (ξ,b) such that they satisfy 3.5 for our original map as constructed in 2.4 and 2.5.

Consider the diagram (K denotes complex K-theory.)

$$\begin{array}{ccc} K_G(*) & \xrightarrow{\phi_0} & K_H(*) \times K_K(*) \\ \downarrow & & \downarrow \psi \\ \hat{K}_G(*) & \xrightarrow{\phi} & \hat{K}_H(*) \times \hat{K}_K(*). \end{array}$$

We use the cartesian square in [Ma 1, Cor. 1.9] to describe the image of ϕ. Also notice that ψ is an injection, it is the product of two completions. Each of these completions is an injection as the group involved is of prime order. With this information it follows that

(3.6) $\operatorname{im}\phi = \{(\hat{U},\hat{U}') \mid (U,U') \in K_H(*) \times K_K(*)$ such that $\dim U = \dim U'$ and U' is invariant under the action of NK/K, given by reversing the direction of rotation$\}$.

Pick the pair (U,U') to be $(\mathrm{Res}_H^2U, \mathrm{Res}_K^3U)$ from 2.3 (this is a good choice as Res_K^3U has an even number of nontrivial irreducible representations as summands). This pair provides us with a bundle ζ_0 in $\hat{K}_G(*)$. Let $\pi_G^Y\colon \hat{Y} \to BG = \hat{*}$. Then $\zeta = (\pi_G^Y)^*\zeta_0$. By the definition of ^2M and ^3M (after 2.3) it follows that $\widehat{TM} \oplus \varepsilon \cong i\hat{f}^*\zeta \cong (\pi_G^M)^*\zeta_0$. Pick an isomorphism a.

Let $\rho\colon BK \to BG$ be the projection. Then $\rho^*\zeta_0 = {}^3\hat{U}$. Choose $\xi = Y \times {}^3U \oplus \varepsilon'$ and b a product, i.e. constant on fibres. Note that $T^3M \oplus \varepsilon \cong {}^3M \times {}^3U \oplus \varepsilon'$. As $^3M = G \times_K S(^3U)$, $\pi_G^{^3M} = \rho \circ \pi_K^{^3M}$. Hence we can impose the compatibility condition on (ζ,a) and (ξ,b) expressed in 3.5. This completes the proof of 2.6.

REMARK. To make this principle even clearer note that the bundle data (restricted to ${}^3M$) pulled back from BK, and this was precisely the completion of the bundle data (ξ,b), which were pulled back from a point. Now each surgery step on the K fixed set attaches $G\times_K D^{k+1}\times D(\mathbb{R}^{n-k}\times W)$ over $G\times_K S^k\times D(\mathbb{R}^{n-k}\times W)$. W is the slice of N^K in N. Considering the map $\pi_G^{{}^3M}\colon {}^3\hat{M}\to BG$, handles, attaching maps, and bundle data with compatibility, then everything factors over BK. Hence surgery preserves 3.5.

Hence we find

(3.7) Suppose $f\colon M\to Y$ is as in 2.6 and $M\to *$ induces $\pi_G^M\colon \hat{M}\to BG$. There exists a bundle ξ_0 over BG and an isomorphism
$a\colon \hat{TM}\oplus\varepsilon\to(\pi_G^M)^*\xi_0$.

<u>Proof of Lemma 2.7</u>. Suppose we are given $f\colon M\to Y$ with completed bundle data (ζ,a). Consider a class $\mu\in\pi_{k+1}(f)=\pi_{k+1}(M_f,M)$, M_f is the mapping cylinder of f. A representative of μ is a diagram

$$\mu\colon\quad\begin{array}{ccc} S^k & \xrightarrow{\iota} & M\\ \downarrow & & \downarrow f\\ D^{k+1} & \longrightarrow & Y.\end{array}$$

We can extend this diagram to

$$\mathcal{U}\colon\quad\begin{array}{ccc} G\times\mathbb{D}_0 & \xrightarrow{i} & M\\ \downarrow & & \downarrow f\\ G\times\mathbb{D} & \xrightarrow{k} & Y\end{array}$$

where $\mathbb{D}_0=S^k\times D^{n-k}$ and $n=\dim M$. We find the following diagram

$$\begin{array}{ccc} (i^*TM)^{\wedge} & \xrightarrow{\hat{i}^*a} & \hat{i}^*\hat{f}^*\xi\\ \uparrow & & \uparrow\\ i^*TM|_{\mathbb{D}_0} & \xrightarrow{A} & \hat{i}^*\hat{f}^*\xi|_{\mathbb{D}_0}\end{array}$$

The bundles in the top row are over $E\times_G G\times\mathbb{D}_0$, the ones in the bottom row are over $\mathbb{D}_0$. The vertical maps are obtained from the inclusion $\mathbb{D}_0\to(G\times\mathbb{D}_0)^{\wedge}$ which is a homotopy equivalence, and A is obtained by restriction.

Choose a regular homotopy class of equivariant immersions $G\times\mathbb{D}_0\to M$ such that its differential restricted over $\mathbb{D}_0$ is A, this is possible by Hirsch's Lemma [W1]. Assume that this immersion can be represented within its regular homotopy class by an imbedding, this is an algebraic problem and it has to be checked when we do surgery. Then we can attach a handle $\mathbb{D}$,

$\mathbb{D} = D^{k+1} \times D^{n-k}$, along $\mathbb{D}_0$ and A extends to $\bar{A}$ over $\mathbb{D}$. Let j be a homotopy inverse of the inclusion $\mathbb{D}_0 \to (G \times \mathbb{D}_0)^\wedge$. The bundle isomorphisms j^*A and $\hat{i}^*a$ are homotopic, so change a via a homotopy such that $j^*A = \hat{i}^*a$. Set $W = M \times I \cup G \times \mathbb{D}$, the G handle is attached via i along $G \times \mathbb{D}_0$. Set $F = f \times \mathrm{Id} \cup_i k$. Then $a \cup_{\hat{i}^*} j^*\bar{A}$: $(TW)^\wedge \to \hat{F}^*\xi$ provides bundle data for $F: W \to Y \times I$. One of the boundary components of W is the manifold constructed by surgery and the underlying function and bundle data for the resulting weak normal map are obtained by restriction.

4. Surgery obstructions, semicharacteristics, and Reidemeister torsion

In this section we describe a computation of $L_3^h(\mathbb{Z}[G],1)$, $G = S_3$. To recognize elements in $L_3^h(\mathbb{Z}[G],1)$, this is a nontrivial group, we have to give geometric invariants in terms of the normal map for which we want to compute the surgery obstruction. I am in debt of W. Pardon for some advice on the study of this group. G. Carlsson and J. Milgram explained their calculation of semicharacteristics to me. I want to thank I. Hambleton who told me much about the strategy to define Reidemeister torsion invariants which detect surgery obstructions.

For each prime p we have a cartesian square

$$\begin{array}{ccc} \mathbb{Z} & \longrightarrow & \hat{\mathbb{Z}}_{(p)} \\ \downarrow & & \downarrow \\ \mathbb{Q} & \longrightarrow & \hat{\mathbb{Q}}_{(p)} \end{array}$$

The groups $L_3^h(\mathbb{Q}[G],1)$ and $L_3(\hat{\mathbb{Z}}_{(p)}[G],1)$, $p \neq 2$, vanish. Elements in $L_3(\hat{\mathbb{Z}}_{(2)}[G],1)$ are detected as semicharacteristics. Set $\hat{\mathbb{Z}} = \lim \prod \hat{\mathbb{Z}}_{\{p_i\}}$, where the product ranges over a finite set of primes $\{p_i\}_{1\leq i\leq k}$, and the limit is taken over the partially ordered set of finite sets of primes. Set $\hat{\mathbb{Q}} = \hat{\mathbb{Z}} \otimes \mathbb{Q}$.

Start out with diagram

(4.1)

$$\begin{array}{ccccccc} & & & & & & 0 \\ & & & & & & \downarrow \\ & & & & & \overset{\text{S.C.}}{\dashrightarrow} & SC(G) \cong \mathbb{Z}_2 \\ & & & & & & \downarrow \\ 0 \longrightarrow & \bar{L}^h_3(\mathbb{Z}[G],1) & \longrightarrow & L^h_3(\mathbb{Z}[G],1) & \xrightarrow{\beta} & & L^h_3(\hat{\mathbb{Z}}_{(2)}[G],1) \cong \mathbb{Z}_2 \\ & & & & & & \downarrow \alpha = 0 \\ & & & & & & L^h_3(\hat{\mathbb{Q}}_{(2)}[G],1) \end{array}$$

The group of semicharacteristics, $SC(G)$, is defined as the kernel of α (see [CM,Def.1.2]). It follows from Theorem 7(b) loc.sit. that $SC(G) \cong \mathbb{Z}_2$, and from Lemma 9 loc. sit. that $L^h_3(\hat{\mathbb{Z}}_{(2)}[G],1) \cong \mathbb{Z}_2$. The kernel of β is denoted by $\bar{L}_3(\mathbb{Z}[G],1)$.

Set $K_1 = K_1(\hat{\mathbb{Q}}[G])$, and $\bar{K}_1(G) = \mathrm{Im}\{K_1(\hat{\mathbb{Z}}[G]) \oplus K_1(\mathbb{Q}[G]) \longrightarrow K_1(\hat{\mathbb{Q}}[G])\}$. There is an exact sequence

$$0 \longrightarrow \bar{K}_1(G) \longrightarrow K_1(\hat{\mathbb{Q}}[G]) \longrightarrow \tilde{K}_0(\mathbb{Z}[G]) \longrightarrow 0.$$

This and the following diagram and statement can be found in [HM 2.9 and 3.6].

(4.2)

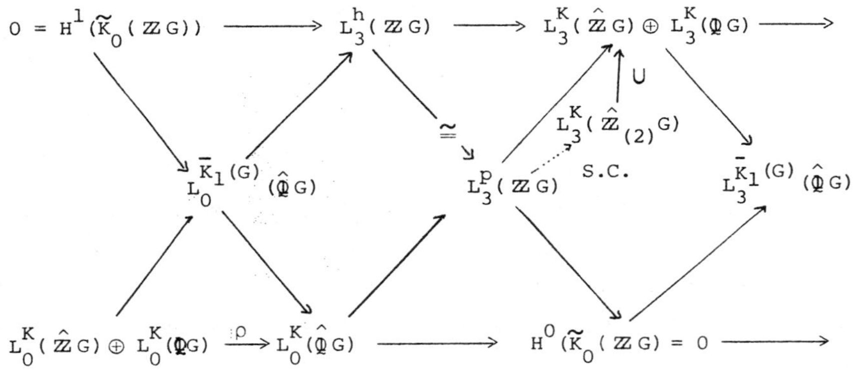

Elements in $L^p_3(\mathbb{Z}G)$ are detected by the semicharacteristic and the Reidemeister torsion defined in [Ma 2].

We conclude this section by computing the semicharacteristic for our surgery problem.

The semicharacteristic is carried by 2-blocks, see [CM, introduction]. In our situation such a 2-block is represented by $\hat{\mathbb{Z}}_{(2)}[NK/K]$, $K \cong \mathbb{Z}_3$. For a manifold M^{2k+1} the semicharacteristic $\chi_{\frac{1}{2}}(M)$ is defined as usual by adding up homology groups up to the middle dimension with a sign, using appropriate coefficients and an equivalence relation as in [L]. This definition also applies to Poincaré Duality spaces. For a map $f: M \to Y$ we set

$$\chi_{\frac{1}{2}}(f) = \chi_{\frac{1}{2}}(M) - \chi_{\frac{1}{2}}(Y).$$

LEMMA 4.3 Let $f: M \to Y$ be as above. The semicharacteristic $\chi_{\frac{1}{2}}(f)$ is

(i) unchanged under surgery in the free part

(ii) is changed in every surgery step on the K fixed set.

PROOF: (i) It is standard that the semicharacteristic is unchanged under free cobordism, and surgery in the free part gives rise to a cobordism which is relative to the fixed sets for all H, $1 \neq H \subset G$. (ii) It is also obvious that surgery on the K fixed set changes $\chi_{\frac{1}{2}}(M)$ by the class of $\hat{\mathbb{Z}}_{(2)}[NK/K]$.

Suppose $f: M \to Y$ is the map in 2.5, and after doing surgery on the K fixed set we constructed $f^e: M^e \to Y$ (e for eventual). So f^e is the map which induces a mod 3 homology equivalence on the K fixed set.

LEMMA 4.4 $\chi_{\frac{1}{2}}(f^e) = 0$.

PROOF: Let $f_0: M_0 \to Y_0$ be the map as defined in the beginning of 2a.) The only homology for M, M_0, Y, and Y_0 below their respective middle dimension is in dimension 0, and as G modules $H_0(M) = H_0(M_0)$ and $H_0(Y) = H_0(Y_0)$. Hence for the initial maps $\chi_{\frac{1}{2}}(f) = \chi_{\frac{1}{2}}(f_0)$. For the eventual map f^e and f_0^e we have the following computation. As we were able to convert f_0^e by surgery in the free part into a pseudo equivalence, it follows that $\chi_{\frac{1}{2}}(f_0^e) = 0$. As $f_0^K = f^K$ we do the same number of surgery steps on f_0 and f to convert these maps into maps f_0^e and f^e i.e. the maps inducing mod 3 homology equivalences on K-fixed sets. As the change of the semicharacteristic of either map depended only on the number of surgery steps on the K fixed set, this implies that also $\chi_{\frac{1}{2}}(f^e) = 0$.

5. The torsion of a simple Poincare duality chain complex

This section reviews material developed in [Ma, §2] and MH, §§2 and 3] where the reader also finds references to original sources like [W1], [R1,2], [LM]. We include it for the convenience of the reader, to establish our notation, and to give some background.

Let (R,α,u) be an anti-structure, i.e. α is an anti involution on the ring R and $u \in \{\pm 1\}$. Let R be a semisimple K algebra, char(K) $\neq$ 2. Such anti-structures can be decomposed into simple (anti) involutive rings

$$(5.1)\quad (R,\alpha,u) = \prod_{i=1}^{k} (M_{n_i}(E_i),\alpha,u) \times \prod_{j=1}^{\ell} (M_{n_j}(E_j) \times M_{n_j}(E_j),\alpha,u)$$

where in the second product α acts by permuting the factors (type GL anti-structures), and the E_i are division rings.

The Reidemeistertorsion invariants we want to consider will lie in $H^*(\mathbb{Z}_2, K_1(R,\alpha,u))$. Because of the splitting 5.1 the invariants split into invariants in $H^*(\mathbb{Z}_2, K_1(M_{n_i}(E_i),\alpha,u))$; the second product of rings does not contribute as their Tate cohomology groups vanish, see [Ma 2]. There exist anti-involutions α_i on E_i which agree with α on the centers F_i, and we determine $u_i \in \{\pm 1\}$ such that $\{E_i,\alpha_i,u_i\}$ and $\{M_{n_i}(E_i),\alpha,u\}$ have the same type. Types of anti-structures (S,α,u) over simple algebras A (e.g. over skew fields) are defined as follows. If F is the center then $|S:F| = m^2$. (S,α,u) has type U if $\alpha|F \neq \mathrm{Id}$. It has type 0 (resp. type Sp) if $\alpha|F = \mathrm{Id}$ and $\dim_F S^\alpha = \frac{1}{2}(m^2+um)$ (resp $\dim_F S^\alpha = \frac{1}{2}(m^2-um)$). Note that type 0 and type Sp are interchanged when u is replaced by $-u$. See [Wl,p.5 or Ma]. As usual we denote $H^*(\mathbb{Z}_2, K_1(S,\alpha,u))$ by $H^*(K_1(S))$.

Since

$$(5.2) \qquad H^*(K_1(R)) = \bigoplus_{i=1}^{k} H^*(K_1(E_i))$$

we shall deal with invariants $\tau_i \in H^*(K_1(E_i))$, $i \in \{1,\dots,k\}$.

Let C_* be an (E,α,u) chaincomplex, where E is a division algebra and $u = 1$. Let C^* be the complex consisting of the dual modules $C^i = \mathrm{Hom}_E(C_i,E)$. This is an E complex with $(f\cdot e)(m) = \alpha(e)f(m)$. A <u>Poincaré duality chaincomplex</u> over E is a pair (C_*,φ) where C_* is a finitely generated E chain complex and $\varphi\colon C^* \to C_*$ is a chain homotopy equivalence. If φ is of degree n (i.e. $\varphi\colon C^i \to C_{n-i}$) we say that (C_*,φ) has <u>formal dimension</u> n.

<u>Provisional assumption 2.3</u> (compare [Ma,2.4]) All chain complexes are even dimensional in the sense that $\dim_E C_i \equiv 0 \pmod 2$ and $\dim_E H_i \equiv 0 \pmod 2$ for all i.

This will allow us to define an invariant in $K_1(E)$ instead of $\overline{K}_1(E) = K_1(E)/\det(-1)$ with the same properties as in [Ml]. Madsen [Ma] also shows that the provisional assumption is not restrictive as we are able to add complexes X_* to a given Poincaré duality chain complex to make a complex even. This is done without changing the class in $H^*(\overline{K}_1(E))$. The invariant will then be well defined by defining the invariant of C_* to be the one of $C_* \oplus X_*$.

We recall Milnor's definition of torsion. Let $\underline{b}_1$ and $\underline{b}_2$ be bases for a finitely generated E module M. Then $\underline{b}_1 = \underline{b}_2 \cdot B$ for some non-singular E-matrix B. The element of $K_1(E)$ associated to B is denoted by $[\underline{b}_1/\underline{b}_2]$.

Let $M^* = \mathrm{Hom}_E(M,E)$ be the dual E-module and let $\underline{b}_i^*$ be the base which is dual to $\underline{b}_i$. Then

$$[\underline{b}_1^*,\underline{b}_2^*] = -\overline{[\underline{b}_1/\underline{b}_2]}.$$

Here the bar indicates the involution on $K_1(E)$ induced from α; on representing matrices it is given by the α-conjugate transpose operation.

Let C_* be a based PD chaincomplex of formal dimension n. The given base for C_p is denoted by $\underline{c}_p$. There is a standard decomposition of chaingroups $C_p \cong B_p \oplus H_p \oplus B_{p-1}$ which becomes an isomorphism of chain complexes when we use the differential $(b_p, h_p, b_{p-1}) \mapsto (b_{p-1}, 0, 0)$. Since we assume C_i and H_i to have even dimension over E, so does B_i for all i. Let $\underline{h}_p$ be a basis for $H_p = H_p(C_*)$. Milnor defines the torsion

$$\tau(C_*, \underline{h}_*) = \Sigma(-1)^i[\underline{b}_i\underline{h}_i\underline{b}_{i-1}/\underline{c}_i] \in K_1(E).$$

Recall that a chain homotopy equivalence (chain homology equivalence) between based complexes $f\colon C_* \to C'_*$ is called simple if the mapping cone complex $C_*(f)$ has vanishing torsion. Similarly, a based PD chaincomplex is simple if the chainmap φ (the duality map) is simple when C^* is given the dual basis.

For such a simple PD chaincomplex we can sometimes define a torsion even if the homology does not come with a preferred basis. Suppose C_* has even formal dimension $n = 2m$. Then $\varphi_m\colon H^m \to H_m$ defines a $(-1)^m$ u-hermitian form (H_m, φ_m) on H_m. Such a form is called hyperbolic if it admits a symplectic basis. In this case $\dim_E H_m = 2p$, and expressed in the symplectic basis, $\varphi_m = J_p$ where

$$J_p = \begin{pmatrix} 0 & Id_p \\ \varepsilon Id_p & 0 \end{pmatrix}, \quad \varepsilon = (-1)^m u$$

Definition 5.3: [Ma]. Let C_* be a based PD chaincomplex of formal dimension n over (E,α,u). A family $\underline{h}_* = \{\underline{h}_i\}$ of bases $\underline{h}_i$ of $H_i(C_*)$ is called a PD basis if $\varphi_i\colon H^{n-i} \to H_i$ is simple for $i \neq n/2$ with respect to the bases $\underline{h}^{n-i} = \underline{h}^*_{n-i}$ and $\underline{h}_i$. If $n = 2m$ we suppose further that $\underline{h}_m$ is a symplectic basis for (H_m, φ_m).

Madsen shows that a simple Poincaré Duality complex of dim n with a Poincaré Duality basis has a well defined torsion in $H^n(K_1(E))$.

Suppose we are given a P.D. chaincomplex (C_*, φ) of formal dimension n over the anti-structure (R,α,u). For each factor $(M_{n_i}(E_i),\alpha,u)$ in 5.1 we obtain a P.D. chaincomplex $C^* \otimes_R M_{n_i}(E_i)$ (we have to take the antiinvolution α into account to see that this is a $M_{n_i}(E_i)$ complex) which we denote by $C_*^{E_i}$. Its torsion in $H^n(K_1(E_i))$ is denoted by $\tau_i(C_*,\varphi)$ - if $(C_*^{E_i}, \varphi^{E_i})$ is a simple $M_{n_i}(E_i)$ P.D. chaincomplex. The collection of the invariants $\tau_i(C_*,\varphi)$ defines $\tau(C_*,\varphi)$ by using 5.2. Madsen studies the algebraic bordism properties of the invariant τ. Let D_* and C_* be chaincomplexes and $i\colon C_* \to D_*$ an inclusion. Suppose there is a diagram of chainmaps

(5.4)

$$\begin{array}{ccccccccc}
0 & \longrightarrow & D_* & \longrightarrow & C_*(i) & \longrightarrow & C_{*-1} & \longrightarrow & 0 \\
 & & \uparrow \Phi & & \uparrow \tilde{\Phi} & & \uparrow \varphi & & \\
0 & \longrightarrow & D^*(i) & \longrightarrow & D^* & \xrightarrow{i^*} & C^* & \longrightarrow & 0
\end{array}$$

where $C_*(i)$ is the mapping cone of the inclusion $i: C_* \to D_*$ and $D^*(i) = \mathrm{Ker}\, i^*$. Each of the complexes has a distinguished class of basis if C_* and D_* do, and with these basis the sequences are based. The chain map $\tilde{\Phi}$ is induced by Φ, it is Φ^* composed with some natural chain homotopies. The pair (D_*,C_*) is called simple if in 5.4 each of the vertical maps is simple.

Define $\Delta_n(E) \subseteq H^n(K_1(E))$ as follows: $\Delta_{2m}(E)$ is the subgroup of discriminants of $(-1)^m$ u-hermitian forms and $\Delta_{2m+1}(E)$ is the subgroup formed by determinantes of automorphisms of the $(-1)^m$ u-hyperbolic form (E^p,J_p).

With these definitions Madsen shows:

<u>LEMMA 5.5 ([Ma 2, Lemma 2.10])</u>. Suppose (D_*,C_*) is a simple PD pair of formal dimension $n+1$, and if n is even that $H_*(C_*)$ admits a P.D. base. Then $\tau(C_*) \in \Delta_{n+1}$.

<u>Definition 5.6</u> If $\tau(C_*) \in H^n(K_1(R))$ we set $\Delta(C_*)$ to be the class of $\tau(C_*)$ in $H^n(K_1(R))/\Delta_n$. Here each summand $H^n(K_1(E))$ of $H^n(K_1(R))$ is factored by the appropriate $\Delta_n(E)$.

We turn to the geometric aspect of torsion.

We shall use the following notion of a G Poincaré Duality complex. Let X be a finite G CW complex and $C_*(X)$ its cellular chain complex. If $\xi \in C_n(X) \otimes_{\mathbb{Z}[G]} \mathbb{Z}$ represents a homology class in $H_n^t(X,\mathbb{Z})$ then

$$\varphi = \xi : C^*(X) \to C_*(X)$$

is defined, see [W2,p.22]. Because of the invariant choice of ξ this is an equivariant map of chaincomplexes. If ξ is a chain homotopy equivalence we call $[\xi] = [X]$ fundamental class of X and $(X,[X])$ (abbreviated as X) is called a G Poincare Duality complex. G Poincare Duality pairs are defined accordingly. Let (R,α,u) be a semisimple antistructure and $j: \mathbb{Z}[G] \to R$ a homomorphism of antistructures. Every $\mathbb{Z}[G]$ module can be tensored with R and hence turned into an R module. Interesting examples are $R = F[G]$, where F is a field and $(\mathrm{char}(F),|G|) = 1$, or $R = \hat{\mathbb{Q}}$. A G CW complex X will then give rise to a based complex $C_*(X)$ over R and the question whether $(C_*(X),\varphi)$ is a simple PD chaincomplex can be studied. If this chaincomplex is simple we say that $(X,[X])$ is a simple P.D. complex.

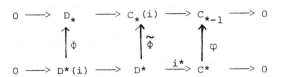

The following is well known:

Theorem 5.7 A closed smooth G manifold M with free action is a simple Poincare Duality complex. A compact smooth G manifold $(M,\partial M)$ with free action gives rise to a simple Poincare Duality pair.

This theorem is proved using the classical proof of Poincare (or Lefschetz) duality, see [W,L]. Abbreviate $\Delta(C_*(X))$ by $\Delta(X)$.

Corollary 5.8 Let M_0 and M_1 be closed smooth G manifolds with free action. If they are freely cobordant and $\Delta(M_i)$ is defined, then $\Delta(M_0) = \Delta(M_1)$.

Proof: Let W be the cobordism between M_0 and M_1, so $\partial W = M_0 \amalg -M_1$. As $\Delta(W) = 0$, $\Delta(M_0) = \Delta(M_1)$.

6. Nonfree group actions and Reidemeister torsion

In the previous section we dealt with simple Poincare duality chaincomplexes and chaincomplexes which arise from free actions on smooth manifolds. So in particular 5.5 is a purely algebraic Lemma whose assumptions are satisfied for $(C_*(M), C_*(\partial M))$ if M is a smooth compact G manifold. In this section we use the algebra of the previous section but we consider nonfree actions on manifolds.

In comparison to 5.7 we would like to show that a closed smooth G manifold (with nonfree action) of dim n is a simple $F[G]$ Poincaré duality complex; F is a field whose characteristic is prime to $|G|$ or $F = \hat{\mathbb{Q}}$. It is not but we are able to compute the torsion of the duality map.

For some Poincare duality complexes (chaincomplexes) of formal dimension n it will be possible to consider a torsion invariant in $K_1(E)/\{M+(-1)^n M^* \mid M \in K_1(E)\}$. I.e. we find an invariant in this group if we start out with a PD complex and impose a P.D. basis on homology.

Let $D(V)$ be the disk in an orthogonal representation. Assume the following two decompositions of $D(V)$. Let $S(V)$ be decomposed as a G CW complex. In D_p we consider the entire interior as one cell, in D_s we use the radial decomposition based on the subdivision of $S(V)$. There is a cellular map $\mathrm{Id}\colon D_p \to D_s$. Let us compute the torsion of this map. To do so we write out the chaincomplex of (M_{Id}, D_p) where M_{Id} is the mapping cylinder of Id. Let $0 \to C_n \to C_{n-1} \to \dots \to C_0 \to 0$ be the chain complex of $S(V)$. Then it is easy to check that the complex for the pair (M_{Id}, D_p) is

$$0 \to \mathbb{Z} \to C_n \to C_{n-1} \to \dots \to C_1 \to C_0 \to \mathbb{Z} \to 0$$

Consider coefficients in $F[G]$, but keep the names of the chaingroups, we obtain the chaincomplex

$$0 \to F \to F \oplus B_{n-1} \to C_{n-1} \to \ldots \to C_1 \to B_0 \oplus F \to F \to 0.$$

Considering that F plays the role of $H_0(S(V))$ and $H_n(S(V))$ respectively we find that

(6.1) $\tau(\mathrm{Id}) = -\tau(S(V))$.

The second computation we need is

(6.2) $\tau(S(V \otimes R)) = -\tau(S(V))$.

This is obvious from the sequence

$$0 \to C_*(S(V)) \to C_*(D(V)) \to C_*(D(V),\ S(V)) \to 0$$

and as the augmented chaincomplex of $S(V \oplus R)$ and of the pair $(D(V),\ S(V))$ are the same up to simple chain homotopy.

To avoid excessive notation in our next theorem we assume that the smooth compact G manifold M has a fixed point p and that T_xM is obtained from T_pM by restricting the group action to G_x. Let the class of M in the Burnside ring be denoted by $[M]$. Define a_H by the equation $[M] = \Sigma\, a_H[G/H]$. For any τ in $K_1(F[G])$ set

(6.3) $[M] \cdot \tau = \Sigma\, a_H\, \mathrm{Ind}_H^G\, \mathrm{Res}_H^G\, \tau$.

This multiplication makes $K_1(F[G])$ into a Frobenius module over the Burnside ring. Res_H^G and Ind_H^G are the natural restriction and induction maps.

Theorem 6.4. With above assumptions on M the duality map $C^*(M) \to C_*(M, \partial M)$ has torsion $[M] \cdot \tau(S(T_pM))$. If we do not assume a fixed point but assume that $\tau(S(T_xM)) = 0$ for all $x \in M$, then the duality map has zero torsion.

PROOF: Remember the classical proof of Lefschetz duality. There we take a (fine) simplicial decomposition of M. To each simplex σ we construct a dual cell D_σ, and these simplices and cells are in equivariant 1-1 correspondence. Say $C_*(M, \partial M)$ is constructed from the simplicial decomposition, $C_f^*(M)$ is modelled on the dual cells. Hence we have a simple isomorphism $C_f^*(M) \to C_*(M, \partial M)$. But $C_f^*(M)$ has two deficiencies. The first one is not serious. The underlying space of the dual cells is $M_0 = M$ - open collar of the boundary. But M_0 and M are diffeomorphic. The serious deficiency is that the dual cells which we constructed are not G cells. They are of the form $D(V)$; if σ was a simplex of the form $G/H \times \Delta^k$, n is the dimension of M^H, and W is the slice representation of M^H in M, then $V = D(\mathbb{R}^{n-k} \times W)$. We call $D(V)$ a funny cell (subscript f). Let $C^*(M)$ be the dual of $C_*(M)$ for some G CW decomposition of M. A subdivision of a G CW decomposition into a finer one has zero torsion. So to compute

$\tau(\varphi)$, $\varphi: C^*(M) \to C_*(M,\partial M)$ we have to subtract the sum of the contributions from subdividing funny cells into G CW cells. The torsion $\tau(S(W))$ is such a single contribution. We had one such torsion for each simplex. Its sign depends on the dimension of the simplex. This gives us the formula we claimed.

Corollary 6.5 Suppose M is an odd dimensional closed compact smooth G manifold. Then M is a simple Poincaré Duality complex.

Theorem 6.4 points out one important fact. We cannot expect Reidemeister torsion to be a bordism invariant for nonfree actions. This is not surprising. The lens-spaces considered by Milnor (spheres with free $\mathbb{Z}_p$ action) have nonvanishing Reidemeister torsion (defined slightly differently) and they bound but nonfreely. Nevertheless there are some cases when we can compute $\tau(S(V))$.

LEMMA 6.6

(i) Suppose $G = \mathbb{Z}_2$ or $\mathbb{Z}_3$ and V is a G representation,
or:

(ii) Suppose $V = 4\,\mathbb{R}[A]$ where A is a finite G set and $\mathbb{R}[A]$ the permutation representation.

Then $\tau(S(V)) = 0$.

Proof: In case (i) the result is checked easily for irreducible representations. The general case follows from induction. With our above choices of basis for chain groups and homology we can write

$$S(V_1 \oplus V_2) = S(V_1) \times S(V_2 \oplus \mathbb{R}) / S(V_1) \vee S(V_2 \oplus \mathbb{R}).$$

To show (ii) note that there exists a 4 dimensional closed manifold N with Euler characteristic one. Then $S(V) = \partial(N[A] - \mathring{D}(V))$ where the disk is removed about a fixed point in N[A] which is constructed similarly to $\mathbb{R}[A]$. This is: we use one factor of N for each element in A, and G permutes them according to the action on A. As $\chi(N) = 1$ we find that S(V) is the boundary of a manifold Z for which $\chi(Z^H) = 0$ for all $H \subset G$.

Let us conclude this section by computing the Reidemeister torsion contribution to the surgery obstructions $\sigma_1(f^e)$ and $\sigma_1(f_0^e)$. Again f^e and f_0^e are obtained from f (see 2.5) and f_0 (see 2.a) by converting the induced map on the $\mathbb{Z}_3$ fixed set into mod 3 homology equivalences. Instead of defining $\Delta(f^e)$ and $\Delta(f_0^e)$ we make the following observation. It suffices to compute $\Delta(M)$ and $\Delta(M_0)$ where $f: M \to Y$ and $f_0: M_0 \to Y_0$. If C_* is the chaincomplex of M (or M_0 respectively) then it suffices to kill the homology groups of

$$(6.7) \qquad \overline{C}_*: \ 0 \to \mathbb{Z} \to C_n \to C_{n-1} \to \dots \to C_0 \to \mathbb{Z} \to 0$$

and to make M (and M_0 respectively) 1-connected. The surgery kernels for f, i.e. $K_*(M)$ are just the homology groups of $\overline{C}_*$. Using the Hurewicz theorem that will imply that we kill $\pi_k(M)$ for $k < \dim M$.

The reference to Y and Y_0 was needed only unessentially. Usually it is needed to compare homologies (or homotopy groups) but this is taken care of in 6.7. Secondly we use Y as space from where we pull our bundle data back. So Y is not needed for this purpose as the bundle data for f_0 pull back from a point, and for f they pull back from BG (see 3.8).

Hence we can try to kill the homology of the complex $\overline{C}_*$, the complex in 6.6, by surgery. We showed that $\chi_{1/2}(\overline{C}_*)$ vanishes after we did surgery on the $\mathbb{Z}_3$ fixed set of f and f_0 respectively. Then $\overline{C}_*$ is a simple P.D. chaincomplex with vanishing torsion as the underlying space for C_* is $G \times_H S(^2U) \amalg - G \times_K S(^3U)$, see 2.c. This follows from 6.5 and 6.6(i). We do only surgeries on the $\mathbb{Z}_3$ fixed set and in the free part.
Also $\mathrm{Iso}(M) = \mathrm{Iso}(M_0) = \{1, \mathbb{Z}_3, \mathbb{Z}_2\}$ (up to conjugacy) so any cobordism under consideration has only these isotropy groups. Hence it gives rise to a simple P.D. pair. So 5.5 shows the invariance of Δ for the domains of the normal maps under consideration. So as it was zero to begin with and did not change we have

Theorem 6.8 The Reidemeister torsion contribution to $\sigma_1(f)$ vanishes.

Remark 6.9 Just as in 6.8 we could also show that $\sigma_1(f_0)$ vanishes. But 6.6(ii) gives a second argument. In [DP1] it was shown that both source and target of the normal map f_0 were of the form stated in 6.6(ii). This also implies that the Reidemeister torsion contribution $\Delta(f_0)$ to $\sigma_1(f_0)$ vanishes. In this case we can easily use $\Delta(f_0) = \Delta(Y_0) - \Delta(X_0)$, and each summand vanishes.

Purdue University
West Lafayette, IN 47907

BIBLIOGRAPHY

[A] E. Artin, Zur Theorie der Reihen mit allgemeinen Gruppencharakteren, Hamburger Abhandlungen 8 (1931), 292-306.

[B] A. Borel, Seminar on transformation groups, Ann. of Math. Studies 46, Princeton University Press, (1960).

[C] G. Carlsson and J. Milgram, The oriented odd L-groups of finite groups, preprint.

[tD] T. tom Dieck, Semilinear group actions on spheres: Dimension functions, in Algebraic Topology, Aarhus 1978, Lecture Notes in Mathematics 763, 1980.

[DH] R. Dotzel and G. Hamrick, p-group actions on homology spheres, Invent. Math. 62 (1981) no.3, 437-442.

[DP1] K. H. Dovermann and Ted Petrie, Artin relation for smooth representations, Proceedings National Academy of Science U.S., 77(1980), 5620-5621.

[DP2] K. H. Dovermann and Ted Petrie, G. Surgery II, Memoirs of the AMS, to appear as vol.259.

[DP3] K. H. Dovermann and Ted Petrie, An induction theorem in equivariant surgery theory-G Surgery III, preprint (1981).

[HM] I. Hambleton and I. Madsen, Semifree actions on $\mathbb{R}^n$ with one fixed point (in this volume).

[L] R. Lee, Semicharacteristic classes, Topolgoy 12 (1973), 183-199.

[Le] Lefschetz, Introduction to topology, Princeton, 1949.

[Ma 1] I. Madsen, Smooth spherical space forms, in: Geometric applications of homotopy theory I, Proceedings, Evanston 1977, Springer Lecture Notes in Mathematics Vol.657, 1976, 303-352.

[Ma 2] I. Madsen, Reidemeister torsion, surgery invariants, and spherical space forms, preprint (1981), Aarhus Universitet.

[M1] J. Milnor, Groups which act on S^n without fixed points, Amer. J. Math. 79 (1957) 623-630.

[M2] J. Milnor, Whitehead torsion, Bull. AMS 72 (1966), 358-426.

[MY1] D. Montgomery and C. T. Yang, A generalization of Milnor's theorem and differentiable dihedral transformation groups, preprint.

[MY2] D. Montgomery and C. T. Yang, Dihedral group actions I, in: General Topology and Modern Analysis (edited by L. F. McAuley and M. M. Rao), Academic Press (1980), 295-304.

[MY3] D. Montgomery and C. T. Yang, Dihedral group actions II, preprint.

[Pa] W. Pardon, Mod 2 semicharacteristics and the converse to a theorem of Milnor, Math, Zeitschrift 171 (1980), 247-268.

[R1] A. Ranicki, Algebraic L theory I: Foundations, Proc. London Math. Soc. (3) 27 (1973), 101-125.

[R2] A. Ranicki, Exact sequences in the algebraic theory of surgery, Princeton University Press, Mathematical Notes, Vol.26.

[St] E. Straume, Dihedral transformation groups on homology spheres, Journal of Pure and Applied Algebra 21 (1981), 51-74.

[S] R. Swan, Periodic resolutions for finite groups, Ann. of Math. Vol.72, (1960), 267-291.

[S2] R. Swan, K theory of finite groups and orders, Lecture Notes in Math 149, Springer, 1970.

[W1] C. T. C. Wall, Surgery on compact manifolds, Academic Press, 1970.

[W2] C. T. C. Wall, On the classification of hermitian forms VI, group rings, Ann. of Math. 103 (1976) 1-80.

Canadian Mathematical Society
Conference Proceedings
Volume 2, Part 2 (1982)

Semi-Free Actions on $\mathbb{R}^n$ with One Fixed Point

Ian Hambleton(*) and Ib Madsen

If π is a finite group and $\rho: \pi \times \mathbb{R}^n \to \mathbb{R}^n$ is a smooth semi-free action with fixed point set the origin, then $(\mathbb{R}^n,\rho)$ is equivariantly diffeomorphic to a free representation space for π. After the solution of the space-form problem [MTW] we know that this is not true for topological actions: the open cone over a non-linear free action on S^{n-1} is an example. In this paper we give necessary and sufficient conditions for π to have a proper topological semi-free action on $\mathbb{R}^n$ with Fix $(\mathbb{R}^n,\pi) = 0$ in the special case when Swan's finiteness obstruction for π vanishes. One consequence is the existence of many such actions with no invariant $S^{n-1} \subset \mathbb{R}^n - \{0\}$. All actions will be assumed to be topological and orientation-preserving.

By a proper semi-free action on $\mathbb{R}^n$ we mean that $\mathbb{R}^n - \{0\}/\pi$ is an open topological manifold which is the infinite cyclic covering of a compact manifold (Definition 1.2). In Section 1 we observe that Ronnie Lee's semi-characteristic invariant [L] is defined for such actions and leads to restrictions on π (Prop. 1.3, 1.13). Our main assumption, that $X = \mathbb{R}^n - \{0\}/\pi$ be homotopy equivalent to a finite complex, amounts to assuming that the Swan obstruction $\sigma_n(\pi) \in \tilde{K}_0(\mathbb{Z}\pi)/T$ vanishes (where T is the Swan subgroup [S], [TW]. Actually, in this paper we only study actions of the above type where homotopy type is restricted to be a linear space form on certain subgroups. For the rest of the paper we let $\sigma_n(\pi)$ denote the finiteness obstruction of the corresponding (almost linear) k-invariant.

The condition $\sigma_n(\pi) = 0$ has been extensively studied in [Mg1], [Mg2].

(*) Research partially supported by NSERC (Canada) Grant A4000
© 1982 American Mathematical Society
0731-1036/82/0000-0462/$12.25

In particular, examples are known of finite groups π such that $\sigma_n(\pi) \neq 0$ to which the methods here do not apply directly.

Let a, b be coprime odd numbers with prime decompositions $a = p_1^{i_1} \cdots p_r^{i_r}$, $b = p_{r+1}^{i_{r+1}} \cdots p_m^{i_m}$. Write $\underline{\underline{a}}$, $\underline{\underline{b}}$ for the sets of prime powers which occur in a, b,

$$\underline{\underline{a}} = \{p_1^{i_1}, \ldots, p_r^{i_r}\}, \quad \underline{\underline{b}} = \{p_{r+1}^{i_{r+1}}, \ldots p_m^{i_m}\}$$

We write ζ_r for any primitive r-th root of 1 and $\eta_r = \zeta_r + \zeta_r^{-1}$. Let $A = Z[\eta_a, \eta_b]$ and consider the reduction homomorphisms

$$\Phi_{a,b} : A^\times \to (A/aA)^\times_{(2)} \times (A/bA)^\times_{(2)}$$

(0.1)

$$\phi_{a,b} : A^\times \to (A/4A\ )^\times_{(2)} = A/2A,$$

where the subscript (2) indicates the 2-primary component.

Note that $A/aA = \prod_{p|a} A/pA$ (p prime) and that $A/pA = Z[\eta_{\bar{a}}, \eta_b]/p\, Z[\eta_{\bar{a}}, b]$ where $\bar{a} = a$ if $p \nmid a$ and $\bar{a} = a/p_\nu^{i_\nu}$ if $p|a$ and $p_\nu = p$. Moreover, each A/pA splits up in a number of isomorphic finite fields of characteristic p.

In $(A/aA)^\times_{(2)} \times (A/bA)^\times_{(2)}$ we consider the element

$$\bar{V}(a,b) = ((\bar{V}_1(a,b), \ldots, \bar{V}_r(a,b)), (\bar{V}_{r+1}(a,b), \ldots, \bar{V}_n(a,b))$$

where $\bar{V}_i(a,b) = \prod_{\substack{k=1 \\ k\neq i}}^{m} (\eta_{a_k} - 2)$ and $a_k = p_k^{i_k}$.

For example, if a and b are primes then $\bar{V}(a,b) = (\eta_a - 2, \eta_b - 2)$.

Theorem A. Let π be a finite group with periodic cohomology of period 2d and $n \geq 3$ a multiple of d such that $\sigma_{2n}(\pi) = 0$. Then π has a proper semi-free action on $\mathbb{R}^{2n}$ with fixed point set the origin if and only if

(i) Lee's semi-characteristic vanishes, and

(ii) for each group $Q(8a,b,1) \subset \pi$, $\bar{V}(a,b) \in \text{Image}\,(\Phi_{a,b}| \text{Ker}\,\phi_{a,b})$

In the above Q(8a,b,1) is the extension of $\mathbb{Z}/ab$ by the quaternion group Q8 presented as

(0.2) $$Q(8a,b,1) = \langle X, Y, A, B \mid A^a = B^b = 1,\ AB = BA,\ XAX^{-1} = A^{-1},\ YBY^{-1} = B^{-1},\ XB = BX,\ YA = AY \rangle.$$

From this result if π has such an action then every subgroup of order p^2 or 2p is cyclic (p prime). The most interesting case occurs if $n = 4k + 2$, when it follows that π has no subgroups τ of type IIL: an extension

$$1 \to \mathbb{Z}/m \to \tau \to Q2^r \to 1$$

with centralizer of $\mathbb{Z}/m$ of index 4 and $r \geq 4$. Section 1 contains, in addition to the properties of the semi-characteristic, a discussion of proper surgery theory resulting in a proof that the conditions stated in Theorem A are necessary for the existence of such actions on $\mathbb{R}^{2n}$ (even when

$\sigma_{2n}(\pi) \neq 0)$. Section 2 gives the basic arithmetic sequences from [W6] and [R3] needed to compute the various surgery obstruction groups and certain braid diagrams (2.8, 2.9, 2.10) relating them. From these we can give an alternate proof of the existence statement of Theorem A in a special case (2.11). Some specific groups which act semifreely on $\mathbb{R}^{8n+4}$ with no invariant S^{8n+3} are listed here. In this case and in the general situation of Theorem A, the fact that $\sigma_{2n}(\pi) = 0$ is used to obtain a surgery problem

$$(f,b): M \to X$$

where X is a finite Swan complex for π in dimension $2n - 1$. The surgery obstruction $\lambda^Y(f,b) \in L^Y_{2n-1}(\mathbb{Z}\pi)$ is then defined, and by (1.8) we must show that its image is zero in $L^P_{2n-1}(\mathbb{Z}\pi)$. The method used for determining $\lambda^Y(f,b)$ is in [M] and almost all of the explicit calculations of Sections 3 and 4 are already contained in that paper. To examine $\lambda^P(f,b)$ we introduce the δ-invariant of Section 3 and show (Corollary 3.8) that it together with the semi-characteristic χ detect all elements in $L^P_3(\mathbb{Z}\pi)$. In (4.6) and (4.11 -4.15) we show how $\delta(\lambda^Y(f,b))$ is determined by the Reidemeister torsion invariants of [M]. From (4.7) and (4.8) we obtain a condition under which $\delta(\lambda^Y(f,b)) = 0$ that can be checked by calculation in the necessary cases. The details are carried out in (4.16) - (4.19).

Finally we remark that the problem considered here has also been studied by Jim Milgram [Mg2] without the assumption $\sigma_{2n}(\pi) = 0$ by completely different methods, (see the discussion at the end of §4). We wish to thank him for many useful conversations on this subject.

§1. Necessary Conditions for Semi-Free Actions on $\mathbb{R}^{2n}$

In this section we will discuss proper actions on $\mathbb{R}^{2n}$ and proper surgery theory following the account in [PR], in order to prove the necessity of the conditions stated in Theorem A.

Definition 1.1 ([PR, §5]) A proper k-dimensional Top manifold consists of (i) an open (k+1)-dimensional Top manifold W, (ii) a free $\mathbb{Z}$-action on W such that the orbit space $W/\mathbb{Z}$ is compact, and (iii) a homotopy retraction $r: W/\mathbb{Z} \to W$ of the projection $W \to W/\mathbb{Z}$.

Under these conditions $r \times c: W/\mathbb{Z} \to W \times S^1$ is a homotopy equivalence where $c: W/\mathbb{Z} \to B\mathbb{Z} = S^1$ classifies the free $\mathbb{Z}$-action and W is a finitely dominated k-dimensional Poincaré complex.

Definition 1.2 A semi-free topological action of a finite group π on $\mathbb{R}^k$ is proper if $\mathrm{Fix}(\mathbb{R}^k,\pi) = \{0\}$ and $W = (\mathbb{R}^k - \{0\})/\pi$ a proper k-dimensional Top manifold.

Let $\mathcal{N}^p_k(\pi)$ denote the unoriented bordism group of free proper π-actions or equivalently of pairs (W,ϕ) where W is a proper k-dimensional Top manifold and $\phi: W \to B\pi$ a continuous map. From (1.1) it follows that $\mathcal{N}^p_k(\pi)$ is a subgroup of $\mathcal{N}_{k+1}(\pi \times \mathbb{Z})$ and so is detected by characteristic numbers.

Proposition 1.3 Suppose that a finite group π has a proper semi-free action on $\mathbb{R}^{2n}$. Then π satisfies the p^2 and 2p conditions for all primes p (i.e. all subgroups of these orders are cyclic). If in addition $n = 4\ell + 2$, then π does not contain a subgroup τ of type IIL.

Proof. Since W is a finitely dominated complex with $\tilde{W} \simeq S^{2n-1}$ the group π must have periodic cohomology and therefore satisfies the p^2-conditions from results of Cartan and Eilenberg. (See [MTW] for a survey of the space-form problem.) To prove the second part observe that the semi-characteristic invariant of R. Lee:

(1.4)
$$\chi_{\frac{1}{2}} : \mathcal{N}^p_{2n}(\pi) \to \tilde{R}_{GL,ev}(\pi;F)$$

is defined for any field F of characteristic 2 by the formula

(1.5)
$$\chi_{\frac{1}{2}}(W,\phi) = \sum_{i=0}^{n-1} (-1)^i [H_i(\tilde{W};F)]$$

exactly as in [L, p.192] and has the formal properties expressed in [L, 2.4, 2.7, 3.8, 4.10, 4.11]. In our non-compact setting the full strength of Bredon's result [B, 7.4] is needed to show that even forms can be factored out of $\tilde{R}_F(\pi)$. It then follows as in [L, 4.14] that π satisfies the 2p-conditions and from [L, 4.15] that the type IIL subgroups are eliminated when π acts on $\mathbb{R}^{8\ell+4}$.

In order to provide a setting for the existence part of Theorem A we must now review some results from proper surgery theory due to Maumary [Ma], Taylor [T] and Pedersen-Ranicki [PR]. Let $L^p_n(\mathbb{Z}\pi)$ denote the projective surgery groups (called $U_n(\mathbb{Z}\pi)$ in [R1]) defined using forms and formations on finitely-generated projective $\mathbb{Z}\pi$-modules. For any CW complex K, let $L^{1,p}_n(K)$ denote the bordism group of normal maps from compact n-manifolds to finitely dominated Poincaré complexes equipped with a reference map to K. Finally let $L^{h,\,open}_{n+1}(K \times \mathbb{R})$ denote the bordism group of proper normal maps from (n + 1)-dimensional paracompact open manifolds to open Poincaré complexes with reference map to $K \times \mathbb{R}$.

Theorem 1.6 (see [PR, 2.1 and 7.4])

(a) If K has a finite 2-skeleton and $n \geq 5$,

$$L_n^{1,p}(K) \cong L_n^p(\mathbb{Z}\pi_1 K)$$

(b) If K is a finite complex and $n \geq 7$,

$$L_n^{1,p}(K) \cong L_n^p(\mathbb{Z}\pi_1 K) \cong L_{n+1}^{h,\text{ open}}(K \times \mathbb{R}).$$

From this result it is clear that the existence of the required proper semi-free actions of π on $\mathbb{R}^{2n}$ can be studied by following the method of [S], [TW]. Let π be a finite group with periodic cohomology of period 2d and let X^{2n-1} be a finitely dominated Poincaré complex with n a multiple of d, $\pi_1 X \cong \pi$ and $\tilde{X} \simeq S^{2n-1}$. Any such X will be called a Swan complex for π. As in [MTW], there exists a normal invariant for X and so a normal map,

$$(1.7) \qquad (f,b)\colon\ M^{2n-1} \to X$$

where M is compact. If $n \geq 4$, the surgery obstruction $\lambda^p(f,b) = 0$ in $L_{2n-1}^p(\mathbb{Z}\pi)$ if and only if $M \times S^1$ is normally cobordant to a manifold N homotopy equivalent to $X \times S^1$ or ([PR, 5.1]) if and only if X has the homotopy type of a proper 2n-dimensional Top manifold. In this case let W be the infinite cyclic cover of N and then π has a free proper action on the covering space $\tilde{W}$.

Lemma 1.8 Suppose that $\lambda^p(f,b) = 0$ for some normal map with target X. Then π has a semi-free proper action on $\mathbb{R}^{2n}$ with $\mathrm{Fix}(\mathbb{R}^{2n},\pi) = \{0\}$.

Proof From the homotopy equivalence $N \simeq X \times S^1$ we obtain a homotopy equivalence $\tilde{N} \simeq S^{2n-1} \times S^1$ which is proper since both sides are compact. Then $\tilde{W}$ is properly homotopy equivalent and therefore homomorphic to $S^{2n-1} \times \mathbb{R}$. Any finite group action on $S^{2n-1} \times \mathbb{R}$ (preserving the ends) extends to the space obtained by compactifying one end, but this is $\mathbb{R}^{2n}$.

In the existence argument of §§3, 4 the space X is assumed to be a finite complex of dimension 2n - 1. Equivalently we assume that Swan's finiteness obstruction $\sigma_{2n}(\pi) = 0$ in $\tilde{K}_0(\mathbb{Z}\pi)/T$ where $T = \{\langle\Sigma,r\rangle \mid r\varepsilon(\mathbb{Z}/|\pi|)^\times\}$ is the Swan subgroup [S, p.278]. Under this assumption the surgery obstruction of (1.7) actually lies in the group $L^h_{2n-1}(\mathbb{Z}\pi)$. The surgery groups are related by the sequence [R1, 4.3]:

$$(1.9) \quad \cdots \to H^1(\tilde{K}_0(\mathbb{Z}\pi)) \to L^h_{2n-1}(\mathbb{Z}\pi) \to L^p_{2n-1}(\mathbb{Z}\pi) \xrightarrow{\sigma} H^0(\tilde{K}_0(\mathbb{Z}\pi)) \to \cdots$$

where $\tilde{K}_0(\mathbb{Z}\pi)$ is a considered to be a $\mathbb{Z}/2$-module with the involution $([P] \to -[P^*])$ and for any $\mathbb{Z}/2$-module A, $H^*(A) \equiv H^*(\mathbb{Z}/2; A)$. From [PR, 2.1] if $(f,b)\colon M \to X$ is a normal map, $\sigma(\lambda^p(f,b)) \in H^0(\tilde{K}_0(\mathbb{Z}\pi))$ is just the image of the Wall finiteness obstruction $\sigma(X)$.

The semi-characteristic invariant of a surgery problem:

$$\chi_{\frac{1}{2}}(f,b) = \chi_{\frac{1}{2}}(X) - f_*\chi_{\frac{1}{2}}(M)$$

defines homomorphisms (for each field F of characteristic 2)

$$L^p_{2n-1}(\mathbb{Z}\pi) \to \tilde{R}_{GL,ev}(\pi;F)$$

In fact the image of this homomorphism lies in the subgroup $\{[V] \in \tilde{R}_{GL,ev}(F) | [V] = [V^*]\}$ and the relevant fields F are those which occur in the splitting of $\hat{\mathbb{Z}}_2\pi/\mathrm{Rad}$ as a direct sum of matrix rings over finite fields F of characteristic 2. The following reformulation of this invariant is due to Pardon [P] in a special case and J. Davis [Stanford Ph.D. Thesis, 1981] in general. We include a sketch of the proof.

Lemma 1.10 (a) The semi-characteristic invariants induce a homomorphism

$$\chi_{\frac{1}{2}}: L^p_{2n-1}(\mathbb{Z}\pi) \to H^1(\tilde{K}_0(\hat{\mathbb{Z}}_2\pi))/\mathrm{Im}(L^p_{2n}(\hat{\mathbb{Z}}_2\pi))$$

(b) This homomorphism may be identified with the natural map:

$$\chi: L^p_{2n-1}(\mathbb{Z}\pi) \to L^h_{2n-1}(\hat{\mathbb{Z}}_2\pi).$$

Proof

(a) Under the splitting $\hat{\mathbb{Z}}_2\pi/\mathrm{Rad} = \prod_{i=1}^{t} M(n_i,F_i)$, $K_0(\hat{\mathbb{Z}}_2\pi) = \mathbb{Z} \oplus (\mathbb{Z})^t$ and $L^p_{2n}(\hat{\mathbb{Z}}_2\pi) = \prod_{i=1}^{t} L^p_{2n}(M(n_i,F_i))$. The projective class of the simple module over $M(n_i,F_i)$ will be in the image of the map (from (1.9))

$$d_{2n}: L^p_{2n}(\hat{\mathbb{Z}}_2\pi) \to H^1(\tilde{K}_0(\hat{\mathbb{Z}}_2\pi))$$

when the simple module admits an even bilinear form. Since $L^p_{2n-1}(\hat{\mathbb{Z}}_2\pi) = 0$ from [CM], we obtain

$$H^1(\tilde{K}_0(\hat{\mathbb{Z}}_2\pi))/\mathrm{Im}(L^p_{2n}(\hat{\mathbb{Z}}_2\pi)) \cong L^h_{2n-1}(\hat{\mathbb{Z}}_2\pi).$$

It remains to show that under this isomorphism the image of the semi-characteristic is just the natural (change of co-efficients) map in (b). For this we use Ranicki's algebraic theory of surgery [R1]. If $(f,b)\colon M \to X$ is a highly-connected geometric surgery problem let (C_*,∂_*) denote the chain complex of surgery kernels over $\mathbb{Z}\pi$, (C^*,δ^*) the dual co-chain complex and $\psi\colon C_* \to C^*$ the duality chain isomorphism. The surgery obstruction $\lambda^p(f,b)$ is then represented by the formation

$$(C_{n-1} \oplus C^{n-1};\ C_{n-1},\ K)$$

where $K = \{(\partial_n(x),\ \psi_n(x) + \delta(y)) \mid x \in C_n,\ y \in C^{n-2}\}$ and the surgery kernel in dimension $n - 1$ is

$$H_{n-1}(C_*) = C_{n-1}/C_{n-1} \cap K.$$

Now if the coefficients are changed to $\hat{\mathbb{Z}}_2\pi$ and then (by projection onto a simple factor) to $M(n_i,F_i)$, the component of $\chi(\lambda^p(f,b))$ in the expression (1.11) is just $[C_{n-1} \cap K] = [H_{n-1}(C_*)] = \chi_{\frac{1}{2}}(f,b)_{(i)}$.

<u>Definition 1.12</u> Let π be a finite group with periodic cohomology of period $2d$ and $n \geq 3$ a multiple of d. Then

$$\chi_{2n-1}(\pi) = \{\chi(\lambda^P(f,b)) \subset L^h_{2n-1}(\hat{\mathbb{Z}}_2\pi)\}$$

is the set obtained as X varies over all Swan complexes for π in dimension 2n-1 and (f,b): $M \to X$ is any normal map.

This is just the subset of all possible semi-characteristic invariants for spherical space form problems. Since the map χ factors through $L^h_{2n-1}(\mathbb{Z}_{(2)}\pi)$ the size of this subset can often be determined. Alternatively the map χ has been completely computed by Carlsson-Milgram [CM].

Proposition 1.13 ([CM], [P]) Suppose π satisfies the p^2-conditions

(a) For n odd, $\chi_{2n-1}(\pi) = 0$

(b) For n = 4k, $\chi_{2n-1}(\pi) = 0$ if and only if π satisfies the 2p-conditions

(c) For n = 4k + 2, $\chi_{2n-1}(\pi) = 0$ if and only if π satisfies the 2p-conditions and has no type IIL subgroups.

Proof (a) Since $L^h_1(\hat{\mathbb{Z}}_2\pi) \to L^h_1(\hat{Q}_2\pi)$ is injective [CM], the map χ: $L^P_1(\mathbb{Z}\pi) \to L^h_1(\mathbb{Z}_{(2)}\pi)$ is zero from the localization sequence.

(b) Pardon shows that $L^h_3(\mathbb{Z}_{(2)}\pi) = 0$ if the Sylow 2-subgroup π_2 is cyclic and π satisfies the 2p-conditions. (If π has any dihedral subgroups we know from Lee's examples that $\chi_{2n-1}(\pi)$ does not contain 0). If π_2 is generalized quaternion, although $L^h_3(\mathbb{Z}_{(2)}\pi) \neq 0$ the obstructions of space form problems vanish [P, 5.1].

(c) The group $L^h_3(\mathbb{Z}_{(2)}\pi) = 0$ again when π_2 is quaternion of order 8 [P, 6.1]. If π contains a type IIL subgroup $\chi_{2n-1}(\pi)$ does not contain 0.

§2. Arithmetic Sequences in K- and L-Theory

The calculation of the relevant surgery obstruction groups in §§3, 4 will be based on the arithmetic sequences introduced by Wall (see [W5] for a more detailed account and complete references). For $\Lambda = \mathbb{Z}, \hat{\mathbb{Z}}$ let $X(\Lambda\pi) = SK_1(\Lambda\pi)$ and $Y(\Lambda\pi) = \Lambda^\times \oplus \pi/\pi' \oplus X(\Lambda\pi)$. The main exact sequence is:

$$(2.1) \qquad \cdots \to L^s_{n+1}(\hat{Q}\pi) \to L^X_n(\mathbb{Z}\pi) \to L^X_n(\hat{\mathbb{Z}}\pi) \oplus L^s_n(Q\pi) \to L^s_n(\hat{Q}\pi) \to \cdots$$

The intermediate groups needed for surgery are $L^Y_n(\mathbb{Z}\pi)$ which are related to $L^X_n(\mathbb{Z}\pi)$ by the Rothenberg sequence:

$$(2.2) \qquad \cdots \to H^{n+1}(Y/X) \to L^X_n(\mathbb{Z}\pi) \to L^Y_n(\mathbb{Z}\pi) \to H^n(Y/X) \to \cdots$$

Another case of this sequence links these to $L^K_n(\mathbb{Z}\pi)$:

$$(2.3) \qquad \cdots \to H^{n+1}(Wh'(\mathbb{Z}\pi)) \to L^Y_n(\mathbb{Z}\pi) \to L^K_n(\mathbb{Z}\pi) \to H^n(Wh'(\mathbb{Z}\pi)) \to \cdots$$

where $Wh'(\Lambda\pi) = K_1'(\Lambda\pi)/\Lambda^\times \oplus \pi/\pi'$ and $K_1'(\Lambda\pi) = K_1(\Lambda\pi)/X(\Lambda\pi)$. Actually to get the surgery obstruction groups in odd dimensions we must factor out the class of $\tau = \begin{pmatrix} 0 & 1 \\ \pm 1 & 0 \end{pmatrix}$ from L^Y and L^K to obtain $L'_n(\mathbb{Z}\pi)$ and $L^h_n(\mathbb{Z}\pi)$ respectively. However, the L^Y and L^K groups have better formal properties and will be used throughout.

In order to state the analogues of (2.1) for L^K and L^p we need the K-theory arithmetic sequences.

$$(2.4) \qquad 0 \to K_1'(\mathbb{Z}\pi) \to K_1'(\hat{\mathbb{Z}}\pi) \oplus K_1(Q\pi) \to K_1(\hat{Q}\pi) \to \tilde{K}_0(\mathbb{Z}\pi) \to 0$$

or

$$(2.5) \qquad 0 \to Wh'(\mathbb{Z}\pi) \to Wh'(\hat{\mathbb{Z}}\pi) \oplus Wh(Q\pi) \to Wh(\hat{Q}\pi) \to \tilde{K}_0(\mathbb{Z}\pi) \to 0$$

It is convenient to break (2.4) up into two short exact sequences by letting $\bar{K}_1(\pi) = \ker\ (K_1(\hat{Q}\pi) \to \tilde{K}_0(\mathbb{Z}\pi))$:

$$(2.6) \qquad 0 \to K_1'(\mathbb{Z}\pi) \to K_1'(\hat{\mathbb{Z}}\pi) \oplus K_1(Q\pi) \to \bar{K}_1(\pi) \to 0$$

and

$$(2.7) \qquad 0 \to \bar{K}_1(\pi) \to K_1(\hat{Q}\pi) \to \tilde{K}_0(\mathbb{Z}\pi) \to 0.$$

The cohomology sequence arising from (2.6) fits with (2.1) to form a commutative ladder.

Proposition 2.8 There is a commutative braid diagram of exact sequences:

$$\begin{array}{ccccccc}
H^0(K_1'(\mathbb{Z}\pi)) & & L_3^X(\mathbb{Z}\pi) & & L_3^X(\mathbb{Z}\pi) \oplus L_3^s(Q\pi) & & L_3^K(\mathbb{Z}\pi) \oplus L_3^K(Q\pi) \\
 & & & & & & \\
L_0^s(\hat{Q}\pi) & & H^0(K_1'(\hat{\mathbb{Z}}\pi) \oplus K_1(Q\pi)) & & L_3^K(\mathbb{Z}\pi) & & L_3^s(\hat{Q}\pi) \\
 & & & & & & \\
L_0^K(\hat{\mathbb{Z}}\pi) \oplus L_0^K(Q\pi) & & L_0^{\bar{K}_1(\pi)}(Q\,\pi) & & H^0(\bar{K}_1(\pi)) & & H^1(K_1'(\mathbb{Z}\pi))
\end{array}$$

The proof follows by standard arguments from the exactness of the various sequences. The arithmetic sequence for L^K appears in [R3].

There is also a similar diagram containing the arithmetic sequences for L^P and (1.9). In both cases we show only the portion needed for the applications.

Proposition 2.9 There is an commutative diagram of exact sequences:

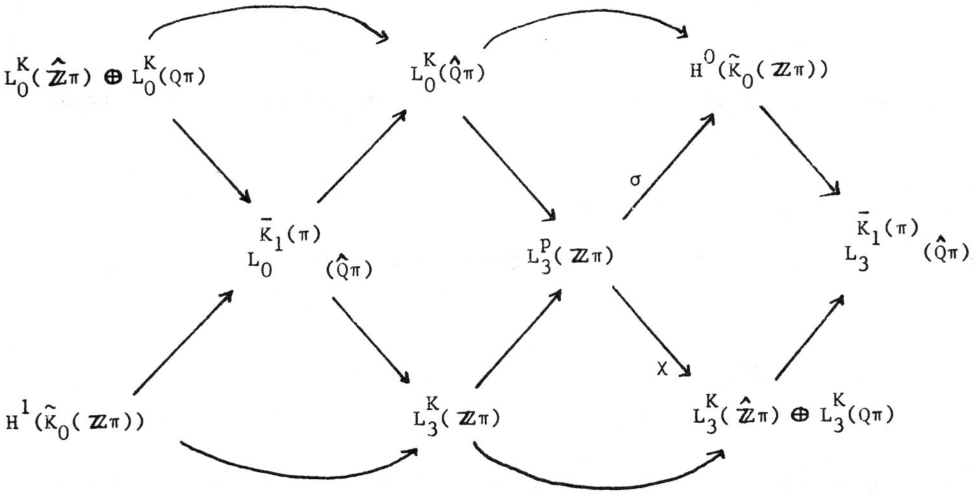

In this diagram we can see the relationship between the finiteness obstruction of an L^p-surgery problem and the semi-characteristics. In addition, there is a diagram involving (2.7).

Proposition 2.10 There is a commutative exact diagram:

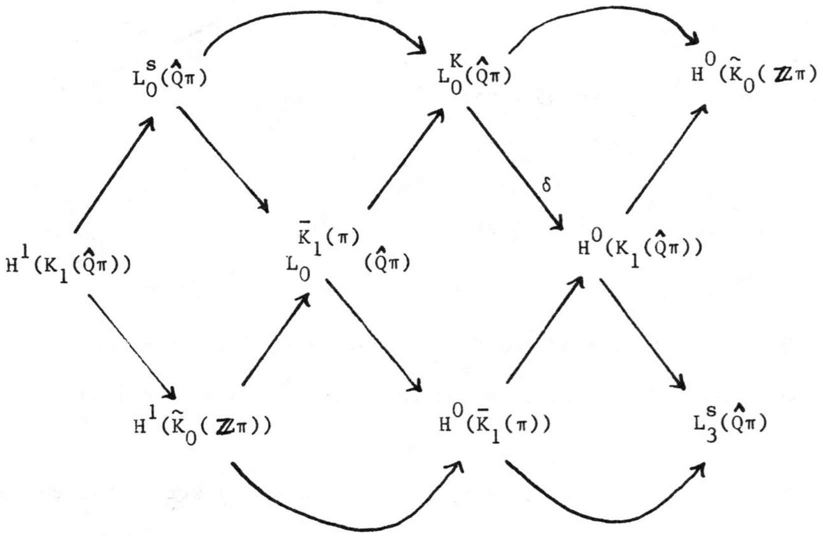

__Corollary 2.11__ Let $\pi = Q(8p,q,1)$ where p and q are odd primes and $p \equiv 3(4)$ or $q \equiv 3(4)$. Suppose that

(i) $\sigma_4(\pi) = 0$,

(ii) $(\eta_p - 2, \eta_q - 2) \in \mathrm{Image}(\Phi_{p,q}|\ker \phi_{p,q})$

Then π admits a proper semi-free action on $\mathbb{R}^{8k+4}$ for any $k > 0$.

__Proof__ In this situation conditions 7.5(i) and 7.5(ii) of [M, Theorem 7.5] are satisfied for a suitable normal map $(f,b)\colon M \to X^{8k+3}$. It follows from [M, 5.8(i)] that $\lambda^h(f,b) \in L_3^h(\mathbb{Z}\pi)$ lies in the image of the composite

$$H^1(K_1(\hat{Q}\pi)) \to L_0^S(\hat{Q}\pi) \to L_3^X(\mathbb{Z}\pi) \to L_3^Y(\mathbb{Z}\pi).$$

The result now follows from the commutative diagram:

$$\begin{array}{ccccc}
H^1(K_1(\hat{Q}\pi)) & \longrightarrow & L_0^s(\hat{Q}\pi) & \longrightarrow & L_3^X(\mathbb{Z}\pi) & & \\
\downarrow & & | & & \downarrow & & \\
 & & L_0^{\bar{K}_1(\pi)}(\hat{Q}\pi) & & & & \\
 & \nearrow & & \searrow & & & \\
H^1(\tilde{K}_0(\mathbb{Z}\pi)) & & \longrightarrow & & L_3^h(\mathbb{Z}\pi) & \longrightarrow & L_3^P(\mathbb{Z}\pi),
\end{array}$$

which shows that $\lambda^P(f,b) = 0$.

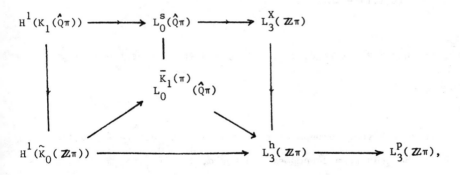

__Remark__ This is a separate proof of the existence part of Theorem A in these special cases. If $p \equiv 3(4)$ then $\sigma_4(\pi) = 0$ if and only if $q \equiv 1(8)$ or $q \equiv 5(8)$ and the Legendre symbol $(\frac{p}{q}) = +1$ by results from [Mg2]; there exists a free action on S^{8k+3} if and only if $q \equiv 1(8)$ and 2.11(ii) is satisfied [M, 7.6].

§3. The δ-invariant

In this section we use standard facts from L-theory to examine the Mayer-Vietoris sequence from (2.9) for calculating $L^p_{2n+1}(\mathbb{Z}\pi)$. In our applications we need only information about $L^p_3(\mathbb{Z}\pi)$, so concentrate on this case.

Let D be a division algebra with center E and (D,α,u) any anti-structure over D. We recall from [W1]

(3.1) $$L^K_1(D,\alpha,u) = 0 \quad \text{if} \quad (D,\alpha,u) \neq (E,1,1)$$

$$L^K_1(E,1,1) = \mathbb{Z}/2$$

Moreover, if char E ≠ 2 then $d_1 : L^K_1(E,1,1) \to H^1(E^\times)$ gives the isomorphism.

We write $L^U_*(\Lambda\pi)$ for the L-groups of the usual anti-structure $(\Lambda\pi,\alpha,1)$. For $\Lambda = Q$ we have the Weddeburn decomposition $(Q\pi,\alpha,1) = (S_i,\alpha_i,1)$ where the S_i are simple algebras over number fields. More precisely, each S_i is associated to an irreducible character χ_i of π and $Z(S_i) = Q(\chi_i)$. The anti-structure has type O, type Sp or type U if the Frobenius-Schur sum $\sum_{g\in\pi} \chi_i(g^2)$ is > 0, < 0 or $= 0$.

<u>Corollary 3.2</u> For every finite group π, $L^K_3(Q\pi) = 0$.

Proof The Wedderburn decomposition of $(Q\pi,\alpha,1)$ gives $L_3^K(Q\pi) \cong \bigoplus L_3^K(S_i,\alpha_i,1)$. For each i we have a scaling equivalence of anti-structures

$$(S_i,\alpha_i,1) \cong (M_n(D_i),\alpha_i,u_iI_n)$$

Morita equivalence gives

$$L_3^K(S_i,\alpha_i,1) \cong L_3^K(D_i,\alpha_i,u_i) \cong L_1^K(D_i,\alpha_i, -u_i).$$

By (3.1), the group is zero unless $(D_i,\alpha_i,u_i) = (E_i,\alpha_i - 1)$. But the exceptional case cannot occur for any i. If it did, the corresponding simple component S_i would have type Sp, hence be associated to an irreducible (complex) character χ_i with $\Sigma\chi_i(g^2) < 0$. Let $E_i \subset \mathbb{R}$ be a completion of E_i. Then $S_i \otimes_{E_i} \mathbb{R}$ is the simple component of $\mathbb{R}\pi$ associated to χ_i, and since the Frobenius-Schur sum is negative $S_i \otimes_{E_i} \mathbb{R} = M_n(\mathbb{H})$. This contradicts the fact that $D_i = E_i$.

Recall from [W3] that if R is a complete semi-local ring and $J \subset R$ is an ideal so that R is J-radically complete then

(3.3) $$L_n^K(R,\alpha,u) \cong L_n^K(R/J,\bar{\alpha},\bar{u})$$

The isomorphism is induced from the projection of anti-structures.

Corollary 3.4 For all odd prime p,

$$L_{2n+1}^K(\hat{\mathbb{Z}}_p\pi) \to L_{2n+1}^K(\hat{Q}_p\pi)$$

is injective.

Proof We may assume by induction that π is 2-hyperelementary, say $\pi = (\mathbb{Z}/p^r \times \mathbb{Z}/m) \tilde{\times} \sigma$ with $\sigma \in \mathrm{Syl}_2(\pi)$. Let $T \in \mathbb{Z}/p^r$ denote a generator. Consider the ideal $J = (1 - T) \subset \hat{\mathbb{Z}}_p\pi$. It satisfies the conditions of (3.3), and $\hat{\mathbb{Z}}_p\pi/J = \hat{\mathbb{Z}}_p[\mathbb{Z}/m \tilde{\times} \sigma]$ where p is not a divisor of m. We have

$$(\hat{\mathbb{Z}}_p[\mathbb{Z}/m \tilde{\times} \sigma], \alpha, 1) \cong \prod_{d|m} (\hat{\mathbb{Z}}_p[\zeta_d]^t\sigma, \alpha, 1)$$

$$(\hat{Q}_p[\mathbb{Z}/m \tilde{\times} \sigma], \alpha, 1) \cong \prod_{d|m} (\hat{Q}_p[\zeta_d]^t\sigma, \alpha, 1)$$

Since $p \nmid 2d$, $\hat{Q}_p[S_d]^t\sigma$ decomposes into matrix rings over local fields E_i and $\hat{\mathbb{Z}}_p[\zeta_d]^t\sigma$ decomposes into matrix rings over the integers A_i in E_i. Using Morita-equivalence we are reduced to showing that

$$L_1^K(A_i, \alpha_i, u_i) \to L_1^K(E_i, \alpha_i, u_i)$$

is injective when A_i is an unramified p-ring and E_i is its fraction field. This follows from (3.1) and (3.3).

Next, we use some classical facts about quadratic forms over division algebras to get information about the left hand part of the Mayer-Vietoris sequence:

$$L_0^K(\hat{\mathbb{Z}}\pi) \oplus L_0^K(Q\pi) \to L_0^K(\hat{Q}\pi) \to L_3^P(\mathbb{Z}\pi)$$

(see [W2], [W4] or [LM], §2).

Consider an anti-structure (S,α,u) over a central, simple E-algebra. If E is local, then $K_1(S) = E^{\times}$ and

$$d_0\colon\ L_0^K(S,\alpha,u) \to H^0(E^{\times})$$

is injective unless (S,α,u) has type 0 and $S = M_n(E)$ where $\operatorname{Ker} d_0 = L_0^X(S,\alpha,u) = \mathbb{Z}/2$. Moreover, if E is global and (S,α,u) has type 0 then there is an exact sequence

$$(3.5)\qquad 0 \to L_0^X(S,\alpha,u) \to L_0^X(S_A,\alpha,u) \to \mathbb{Z}/2 \to 0$$

where $S_A = S \otimes Q_A$, $Q_A = \hat{Q} \oplus \mathbb{R}$. (See [W2], [W4].)

<u>Proposition 3.6</u> The sequence

$$L_0^K(\hat{\mathbb{Z}}\pi) \oplus L_0^K(Q\pi) \to L_0^K(\hat{Q}\pi) \xrightarrow{d_0} H^0(K_1(\hat{Q}\pi)$$

is exact.

<u>Proof</u> As a first step we show the weaker statement that

$$L_0^K(Q\pi) \to L_0^K(\hat{Q}\pi) \xrightarrow{d_0^X} H^0(K_1(\hat{Q}\pi))$$

is exact. According to the discussion above

$$\operatorname{Ker} d_0^X = L_0^X(S_i,\ \alpha_i,\ u_i)$$

where the product ranges over the type 0 anti-structures in the Wedderburn decomposition for which S_i is split (i.e. $S_i = M_k(E_i)$). We show that

$$L_0^X(S_i,\alpha_i,u_i) \to L_0^X(\hat{S}_i,\hat{\alpha}_i,\hat{u}_i)$$

is surjective for each such i. This follows from (3.5). Indeed,

$$(S_i)_A = \hat{S}_i \times M_n(\mathbb{R} \otimes S_i) \quad \text{and} \quad L_0^X((S_i)_A) = L_0^X(\hat{S}_i) \oplus (4\mathbb{Z})^{\oplus g}$$

where g is the number of real primes of E_i. Each of the g components $4\mathbb{Z}$ surjects onto the $\mathbb{Z}/2$ in (3.5), so $L_0^X(S_i) \to L_0^X(\hat{S}_i)$ is surjective as claimed, and $L_0^K(Q\pi)$ maps onto the kernel of d_0^X: $L_0^K(\hat{Q}\pi) \to H^0(K_1(\hat{Q}\pi))$.

The kernel of d_0: $L_0^K(\hat{Q}\pi) \to H^0(Wh(\hat{Q}\pi))$ is the image of $L_0^Y(\hat{Q}\pi)$ in $L_0^K(\hat{Q}\pi)$. It follows from the Rothenberg sequence relating L_0^X and L_0^Y that we are done if we can show that the two compositions

$$L_0^Y(\hat{\mathbb{Z}}\pi) \oplus L_0^Y(Q\pi) \to L_0^Y(\hat{Q}\pi) \to H^0(\hat{Q}^\times) \oplus H^0(\pi/\pi')$$

$$L_0^Y(\hat{Q}\pi) \to H^0(\hat{Q}^\times) \oplus H^0(\pi/\pi')$$

have the same images.

We write $L_0^Y(\Lambda\pi) = L_0^K(\Lambda) \oplus \tilde{L}_0^Y(\Lambda\pi)$ and first show that $L_0^K(\hat{\mathbb{Z}}) \oplus L_0^K(Q) \to H^0(\hat{Q}^\times)$ is surjective. Indeed, $L_0^K(\hat{\mathbb{Z}}) \to H^0(\hat{\mathbb{Z}}^\times)$ has image of index 2; and $\langle -1\rangle \in H^0(\hat{\mathbb{Z}}^\times)$ lies outside the image since

$-1 \notin$ Image $(L_0^K(\hat{\mathbb{Z}}_2) \to H^0(\hat{\mathbb{Z}}_2^\times))$. The maps $L_0^K(Q) \to H^0(Q^\times)$ and $L_0^K(\hat{Q}) \to H^0(\hat{Q}^\times)$ are surjective, and the surjectivity of $L_0^K(\hat{\mathbb{Z}}) \oplus L_0^K(Q) \to H^0(\hat{Q}^\times)$ follows from the exact sequence

$$0 \to H^0(\mathbb{Z}^\times) \to H^0(\hat{\mathbb{Z}}^\times) \oplus H^0(Q^\times) \to H^0(\hat{Q}^\times) \to 0$$

Finally, $L_3^X(Q\pi) \to L_3^X(\hat{Q}\pi) = L_3^X(Q_A\pi)$ is injective, and the diagram

$$\begin{array}{ccccc}
\tilde{L}_0^Y(Q\pi) & \xrightarrow{d_0} & H^0(\pi/\pi') & \longrightarrow & L_3^X(Q\pi) \\
\downarrow & & \| & & \downarrow \\
\tilde{L}_0^Y(\hat{Q}\pi) & \xrightarrow{d_0} & H^0(\pi/\pi') & \longrightarrow & L_3^X(\hat{Q}\pi)
\end{array}$$

shows that $\tilde{L}_0^Y(Q\pi)$ and $\tilde{L}_0^Y(\hat{Q}\pi)$ have the same images in $H^0(\pi/\pi')$.

<u>Definition 3.7</u> The discriminant d_0 defines the invariant

$$\delta:\ \text{Image } (L_0^K(\hat{Q}\pi) \to L_3^P(\mathbb{Z}\pi)) \to H^0(K_1(\hat{Q}\pi))/L_0^K(\hat{\mathbb{Z}}\pi) \oplus L_0^K(Q\pi)$$

Note from the Mayer-Vietoris sequence that δ is defined on the kernel of

$$L_3^P(\mathbb{Z}\pi) \to L_3^K(\hat{\mathbb{Z}}\pi) \oplus L_3^K(Q\pi)$$

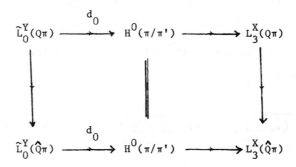

This kernel is by (3.2) and (3.4) the same as the kernel of $L_3^p(\mathbb{Z}\pi) \to L_3^K(\hat{\mathbb{Z}}_2\pi)$, or what is equivalent the kernel of the semi-characteristic χ. We have from (3.6):

<u>Corollary 3.8</u> The invariants χ and δ detect $L_3^p(\mathbb{Z}\pi)$.

There is a decomposition

$$K_1(Q\pi) = K_1(Q\pi)_+ \oplus K_1(Q\pi)_- \oplus K_1(Q\pi)_0$$

associated to the decomposition of $(Q\pi,\alpha,1)$ into type O, type Sp and type U parts. In other words

$$K_1(Q\pi)_+ = \prod Q(\chi_i^+)^\times,\ K_1(Q\pi)_- = \prod Q(\chi_i^-)^*,\ K_1(Q\pi)_0 = \prod Q(\chi_i^0)^\times$$

where χ_i^+, χ_i^- and χ_i^0 are irreducible (complex) characters with $\Sigma\chi_i^+(g^2) > 0$, < 0, and $= 0$, respectively ($Q(\chi)^* \subset Q(\chi)^\times$ is the subgroup of totally positive elements). There is a similar decomposition in the case of complete group rings $K_1(\hat{Q}_p\pi)$ (placing type GL in $K_1(\hat{Q}_p\pi)_0$). If $\ell \mid |\pi|$ then

$$K_1(\hat{\mathbb{Z}}_\ell\pi) = K_1(\hat{\mathbb{Z}}_\ell\pi)_+ \oplus K_1(\hat{\mathbb{Z}}_\ell\pi)_- \oplus K_1(\hat{\mathbb{Z}}_\ell\pi)_0$$

with $K_1(\hat{\mathbb{Z}}_\ell\pi)_\pm = \prod U(Q(\chi_i^\pm)\hat{}_\ell)$ and $K_1(\hat{\mathbb{Z}}_\ell\pi)_0 = \prod U(Q(\chi_i^0)\hat{}_\ell)$. Here $Q(\chi)\hat{}_\ell = \prod_{y|\ell} Q(\chi)\hat{}_y$ and U indicates units (of the algebraic integers). If ℓ is odd and ℓ divides $|\pi|$ then $H^0(K_1(\hat{\mathbb{Z}}_\ell\pi)) \cong H^0(K_1(\hat{\mathbb{Z}}_\ell\pi/\mathrm{Rad}))$. The induced anti-structure over $\hat{\mathbb{Z}}_\ell\pi/\mathrm{Rad}$ decomposes into summands whose underlying rings are matrix rings over finite fields. Thus we have a decomposition of

$K_1(\hat{\mathbb{Z}}_\ell\pi/\text{Rad})$ as above, inducing a decomposition of $H^0(K_1(\hat{\mathbb{Z}}_\ell\pi))$. We note that $H^0(K_1(\hat{\mathbb{Z}}_\ell\pi)_0) = 0$ (because the norm homormophisms is surjective for finite fields.

Proposition 3.9 The range of δ is contained in

$$H^0(Wh(\hat{Q}\pi)+)/(H^0(Wh(Q\pi)_+) \oplus H^0(Wh(\hat{\mathbb{Z}}_{odd}\pi)_+) \oplus L_0^K(\hat{\mathbb{Z}}_2\pi))$$

Proof Let (S,α,u) be a type Sp anti-structure over a central, simple E-algebra and suppose char $E \neq 2$. From [W2], [W4] we have

(i) $L_0^K(S,\alpha,u) = 0$, E local or finite

(ii) $L_0^K(S,\alpha,u) \cong \Sigma$, E global

where $\Sigma \quad \oplus \quad (2\mathbb{Z})$ is a signature group. Since $L_0^K(\hat{\mathbb{Z}}_\ell\pi) \cong L_0^K(\hat{\mathbb{Z}}_\ell\pi/J)$ and $L_0^K(\hat{Q}\pi)$ is the restricted product of $L_0^K(\hat{Q}_\ell\pi)$ and $L_0^K(\hat{\mathbb{Z}}_\ell\pi)$, it follows that $d_0 : L_0^K(\Lambda\pi) \to H^0(K_1(\Lambda\pi))$ has image contained in $H^0(K_1(\Lambda\pi)_+) \oplus H^0(K_1(\Lambda\pi)_0)$ for $\Lambda = Q, \hat{\mathbb{Z}}_{odd}$ and $\hat{Q}$.

If (S,α,u) has type O or type U then

$$d_o: \ L_0^K(S,\alpha,u) \to H^0(K_1(S))$$

is surjective where E is finite, local or global. Moreover, if (S,α,u) has type U and E is global then $H^0(K_1(S)) \to H^0(K_1(\hat{S}))$ is subjective by class field theory so $H^0(K_1(\hat{Q}\pi)_0)/H^0(K_1(Q\pi)_0) = 0$.

Finally, d_0: $L_0^K(\hat{\mathbb{Z}}_2\pi) \to H^0(K_1(\hat{\mathbb{Z}}_2\pi)) \to H^0(K_1(\hat{Q}\pi)$ has image contained in $H^0(K_1(\hat{Q}\pi)_+)$ because it factors through

$$L_0^K(\hat{Q}_2\pi) \to H^0(K_1(\hat{Q}_2\pi)) \text{ and } L_0^K(\text{type Sp}) = 0$$

in the local case.

§4. The Proof of Theorem A

Recall that the groups π which satisfy the p^2- and 2p-conditions are classified into six types and that the results of [MTW], [W] imply a free action of π on S^{2n-1} where 2n is the period of π except in some cases where π has type II. Since a free action on S^{2n-1} gives a semi-free action on $\mathbb{R}^{2n}$ (free outside 0) by coning, we have left only to prove Theorem A for groups of type II. These are extensions

$$(4.1) \qquad 0 \to \mathbb{Z}/m \to \tau \to Q2^r \times \mathbb{Z}/k \to 1$$

where m and k are odd coprime integers. The period of $H^*(\tau;\mathbb{Z})$ is 4n, where n depends on the action of $\mathbb{Z}/k$ on $\mathbb{Z}/m$.

Let $g \in H^{4n}(\tau;\mathbb{Z})$ be any generator with vanishing finiteness obstruction,

$$\sigma_{4n}(g) = 0 \qquad \text{in} \qquad \tilde{K}_0(\mathbb{Z}\tau)$$

We pick a finite CW complex $\Sigma \simeq S^{4n-1}$ with a free action of τ such that the first k-invariant of Σ/τ is g. Let $\tau_p \in \mathrm{Syl}_p(\tau)$. Every generator of

$H^*(\tau_p; \mathbb{Z})$ with vanishing finiteness obstruction is realized by a free linear action: there are free representations V_p such that

$$(\Sigma, \tau_p) \simeq (S(V_p), \tau_p)$$

Using induction on normal invariants (cf. [M], §7) we get a normal map

$$(4.2) \qquad f: M^{4n-1} \to \Sigma/\tau, \ b: \nu_M \to \zeta$$

inducing the homotopy equivalences $S(V_p)/\tau_p \simeq \Sigma/\tau_p$ (up to normal cobordism).

From (1.6), (1.8) there exists a proper semi-free action of τ on $(\mathbb{R}^{4n}, 0)$ which is free on $\mathbb{R}^{4n} - \{0\}$ and with $(\mathbb{R}^{4n} - \{0\}, \tau) \simeq (\Sigma, \tau)$ if and only if

$$\lambda^p(f,b) = 0 \quad \text{in} \quad L_3^p(\mathbb{Z}\pi)$$

By Dress' induction theorem $L_3^p(\mathbb{Z}\pi)$ injects into $L_3^p(\mathbb{Z}G)$ where $G \subset \tau$ is the subgroup

$$(4.3) \qquad 0 \to \mathbb{Z}/m \to G \to Q2^r \to 1$$

Thus we need only consider the case where $k = 1$ in (4.1) (where we write G instead of τ). If $r > 3$ in (4.2) and the centralizer of $\mathbb{Z}/m$ has index 4 in G then $\chi(f,b) \neq 0$ from (1.13). In particular, $\lambda_3^p(f,b) \neq 0$. If the centralizer of $\mathbb{Z}/m$ in G has index 2 then G admits a free representation in degree $4n$. Thus to complete the proof of Theorem A we have left to construct actions on $(\mathbb{R}^{8k+4}, 0)$ of the groups in (4.3) with $r = 3$ and $C_G(\mathbb{Z}/m) = \mathbb{Z}/2m$.

Using notation from [Mil] these are the groups $Q(8a,b,c)$ with at least two of the three coprime integers a, b and c larger than 1. The numbers a, b and c can be permuted without changing the isomorphism class, so we may assume $a \geq b \geq c$. The required actions on $(\mathbb{R}^{8k+4},0)$ follow from (1.8) and

<u>Theorem 4.4</u> Let $\pi = Q(8a,b,c)$, $a \geq b \geq c$, where $b > 1$. Suppose that $\sigma(g) = 0$ for a generator $g \in H^{8k+4}(\pi: \mathbb{Z})$ which restricts to the generator of a linear space form on the subgroup $Q(4a,b,c)$. Let (f,b) be the normal map from (4.2) and $k \geq 1$. Then $\lambda^P(f,b) = 0$ in $L_3^P(\mathbb{Z}\pi)$ if and only if the condition (ii) of Theorem A is satisfied (for all subgroups τ of π isomorphic to $Q(8\alpha,\beta,1)$).

We begin the proof of (4.4) with a general result which restricts the calculation of $\lambda^P(f,b)$ when (f,b) is a surgery problem over a weakly simple Poincaré complex with fundamental group π. In this situation we have $\lambda^Y(f,b) \in L_3^Y(\mathbb{Z}\pi)$ and $\lambda^Y(f,b)$ maps to $\lambda^P(f,b)$ under the natural map $L_3^Y(\mathbb{Z}\pi) \to L_3^P(\mathbb{Z}\pi)$.

Consider the homomorphism:

(4.5) $$\delta^Y\colon \operatorname{Ker}(L_{2n-1}^Y(\mathbb{Z}\pi) \to L_{2n-1}^K(\hat{\mathbb{Z}}_2\pi)) \to H^0(Wh(\hat{Q}\pi))/(L_{2n}^K(Q\pi) \oplus L_{2n}^K(\mathbb{Z}\pi))$$

induced from the additive relation in diagram (2.8) replacing X by Y and using the cohomology sequence of (2.5) instead:

$$\begin{array}{ccc} L_{2n-1}^Y(\mathbb{Z}\pi) \to L_{2n-1}^Y(\hat{\mathbb{Z}}\pi) & \oplus & L_{2n-1}^Y(Q\pi) \\ & \uparrow & \\ H^0(Wh(\hat{\mathbb{Z}}\pi)) & \oplus & H^0(Wh(Q\pi)) \to H^0(Wh(\hat{Q}\pi)) \end{array}$$

We compare δ^Y with the δ-invariant from (3.7):

Proposition 4.6 Given a surgery problem (f,b) over an odd dimensional weakly simple Poincaré complex. If $\chi(f) = 0$ then $\delta^Y(\lambda^Y(f)) = \delta(\lambda^P(f))$.

Proof The argument is a variant of arguments used in [M, 2.10 and 3.10] and is based on the algebraic surgery technique from [R1]. Let C_* be the quadratic Poincaré chain complexes representing $\lambda^Y(f) \in L^Y_{2n-1}(\mathbb{Z}\pi)$. Then $C_* \otimes \hat{\mathbb{Z}}$ and $C_* \otimes Q$ represent the images in $L^Y_{2n-1}(\hat{\mathbb{Z}}\pi)$ and $L^Y_{2n-1}(Q\pi)$, so there exists quadratic algebraic cobordisms $\hat{D}_*$ and D^0_*,

$$\partial\hat{D}_* = C_* \otimes \hat{\mathbb{Z}} - \hat{C}_*, \qquad \partial D^0_* = C_* \otimes Q - C^0_*$$

with $\hat{C}_*$ and C^0_* acyclic complexes. Let $\Delta(\hat{C}_*)$ and $\Delta(C^0_*)$ be the Whitehead torsion invariants in $H^0(Wh(\hat{\mathbb{Z}}\pi))$ and $H^0(Wh(Q\pi))$. The difference of their images in $H^0(Wh(\hat{Q}\pi))$ represents $\delta^Y(\lambda^Y(f))$ according to [M, 3.10]:

$$\Delta(\hat{C}_* \otimes_{\hat{\mathbb{Z}}} \hat{Q}) - \Delta(C^0_* \otimes_Q \hat{Q}) = \delta^Y(\lambda^Y(f))$$

On the other hand we can glue $\hat{D}_* \otimes_{\mathbb{Z}} Q$ and $D^0_* \otimes_Q \hat{Q}$ together along $C_* \otimes_{\mathbb{Z}} \hat{Q}$ to get

$$D_* = \hat{D}_* \otimes_{\mathbb{Z}} \hat{Q} \cup D^0_* \otimes_Q \hat{Q}$$

with $\partial D_* = \hat{C}_* \otimes_{\hat{\mathbb{Z}}} Q - C^0_* \otimes_Q \hat{Q}$. The boundary is acyclic, so D_* represents an element in $L^K_{2n}(\hat{Q}\pi)$ which maps to $\lambda^P(f)$ in $L^P_{2n-1}(\mathbb{Z}\pi)$. This follows from the cobordism interpretation of the Mayer-Vietoris sequence for calculating $L_*^P(\mathbb{Z}\pi)$. Finally, using [M, 2.10]

$$d_{2n}: L_{2n}^K(\hat{Q}\pi) \to H^0(Wh(\hat{Q}\pi))$$

can be calculated from:

$$d_{2n}([D_*]) = \Delta(\partial D_*) = \Delta(\hat{C}_* \otimes_{\hat{\mathbb{Z}}} \hat{Q}) - \Delta(C_*^0 \otimes_Q \hat{Q}).$$

For ℓ an odd prime,

$$H^0(Wh(\hat{\mathbb{Z}}_\ell\pi)_+) \to L_3^Y(\hat{\mathbb{Z}}_\ell\pi)$$

$$H^0(Wh(Q\pi)_-) \to L_3^Y(Q\pi)$$

are surjective. Thus the components of $\delta^Y(f)$ in $H^0(Wh(\hat{Q}_\ell\pi))$ lie in $H^0(Wh(\hat{Q}_\ell\pi)_-)$. On the other hand, by (3.9), the range of δ is a quotient of $H^0(Wh(\hat{Q}\pi)_+$. Hence we have

<u>Corollary 4.7</u> (i) In the situation of (4.6), $\delta(\lambda^P(f))$ belongs to the image of

$$H^0(Wh(\hat{Q}_2\pi)_+)/L_0^K(\hat{\mathbb{Z}}_2\pi) \to H^0(Wh(\hat{Q}\pi)_+)/I$$

where $I = H^0(Wh(Q\pi)_+) \oplus H^0(Wh(\hat{\mathbb{Z}}_{odd}\pi)_+) \oplus L_0^K(\hat{\mathbb{Z}}_2\pi)$.

(ii) In fact, $\delta(\lambda^P(f)) = \delta_2^Y(\lambda^Y(f))$ where δ_2^Y is the composition

$$L_3^Y(\mathbb{Z}\pi) \to L_3^Y(\hat{\mathbb{Z}}_2\pi) \leftarrow H^0(Wh(\hat{\mathbb{Z}}_2\pi))/L_0^K(\hat{\mathbb{Z}}_2\pi) \to H^0(Wh(\hat{Q}_2\pi)_+)/L_0^K(\hat{\mathbb{Z}}_2\pi)$$

At the moment we do not know a satisfactory direct description of the invariant $\delta_2^Y(\lambda^Y(f))$ in general. This appears to be a very interesting problem.

Lemma 4.8 In the situation of (4.7), suppose further that

$$\delta_2^Y(\lambda^Y(f)) \in \text{Image } (H^0(Wh'(\mathbb{Z}\pi)) \to H^0(Wh(\hat{Q}_2\pi)_+) / L_0^K(\hat{\mathbb{Z}}_2\pi))$$

Then $\lambda^P(f) = 0$.

Proof This follows from the exact sequence

$$0 \to Wh'(\mathbb{Z}\pi) \to Wh'(\hat{\mathbb{Z}}\pi) \oplus Wh(Q\pi) \to Wh(\hat{Q}\pi) \to \tilde{K}_0(\mathbb{Z}\pi) \to 0$$

Indeed, under the stated assumptions $\delta^Y(\lambda^Y(f))$ in $H^0(Wh(\hat{Q}_2\pi))/L_0^K(\hat{\mathbb{Z}}_2\pi)$ maps trivially to $H^0(Wh(\hat{Q}\pi))/I$. Hence $\lambda^P(f) = 0$ by (4.7).

We need one more general result. Suppose $\pi = \mathbb{Z}/m \tilde{\times} \sigma$ is an arbitrary 2-hyperelementary group. There are decompositions

$$L_*^P(\mathbb{Z}\pi) = \prod_{d|m} L_*^P(\mathbb{Z}\pi)(d), \quad L_*^Y(\mathbb{Z}\pi) = \prod_{d|m} L_*^Y(\mathbb{Z}\pi)(d)$$

and for $d > 1$, $L_*^Y(\mathbb{Z}\pi)(d) = L_*^X(\mathbb{Z}\pi)(d)$. Each factor can be calculated from the exact sequences

$$\cdots \to L_0^X(\hat{S}(d)) \to L_3^X(\mathbb{Z}\pi)(d) \to \prod_{\ell \nmid d} L_3^X(\hat{R}_\ell(d)) \oplus L_3^X(S(d)) \to \cdots$$

(4.9)

$$\cdots \to L_0^K(\hat{S}(d)) \to L_3^p(\mathbb{Z}\pi)(d) \to L_3^K(\hat{R}_2(d)) \to \cdots$$

where $R(d) = \mathbb{Z}[\zeta_d]^t[\sigma]$, $S(d) = Q(\zeta_d)^t[\sigma]$ and $\hat{R}_\ell(d) = \hat{\mathbb{Z}}_\ell \otimes R(d)$. (The result for L_*^X can be found in [W6] and the result for L_*^p is similar.)

With these preparations we return to the proof of Theorem 4.4, so let $\pi = Q(8a,b,c)$.

<u>Lemma 4.10</u> If $(d,a) \neq 1$, $(d,b) \neq 1$ and $(d,c) \neq 1$ then $L_3^p(\mathbb{Z}\pi)(d) = 0$ and $L_3^X(\mathbb{Z}\pi)(d) = 0$.

<u>Proof</u> Each factor in $S(d)$ and in $\hat{R}_2(d)/J$ has type U (cf. [LM], §3). It follows that $L_3^K(\hat{R}_2(d)) = 0$ and that $C\ell_0^K(S(d)) = \oplus\mathbb{Z}/2$ with one summand $\mathbb{Z}/2$ for each simple factor in $S(d)$, (cf. [W4], §5). Moreover, $L_0^K(S(d) \otimes_Q \mathbb{R})$ surjects onto $C\ell_0^K(S(d))$ so $L_0^K(S(d)) \to L_0^K(\hat{S}(d))$ must be surjective. Hence $L_3^p(\mathbb{Z}\pi)(d) = 0$ by (4.9). The argument for $L_3^X(\mathbb{Z}\pi)(d)$ is similar.

It suffices to prove (4.5) when precisely one of the three numbers a, b and c is equal to 1. Indeed, if $d = \alpha\beta$ with $\alpha|a$, $\beta|b$ then we have an isomorphism

$$\text{Res: } L_3^p(\mathbb{Z}[Q(8a,b,c)])(d) \xrightarrow{\cong} L_3^p(\mathbb{Z}[Q(8\alpha,\beta,1)])(d)$$

(cf. [M], 4.17.) Moreover, if in addition $\beta = 1$ then $Q(8\alpha,\beta,1) = Q(8\alpha)$

admits free linear representations in degrees 8k+4, and the covering of (f,b) induced from $Q(8\alpha) \subset Q(8a,b,c)$ is normally cobordant to a homotopy equivalence $S(V)/Q(8\alpha) \simeq \Sigma/Q(8\alpha)$. Hence $\mathrm{Res}(\lambda^P(f)(d)) = 0$ in this case, so $\lambda^P(f)(d) = 0$.

For the rest of the section we write

$$G = Q(8a,b,1),\ d = ab$$

with a and b odd coprime numbers. We evaluate $\lambda^P(f) = 0$ by arguments which are similar to the arguments used in [M] to examine $\lambda^K(f)$ in the special cases where a and b are prime numbers.

Consider the 'diagonal' subgroup H G,

$$H = \mathbb{Z}/ab \tilde{\times} \mathbb{Z}/4$$

where $\mathbb{Z}/4$ acts by inversion on $\mathbb{Z}/ab$. The centralizer $C = C_H(\mathbb{Z}/ab)$ is cyclic of order 2ab, and any free representation V of C, $W = \mathrm{Ind}_H^G(V)$ is a free representation of G. Thus H acts freely S(W) (in all dimensions 4n-1).

Different generators $g \in H^{8k+4}(G;\mathbb{Z})$ usually have different finiteness obstructions $\sigma_4(g) \in \tilde{K}_0(\mathbb{Z}G)$. The variation lies in the subgroup $T(\mathbb{Z}G)$ generated by the projective ideals (Σ,r) mentioned above.

Let $g \in H^{8k+4}(G;\ \mathbb{Z})$ be a generator with

(i) $\sigma(g) = 0$

(ii) $\mathrm{Res}_H^G(g) = c_{4k+2}(W)$, $\quad \mathrm{Res}_H^G\colon H^*(G;\mathbb{Z}) \to H^*(H;\mathbb{Z})$

where $W: H \to U(4k+2)$ is a free representation and c_{4k+2} denotes its Chern class.

The Chern class is equal to the first k-invariant of $S(W)/H$ and determines uniquely its homotopy type. It follows that there is a surgery problem

$$f_H: \quad S(W)/H \to \Sigma/H, \ b: \nu \to \zeta$$

where Σ/H is the covering of the finite Poincaré complex Σ/G determined by $g \in H^{8k+4}(G; \mathbb{Z})$.

Induction theorems for normal invariants shows that (up to normal cobordism) f_H covers a surgery problem

$$f: \quad M \to \Sigma/G, \ b: \quad \nu_M \to \zeta$$

(cf. [M], §7). This is the normal map whose λ^P-obstruction we shall now calculate.

It follows from duality of Whitehead torsions that Σ/G is a weakly simple Poincaré complex. Thus there is an obstruction

$$\lambda^Y(f,b) = \lambda^Y(f) \in L_3^Y(\mathbb{Z}G)$$

which projects onto $\lambda^K(f)$ and $\lambda^P(f)$. By construction $\lambda^K(f_H) = \mathrm{Res}_H^G(\lambda^K(f)) = 0$. More precisely we have the exact sequence

$$Wh'(\mathbb{Z}H) \otimes \mathbb{Z}/2 \quad \to \quad L_3^Y(\mathbb{Z}H) \to L_3^K(\mathbb{Z}H) \to 0$$

and

$$(4.11) \qquad \mathrm{Res}_H^G(\lambda^Y(f)) = \Delta(S(W)/H) \ / \ \Delta(\Sigma/H)$$

where $\mathrm{Res}_H^G\colon\ L_3^Y(\mathbb{Z}G) \to L_3^Y(\mathbb{Z}H)$ and Δ denotes the Reidemeister torsion, considered as an element of $Wh(QH) \subset \mathbb{Z}(QH)^\times$ (inclusion via reduced norm), cf. [M, 3.10].

Actually, we are only interested in the top component $\lambda^Y(f)(d) = \lambda^X(f)(d)$, $d = ab$. The corresponding part of $Wh(QH)$ is $K_1(Q(\zeta_d)^t[\mathbb{Z}/4])$, and

$$K_1(Q(\zeta_d)^t[\mathbb{Z}/4]) \cong K_1(Q(\zeta_d)^t[\mathbb{Z}/4])_+ \times K_1(Q(\zeta_d)^t[\mathbb{Z}/4])_-$$

$$\cong Q(\zeta_d + \zeta_d^{-1})^\times \times Q(\zeta_d + \zeta_d^{-1})^*$$

In general (4.11) is not enough to calculate $\lambda^X(f)(d)$, since $\mathrm{Res}_H^G\colon L_3^X(\mathbb{Z}G)(d) \to L_3^X(\mathbb{Z}H)(d)$ is not injective, but it does determine the image $\lambda_2^X(f)(d) \in L_3^X(\hat{\mathbb{Z}}_2G)$. This follows from the next lemma. Let

$$A = \mathbb{Z}[\zeta_a + \zeta_b^{-1},\ \zeta_b + \zeta_b^{-1}] = \mathbb{Z}[\eta_a, \eta_b]$$

$$B = \mathbb{Z}[\zeta_a\zeta_b + \zeta_a^{-1}\zeta_b^{-1}] = \mathbb{Z}[\eta_{ab}]$$

<u>Lemma 4.12</u> Let $d = ab$.

$$\mathrm{Im}(L_3^X(\mathbb{Z}G)(d) \to L_3^X(\hat{\mathbb{Z}}_2 \otimes \mathbb{Z}(\zeta_d)^t[Q8]) \cong A/2A$$

$$\operatorname{Im}(L_3^X(\mathbb{Z}H)(d) \to L_3^X(\hat{\mathbb{Z}}_2 \otimes \mathbb{Z}(\zeta_d)^t[\mathbb{Z}/4])) \cong B/2B$$

and Res_H^G: $A/2A \to B/2B$ is the natural inclusion. (cf [M], §4 or [LM], §3.)

The proof of 4.12 is in part based on the isomorphisms

$$(4.13) \quad \hat{\mathbb{Z}}_2 \otimes \mathbb{Z}[\zeta_d]^t Q8 \cong M_2(\hat{A}_2[\mathbb{Z}/2]); \quad \hat{\mathbb{Z}}_2 \otimes \mathbb{Z}[\zeta_d]^t \mathbb{Z}/4 \cong M_2(\hat{B}_2[\mathbb{Z}_2])$$

where $\hat{A}_2 = \hat{\mathbb{Z}}_2 \otimes A$, $\hat{B}_2 = \hat{\mathbb{Z}}_2 \otimes B$. Note from these that

$$K_1(\hat{\mathbb{Z}}_2G)(d) \cong \hat{A}_2[\mathbb{Z}/2]^\times; \quad K_1(\hat{\mathbb{Z}}_2H)(d) \cong \hat{B}_2[\mathbb{Z}_2]^\times$$

Let $\phi : A^\times \to (A/4A)^\times \otimes \hat{\mathbb{Z}}_2 \cong A/2A$ denote reduction modulo 4. The next result is an addendum to the proof of (4.12).

Lemma 4.14 The homomorphism

$$t: \; H^0(K_1(\hat{\mathbb{Z}}_2G)(d)) \to L_3^X(\hat{\mathbb{Z}}_2G)(d)$$

in the Rothenberg sequence maps $a_0 + a_1T \in \hat{A}_2[\mathbb{Z}_2]^\times$ to $\phi(a_0 + a_1) \in A/2A$. There is a similar result for $H^0(K_1(\hat{\mathbb{Z}}_2H)(d)) \to L_3^X(\hat{\mathbb{Z}}_2H)(d)$.

The isomorphisms in (4.13) induce isomorphisms of corresponding summands in $\hat{Q}_2G$ and $\hat{Q}_2H$. Let FA and FB denote the field of fractions of A and B. Then

$$K_1(\hat{Q}_2G)_\pm(d) \cong (F\hat{A}_2)^\times, \quad K_1(\hat{Q}_2H)_\pm(d) \cong (F\hat{B}_2)^\times$$

and the transfer map Res: $K_1(\hat{Q}_2G)_\pm(d) \to K_1(\hat{Q}_2H)_\pm(d)$ becomes the inclusion $(F\hat{A}_2)^\times \subset (F\hat{B}_2)^\times$.

Lemma 4.15 The natural homomorphism

$$K_1(\hat{\mathbb{Z}}_2G)(d) \to K_1(\hat{Q}_2G)_+(d)$$

sends $a_0 + a_1T$ to $a_0 + a_1$ and similarly for $K_1(\hat{\mathbb{Z}}_2H)(d) \to K_1(\hat{Q}_2G)_+(d)$.

Proof In (4.13) we have not kept track of the anti-structure. Actually it is twisted: there is a Morita equivalence:

$$(\hat{\mathbb{Z}}_2 \otimes \mathbb{Z}[\zeta_d]^t Q8, \alpha, 1) \sim (\hat{A}_2[\mathbb{Z}/_2], 1, T)$$

Thus the type 0 summand (in the corresponding rational structure) corresponds to setting $T = +1$.

The results (4.11) - (4.15) imply a calculation of $\delta_2^Y(\lambda^Y(f)(d))$ in terms of the Reidemeister torsion invariants $\Delta(S(W)/H)_+(d)$ and $\Delta(\Sigma/H)_+(d)$ in $H^0(Wh(\hat{Q}_2H)_+(d))$. By (4.7) this amounts to a calculation of $\delta(\lambda^P(f)(d))$. There is a well-known formula for calculating $\Delta(S(W)/H)$ in terms of the characters of the representation W, see e.g. [Mil 2]; it is more problematic to calculate $\Delta(\Sigma/H)$.

We have

$$\Delta(\Sigma/H) = \mathrm{Res}_H^G(\Delta(\Sigma/G))$$

and Res_H^G: $Wh(QG) \to Wh(QH)$ corresponds to the inclusion of FA in FB on the top component. Moreover, the image of $\Delta(\Sigma/G)$ in the quotient group $Wh(QG)/Wh'(\mathbb{Z}G)$ only depends on the homotopy type of Σ/G, or on the k-invariant $g \in H^{8k+4}(G;\mathbb{Z})$. In fact, $\Delta(\Sigma/G)$ represents the 'reason' that $\sigma(g) = 0$, (cf. [M], §6).

We only need information about $\Delta(\Sigma/G)(d) \in FA^{\times}$. Recall the notations $\underline{\underline{a}}$, $\underline{\underline{b}}$ for the sets of full prime power divisors of a and b from §0. We shall consider subsets S of $\underline{\underline{a}} \cup \underline{\underline{b}}$ and write

$$\eta_S = \zeta_r + \zeta_r^{-1} \quad \text{where} \quad r = \prod_{q \in S} q$$

$$\eta_{\emptyset} = 2$$

The cardinality of S is denoted $|S|$. Without proof we list

<u>Proposition 4.16</u>. Let $G = Q(8a,b,1)$, $d = ab$.
If $\sigma(g) = 0$ with g as above then

$$\Phi_{a,b}\colon A^{\times} \to (A/aA)^{\times}_{(2)} \times (A/bA)^{\times}_{(2)}$$

contains $(-1,1)$ in its image. For the resulting finite Σ/G

$$\Delta(\Sigma/G)_+(d) = \prod_{q \in \underline{\underline{a}} \cup \underline{\underline{b}}} (2 - \eta_q)^{n-1} V(d)\, U_+$$

in $H^0(Wh(QG))/H^0(Wh'(\mathbb{Z}G))$. Here $\Phi(U_+) = ((-1)^{r-1}, (-1)^r)$

where $r = |\underline{\underline{a}}|$ and

$$V(d) = \prod_{i=2}^{n} \prod_{|S|=i} (n_{S \cap \underline{\underline{a}}} - n_{S \cap \underline{\underline{b}}})^{(-1)^i}$$

This result is proved in [M, §6] when a and b are primes. The general result is similar in spirit but details are cumbersome, cf. [B].

For any unramified 2-ring A_y, $H^0(A_y) \cong A_y/2A_y \oplus \mathbb{Z}/2$ where $\mathbb{Z}/2$ is generated by $1 + 4\beta$ with β any element which is not of the form $x + x^2$ in the residue field. In terms of the Rothenberg exact sequence

$$L_0^K(A_y, 1, 1) \to H^0(A_y^\times)$$

maps onto $\mathbb{Z}/2$ with cokernel $A_y/2A_y$. We have $A_2 = \prod_{y|2} A_y$ and get the exact sequence

$$0 \to L_0^K(A_2, 1, 1) \xrightarrow{d_0} H^0(A_2^\times) \xrightarrow{\phi} A/2A \to 0$$

In particular (4.13) and (4.15) give

$$H^0(Wh(\hat{\mathbb{Z}}_2G)_+(d))/L_0^K(\hat{\mathbb{Z}}_2G)(d) \cong A/2A$$

and A/2A included into $H^0(Wh(\hat{Q}_2G)_+(d))/L_0^K(\hat{\mathbb{Z}}_2G)(d)$ via the natural inclusion $\hat{A}_2 \subset F\hat{A}_2$. We summarize our calculations in

Proposition 4.17 Let $G = Q(8a, b, 1)$, $d = ab$. If $\sigma(g) = 0$ then there exists a surgery problem $f\colon M \to \Sigma/G$ with

$$\delta^{Y}(\lambda^{Y}(f)(d)) = \phi(V(d)U_{+})$$

where $\phi_A = \phi_{a,b}\colon A^{\times} \to (A/4A)^{\times}_{(2)}$, and U_+ and $V(d)$ are from (4.16).

Proof. We use the surgery problem $(f,b)\colon M \to \Sigma/G$ described above with covering $f_H\colon S(W)/H \to \Sigma/H$.

The Reidemeister torsion of $S(W)/H$ has top component

$$\Delta(S(W)/H)(d) = \prod_{i=1}^{n} \prod_{|S|=i} (2 - \eta_S)^{(-1)^{n-i}}$$

For $|S| > 2$, $\eta_S - 2$ is a square in $B^{\times}$ so calculating modulo squares,

$$\Delta(\Sigma/H)_{+}(d) \;/\; \Delta(S(W)/H)_{+}(d) \equiv (-1)^{n+1} V(d)U_{+}$$

From (4.12) we get

$$\lambda_2^{Y}(f_H) = \phi_B(V(d)U_{+}) \in \mathrm{Im}(L_3^{X}(\mathbb{Z}H)(d) \to L_3^{X}(\hat{\mathbb{Z}}_2H)(d))$$

where $\phi_B\colon B^{\times} \to (B/4B)^{\times}_{(2)} = B/2B$.

Since $\lambda_2^{Y}(f_H) = \mathrm{Res}_H^G(\lambda_2^{Y}(f))$ and $A/2A \subset B/2B$,

$$\lambda_2^{Y}(f) = \phi_A(U_{+}V(d)) \in \mathrm{Im}(L_3^{X}(\mathbb{Z}G)(d) \to L_3^{X}(\hat{\mathbb{Z}}_2G)(d))$$

The rest follows from (4.14), (4.15) and (4.7).

We complete the proof of Theorem (4.4) by showing that $\phi(V(d)U_+)$ in (4.17) actually belongs to the image of

$$H^0(Wh'(\mathbb{Z}G)) \to H^0(Wh'(\hat{Q}_2G)_+)/L_0^K(\mathbb{Z}_2G) \tag{4.18}$$

precisely when the condition of Theorem A is satisfied. In fact, we calculate the map (4.18) and (4.8) implies our results on $\delta(\lambda^P(f))$.

Consider the exact diagram:

$$\begin{array}{ccccccccc}
 & & & & 0 & & & & \\
 & & & & \uparrow & & & & \\
0 & \to & \hat{D}(G)_+(d) & \longrightarrow & D(G)_+(d) & \longrightarrow & 0 & & \\
 & & \uparrow{\scriptstyle \Phi} & & \uparrow & & \uparrow & & \\
0 & \to & A^\times & \longrightarrow & (FA)^\times \times A^\times & \longrightarrow & (F\hat{A})^\times & \to \tilde{K}_0(A) & \to 0 \\
 & & \uparrow & & \uparrow & & \uparrow & \uparrow & \\
0 & \to & Wh'(\mathbb{Z}G)_+(d) & \to & Wh(QG)_+(d) \times Wh'(\hat{\mathbb{Z}}G)_+(d) & \to & Wh(\hat{Q}G)_+(d) & \to \tilde{K}_0(\mathbb{Z}G)_+(d) & \\
 & & \uparrow & & \uparrow & & \uparrow & & \\
0 & \text{---} & Ker & \longrightarrow & Ker & \longrightarrow & 0 & & \\
 & & \uparrow & & \uparrow & & & & \\
 & & 0 & & 0 & & & &
\end{array}$$

where $A = \mathbb{Z}[\eta_a, \eta_b]$ and $Wh'(\mathbb{Z}G)_+(d) \subset Wh'(\mathbb{Z}G)$ is defined by the diagram.

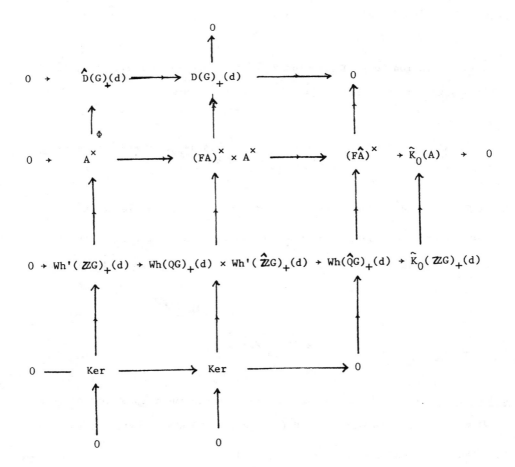

The group $D(G)_+(d) = \mathrm{cok}(Wh'(\mathbb{Z}G)_+(d) \to A^\times)$ was calculated in [M, 5.3]: its 2-primary part is

$$D(G)_+(d)_{(2)} \cong (A/2ab)^\times_{(2)}/\langle -1\rangle$$

Moreover, the kernels Ker in the diagram are products of profinite ℓ-groups with $\ell | ab$, and can be disregarded since ab is odd: 2-locally there is an exact sequence

$$0 \to Wh'(\mathbb{Z}G)_+(d) \to A^\times \xrightarrow{\Phi} (A/ab)^\times/\langle -1\rangle \to \tilde{K}_0(\mathbb{Z}\pi)_+(d) \to \tilde{K}_0(A) \to 0$$

In conclusion, $\phi(V(d)U_+)$ belongs to the image in (4.18) if and only if $\Phi(V(d)U_+)$ is $(\pm 1)\cdot(\text{square})$ in $(A/aA)^\times \times (A/bA)^\times$. Now

$$(A/aA)^\times_{(2)} \times (A/bA)^\times_{(2)} = \prod_{p|a}(A/p)^\times_{(2)} \times \prod_{q|b}(A/q)^\times_{(2)}, \text{ p and q primes.}$$

Since $\mathbb{F}_p^\times \subset (A/p)^\times$ and $-1 \in (\mathbb{F}_p^\times)^2$ precisely when $p \equiv 1(4)$ we check that either all prime divisors of a or all prime divisors of b are congruent to 1 (mod 4). It then follows that $\phi(V(d)U_+)$ is in the image in (4.18) if and only if $\Phi_{a,b}(V(d))$ belongs to

$$\mathrm{Image}\ (\Phi_{a,b}|\mathrm{Ker}\ \phi_{a,b}).$$

<u>Lemma 4.19</u>. If p and q are prime numbers both congruent to 3 (mod 4) then $\sigma_4(g) \neq 0$ for each generator $g \in H^4(Q(8p,q,1); \mathbb{Z})$ which restricts to a k-invariant of a linear space form on $Q(4p,q)$.

<u>Proof</u> The Galois group of $A/\mathbb{Z}$ has odd order. Consider the diagram

$$\begin{array}{ccc} A^{\times} & \xrightarrow{\Phi_A} & (A/pA)^{\times} \times (A/qA)^{\times} \\ \downarrow N & & \downarrow N \\ \mathbb{Z}^{\times} & \xrightarrow{\Phi_{\mathbb{Z}}} & \mathbb{F}_p^{\times} \times \mathbb{F}_q^{\times} \end{array}$$

where N is the norm. Since the Galois group has odd order, N is a 2-local isomorphism. But $(-1,1) \in \mathbb{F}_p^{\times} \times \mathbb{F}_q^{\times}$ is not in the image of $\Phi_{\mathbb{Z}}$, and the result follows from (4.16).

Since $\sigma(Q(8a,b,1)) = 0$ implies that $\sigma(Q(8p,q,1) = 0$ for prime divisors $p|a$ and $q|b$ it follows that either a or b is a product of primes $\equiv 1(4)$. This ends the proof of Theorem 4.4 and thus also of Theorem A.

We end with a discussion of $\lambda^p(f)$ for the group $Q(8p,q,1)$ where p and q are odd primes. This is the case also considered by Jim Milgram in [Mg2], but unfortunately our results do not completely agree.

First we note that a necessary condition that $\lambda^p(f) = 0$ is that the Legendre symbol $\left(\frac{p}{q}\right) = +1$. This follows from the diagram

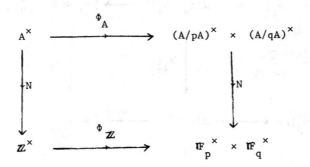

$$
(4.20)\qquad
\begin{array}{ccccc}
\mathbb{Z}[\eta_p, \eta_q]/2 & \xleftarrow{\ \phi\ } & \mathbb{Z}[\eta_p, \eta_q]^{\times} & \longrightarrow & (\mathbb{F}_p \otimes \mathbb{Z}[\eta_q])^{\times}_{(2)} \\
\downarrow{\scriptstyle T} & & \downarrow{\scriptstyle N_q} & & \downarrow{\scriptstyle N_q} \\
\mathbb{Z}[\eta_p]/2 & \xleftarrow{\ \phi\ } & \mathbb{Z}[\eta_p]^{\times} & \longrightarrow & (\mathbb{F}_p^{\times})_{(2)} \\
\downarrow{\scriptstyle T_0} & & \downarrow{\scriptstyle N_0} & & \downarrow{\scriptstyle \cong} \\
\mathbb{Z}[\eta_p]_0/2 & \underset{\approx}{\xleftarrow{\ \phi_0\ }} & \mathbb{Z}[\eta_p]_0^{\times} & \longrightarrow & (\mathbb{F}_p^{\times})_{(2)} \\
\downarrow & & \downarrow{\scriptstyle N_2} & & \downarrow \\
\mathbb{F}_2 & \xleftarrow{\ \phi_2\ } & \mathbb{Z}^{\times} & \longrightarrow & (\mathbb{F}_p^{\times})_{(2)}
\end{array}
$$

Here the N's are norms and the T's are traces; $\mathbb{Z}[\eta_p]_0 = \mathbb{Z}[\eta_p]^{G_0}$ where G_0 is the subgroup of the elements of odd order in $(\mathbb{Z}/p)^{\times}/\langle -1 \rangle$. The map ϕ_0 is an isomorphism because if ε in $\mathbb{Z}[\eta_p]_0^{\times}$ is a unit with $N_2(\varepsilon) = -1$ then by (4.20), $\phi(\varepsilon)$ has non-zero trace to $\mathbb{Z}/2$. Since $(\mathbb{Z}[\eta_p]_0/2 : \mathbb{F}_2)$ is a power of 2 it follows that $\phi_0(\varepsilon)$ is a normal basis of $\mathbb{Z}[\eta_p]_0/2$. Hence ϕ_0 is subjective and therefore an isomorphism.

We have proved

<u>Lemma</u> 4.21. If $\pi = Q(8p,q,1)$ and $\left(\frac{p}{q}\right) = -1$ then $\lambda^P(f) \neq 0$.

This result has also been proved by Jim Milgram. In a previous version of this paper we stated that $\sigma(Q(8p,q,1)) = 0$ implies that $\lambda^P(f) = 0$, contradicting Milgram's results. The mistake appeared in our formulation of 4.17, where we falsely asserted that $\phi_{p,q}(V(pq)) = 0$.

In [Mg 2] it is asserted that $\left(\frac{p}{q}\right) = +1$ is a sufficient condition for the vanishing of the projective surgery obstruction. This contradicts our Theorem A as shown by the following

Example 4.22 For $(p, q) = (19, 5)$, $\lambda^P(f) \neq 0$. Indeed, the finiteness obstruction vanishes and condition (i) in Theorem A is satisfied. Thus the question is whether or not

$$(\eta_5 - 2, \eta_{19} - 2) \in (\mathbb{F}_{19} \otimes \mathbb{Z}[\eta_5])^{\times}_{(2)} \times (\mathbb{F}_5 \otimes \mathbb{Z}[\eta_{19}])^{\times}_{(2)}$$

belongs to the image of $\Phi_{19,5} \mid \operatorname{Ker} \phi_{19,5}$.

We have $\mathbb{F}_{19} \otimes \mathbb{Z}[\eta_5] = \mathbb{F}_{19} \times \mathbb{F}_{19}$ and $\mathbb{F}_5 \otimes \mathbb{Z}[\eta_{19}] = \mathbb{F}_5{}^9$. A short calculation in $\mathbb{Z}[\eta_5]$ shows that the reduction of $\eta_5 - 2$ gives a pair of non-squares in $\mathbb{F}_{19} \times \mathbb{F}_{19}$, namely $(2, -7)$. Finally if $(\eta_5 - 2, \eta_{19} - 2) \in$ Image $(\Phi \mid \operatorname{Ker} \phi)$ then by (4.20)

$$\Phi_0 \colon \mathbb{Z}[\eta_5]^{\times 2} \to \mathbb{F}_{19} \times \mathbb{F}_{19} \times \mathbb{F}_5$$

would contain $((\eta_5 - 2)^5, 1)$ whence a non-square, in its image.

References

[B] S. Bentzen, Finiteness obstructions of periodic groups (to appear).

[BR] G.E. Bredon, "Introduction to Compact Transformation Groups", Academic Press, New York, 1972.

[CM] G. Carlsson and R.J. Milgram, "The oriented odd L-groups of finite groups", Preprint, Stanford, 1980.

[L] R. Lee, "Semi-characteristic classes", Topology 12 (1973), 183-200.

[LM] E. Laitinen, I. Madsen, "L-theory of groups with periodic cohomology I", preprint, Aarhus 1981.

[M] I. Madsen, "Reidemeister torsion, surgery invariants and spherical space forms", preprint, Aarhus 1981.

[M1] I. Madsen, "Spherical space forms in the period dimension", (revised) preprint, Aarhus 1982. To appear in Proc. London Math. Soc.

[M2] I. Madsen, "Spherical space forms: A survey", 18th Scandinavian Congress of Mathematicians, Proceedings, Birkhauer, Boston, 1981.

[MTW] I. Madsen, C.B. Thomas and C.T.C. Wall, "The topological spherical space form problem II", Topology 15 (1976), 375-382.

[Ma] S. Maumary, "Proper surgery groups and Wall-Novikov groups", Proc. of Battelle Conf. on Algebraic K-Theory, Springer Lecture Notes 343 (1973), 526-539.

[Mg] R.J. Milgram, "Odd index subgroups of units in cyclotomic fields and applications", preprint, Stanford (1980).

[Mg2] R.J. Milgram, "Patching techniques in surgery and the solution of the compact space form problem", preprint, Stanford, 1981.

[Mil] J. Milnor, "Groups which act on S^n without fixed points", Amer. J. Math, 79 (1957), 623-630.

[Mil2] J. Milnor, "Whitehead torsion", Bull. AMS 72 (1966), 358-426.

[P] W. Pardon, "Mod 2 semi-characteristics and the converse to a theorem of Milnor", Math. Z. 171 (1980), 247-268.

[PR] E.K. Pedersen and A.A. Ranicki, "Projective surgery theory", Topology 19(1980), 239-254.

[R1] A.A. Ranicki, "The algebraic theory of surgery I, II", Proc. London Math. Soc. (3) 40(1980), 87-192, 193-283.

[R2] A.A. Ranicki, "Algebraic L-theory I. Foundations". Proc. Lond. Math. Soc (3) 27 (1973), 101-125.

[R3] A.A. Ranicki, "Exact sequences in the algebraic theory of surgery", Math Notes 26, Princeton University Press, Princeton, 1981.

[S] R. Swan, "Periodic resolutions for finite groups", Ann. of Math. 72 (1960), 167-191.

[T] L. Taylor, "Surgery on paracompact manifolds". Berkeley Ph.D. Thesis (1972).

[TW] C.B. Thomas and C.T.C. Wall, "The topological spherical space form problem I", Comp. Math. 23 (1971), 101-114.

[W] C.T.C. Wall, "Periodic projective resolutions", Proc. London Math. Soc. (3)39(1979), 509-553.

[W1] C.T.C. Wall, "Foundations of algebraic L-theory", Algebraic K-theory III, Battelle Institute Conference, 1972, Springer Lecture Notes 343.

[W2] C.T.C. Wall, "The classification of hermitian forms II", Invent. Math. 18 (1972), 119-141.

[W3] C.T.C. Wall, "The classification of hermitian forms III", Invent. Math. 19 (1973), 59-71.

[W4] C.T.C. Wall, "The classification of hermitian forms IV", Invent. Math 23 (1974), 241-260.

[W5] C.T.C. Wall, "The classification of hermitian forms VI", Ann. of Math. 103 (1976), 1-80.

Ian Hambleton
Department of Mathematical Sciences
McMaster University
Hamilton, Ontario L8S 4K1
Canada

Ib Madsen
Matematisk Institut
Aarhus Universitet
8000 Aarhus C.
Denmark

April 30, 1982
IH.1/A/IH2.1/dm

Canadian Mathematical Society
Conference Proceedings
Volume 2, Part 2 (1982)

REPRESENTATIONS AT FIXED POINTS OF ACTIONS OF FINITE GROUPS ON SPHERES

Sören Illman

1. INTRODUCTION. Let G be a finite group acting smoothly on a sphere S^n. In [14] (see footnote on page 406), P.A. Smith asked the question: Assuming that the fixed point set $(S^n)^G$ consists of two points x and y, does it follow that the linear representations of G on the tangent spaces at x and y are isomorphic, i.e., linearly equivalent? As commonly happens, time changed a possible affirmative answer to the above question into a conjecture. Thus we arrive at the following "conjecture" of P.A. Smith: Let G be a finite group acting smoothly on a (homology) sphere with precisely two fixed points x and y. Then the representations of G on the tangent spaces at x and y are linearly equivalent.

In [10], Theorem B, Petrie announces the following result:

P. Every odd order abelian group G with at least 4 non cyclic Sylow subgroups acts smoothly on a homotopy sphere such that the fixed point set of G consists of two points and the representations at these points are not linearly equivalent.

In [4], Corollary 1A, Cappell and Shaneson annouce the following result:

C.S.I. For every $n = 2^a b$ with $a \geq 2$ and b odd and $b > 1$, there is a smooth action of Z_n on the sphere S^{2k+1}, for each $k \geq 4$, with precisely two fixed points and with the representations of Z_n on the $(2k+1)$-dimensional tangent spaces at the fixed points being linearly inequivalent. Moreover these representations are pseudo-free and the smooth Z_n action on S^{2k+1} is free outside of a circle.

For $n = 4q$, $q > 1$, they have counterexamples given by smooth Z_n actions on 9-dimensional homotopy spheres, [4], Theorem 2.

Before the occurrence of these counterexamples, the conjecture of P.A. Smith had been proved in many important cases. Atiyah and Bott [1] proved the conjecture for actions of cyclic groups of odd prime order on homology spheres.

1980 Mathematics Subject Classification. 57S17, 57S25.

© 1982 American Mathematical Society
0731-1036/82/0000-0463/$06.25

Milnor (Theorem 12.11 in [9], Theorem 7.27 in [1]) extended this to all finite groups (in fact all compact Lie groups) in the case when the action is free outside the two fixed points. Sanchez [11] proved the conjecture for actions of Z_n, where $n = p^s$ and p is an odd prime, on orientable Z_p-homology spheres. By Bredon [2a] it follows that the conjecture is true for actions of Z_n, where $n = 2^s$, on homotopy spheres Σ^m, provided m is sufficiently large with respect to n.

Although the representations at two fixed points need not, in general, be linearly equivalent one may hope for some other kind of equivalence between them. In this direction Cappell and Shaneson announce in [4], (see Theorem 1), the following result:

C.S.II. If a cyclic group Z_n acts smoothly on a (homology) sphere such that the action is free outside of a set of dimension at most 1, then the representations of Z_n on the tangent spaces of two fixed points are topologically equivalent. (These representations are pseudo-free.)

Their result lead them to suggest the following revised form of the conjecture of P.A. Smith.

CONJECTURE (Cappell and Shaneson). Let G be a finite group acting smoothly on a (mod 2) homology sphere with each cyclic subgroup having fixed points which are finite or connected. Then the representations of G on the tangent spaces of the fixed points are topologically equivalent.

Recall that by Cappell and Shaneson [3] and [5] topological equivalence of linear representations does not in general imply linear equivalence.

Observe that we do not have at our disposal any means of concluding topological equivalence of two representations V and W of G from an assumed topological equivalence of the induced representations $V|_K$ and $W|_K$, for all cyclic subgroups K of G. Hence from a possible affirmative solution of the above conjecture for all finite cyclic groups we are not able to conclude an affirmative answer for all finite groups.

We say that a representation of a finite group G on a linear vector space V is <u>n-pseudo-free</u> if the action of G is free outside a set of dimension at most n. Thus an n-pseudo-free representation is k-pseudo-free for every $k \geq n$. A 0-pseudo-free representation is free outside the origin, and such representations are usually called free. The 1-pseudo-free representations are usually called pseudo-free. We shall be concerned with 2-pseudo-free and pseudo-free representations.

The main results of this paper are Theorems A and C and their respective Corollaries B and D given below.

THEOREM A. Let G be a finite solvable group acting smoothly on a homotopy sphere Σ^m, where $m \geq 5$, such that there is a fixed point $a \in \Sigma^G$ at which the representation of G is 2-pseudo-free. Then the G-space Σ^m is

G-homeomorphic to a linear representation sphere $S^m(\rho)$.

COROLLARY B. Let G be a finite solvable group acting smoothly on a homotopy sphere Σ^m, where $m \geq 5$, such that there is a fixed point $a \in \Sigma^G$ at which the representation of G is 2-pseudo-free. Then the representations of G at any two fixed points are topologically equivalent.

The assumption that G is solvable is used only in a preliminary result, Proposition 1, which says that from the assumptions in Theorem A (the condition $m \geq 5$ is not needed here) it follows that for any non-trivial subgroup H of G we have $\Sigma^H \cong S^k$, where $k = 0, 1$ or 2. In Theorem 5 we then prove that for actions of arbitrary finite groups this situation (where now the condition $m \geq 5$ is used) implies that the given action is topologically equivalent to a linear action. Observe that in Theorem A we only assume the existence of one fixed point. In this connection one should keep in mind that there exist smooth actions of finite groups on spheres with exactly one fixed point. A smooth action of the binary icosahedral group on S^7 with exactly one fixed point is given in Stein [15], Proposition 4.3. In this example the representation at the fixed point is 3-pseudo-free, and the binary icosahedral group is perfect and hence in particular not solvable.

But, for the problem of comparing the representations at two fixed points, one may of course as well assume that there are at least two fixed points. An argument due to Gary Hamrick shows that if an arbitrary finite group acts on a homotopy sphere such that there exist at least two fixed points and the representation of G at one of the fixed points is pseudo-free, then for any non-trivial subgroup H of G we have $\Sigma^H \cong S^k$, where $k = 0$ or 1. See Proposition 3. Combining Proposition 3 and Theorem 5 we obtain.

THEOREM C. Let G be a finite group acting smoothly on a homotopy sphere Σ^m, where $m \geq 5$, such that there exist at least two fixed points $a, b \in \Sigma^G$, and the representation of G at a is pseudo-free. Then the G-space Σ^m is G-homeomorphic to a linear representation sphere $S^m(\rho)$.

COROLLARY D. Let G be a finite group acting smoothly on a homotopy sphere Σ^m, $m \geq 5$, such that there is a fixed point at which the representation of G is pseudo-free. Then the representations of G at fixed points are topologically equivalent to each other.

Corollary B extends, for actions on homotopy spheres of dimension at least 5, the above mentioned result C.S.II of Cappell and Shaneson in two ways. It replaces cyclic groups by solvable groups and it extends the condition of the representations being pseudo-free to the more general assumption that the representations are 2-pseudo-free. Moreover, Theorem A proves more, namely that from the same assumptions as in Corollary B it in fact follows that the given action itself is topologically equivalent to a linear action. Corollary D generalizes, in the case of actions on homotopy spheres of dimension at least 5, the result C.S.II of Cappell and Shaneson by replacing cyclic groups by arbitrary finite groups. Thus Corollary D proves the conjecture of Cappell and Shaneson in the case of actions on homotopy spheres of dimension at least 5 with pseudo-free representations at fixed points.

Also observe that Theorem A shows that the smooth actions of cyclic groups on spheres S^{2k+1}, $k \geq 4$, given in C.S.I are G-homeomorphic to linear representation spheres, but they cannot be G-diffeomorphic to linear representation spheres, since in that case the representations at fixed points would have to be linearly equivalent.

The above results also raise the question of which finite groups admit pseudo-free and 2-pseudo-free representations. In Laitinen [8] a complete determination of which groups admit pseudo-free representations is given. Except for a few low dimensional examples the pseudo-free representations reduce to free ones. In particular he shows that if $\rho: G \to O(m)$, where $m \geq 4$, is a pseudo-free representation, which is not free, then there exist a free representation $\psi: G \to O(m-1)$ and homomorphism $\chi: G \to O(1) = Z_2$ such that $\rho = \psi + \chi$. The groups that admit free representations are known, see Wolf [17], Theorems 6.1.11 and 6.3.1. Since in Theorem C and Corollary D we have the case when $m \geq 5$, it follows that the finite group G in Theorem C and Corollary D must in fact be one of the groups listed in Wolf [17], 6.1.11,3 and 6.3.1.

Next we consider the question of which solvable groups can occur in Theorem A and Corollary B, i.e., which solvable groups admit 2-pseudo-free representations of real degree at least 5. We will here be content with pointing out that there exist infinitely many finite solvable groups that admit 2-pseudo-free representations, of real degree at least 5, but do not admit any free representations, i.e., they do not belong to the family of finite solvable groups given in Theorem 6.1.11,3 in Wolf [17]. The groups in the example given below are metacyclic groups of order pq, where p and q are distinct odd primes, and hence it follows by the full result in Laitinen [8] that these groups do not admit any pseudo-free representations, except of course the trivial representation of real degree 1. Let p and q be odd primes such that $q|p-1$. The automorphism group of Z_p is cyclic of order $p-1$, and hence there exists a nontrivial homomorphism

$$\varphi: Z_q \to \text{Aut}(Z_p).$$

Let φ be such a non-trivial homomorphism and form the semi-direct product

$$G = Z_p \times_\varphi Z_q .$$

Then G is a non-cyclic solvable group of order pq. Thus G does not satisfy the pq-condition and hence does not admit any free representations, see Theorem 5.3.1 in Wolf [17]. A 2-pseudo-free representation of G can be constructed as follows. Let W denote the euclidean plane $\mathbb{R}^2$ with standard free representation of Z_p. Let H denote the normal subgroup of G which is the image of Z_p in G under the standard inclusion. Let $V = \mathrm{Ind}_H^G(W)$ be the representation of G induced by the representation W of $H = Z_p$. Then V is a 2-pseudo-free representation of G of real degree $2q$. In fact $V^H = \{0\}$ and if K is any subgroup of order q of G we have $V^K \cong \mathbb{R}^2$. (This follows easily from Proposition 22 on page 58 in Serre [13].) Observe that by Dirichlet's theorem there exists, for any given prime q, an infinite number of primes p satisfying $p \equiv 1 \pmod q$, i.e., such that $q|p-1$. This shows that there are infinitely many solvable groups that do not admit any free representations but do admit 2-pseudo-free representations.

The contents of this paper can in short be described as follows. In Proposition 1 we show that, if a finite solvable group G acts smoothly on a homotopy sphere Σ^m such that there is a fixed point at which the representation of G is 2-speudo-free, then for any non-trivial subgroup $H \neq \{e\}$ of G the fixed point set Σ^H is diffeomorphic to S^0, S^1 or S^2. In Proposition 3, due to Gary Hamrick, it is shown that if an arbitrary finite group G acts smoothly on a homotopy sphere Σ^m such that there are at least two fixed points and the representation of G at one of these fixed points is pseudo-free, then for any non-trivial subgroup $H \neq \{e\}$ the fixed point set Σ^H is diffeomorphic to S^0 or S^1. In order to conclude from the situations arising in Propositions 1 and 3 that the action of G on Σ^m, in the case when $m \geq 5$, is topologically equivalent to a linear action we need a suitable recognition of linear actions result. Although the situation here in many respects is simpler that the one in the general case given in [7], it is not covered by the result in [7]. Hence we prove in Theorem 5, for actions of arbitrary finite groups, the version needed here. The proof of Theorem 5 relies on results and techniques from [7]. Theorem A follows directly from Proposition 1 and Theorem 5, and Theorem C follows from Proposition 3 and Theorem 5. The Corollaries B and D follow from Theorems A and C, respectively, after an application of the conical orbit structure theorem in Bredon [2].

2. TERMINOLOGY AND NOTATION. Two representations of a finite group on finite dimensional real vector spaces V and W are said to be topologically equiva-

lent if there exists a G-homeomorphism $h: V \to W$, which we may assume to satisfy $h(0) = 0$. A representation space V is called n-pseudo-free if the action of G on V is free outside a set of dimension at most n. Thus the 0-pseudo-free representations are exactly the free representations and the 1-pseudo-free representations are the pseudo-free representations.

We use the notation $\mathbb{R}^m(\rho)$ for the orthogonal representation space determined by the representation $\rho: G \to O(m)$. Furthermore we denote

$$B^m(\rho) = \{x \in \mathbb{R}^m(\rho) \mid \|x\| \leq 1\}$$

$$\overset{\circ}{B}{}^m(\rho) = \{x \in \mathbb{R}^m(\rho) \mid \|x\| < 1\}$$

$$S^{m-1}(\rho) = \{x \in \mathbb{R}^m(\rho) \mid \|x\| = 1\}.$$

We call $B^m(\rho)$ for a linear representation ball and $S^{m-1}(\rho)$ for a linear representation sphere. By $B^m_r(\rho)$, $0 < r < \infty$, we denote a linear representation ball of radius r.

Let M be a smooth G-manifold and $x \in M^G$ a fixed point of G. Then the tangent space to M at x is a representation space for G. We usually call such a representation of G on the tangent space at x simply for the representation of G at x.

If X is a G-space and H is a subgroup of G we denote

$$X_H = \{x \in X \mid G_x = H\}.$$

3. ACTIONS OF SOLVABLE GROUPS WITH A 2-PSEUDO-FREE REPRESENTATION AT A FIXED POINT.

PROPOSITION 1. Let G be a finite solvable group acting smoothly on a homotopy sphere Σ^m, $m \geq 0$. Let $a \in \Sigma^G$ and assume that the representation $\mathbb{R}^m(\rho_a)$ of G at a is 2-pseudo-free. Then for any non-trivial subgroup $H \neq \{e\}$ of G the fixed point set Σ^H is diffeomorphic to S^k, where $k = 0$, 1 or 2.

PROOF. Let us first consider the case when H is a cyclic subgroup of order p, where p is a prime. By P.A. Smith theory we have that Σ^H is a Z_p-homology sphere. In case p is odd it follows by another result of P.A. Smith (see Bredon [2], Theorem IV.2.1) that each component of Σ^H is orientable. In case p equals 2 each component of Σ^H is of course Z_2-orientable. Hence, in both cases, it follows that Σ^H equals two points or is connected. Since $a \in \Sigma^H$ and the representation of G at a is 2-pseudo-free we have

$$\dim \Sigma^H \le 2.$$

If $\dim \Sigma^H = k \le 1$ it follows directly that $\Sigma^H \cong S^k$, where $k = 0$ or 1. If $\dim \Sigma^H = 2$ it follows that $\Sigma^H \cong S^2$, because S^2 is the only compact, connected, orientable 2-manifold which is a homology Z_p-sphere.

Now let H be an arbitrary non-trivial subgroup of G. Then H is solvable and hence there exists a tower

$$\{e\} = H_0 \triangleleft H_1 \triangleleft \ldots \triangleleft H_n = H,$$

where H_{i-1} is normal in H_i and H_i/H_{i-1} is abelian, $1 \le i \le n$, and H_1 is cyclic of prime order. By what we showed above we have

$$\Sigma^{H_1} \cong S^k, \quad k = 0, 1 \text{ or } 2.$$

The quotient group H_{i+1}/H_i acts smoothly on Σ^{H_i} and we have

$$(1) \qquad \Sigma^{H_{i+1}} = (\Sigma^{H_i})^{H_{i+1}/H_i}, \quad 1 \le i \le n-1.$$

Let $1 \le i \le n-1$ and assume inductively that $\Sigma^{H_i} \cong S^k$, where $k = 0, 1$ or 2. Since $a \in \Sigma^G \subset \Sigma^{H_i}$ it follows that H_{i+1}/H_i acts smoothly on $\Sigma^{H_i} - \{a\} \cong \mathbb{R}^k$. Since any smooth action on $\mathbb{R}^k$, $0 \le k \le 2$, is topologically equivalent to a linear action (see Bredon [2], Theorems IV.8.1 and IV.8.5) it follows that the action of H_{i+1}/H_i on $\Sigma^{H_i} \cong S^k$ is topologically equivalent to a linear action. Hence by (1) we have that $\Sigma^{H_{i+1}} \cong S^{k'}$, where $k' = 0, 1$ or 2. Thus induction shows that the fixed point set $\Sigma^H = \Sigma^{H_n}$ is diffeomorphic to S^0, S^1 or S^2. □

4. ACTIONS OF FINITE GROUPS WITH PSEUDO-FREE REPRESENTATIONS AT FIXED POINTS.

The material in this section is due to Gary Hamrick.

LEMMA 2. Let G be a simple and non-cyclic finite group. Then there exist two distinct maximal proper subgroups H_1 and H_2 of G such that $H_1 \cap H_2 \neq \{e\}$.

PROOF. Since G is non-cyclic it follows that every element of G is contained in some maximal proper subgroup of G, and that there exist at least two distinct maximal proper subgroups of G, and that the trivial subgroup $\{e\}$

is not a maximal proper subgroup of G.

Assume that the intersection of any two distinct maximal proper subgroups of G equals $\{e\}$. We shall show that this assumption leads to a contradiction. Let H be a maximal proper subgroup of G, and denote

$$|H| = r > 1, \quad |G/H| = q \quad \text{and} \quad |G| = qr = n.$$

Since H is maximal and G is simple we have $N(H) = H$. Thus the number of distinct conjugates of H in G equals $|G/H| = q$. By our assumption the intersection of any two distinct maximal proper subgroups equals $\{e\}$. Hence it follows that the number of elements different from the identity in the union of all conjugates of H equals

$$u = q(r - 1) = n - q.$$

Since $q \leq n/2$, we have $u \geq n/2$. Therefore the existence of a maximal proper subgroup K not conjugate to H would imply that there are at least $n/2 + n/2 + 1 = n + 1$ elements in G, which is impossible. Hence every maximal proper subgroup of G is conjugate to H. Since every element of G is contained in a maximal proper subgroup we have

$$G = \bigcup_{g \in G} gHg^{-1}$$

and hence

$$n = u + 1 = n - q + 1.$$

Thus $q = 1$, a contradiction. □

PROPOSITION 3. Let G be a finite group acting smoothly on a homotopy sphere Σ^m, $m \geq 0$. Assume that there exist $a, b \in \Sigma^G$, where $a \neq b$, such that the representation of G at a is pseudo-free. Then for any non-trivial subgroup $H \neq \{e\}$ of G we have $\Sigma^H \cong S^k$, where $k = 0$ or 1.

PROOF. In case $G = Z_p$, where p is a prime, we saw already in the proof of Proposition 1 that $\Sigma^G \cong S^k$, where $k = 0$ or 1. Now let G be an arbitrary finite group and let $\alpha(G)$ denote the total number of prime factors in the order of G, i.e., if $|G| = p_1^{a_1} \cdots p_s^{a_s}$, where p_i, $1 \leq i \leq s$, are primes, then $\alpha(G) = a_1 + \ldots + a_s$. We shall prove our claim by induction in $\alpha(G)$.

If $\alpha(G) = 1$ we have $G = Z_p$, where p is a prime, and the claim is valid. Now assume $\alpha(G) > 1$. By the inductive assumption it is enough to show that $\Sigma^G \cong S^k$, $k = 0$ or 1. If there exists a non-trivial proper normal subgroup H of G the quotient group G/H acts on Σ^H, and by the inductive assumption $\Sigma^H \cong S^k$, $k = 0$ or 1. Thus

$$\Sigma^G = (\Sigma^H)^{G/H} \cong (S^1)^{G/H} \cong S^{k'}, \text{ where } k' = 0 \text{ or } 1.$$

If there exists a non-trivial proper subgroup H of G such that $\Sigma^H \cong S^0$ we have

$$\{a,b\} \subset \Sigma^G \subset \Sigma^H \cong S^0,$$

and hence $\Sigma^G = \{a,b\} \cong S^0$.

Thus it is enough to consider the case when G is simple and non-cyclic and $\Sigma^H \cong S^1$ for every non-trivial subgroup H of G. In this case we have by Lemma 2 that there exist two distinct maximal proper subgroups H_1 and H_2 of G such that $H_1 \cap H_2 \neq \{e\}$. Then $\Sigma^{H_1 \cap H_2} = S^1$ and $\Sigma^{H_i} \cong S^1$, for $i = 1$ and 2. Since moreover $\Sigma^{H_i} \subset \Sigma^{H_1 \cap H_2}$, $i = 1$ and 2, it follows that $\Sigma^{H_1} = \Sigma^{H_1 \cap H_2} = \Sigma^{H_2}$, and hence

$$\Sigma^{<H_1,H_2>} = \Sigma^{H_1} \cap \Sigma^{H_2} = \Sigma^{H_1} \cong S^1.$$

Here $<H_1,H_2>$ denotes the subgroup of G generated by H_1 and H_2. But, since H_1 and H_2 are distinct and maximal, we have $<H_1,H_2> = G$, and hence we have shown that $\Sigma^G \cong S^1$. □

5. RECOGNITION OF LINEAR ACTIONS. Let M be a smooth G-manifold with boundary ∂M and interior Int M. Let $N \subset \partial M$ be a smooth G-submanifold of ∂M, such that N is a closed subset of ∂M. Let $\overline{D}$ be a closed G-invariant tubular neighborhood of N in ∂M, and let $D \subset \overline{D}$ denote the corresponding open G-invariant tubular neighborhood. We form the G-space

$$M_1 = M \cup (\overline{D} \times [0,1))$$

where the union is along the set $\overline{D} = \overline{D} \times \{0\}$ in ∂M. The interior of, the topological manifold with boundary, M_1 equals

$$\text{Int } M \cup (D \times [0,1)).$$

LEMMA 4. The spaces $\text{Int } M \cup (D \times [0,1))$ and Int M are G-homeomorphic.

PROOF. Let us denote

$$\overline{A} = \{(x,t) \in \overline{D} \times [0,1) \mid t \le -2\|x\| + 2\}$$

and

$$A = \{(x,t) \in \overline{D} \times [0,1) \mid t < -2\|x\| + 2\}.$$

The map $(x,t) \mapsto (\frac{2-t}{2}x, t)$ gives a G-homeomorphism from $\overline{D} \times [0,1)$ onto $\overline{A}$, which is the identity of $\overline{D} \times \{0\}$ and maps $D \times [0,1)$ onto A.

By the equivariant collaring theorem (Bredon [2], Theorem V.1.5) there exists a G-homeomorphism h of $\partial M \times [0,-1]$ onto a neighborhood of ∂M in M such that $h(x,0) = x$. Using such an equivariant collar of ∂M in M we can by linear pushing in the t-coordinate, where $-1 \le t < 1$, map $M \cup \overline{A}$, (union along $\overline{D} \times \{0\} = \overline{D}$) into M giving us a G-homeomorphism from $\operatorname{Int} M \cup A$ onto $\operatorname{Int} M$. □

THEOREM 5. Let G be a finite group acting smoothly on a homotopy sphere Σ^m, where $m \ge 5$, such that for each non-trivial subgroup $H \ne \{e\}$ we have $\Sigma^H \cong S^k$, where $k = 0$, 1 or 2. Then the G-space Σ^m is G-homeomorphic to a linear representation sphere $S^m(\rho)$.

PROOF. Since $\Sigma^G \cong S^k$, where $0 \le k \le 2$, we have that Σ^G contains at least two points a and b. We consider the smooth G-manifold

$$M = \Sigma - \{b\}.$$

Then $M \cong \mathbb{R}^m$ and for any non-trivial subgroup H of G we have $M^H \cong \mathbb{R}^k$, where $0 \le k \le 2$. We shall prove that M is G-homeomorphic to an orthogonal representation space $\mathbb{R}^m(\rho_a)$ for G. Since Σ is the one-point-compactification of M it then follows that Σ is G-homeomorphic to the linear representation sphere $S^m(\rho_a \oplus 1)$. In order to prove that M is G-homeomorphic to an orthogonal representation space, we prove that M is a monotone union of G-subsets, each of which is G-homeomorphic to a linear representation ball $B^m(\rho_a)$. Thus what we really need to prove is that any compact subset of M is contained in a G-subset which is G-homeomorphic to $B^m(\rho_a)$. The proof of this relies on techniques from [7], and part of the proof is completely analogous to the proof of Proposition 3.2 in [7]. We will indicate in the proof below when we reach the point from which on the proof proceeds as in the proof of Proposition 3.2 in [7].

Let H be any subgroup of G. Since M^H is connected there exists a subgroup $\overline{H}$ of G such that $H \subset \overline{H}$ and $M^H = M^{\overline{H}}$ and $\overline{H}$ occurs in M^H, i.e., there is $x \in M^H$ such that $G_x = \overline{H}$, see the Lemma in § 1 of [7]. Moreover such a subgroup $\overline{H}$ is uniquely determined by the two conditions; $M^{\overline{H}} = M^H$ and $\overline{H}$ occurs in M^H. In this way we get a one-to-one correspondence between the set of all subgroups of G that occur in M and the set of all fixed point sets of subgroups of G. The normalizer $N(H)$ of H in G acts on M^H, and if H occurs in M^H we have that $gM^H \neq M^H$, for all $g \in G \setminus N(H)$.

Let

$$(1) \qquad M^{H_1}, \dots, M^{H_r}$$

be all the 1-dimensional fixed point sets, where we have chosen the subgroups H_i such that H_i occurs in M^{H_i}, $1 \leq i \leq r$. We have

$$M^{H_i} \cong \mathbb{R}, \quad 1 \leq i \leq r.$$

Let

$$(2) \qquad M^{K_1}, \dots, M^{K_s}$$

be all the 2-dimensional fixed point sets, and where moreover K_i occurs in M^{K_i}, $1 \leq i \leq s$. Then we have

$$M^{K_i} \cong \mathbb{R}^2, \quad 1 \leq i \leq s.$$

The intersection of two different sets in (1) equals $\{a\}$ and the intersection of two different sets in (2) equals $\{a\}$ or one of the sets in (1).

Let us denote

$$(3) \qquad R = \bigcup_{F \neq \{e\}} M^F = \bigcup_{i=1}^{r} M^{H_i} \cup \bigcup_{i=1}^{s} M^{K_i}.$$

Then R is a G-subset of M and the action of G on $M - R$ is free. We now proceed to construct a suitable open G-neighborhood U^* of R in M, such that U^* is G-homeomorphic to an orthogonal representation space for G.

We may assume that M is provided with an invariant riemannian metric. Let $\mathbb{R}^m(\rho_a)$ be the orthogonal representation of G on the tangent space at a. Let

$$\alpha: \mathbb{R}^m(\rho_a) \to M$$

be a smooth open G-embedding with $\alpha(0) = a$. We denote

$$\tilde{U} = \alpha(\mathbb{R}^m(\rho_a))$$

$$(4) \qquad \overline{U} = \alpha(B^m(\rho_a))$$

$$U = \alpha(\mathring{B}^m(\rho_a)).$$

Let H be one of the subgroups H_i, $1 \le i \le r$, occuring in (1). Then $M^H \cong \mathbb{R}$, and $N(H)$ acts on M^H. The action of $N(H)$ on $M^H \cong \mathbb{R}$ is topologically equivalent to an orthogonal action. We have

$$\partial\overline{U}^H = \overline{U}^H - U^H = \{x,y\},$$

and $N(H)$ acts on $\{x,y\}$. Since the action of $N(H)$ on M^H is equivalent to an orthogonal action it follows that

$$(5) \qquad M^H - U^H \cong \partial\overline{U}^H \times [1,\infty),$$

where $\cong$ denotes an $N(H)$-homeomorphism which is the identity on $\partial(M^H - U^H) = \partial\overline{U}^H$. (Here $\partial\overline{U}^H \times \{1\}$ is identified with $\partial\overline{U}^H$.)

Let $\overline{V}(M^H - U^H)$ be a closed $N(H)$-invariant tubular neighborhood of $M^H - U^H = (M - U)^H$ in $M - U$. Since (5) holds it follows (Wasserman [16], Corollary 2.5) that the normal bundle η of $(M - U)^H$ in $M - U$ is $N(H)$-isomorphic to the pull-back $r^*(\eta|\partial\overline{U}^H)$, where $r: M^H - U^H \to \partial\overline{U}^H$ is a $N(H)$-retraction and $\eta|\partial\overline{U}^H$ is the normal bundle of $\partial\overline{U}^H$ in $\partial\overline{U}$. Hence it follows that

$$(6) \qquad \overline{V}(M^H - U^H) \cong \overline{V}(\partial\overline{U}^H) \times [1,\infty),$$

where $\cong$ denotes $N(H)$-homeomorphism, which is the identity on $\overline{V}(\partial\overline{U}^H)$. Here $\overline{V}(\partial\overline{U}^H)$ is a closed $N(H)$-invariant tubular neighborhood of $\partial\overline{U}^H = \{x,y\}$ in $\partial\overline{U}$, i.e., $\overline{V}(\partial\overline{U}^H)$ is the disjoint union of a closed H-slice $\overline{S}_x$ at x in $\partial\overline{U}$ and a closed H-slice $\overline{S}_y$ at y in $\partial\overline{U}$. By $V(M^H - U^H)$ we denote the open $N(H)$-invariant tubular neighborhood corresponding to $\overline{V}(M^H - U^H)$. Thus $V(M^H - U^H)$ is the interior of $\overline{V}(M^H - U^H)$ with respect to $M - U$.

For each subgroup H_i, $1 \le i \le r$, we choose $N(H_i)$-invariant tubular neighborhoods

(7)
$$\overline{V}_i = \overline{V}(M^{H_i} - U^{H_i})$$
$$V_i = V(M^{H_i} - U^{H_i}),$$

and by choosing them small enough we may assume that $\overline{V}_i \cap \overline{V}_j = \emptyset$, for $i \neq j$. If H and H' are conjugate in G, say $H' = gHg^{-1}$, we moreover choose the tubular neighborhoods such that $\overline{V}(M^{H'} - U^{H'}) = g\overline{V}(M^H - U^H)$. Then the unions

(8)
$$\overline{V} = \overline{V}_1 \cup \ldots \cup \overline{V}_r$$
$$V = V_1 \cup \ldots \cup V_r$$

are G-subsets of $M - U$. It follows from (6) that

$$\overline{V} \cong (\partial\overline{U} \cap \overline{V}) \times [1,\infty), \tag{9}$$

where $\cong$ denotes a G-homeomorphism, which is the identity on $\partial\overline{U} \cap \overline{V}$.

We now form

$$\overline{U}_1 = \overline{U} \cup \overline{V}, \tag{10}$$

where the union is along $\partial\overline{U} \cap \overline{V}$. The interior of $\overline{U}_1$ equals

$$U_1 = U \cup V. \tag{11}$$

Since (9) holds it follows by Lemma 4 that U_1 is G-homeomorphic to U, i.e.,

$$U_1 \cong U \cong \overset{\circ}{B}{}^m(\rho_a). \tag{12}$$

The corner at $\partial\overline{U} \cap (\overline{V} - V)$ in $\overline{U}_1$ can be smoothed in such a way that $\overline{U}_1$ is a smooth G-manifold with boundary $\partial\overline{U}_1 = \overline{U}_1 - U_1$, and $M - U_1$ is a smooth G-manifold with boundary $\partial(M - U_1) = \partial\overline{U}_1$.

Let K be one of the subgroups K_i, $1 \leq i \leq s$, occurring in (2). The action of $N(K)$ on $M^K \cong \mathbb{R}^2$ is topologically equivalent to an orthogonal action (Bredon [2], Theorem IV.8.5). We have

$$\partial\overline{U}^K = \overline{U}^K - U^K \cong S^{m-1}(\rho_a)^K.$$

The group $N(K)$ acts on $\partial\overline{U}^K$ and $S^{m-1}(\rho_a)^K$ equals the unit circle in the 2-dimensional orthogonal representation space $\mathbb{R}^m(\rho_a)^K$ for $N(K)$. We claim that

(13) $$M^K - U^K \cong \partial\overline{U}^K \times [1,\infty),$$

where $\cong$ denotes an $N(K)$-homeomorphism, which is the identity on $\partial\overline{U}^K$. This can, for example, be seen as follows. The set $\widetilde{U}^K$ is an open $N(K)$-neighborhood of the closed $N(K)$-subset $\overline{U}^K$ in M^K. We have the following sequence of $N(K)$-homeomorphisms

(14) $$\widetilde{U}^K - U^K \cong \mathbb{R}^m(\rho_a)^K - \overset{\circ}{B}{}^m(\rho_a)^K \cong S^{m-1}(\rho_a)^K \times [1,\infty) \cong \partial\overline{U}^K \times [1,\infty).$$

It follows that the $N(K)$-space $M^K/\overline{U}^K$ ($\overline{U}^K$ collapsed to a point) is $N(K)$-homeomorphic to M^K. Since the $N(K)$-action on M^K is equivalent to an orthogonal action the orbit space of the $N(K)$-space M^K has conical orbit structure with vertex any image in the orbit space of a fixed point. Thus the orbit space of the $N(K)$-space $M^K/\overline{U}^K$ has conical orbit structure with vertex the point $\{\overline{U}^K\}^*$ in the orbit space.

The $N(K)$-space $\widetilde{U}^K/\overline{U}^K$ is $N(K)$-homeomorphic to $\mathbb{R}^m(\rho_a)^K$, by an $N(K)$-homeomorphism which takes the point $\{\overline{U}^K\}^*$ into the origin. Hence, in particular the orbit space of the $N(K)$-space $\widetilde{U}^K/\overline{U}^K$ has canonical orbit structure with vertex the point $\{\overline{U}^K\}^*$ in the orbit space. Now, by Theorem II.8.5 in Bredon [2], there exists an $N(K)$-homeomorphism of $\widetilde{U}^K$ onto M^K which is the identity in some neighborhood of $\overline{U}^K$. Thus (14) shows that (13) is valid.

Now we claim that

(15) $$M^K - U_1^K \cong \partial\overline{U}_1^K \times [1,\infty),$$

where $\cong$ denotes an $N(K)$-homeomorphism which is the identity on $\partial\overline{U}_1^K$. By (13) we have $M^K - U^K \cong \partial\overline{U}^K \times [1,\infty) \cong S^1 \times [1,\infty)$, where S^1 is equipped with an orthogonal $N(K)$-action. Assume that the number of 1-dimensional subsets M^H lying in M^K is p, where $1 \leq p \leq r$. (In case $p = 0$ we have $M^K - U_1^K = M^K - U^K$.) Then $M^K - U_1^K$ is obtained by removing from $M^K - U^K \cong S^1 \times [1,\infty)$ 2-dimensional open tubes around sets of the form $\{x_i, y_i\} \times [1,\infty)$, $1 \leq i \leq p$. Thus $M^K - U_1^K$ has $2p$ components and each component C is of the form $\partial C \times [1,\infty)$ with $\partial C \cong \mathbb{R}$. The group $N(K)/K$ acts freely on $M^K - U_1^K$. Thus we see from this description that $\partial\overline{U}_1^K$ has $2p$ components each of which is diffeomorphic to $\mathbb{R}$ and that (15) holds.

Now let $\overline{W}(M^K - U_1^K)$ be a closed $N(K)$-invariant tubular neighborhood of $M^K - U_1^K$ in $M - U_1$. In the same way as (6) followed from (5) it now follows from (15) that

(16) $$\overline{W}(M^K - U_1^K) = \overline{W}(\partial\overline{U}_1^K) \times [1,\infty).$$

Here $\overline{W}(\partial\overline{U}_1^K)$ is a closed N(K)-invariant tubular neighborhood of $\partial\overline{U}_1^K$ in $\partial\overline{U}_1$, and $\cong$ denotes an N(K)-homeomorphism which is the identity on $\overline{W}(\partial\overline{U}_1^K)$.

For each subgroup K_i, $1 \leq i \leq s$, we choose $N(K_i)$-invariant, closed and corresponding open, tubular neighborhoods

$$\overline{W}_i = \overline{W}(M^{K_i} - U_1^{K_i})$$

$$W_i = W(M^{K_i} - U_1^{K_i}).$$

As in the case above, in (7) and (8), we may assume that $\overline{W}_i \cap \overline{W}_j = \emptyset$, if $i \neq j$, and that the unions

$$\overline{W} = \overline{W}_1 \cup \ldots \cup \overline{W}_s$$

$$W = W_1 \cup \ldots \cup W_s$$

are G-subsets of $M - U_1$. From (16) we get

$$(17) \qquad \overline{W} \cong (\partial\overline{U}_1 \cap W) \times [1,\infty),$$

where $\cong$ denotes a G-homeomorphism which is the identity on $\partial\overline{U}_1 \cap \overline{W}$.

We form

$$\overline{U}^* = \overline{U}_1 \cup \overline{W},$$

where the union is along $\partial\overline{U}_1 \cap \overline{W}$. Since (17) holds it follows by Lemma 4 that the interior $U^* = U_1 \cup W$ of $\overline{U}^*$ is G-homeomorphic to U_1. Hence, by combining with (12), we get

$$U^* \cong U_1 \cong U \cong \overset{\circ}{B}{}^m(\rho_a).$$

Moreover U^* is an open G-neighborhood of R, (see (3)), in M.

We also need closed G-invariant neighborhoods $C_{1/2}$ and $C_{1/3}$ of R in M. Here $C_{1/3}$ is contained in the interior of $C_{1/2}$ and $C_{1/2}$ is contained in U^*. The set $C_{1/2}$ is constructed in the same way as U^* but using closed tubular neighborhoods and always using 1/2 of the radius used in constructing U^*. In constructing $C_{1/3}$ we use 1/3 of the radius used in constructing U^*. Thus in constructing $C_{1/2}$ we start with $\overline{U}_{1/2} = \alpha(B^m_{1/2}(\rho))$ and add closed invariant tubular neighborhoods of $M^{H_i} - U_{1/2}^{H_i}$ in $M - U_{1/2}$, $1 \leq i \leq r$, and so on. We now denote

$$C = C_{1/2}$$

$$W^* = M - C_{1/3}.$$

Then $C \subset U^*$ and

$$\operatorname{Int} C \cup W^* = M.$$

It is easy to see that the inclusions

$$U^* - C \hookrightarrow U^* - R$$

and

$$M - C \hookrightarrow M - R$$

are homotopy equivalences. Hence the inclusion

$$\text{(a)} \qquad i\colon (M - C, U^* - C) \to (M - R, U^* - R) = (M_{\{e\}}, U^*_{\{e\}})$$

induces isomorphisms in all degrees.

From now on the proof is completely analogous to the proof of Proposition 3.2 in [7].

Both U^* and M are homeomorphic to $\mathbb{R}^m$, where $m \geq 5$. Since $\dim M^H \leq 2$, for every subgroup $H \neq \{e\}$, it follows, by general position, that

$$i_*\colon \pi_1(U^*_{\{e\}}) \to \pi_1(U^*) = 0$$

and

$$i_*\colon \pi_1(M_{\{e\}}) \to \pi_1(M) = 0$$

are isomorphisms. Thus $U^*_{\{e\}}$ and $M_{\{e\}}$ are simply connected. By excision we have

$$H_q(M_{\{e\}}, U^*_{\{e\}}) \cong H_q(M, U^*) = 0, \text{ all } q \geq 0.$$

Hence it follows that the pair

(b) $(M_{\{e\}}, U^*_{\{e\}})$, is q-connected for all $q \geq 0$.

Let $A \subset M$ be an arbitrary compact subset of M. We claim that there exists a compact G-subset B of M, such that

$$A \cup \{a\} \subset B$$

and denoting $V^* = M - B$ we have that

(c) $(M_{\{e\}}, V^*_{\{e\}})$ is 2-connected.

This is seen as follows. Let V' be a small open G-slice at b in Σ, such that $(A \cup \{a\}) \cap V' = \emptyset$. Then $B = \Sigma - V'$ is a compact G-subset of M with $A \cup \{a\} \subset B$. Let $V^* = V' - \{b\} \subset M$. Then $V^* = M - B$. Let us denote $M' = \Sigma - \{a\}$. Then

(d) $(M_{\{e\}}, V^*_{\{e\}}) = (M'_{\{e\}}, V'_{\{e\}})$.

Both M' and V' are homeomorphic to $\mathbb{R}^m$, where $m \geq 5$, and hence it follows, by general position, that $M'_{\{e\}}$ and $V'_{\{e\}}$ are simply connected. Moreover, it follows by duality (see Lemma 4.1 in [7]) that

$$i_*: H_q(V'_{\{e\}}) \to H_q(M'_{\{e\}})$$

is an isomorphism for all q. Thus it follows that $(M'_{\{e\}}, V'_{\{e\}})$ is q-connected for all q, and hence in particular 2-connected. Thus (d) shows that the claim in (c) is valid.

Now let T be a smooth equivariant triangulation of M, which is subordinate to the open cover $\{\text{Int } C, W^*\}$ of M, [6]. This gives $|T| = M$ the structure of an equivariant p.l. manifold. Let L be the equivariant subcomplex of T consisting of all closed simplexes of T that are contained in W^*. Observe that the only isotropy subgroup that occurs in L is $\{e\}$. Let L^2 denote the 2-skeleton of L. Since $(M_{\{e\}}, V^*_{\{e\}})$ is 2-connected we have that (M_H, V^*_H) is 2-connected for every subgroup H which occurs in L^2. Furthermore $L^2 - V^*$ is compact, and $M_{\{e\}}$ is connected and we have

$$\dim L^2_{\{e\}} \leq \dim M_{\{e\}} - 3.$$

Thus Theorem 2.4 in [7] applies and hence there exist a compact G-subset E'_1

of M and a (p.l.) G-homeomorphism $v: M \to M$ such that

$$L^2 \subset v(V^*) \quad \text{and} \quad v|M - E_1' = \text{id}.$$

Let $E_1 = E_1' \cup B$. Then $M - E_1 \subset M - E_1'$ and $M - E_1 \subset M - B = V^*$.

Now let P be the equivariant subcomplex of T which consists of all closed simplexes of T that are contained in $M - E_1$, and define

$$K = L^2 \cup P.$$

Since $L^2 \subset v(V^*)$ and $P \subset M - E_1 = v(M - E_1) \subset v(V^*)$, it follows that

$$K \subset v(V^*).$$

Now define

$$J = T \div K$$

i.e., J is the maximal subcomplex of the first barycentric subdivision T' of T which does not intersect K. We have

$$J \subset T - K \subset T - P \subset N(E_1;T)$$

and hence J is compact. The only isotropy subgroup that occurs in $J - C$ is $\{e\}$. We claim that

$$\text{(e)} \qquad \dim (J - C)_{\{e\}} \leq \dim M_{\{e\}} - 3.$$

Since $J - C \subset (T \div L^2) - C$ it is enough to prove that

$$\text{(f)} \qquad \dim ((T \div L^2) - C) \leq \dim M_{\{e\}} - 3.$$

Let s' be a simplex of $T \div L^2$ which is not contained in C. Let s be the smallest simplex of T which contains the simplex s' of T'. Then s is not contained in C and therefore $s \subset W^*$, since T is subordinate to $\{\text{Int } C, W^*\}$. Since $x \in W^*$ implies $G_x = \{e\}$ we have $s \subset M_{\{e\}}$. The 2-skeleton s^2 of s is contained in L^2 and hence $s' \cap s^2 = \emptyset$. It now follows that

$$\dim s' \le \dim s - 3 \le \dim M_{\{e\}} - 3.$$

This proves (f) and hence (e) is valid.

By (b) the pair $(M_{\{e\}}, U^*_{\{e\}})$ is q-connected for all q and by (a) the inclusion $(M - C, U^* - C) \to (M_{\{e\}}, U^*_{\{e\}})$ induces isomorphisms in homotopy in all degrees. Hence the pair

$$(M - C, U^* - C) \text{ is q-connected for all } q.$$

Thus Theorem 2.5 in [7] applies and gives us a (p.l.) G-homeomorphism $u: M \to M$ such that $J \subset u(U^*)$.

Since now $K \subset v(V^*)$ and $J \subset u(U^*)$ and $J = T \dot{-} K$ there exists a G-homeomorphism $w: |T'| \to |T'|$, i.e., $w: M \to M$ such that

$$w(u(U^*)) \cup v(V^*) = |T'| = M.$$

Then

$$h = v^{-1} \circ w \circ u: M \to M$$

is a G-homeomorphism such that $h(U^*) \cup V^* = M$, and hence

$$A \subset B \subset h(U^*).$$

Since A is compact and

$$A \subset (h \circ \gamma)(\overset{\circ}{B}{}^m(\rho_a))$$

it follows that there exists $0 < r < 1$ such that

$$A \subset (h \circ \gamma)(B_r^m(\rho_a)) \cong B_r^m(\rho_a).$$

Thus we have proved that every compact subset of M is contained in a G-subset which is G-homeomorphic to $B^m(\rho_a)$. Hence M is a monotone union of G-subsets G-homeomorphic to $B^m(\rho_a)$, and therefore, by Proposition 3.4 in [7], M is G-homeomorphic to $\mathbb{R}^m(\rho_a)$. Since Σ is the one-point compactification of M it now follows that Σ is G-homeomorphic to $S^m(\rho_a \oplus 1)$. □

6. THE MAIN RESULTS

THEOREM A. Let G be a finite solvable group acting smoothly on a homotopy sphere Σ^m, where $m \geq 5$, such that there is a fixed point $a \in \Sigma^G$ at which the representation of G is 2-pseudo-free. Then the G-space Σ^m is G-homeomorphic to a linear representation sphere $S^m(\rho)$.

PROOF. Follows directly from Proposition 1 and Theorem 5. □

The proof of Theorem 5 shows that if $\mathbb{R}^m(\rho_a)$ denotes the, 2-pseudo-free, representation of G at a, then Σ^m is G-homeomorphic to $S^m(\rho_a \oplus 1)$. □

COROLLARY B. Let G and the action of G on Σ^m be as in Theorem A. Then the representations of G at any two fixed points of G are topologically equivalent.

PROOF. By Theorem A there exists a G-homeomorphism $h: \Sigma^m \to S^m(\rho)$. Let $x, y \in \Sigma^G$. An application of the conical orbit structure result, Corollary II.8.4 in Bredon [2], shows that the representation $\mathbb{R}^m(\rho_x)$ of G at x is topologically equivalent to the representation at $h(x)$ in $S^m(\rho)$, (compare with § 3 in Schultz [12]). Likewise the representation $\mathbb{R}^m(\rho_y)$ of G at y is topologically equivalent to the representation at $h(y)$ in $S^m(\rho)$. Since the representations at $h(x)$ and $h(y)$ in $S^m(\rho)$ are, even linearly, equivalent the result follows. □

THEOREM C. Let G be a finite group acting smoothly on a homotopy sphere Σ^m, where $m \geq 5$, such that there exist at last two fixed points $a, b \in \Sigma^G$, and the representation of G at a is pseudo-free. Then the G-space Σ^m is G-homeomorphic to a linear representation sphere $S^m(\rho)$.

PROOF. Follows directly from Proposition 3 and Theorem 5. □

COROLLARY D. Let G be a finite group acting smoothly on a homotopy sphere Σ^m, $m \geq 5$, such that there is a fixed point at which the representation of G is pseudo-free. Then the representations of G at fixed points are topologically equivalent to each other.

PROOF. In case there is only one fixed point of G there is nothing to prove. If there are at least two fixed points Theorem C applies, and Corollary D follows from Theorem C exactly in the same way as Corollary B followed from Theorem A. □

BIBLIOGRAPHY

1. M.F. Atiyah and R. Bott, The Lefschetz fixed point theorem for elliptic complexes: II. Applications, Ann. of Math. (2) 88 (1968), 451-491.

2a. G.E. Bredon, Representations at fixed points of smooth actions of compact groups, Ann. of Math. (2) 89 (1969), 515-532.

2. G.E. Bredon, Introduction to Compact Transformation Groups, Academic Press, 1972.

3. S.E. Cappell and J.L. Shaneson, Linear algebra and topology, Bull. Amer. Math. Soc. (N.S.) 1 (1979), 685-687.

4. S.E. Cappell and J.L. Shaneson, Fixed points of periodic maps, Proc. Nat. Acad. Sci. USA 77 (1980), 5052-5054.

5. S.E. Cappell and J.L. Shaneson, Non-linear similarity, Ann. of Math. (2) 113 (1981), 315-355.

6. S. Illman, Smooth equivariant triangulations of G-manifolds for G a finite group, Math. Ann. 233 (1978), 199-220.

7. S. Illman, Recognition of linear actions on spheres. (To appear in Trans. Amer. Math. Soc.)

8. E. Laitinen, Pseudo-free representations of finite groups. (To appear.)

9. J.W. Milnor, Whitehead torsion, Bull. Amer. Math. Soc. 72 (1966), 358-426.

10. T. Petrie, Three theorems in transformation groups, Algebraic Topology, Aarhus 1978, Proceedings, Lecture Notes in Math., Vol. 763, Springer-Verlag (1979), 549-572.

11. C.U. Sanchez, Actions of groups of odd order on compact, orientable manifolds, Proc. Amer. Math. Soc. 54 (1976), 445-448.

12. R. Schultz, On the topological classification of linear representations, Topology 16 (1977), 263-269.

13. J.-P. Serre, Linear Representations of Finite Groups, Springer-Verlag, 1977.

14. P.A. Smith, New results and old problems in finite transformation groups, Bull. Amer. Math. Soc. 66 (1960), 401-415.

15. E. Stein, Surgery on products with finite fundamental group, Topology 16 (1977), 473-493.

16. A.G. Wasserman, Equivariant differential topology, Topology 8 (1969), 127-150.

17. J.A. Wolf, Spaces of Constant Curvature, Fourth Edition, Publish or Perish, Inc., 1977.

DEPARTMENT OF MATHEMATICS
UNIVERSITY OF HELSINKI
HALLITUSKATU 15
00100 HELSINKI 10
FINLAND

Canadian Mathematical Society
Conference Proceedings
Volume 2, Part 2 (1982)

THE CONNER-MILLER CLASSES OF PERIODIC MAPS AND AN EQUIVARIANT POINT INDEX

Steven M. Kahn

ABSTRACT. Let (T,M^m) and (T',V^n) be diffeomorphisms of prime period p on closed manifolds with (T,M) being fixed point free. For any map $f : M \to V$ with $m > (p-1)n$, we obtain an invariant $C(f)$ and show that if $C(f) \neq 0$, then there exists a point $x \in M$ such that $f(T^i x) = T'^i f(x) \; \forall \; 1 \leq i \leq p-1$. This invariant depends on the Conner-Miller and Bredon classes of (T,M) and (T',V) respectively and actually results from the determination of the Conner-Miller classes of the product diffeomorphism $\tau = T \times T'$. Additionally, we obtain another index that deals with the case where both (T,M) and (T',V) have fixed points.

1. INTRODUCTION. In 1965, Conner and Floyd [3] proved a generalization of the Borsuk-Ulam Theorem that states that if T is any fixed point free involution on S^m and V^n is any manifold with $m > n$, then for any map $f : S^m \to V^n$, $\dim \{x \in S^m | f(Tx) = f(x)\} \geq m - n$.

H. J. Munkholm [6] in 1969 considered $\mathbb{Z}_p$-actions on spheres for p any prime and obtained what he called

THE (MOD p) CONNER-FLOYD THEOREM: Let (T,S^m) be any fixed point free diffeomorphism of prime period p, and let V^n be any manifold with $m > (p-1)n$. (For $p > 2$, assume that V is orientable). Then for any map $f : S^m \to V^n$, $\dim \{x \in S^m | f(T^i x) = f(x) \; \forall \; 1 \leq i \leq p-1\} \geq m - (p-1)n$.

1980 Mathematics Subject Classification. 57E99, 57D20.

© 1982 American Mathematical Society
0731-1036/82/0000-0464/$02.75

Let (T,M^m) and (T',V^n) be diffeomorphisms of prime order p on closed manifolds. (For $p > 2$ we assume here and throughout this paper that the manifolds are oriented and that the diffeomorphisms are orientation-preserving). An <u>equivariant point</u> of a map $f : M^m \to V^n$ is a point $x \in M$ such that $f(T^i x) = T'^i f(x) \ \forall\ 1 \le i \le p-1$. In this paper we present an equivariant point index that is defined for any prime p and that yields generalizations of both the Conner and Floyd, and Munkholm results.

In [2], Conner and E. Y. Miller defined certain cohomology classes for manifolds with a free involution. T. Y. Lin [5] has recently extended this definition obtaining Conner-Miller classes $\rho_k(M^m) \in H^k(M;\mathbb{Z}_p)$, $k > \frac{(p-1)}{p}m$, for manifolds with fixed point free diffeomorphisms (T,M) of any prime period p. There are the so-called Bredon classes $B_k(V) \in H^k(V;\mathbb{Z}_p)$ (again defined by Conner and Miller [2] for $p = 2$ and by Lin [5] for $p > 2$) for arbitrary periodic maps (T',V). The definitions of both of these classes are provided in section three. The main result is the following:

<u>THEOREM 1.1</u>. <u>Let</u> (T,M^m) <u>and</u> (T',V^n) <u>be diffeomorphisms of prime period</u> p <u>with</u> (T,M) <u>free</u>. <u>For any map</u> $f : M^m \to V^n$ with $m > (p-1)n$, <u>let</u>

$$C(f) = \Sigma \langle \rho_{m-j}(M) f^* B_j(V) \ , \ [M] \rangle$$

<u>If</u> $C(f) \neq 0$, <u>then</u>

$$\underline{\dim}\ \{x \in M \mid f(T^i x) = T'^i f(x) \ \forall\ 1 \le i \le p-1\} \ge m - (p-1)n \ .$$

<u>REMARK</u>. The Conner-Floyd and Munkholm results follow simply from the fact that for any free periodic map (T,S^m), $\rho_m(S^m) \neq 0$, while for (T',V) the identity map $B_0(V) = 1 \in H^0(V;\mathbb{Z}_p)$.

Theorem 1.1 is an example of the classic Borsuk-Ulam type theorem in which in which at least one of the actions, on the domain or on the range, is taken to be free. Interestingly it leads to another result that deals with the case where neither action is free, that is where both periodic maps have fixed points.

<u>THEOREM 1.2</u>. <u>Let</u> $(T,M^{(p-1)pn})$ <u>and</u> (T',N^{pn}) <u>be diffeomorphisms of prime period</u> p <u>and let</u> $f : M \to N$ <u>be any map</u>. <u>Let</u>

$$B(f) = \langle B_{(p-1)^2 n}(M) f^* B_{(p-1)n}(N) , [M]\rangle$$

_If_ $B(f) \neq 0$, _then_ f _has an equivariant point_.

The results stated above, along with several applications and observations, were given for the case of $p = 2$ in [4]. The proof of Theorem 1.2 parallels the $p = 2$ case proof and so is omitted here. The proof of the following example may be found in [4] as well.

COROLLARY 1.3. _Let_ $(T, \mathbb{K}P(2n))$ _and_ (T', N^{2kn}) _be involutions where_ $\mathbb{K} = \mathbb{R}, \mathbb{C}$ _or_ $\mathbb{H}$ _and_ $k = 1, 2$ _or_ 4 _respectively. Assume that the Euler characteristic_ $\chi(N)$ _is odd. Then any map_ $f : \mathbb{K}P(2n) \to N$ _of odd degree has an equivariant point._

REMARK. There is a generalized index $B(f)$ that can be applied to maps $f : M^m \to N^n$ with $m \leq (p-1)n$. We deal with this index in a future paper.

Throughout this paper, all cohomology will be with $\mathbb{Z}_p$-coefficients.

2. THE BREDON OPERATION. For any fixed point free diffeomorphism (T, M^m) of prime order p there is a cohomology operation $Q : H^k(M) \to H^{pk}(M/T)$, originally defined for $p = 2$ by G. E. Bredon in [1]. (See also [2]). The definition was extended by Lin [5] to include all primes. We list some properties of Q .

(2.1) For any equivariant map $f : (T,X) \to (T',Y)$ inducing a map $F : X/T \to Y/T$, $Q(f^*\alpha) = F^*Q(\alpha)$.

(2.2) $\nu^*Q(\alpha) = \alpha \cdot T^*(\alpha) \cdot T^{2*}(\alpha) \ldots T^{(p-1)*}(\alpha)$ where $\nu : M \to M/T$ is the quotient map

(2.3) $Q(\alpha\beta) = Q(\alpha)Q(\beta)$

(2.4) $Q(1_M) = 1_{M/T} \in H^0(M/T)$

The proof of Theorem 1.1 relies on a so-called separation property of the Bredon operation which Conner and Miller [2] exhibited in the $p = 2$ case. The (mod p) version and two lemmas that essentially make up its proof read as follows:

Let (T,X) be a fixed point free diffeomorphism of prime order p , and let $Y \subset X$ be a closed subset for which $X = Y \cup T(Y) \cup \ldots \cup T^{p-1}(Y)$.

LEMMA 2.5. *The homomorphism*

$$\nu^* : H^*(X/T,\ Y \cap T(Y) \cap \ldots \cap T^{p-1}(Y)/T) \to H^*(X, Y \cap T(Y) \cap \ldots \cap T^{p-1}(Y))$$

is a monomorphism.

LEMMA 2.6. *The Bredon operation* Q *is trivial on the image of* $H^k(X,Y) \to H^k(X, Y \cap T(Y) \cap \ldots \cap T^{p-1}(Y))$.

SEPARATION THEOREM. *If* $\alpha \in H^k(X)$ *lies in the kernel of* $H^k(X) \to H^k(Y)$, *then* $Q(\alpha) = 0$.

3. THE CONNER-MILLER CLASSES AND A PRODUCT FORMULA. We begin this section by recalling the cohomology ring structure of the classifying space $B\mathbb{Z}_p$:

For $p = 2$, $H^*(B\mathbb{Z}_2) = \mathbb{Z}_2[c]$ where $\deg c = 1$

For p odd , $H^*(B\mathbb{Z}_p) = \mathbb{Z}_p[d_2] \otimes \Lambda_p[d_1]$ where $\deg d_1 = 1$, $\deg d_2 = 2$ and d_2 is the (mod p) Bockstein of d_1 .

For notation we write $d_{2j} = d_2^j$ and $d_{2j+1} = d_1 d_2^j$ for p odd and $d_i = c^i$ for $p = 2$.

DEFINITION 3.1. Let (T,M^m) be a fixed point free diffeomorphism of prime period p . For $k > \frac{(p-1)}{p}m$, the Conner-Miller class $\rho_k(M) \in H^k(M)$ is defined as the unique class such that

$$\langle \rho_k(M) \cdot \alpha, [M] \rangle = \langle d_{pk-(p-1)m} Q(\alpha), [M/T] \rangle$$

for all $\alpha \in H^{m-k}(M)$, where $d_j \in H^j(M/T)$ denotes the image of $d_j \in H^j(B\mathbb{Z}_p)$ under the homomorphism induced by the homotopically unique map of M/T into $B\mathbb{Z}_p$.

Let $S^{2N+1} = \{(z_1,\ldots,z_{N+1}) | \Sigma\, z_i \bar{z}_j = 1\}$ and let $\lambda = \exp(2\pi i/p)$. Let A be the map defined by $A(z_1,\ldots z_{N+1}) = (\lambda z_1,\ldots,\lambda z_{N+1})$.

DEFINITION 3.2. Let (T,V^n) be any diffeomorphism of prime period p . With $2N+1 > (p-1)n$, the Bredon class $B_k(V) \in H^k(V)$ is defined as the unique class such that

$$\langle B_k(V) \cdot \alpha, [V] \rangle = \langle d_{pk+2N+1-(p-1)n} Q(\alpha \otimes 1), [V \times S^{2N+1}/\tau] \rangle$$

for all $\alpha \in H^{n-k}(V)$, where $\tau = T \times A$.

Let (T,M^m) and (T',V^n) be periodic maps with (T,M^m) fixed point free. We now consider the product diffeomorphism $\tau = T \times T'$ on $M \times V$. We note that τ is fixed point free and so possesses Conner-Miller classes $\rho_k(M \times V)$. It turns out that the determination of these classes is the key to presenting the invariant $C(f)$ in Theorem 1.1 as a computable index. In [4], we obtained the product formula for the case of $p = 2$. In that the proof for the (mod p) case essentially follows the one given in [4], we omit it here and simply state the result.

THEOREM 3.3. $\rho_k(M \times V) = \Sigma \rho_{k-j}(M) \otimes B_j(V)$

4. PROOF OF THEOREM 1.1.

DEFINITION 4.1. A set $S \subset M^m$ *supports* a cohomology class $\alpha \in H^k(M)$ if for all open sets U, with $S \subset U$, α lies in the kernel of $H^k(M) \to H^k(M-U)$.

REMARK 4.2. Using a simple Poincaré duality argument, it is easy to see that if S supports $\alpha \in H^k(M^m)$ and $\alpha \neq 0$, then $\dim(S) \geq m-k$. (The dimension here as in Theorem 1.1 is cohomological).

PROPOSITION 4.3. *Let* (T,M^m) *be a fixed point free diffeomorphism of prime period* p. *If* $S \subset M$ *is a closed subset that supports a class* $\alpha \in H^k(M)$, *then* $\dfrac{S \cap T(S) \cap \ldots \cap T^{p-1}(S)}{T} \subset M/T$ supports $Q(\alpha)$.

PROOF: Let $\bar{U} \subset M/T$ be any open set containing $\dfrac{S \cap T(S) \cap \ldots \cap T^{p-1}(S)}{T}$.

Let $U = \nu^{-1}(\bar{U}) \subset M$. (Note that $S \cap T(S) \cap \ldots \cap T^{p-1}(S) \subset U$). We must show that $I^*Q(\alpha) = 0 \in H^{pk}(M/T - \bar{U}) = H^{pk}(\frac{M-U}{T})$ where $I : \frac{M-U}{T} \to M/T$ is the inclusion. By (2.1) this is equivalent to showing that $Q(i^*\alpha) = 0$ where $i : M-U \to M$ is the inclusion.

LEMMA 4.4. If U is a T-invariant open set containing $S \cap T(S) \cap \ldots \cap T^{p-1}(S)$, then $U = S' \cap T(S') \cap \ldots \cap T^{p-1}(S')$ for some open set $S' \supset S$.

PROOF OF LEMMA: Consider first the case $p = 2$. Then U is an open set containing $S \cap T(S)$. Since S and therefore $T(S)$ are closed in M , $S-U$ and $T(S) - U$ are disjoint closed sets in M . With M being normal, there are disjoint open sets F_0 and F_1 such that $S-U \subset F_0$ and $T(S) - U \subset F_1$. Let $E_0 = F_0 \cup U$ and let $E_1 = F_1 \cup U$. Then $S \subset E_0$ and $T(S) \subset E_1$ and $E_0 \cap E_1 = U$. Using an induction argument on p (note that $\bigcap_{i=0}^{p-1} E_i = \bigcap_{i=1}^{p-1} (E_0 \cap E_i)$, one sees that for any prime p , $U = E_0 \cap E_1 \cap \ldots \cap E_{p-1}$ where E_i is open and contains $T^i(S)$. Letting $S' = E_0 \cap T^{-1}(E_1) \cap T^{-2}(E_2) \cap \ldots \cap T^{-(p-1)}(E_{p-1})$ completes the proof of the lemma.

Consider the following commutative diagram:

$$\begin{array}{ccc} & M - S' & \\ {}^{k}\swarrow & & \searrow^{j} \\ M-(S' \cap T(S') \cap \ldots \cap T^{p-1}(S')) & \xrightarrow{\ i\ } & M \end{array}$$

where i,j and k are inclusions. Given that S supports $\alpha \in H^k(M)$, we have by definition that $j^*(\alpha) = 0$, and so $k^* i^*(\alpha) = 0$. Now, let

$$X = M-(S' \cap T(S') \cap \ldots \cap T^{p-1}(S')) = M-S' \cap T(M-S') \cap \ldots \cap T^{p-1}(M-S') .$$

Applying the Separation theorem with $Y = M-S'$ yields the result that $Q(i^* \alpha) = 0$. ∎

Let (T, M^m) , (T', V^n) , and $f : M \to V$ be as in Theorem 1.1, and let $(\tau, M \times V)$ be the product diffeomorphism $\tau = T \times T'$. Let $g : M \to M \times V$ be defined by $g(x) = (x, f(x))$ and let G denote the image of g (i.e. G is the graph of f) .

<u>COROLLARY</u> 4.5. <u>Let $g_!(1) \in H^n(M \times V)$ be the cohomology class dual to the graph of</u> f . If $Q(g_!(1)) \neq 0$, <u>then</u>

$$\underline{\dim} \left(\frac{G \cap \tau(G) \cap \ldots \cap \tau^{p-1}(G)}{\tau} \right) \geq m - (p-1)n .$$

PROOF: We simply note that G supports $g_!(1) \in H^n(M \times V)$ and apply Proposition 4.3 along with Remark 4.2. ∎

Theorem 1.1 is now obtained by considering the number $C(f) = \langle d_{m-(p-1)n} Q(g_!(1)), [\frac{M \times V}{T}] \rangle$. By Definition 3.1, $C(f) = \langle \rho_m(M \times V) \cdot g_!(1), [M \times V] \rangle$. The product formula, Theorem 3.3, yields the final result. ∎

BIBLIOGRAPHY

1. G. E. Bredon, Cohomological aspects of transformation groups, Proc. Conf. Transformation Groups (New Orleans, La. 1967) Springer-Verlag, 1968, pp. 245-280.

2. P. E. Conner and E. Y. Miller, Equivariant self-intersection, preprint, 1979.

3. P. E. Conner and E. E. Floyd, Differentiable Periodic Maps, Springer-Verlag, 1964.

4. S. M. Kahn, The Conner-Miller classes of product involutions and Borsuk-Ulam type theorems, to appear in Houston J. Math.

5. T. Y. Lin, On generalization of Conner-Miller classes to $\mathbb{Z}_p$-actions I, preprint.

6. H. J. Munkholm, Borsuk-Ulam type theorems for proper $\mathbb{Z}_p$-actions on (mod p) homology n-spheres, Math. Scand. 24(1969), 167-185.

DEPARTMENT OF MATHEMATICS
WAYNE STATE UNIVERSITY
DETROIT, MICHIGAN 48202

Canadian Mathematical Society
Conference Proceedings
Volume 2, Part 2 (1982)

CLASSIFYING G-SPACES AND THE SEGAL CONJECTURE

by L. G. LEWIS, J. P. MAY, and J. E. McCLURE

In a previous note [11], two of us promised to describe the natural map (induced by $EG \to pt$)

$$(1) \qquad \varepsilon: (S_G)^G \to F(EG^+, S_G)^G \simeq F(BG^+, S) = D(BG^+)$$

in nonequivariant terms. Here S_G is the sphere G-spectrum and $(k_G)^G$ denotes the G-fixed point spectrum associated to a G-spectrum k_G. In section 1, following up and reinterpreting ideas of Adams, Gunawardena, and Miller, we shall generalize the situation and describe the natural map

$$(2) \qquad \varepsilon: [\Sigma_G^\infty B(G,\Pi)^+]^G \to F(EG^+, \Sigma_G^\infty B(G,\Pi)^+)^G \simeq F(BG^+, \Sigma^\infty B\Pi^+)$$

in nonequivariant terms for a finite group G and a compact Lie group Π. Here $B(G,\Pi)$ is the classifying G-space for principal (G,Π)-bundles and Σ_G^∞ denotes the suspension G-spectrum functor. (The equivalence follows from [11; Lemma 12]; see Remarks 7 below for its hypotheses.) More precisely, we shall specify an equivalence from a wedge of spectra $\Sigma^\infty BW\rho^+$ for certain groups $W\rho$ to the domain of ε and shall show that the composite of ε and this equivalence is the wedge sum of appropriate composites of transfer and classifying maps. The reader is referred to Lashof [5] for a good discussion of (G,Π)-bundles and to [8] for a comprehensive study of G-spectra and the equivariant stable category.

The completion conjecture for a G-spectrum k_G asks if $\varepsilon^*: k_G^*(pt) \to k_G^*(EG)$ induces an isomorphism upon completion in the $I(G)$-adic topology, where $I(G)$ is the augmentation ideal of the Burnside ring $A(G)$; see [11] for discussion. The Segal conjecture is the completion conjecture for S_G. In section 2, we shall prove the following result.

THEOREM A. If the Segal conjecture holds for all finite groups G, then the completion conjecture holds for the G-spectra $\Sigma_G^\infty B(G,\Pi)^+$ for all finite groups G and Π.

It would be of interest to know whether or not this result

© 1982 American Mathematical Society
0731-1036/82/0000-0465/$04.75

remains valid for general compact Lie groups Π.

By the results of [11], it suffices to prove the theorem for p-groups G, and in this case the conclusion is equivalent to the assertion that the maps ε of (2) induce equivalences upon completion at p. Given our nonequivariant interpretation of these maps, the conclusion is an observation, suggested to us by Gunawardena and also noticed by Segal, when Π is also a p-group. Indeed, by an argument that is entirely symmetrical in G and Π, one sees that if ε in (1) induces an equivalence upon completion at p for all groups Γ and $G \times \Gamma$, where Γ is a subquotient of Π, then ε in (2) induces an equivalence upon completion at p for all pairs (G,Γ).

The passage from p-groups Π to general finite groups Π is based on the following equivariant generalization of a standard result about transfer. Define a G-cover to be a G-map $\pi: E \to B$ which is also a finite cover. A G-cover π has an associated equivariant transfer map $\tau: \Sigma_G^\infty B^+ \to \Sigma_G^\infty E^+$ [8].

THEOREM B. Let G be a p-group and let $\pi: E \to B$ be a G-cover whose fibre has cardinality prime to p. Then the composite

$$\Sigma_G^\infty B^+ \xrightarrow{\tau} \Sigma_G^\infty E^+ \xrightarrow{\Sigma_G^\infty \pi} \Sigma_G^\infty B^+$$

induces an equivalence upon localization at p, hence $\Sigma_G^\infty B^+$ is p-locally a wedge summand of $\Sigma_G^\infty E^+$.

COROLLARY C. Let G be a p-group and let Λ be a p-Sylow subgroup of a finite group Π. Then $\Sigma_G^\infty B(G,\Pi)^+$ is p-locally a wedge summand of $\Sigma_G^\infty B(G,\Lambda)^+$.

There is an implied splitting of localized G-fixed point spectra, a fact that is not at all obvious from our nonequivariant description of the latter. We are very grateful to George Glauberman for proving a result in finite group theory for us that made clear that such a splitting was plausible; see Remark 14.

The second author wishes to thank Frank Adams, Jeremy Gunawardena, and Haynes Miller for discussions of the Segal conjecture, Dick Lashof for discussions of classifying G-spaces, and George Glauberman and Daniel Frohardt for background information about finite groups.

§1. THE SPLITTING OF $\Sigma_G^\infty B(G,\Pi)^+$

Our nonequivariant description of ε will be based on the following general splitting theorem for stable G-homotopy groups. Let EG be a contractible free left G-space; for a left G-space Y, write $EG\times_G Y$ for the orbit space $(EG\times Y)/G$. For $H \subset G$, let NH denote the normalizer of H in G and let $WH = NH/H$.

THEOREM 1. For a G-space Y, let ξ_H be the composite homomorphism of stable homotopy groups displayed in the following diagram:

$$\begin{array}{ccccc}
\pi_*(EWH\times_{WH}Y^H) & \xrightarrow{\zeta} & \pi_*^G(EWH\times_{WH}Y^H) & \xrightarrow{\tau} & \pi_*^G(G\times_{NH}(EWH\times Y^H)) \\
\downarrow{\scriptstyle \xi_H} & & & & \downarrow{\scriptstyle i_*} \\
\pi_*^G(Y) & \xleftarrow{\varepsilon_*} & \pi_*^G((G\times_{NH}EWH)\times Y) & \cong & \pi_*^G(G\times_{NH}(EWH\times Y))
\end{array}$$

Here ζ is induced by the unit map $S \to (S_G)^G$ and τ is the equivariant transfer associated to the natural G-cover

$$\pi\colon G\times_{NH}(EWH\times Y^H) \to EWH\times_{NH}Y^H = EWH\times_{WH}Y^H$$

with fibre G/H; i and ε are the evident inclusion and projection. Define

$$\xi = \sum_{(H)} \xi_H\colon \sum_{(H)} \pi_*(EWH\times_{WH}Y^H) \to \pi_*^G(Y),$$

where the sum runs over one group H from each conjugacy class (H) of subgroups of G. Then ξ is an isomorphism.

PROOF. Such a splitting is given by Segal [14], Kosniowski [4], and tom Dieck [1], but the present description of the splitting map is not explicit in the literature. We derive it from tom Dieck's proof and diagram chasing. Thus fix H, abbreviate $N = NH$ and $W = WH$, and consider the following commutative diagram:

$$\begin{array}{ccccc}
\pi_*(EW\times_W Y^H) & & \xrightarrow{\quad\zeta\quad} & & \pi_*^G(EW\times_W Y^H) \\
\downarrow{\scriptstyle \zeta} & \searrow{\scriptstyle \zeta} & & \swarrow{\scriptstyle \rho} & \downarrow{\scriptstyle \tau} \\
\pi_*^W(EW\times_W Y^H) & \xrightarrow{\rho} & \pi_*^N(EW\times_N Y^H) & \xrightarrow{\omega} & \pi_*^G(G/N\times(EW\times_N Y^H)) \\
\downarrow{\scriptstyle \tau} & & \downarrow{\scriptstyle \tau} & & \downarrow{\scriptstyle \tau} \\
\pi_*^W(EW\times Y^H) & \xrightarrow{\rho} & \pi_*^N(EW\times Y^H) & \xrightarrow{\omega} & \pi_*^G(G\times_N(EW\times Y^H)) \\
\vdots & & & & \downarrow{\scriptstyle i_*} \\
\pi_*^G(Y) & \xleftarrow{\varepsilon_*} & \pi_*^G((G\times_N EW)\times Y) & \cong & \pi_*^G(G\times_N(EW\times Y))
\end{array}$$

Here the ρ are restriction homomorphisms and the ω are extension of group isomorphisms due to Wirthmuller [12]; see also [8], where $\omega\rho = \tau$, $\tau\rho = \rho\tau$, and $\omega\tau = \tau\omega$ are proven. By the transitivity of transfer [8], the right vertical composite $\tau\circ\tau$ is the transfer referred to in the statement. By [11, Lemma 16], the left vertical composite $\tau\circ\zeta$ is an isomorphism. Tom Dieck [1] proved that the sum over (H) of the dotted arrow composites $\varepsilon_* i_* \omega\rho$ is an isomorphism.

REMARK 2. Except that ω requires reinterpretation and is no longer an isomorphism, the diagram in the proof commutes and tom Dieck's result remains valid for general compact Lie groups G. However, [11, Lemma 16] fails in this generality, hence so does the conclusion of the theorem. In fact, the left vertical composite $\tau\circ\zeta$ is an isomorphism if WH is finite, but the transfer here is zero if WH is not finite (see e.g. [8]). The logic becomes clear if one takes Y to be a point. Here $\pi_0^G(pt)$ is the Burnside ring and is thus the free Abelian group with one generator for each (H) such that WH is finite. However, $\pi_*^G(pt)$ is the direct sum over all (H) of the groups $\pi_*^{WH}(EWH)$, the (H) with WH infinite contributing only to the positive dimensional homotopy groups. Consideration of the completion conjecture here requires consideration of all (H), while the 0-dimensional Segal conjecture requires consideration only of those (H) with WH finite.

With the understandings that our diagram chases will be of interest only when WH is finite, that subgroups are to be closed and homomorphisms continuous, and that G-covers are to be interpreted as G-equivariant bundles, we can allow both G and Π to be arbitrary compact Lie groups until otherwise specified.

We need some notations to describe Y^H when $Y = B(G,\Pi)$.

NOTATIONS 3. Consider a subgroup H of G and a homomorphism $\rho: H \to \Pi$. Define the following associated groups:

$$\Delta\rho = \{(h,\rho(h))\} \subset G \times \Pi$$

$$N\rho = N(\Delta\rho) = \{(g,\sigma) \mid g \in NH,\ \sigma\rho(h)\sigma^{-1} = \rho(ghg^{-1}) \text{ for } h \in H\}$$

$$W\rho = W(\Delta\rho) = N\rho/\Delta\rho$$

$$M\rho = \{g \mid (g,\sigma) \in N\rho \text{ for some } \sigma \in \Pi\} \subset NH$$

$$V\rho = M\rho/H \subset WH$$

$$\Pi^{\rho} = \{\sigma \mid \sigma\rho(h) = \rho(h)\sigma \quad \text{for} \quad h \in H\}$$

Taking $g = e$, we see that Π^{ρ} is a normal subgroup of $N\rho$ with quotient $M\rho$ and a normal subgroup of $W\rho$ with quotient $V\rho$.

DEFINITION 4. Consider the set $R(H,\Pi)$ of Π-conjugacy classes of homomorphisms $\rho\colon H \to \Pi$, $H \subset G$. Define an action of WH on $R(H,\Pi)$ by letting $\bar{g}(\rho)$ be $(g\rho)$, where $g \in NH$ with image $\bar{g} \in WH$ and where $(g\rho)(h) = \rho(g^{-1}hg)$. It is trivial to check that this is well-defined. Observe that the isotropy group of (ρ) is $V\rho$.

PROPOSITION 5. For any subgroup H of G,

$$EWH \times_{WH} B(G,\Pi)^{H} = \coprod_{[(\rho)]} BW\rho,$$

where the union is taken over one ρ in each WH-orbit $[(\rho)]$ of Π-conjugacy classes (ρ) of homomorphisms $H \to \Pi$.

PROOF. Recall that a principal (G,Π)-bundle $p\colon E \to B$ is a principal Π-bundle and a G-map such that the actions of G and Π on E commute. We write the G action on the left and the Π action on the right. Let $p\colon E(G,\Pi) \to B(G,\Pi)$ be a universal principal (G,Π)-bundle. According to Lashof [5], p is characterized by the assertion that the $\Delta\rho$-fixed point space, $E(G,\Pi)^{\rho}$, is non-empty and contractible for all $H \subset G$ and $\rho\colon H \to \Pi$. If $y \in E(G,\Pi)^{\rho}$ and $\sigma \in \Pi$, then $y\sigma$ is fixed by the σ-conjugate $\rho\sigma$ of ρ. In particular, Π^{ρ} acts freely on the contractible space $E(G,\Pi)^{\rho}$, hence we may set

$$B\Pi^{\rho} = E(G,\Pi)^{\rho}/\Pi^{\rho}.$$

It is easily seen that $B(G,\Pi)^{H}$ is the disjoint union over $(\rho) \in R(H,\Pi)$ of the spaces $B\Pi^{\rho}$ and that $p^{-1}(B\Pi^{\rho})$ is the disjoint union of the $E(G,\Pi)^{\rho\sigma}$ as $\rho\sigma$ ranges over the distinct Π-conjugates of ρ. Under the natural action of WH on $B(G,\Pi)^{H}$, $V\rho$ fixes the space $B\Pi^{\rho}$ and

$$B(G,\Pi)^{H} = \coprod_{(\rho)} B\Pi^{\rho} \cong \coprod_{[(\rho)]} WH \times_{V\rho} B\Pi^{\rho}.$$

Therefore

$$EWH \times_{WH} B(G,\Pi)^{H} \cong \coprod_{[(\rho)]} EWH \times_{V\rho} B\Pi^{\rho}.$$

Let $W\rho$ act on EWH via the projection $W\rho \to V\rho \subset WH$ and have its natural action on $E(G,\Pi)^{\rho}$. It is easily checked that the diagonal action of $W\rho$ on $EWH \times E(G,\Pi)^{\rho}$ is free, and we may therefore set

$$EW\rho = EWH \times E(G,\Pi)^{\rho} \quad \text{and} \quad BW\rho = EW\rho/W\rho.$$

The projection $p\colon E(G,\Pi)^{\rho} \to B\Pi^{\rho}$ induces a homeomorphism $BW\rho \cong EWH\times_{V\rho}B\Pi^{\rho}$, and this proves the result.

We retain the notations of the proof in the following result.

PROPOSITION 6. When $Y = B(G,\Pi)$, the restriction ξ_{ρ} of ξ_H to $\pi_*BW\rho$ is the composite

$$\pi_*BW\rho \xrightarrow{\zeta} \pi_*^G BW\rho \xrightarrow{\tau} \pi_*^G\tilde{B}\rho \xrightarrow{\mu_*} \pi_*^G B(G,\Pi).$$

Here

$$\tilde{B}\rho = G\times_{M\rho}(EWH \times B\Pi^{\rho}) \cong [(G \times \Pi)\times_{N\rho}EW\rho]/\Pi,$$

τ is the equivariant transfer associated to the natural G-cover $\pi\colon \tilde{B}\rho \to BW\rho$ with fibre G/H, and μ is the classifying G-map of the natural principal (G,Π)-bundle $p\colon \tilde{E}\rho \to \tilde{B}\rho$, where

$$\tilde{E}\rho = G\times_{M\rho}(EWH \times p^{-1}B\Pi^{\rho}) \cong (G \times \Pi)\times_{N\rho}EW\rho.$$

PROOF. Notice that $\pi\colon \tilde{B}\rho \to BW\rho$ is a restriction of

$$\pi\colon G\times_{NH}(EWH \times B(G,\Pi)^H) \to EWH\times_{WH}B(G,\Pi)^H$$

and that the following squares are pullbacks:

$$\begin{array}{ccccc} G\times_{M\rho}(EWH\times p^{-1}B\Pi^{\rho}) & \xrightarrow{i} & G\times_{NH}(EWH\times E(G,\Pi)) \cong (G\times_{NH}EWH)\times E(G,\Pi) & \xrightarrow{\varepsilon} & E(G,\Pi) \\ \downarrow & & \downarrow \qquad\qquad \downarrow & & \downarrow \\ G\times_{M\rho}(EWH\times B\Pi^{\rho}) & \xrightarrow{i} & G\times_{NH}(EWH\times B(G,\Pi)) \cong (G\times_{NH}EWH)\times B(G,\Pi) & \xrightarrow{\varepsilon} & B(G,\Pi) \end{array}$$

The vertical arrows are all induced by p. The conclusion follows by a simple diagram chase.

REMARKS 7. (i) When $\rho\colon H \to \Pi$ is the trivial homomorphism, $\Pi^{\rho} = \Pi$ and $W\rho = WH \times \Pi$. In particular, with $H = e$, we see that $B(G,\Pi) = B\Pi$ as a nonequivariant space and, with $H = G$, we see that there is an inclusion $\zeta\colon B\Pi \to B(G,\Pi)^G$ whose composite with the inclusion of $B(G,\Pi)^G$ in $B(G,\Pi)$ is a nonequivariant homotopy equivalence. It follows that $\Sigma_G^{\infty}B(G,\Pi)^+$ is a split G-spectrum in the sense of [11, Remarks 11] with splitting map

$$\zeta\colon \Sigma^\infty B\Pi^+ = S \wedge B\Pi^+ \xrightarrow{\zeta\wedge\zeta} (S_G)^G \wedge [B(G,\Pi)^+]^G \to [\Sigma_G^\infty B(G,\Pi)^+]^G.$$

The unlabeled arrow is an instance of the natural map $(k_G)^G \wedge X^G \to (k_G \wedge X)^G$ for G-spectra k_G and based G-spaces X (which need not be an equivalence; see [8]).

(ii) Up to G-homotopy, $B(G,\Pi)$ is a product - preserving functor of Π. If Π is Abelian, this implies that $B(G,\Pi)$ is a Hopf G-space. In fact, for any compact Lie group G, the topological Abelian G-group $F(EG^+, B\Pi)$ is here a model for $B(G,\Pi)$; see Lashof, May, and Segal [6]. From this, we easily deduce that $\Sigma_G^\infty B(G,\Pi)^+$ is a split ring G-spectrum and thus that [11, Proposition 15] applies (its finite generation and $\lim^1$ vanishing hypotheses also being easy to check).

The various homomorphisms of which the ξ_H for general Y and their restrictions ξ_ρ for $Y = B(G,\Pi)$ are composites are all induced by maps in the stable category, and we also write ξ_H and

$$\xi_\rho\colon \Sigma^\infty BW\rho^+ \to [\Sigma_G^\infty B(G,\Pi)^+]^G$$

for the resulting composite maps. Thus the wedge sum of the ξ_ρ is an equivalence (for G finite)

$$(3) \qquad \xi\colon \bigvee_{(H)} \bigvee_{[(\rho)]} \Sigma^\infty BW\rho^+ \to [\Sigma_G^\infty B(G,\Pi)^+]^G.$$

We can now compute the composite of ξ and the map ε of (2) in nonequivariant terms.

<u>THEOREM 8</u>. The adjoint of $\varepsilon\circ\xi_\rho$ is the following composite:

$$\Sigma^\infty BG^+ \wedge \Sigma^\infty BW\rho^+ \cong \Sigma^\infty (BG \times BW\rho)^+ \xrightarrow{\tau} \Sigma^\infty B\rho^+ \xrightarrow{\Sigma^\infty \mu} \Sigma^\infty B\Pi^+.$$

Here $B\rho = EG\times_G \tilde{B}\rho$, τ is the transfer associated to the natural cover $B\rho \to BG \times BW\rho$, and μ is the classifying map of the natural principal Π-bundle $p\colon E\rho \to B\rho$, where $E\rho = EG\times_G \tilde{E}\rho$.

<u>PROOF</u>: Passage to orbits over G gives maps of covers

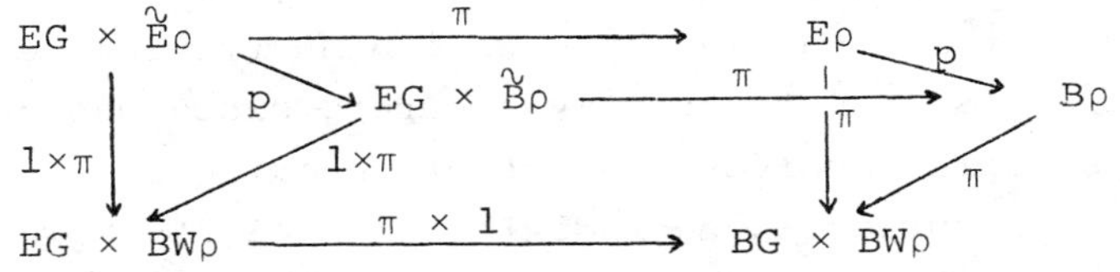

in which the top rectangle is a map of principal Π-bundles. With

G acting trivially on the right hand triangle, we may view these as maps of G-covers in which the top rectangle is a map of principal (G,Π)-bundles. By the naturality of the equivariant transfer and its behavior on products, this implies the following commutative diagram of G-spectra:

$$\begin{array}{ccccc}
EG^+ \wedge \Sigma^\infty_G BW\rho^+ & \cong & \Sigma^\infty_G(EG \times BW\rho)^+ & \xrightarrow{\Sigma^\infty_G(\pi\times 1)} & \Sigma^\infty_G(BG \times BW\rho)^+ \\
\Big\downarrow{\scriptstyle 1\wedge\tau} & & \Big\downarrow{\scriptstyle \tau} & & \Big\downarrow{\scriptstyle \tau} \\
EG^+ \wedge \Sigma^\infty_G \tilde{B}\rho & \cong & \Sigma^\infty_G(EG \times \tilde{B}\rho)^+ & \xrightarrow{\Sigma^\infty_G \pi} & \Sigma^\infty_G B\rho^+
\end{array}$$

Direct inspection of definitions shows that the nonequivariant transfer of a cover $\pi\colon E \to B$ and the equivariant transfer of π regarded as a G-trivial G-cover are related by the commutative diagram

$$\begin{array}{ccc}
\Sigma^\infty B^+ & \xrightarrow{\zeta} & (\Sigma^\infty_G B^+)^G \\
\Big\downarrow{\scriptstyle \tau} & & \Big\downarrow{\scriptstyle \tau^G} \\
\Sigma^\infty E^+ & \xrightarrow{\zeta} & (\Sigma^\infty_G E^+)^G
\end{array}$$

With these preliminaries, we turn to the proof. By Proposition 6 and the definition of ε in [11], we see that the following diagram of spectra must be shown to commute:

$$\begin{array}{ccccc}
\Sigma^\infty BW\rho^+ & \xrightarrow{\mathrm{adj}(\Sigma^\infty \mu\circ\tau)} & F(BG^+, \Sigma^\infty B\Pi^+) & \xrightarrow{F(1,\zeta)} & F(BG^+, \Sigma^\infty_G B(G,\Pi)^+)^G \\
\Big\downarrow{\scriptstyle \zeta} & & \Big\uparrow{\scriptstyle \varepsilon} & \nwarrow{\scriptstyle \cong} & \Big\downarrow{\scriptstyle F(\pi,1)^G} \\
(\Sigma^\infty_G BW\rho^+)^G & \xrightarrow{(\Sigma^\infty \mu\circ\tau)^G} & [\Sigma^\infty_G B(G,\Pi)^+]^G & \xrightarrow{F(\varepsilon,1)} & F(EG^+, \Sigma^\infty_G B(G,\Pi)^+)^G
\end{array}$$

At the top right, we have used that $F(X,E^G) = F(X,E)^G$ for a space X and G-spectrum E, $F(X,E)$ being the function G-spectrum; the equality is immediate from the definitions in [8]. The composite of $F(\pi,1)^G$ and $F(1,\zeta)$ is an equivalence by [11, Lemma 12] and, by definition, ε is the composite of $F(\varepsilon,1)$ and the inverse of this equivalence. The fixed point spectrum functor from G-spectra to spectra has a left adjoint, which we shall denote by Φ. It may be viewed as assigning trivial G action to a spectrum. (This is not strictly accurate since passage to fixed point spectra involves neglect of indexing representations, hence Φ must resurrect such indexing; see [8].) Moreover, Φ commutes with smash products. On passage to adjoints, the outer rectangle of the previous diagram transforms to the outer part of the following diagram of G-spectra; the maps $\tilde{\zeta}$ are adjoint to ζ and are equivalences.

$$\begin{array}{ccccccc}
EG^+ \wedge \Phi\Sigma^\infty BW\rho^+ & \xrightarrow{\pi\wedge 1} & BG^+ \wedge \Phi\Sigma^\infty BW\rho^+ \cong \Phi\Sigma^\infty(BG\times BW\rho)^+ & \xrightarrow{\Phi\tau} & \Phi\Sigma^\infty B\rho^+ & & \\
\downarrow{\scriptstyle 1\wedge\tilde{\zeta}} & & \downarrow{\scriptstyle 1\wedge\tilde{\zeta}} \qquad\qquad \downarrow{\scriptstyle\tilde{\zeta}} & & \downarrow{\scriptstyle\tilde{\zeta}} & \searrow{\scriptstyle\Phi\Sigma^\infty\mu} & \\
EG^+ \wedge \Sigma^\infty_G BW\rho^+ & \xrightarrow{\pi\wedge 1} & BG^+ \wedge \Sigma^\infty_G BW\rho^+ \cong \Sigma^\infty_G(BG\times BW\rho)^+ & \xrightarrow{\tau} & \Sigma^\infty_G B\rho^+ & & \Phi\Sigma^\infty B\Pi^+ \\
\downarrow{\scriptstyle 1\wedge\tau} & & \nearrow{\scriptstyle\Sigma^\infty_G\pi} & & & \searrow{\scriptstyle\Sigma^\infty_G\mu} & \downarrow{\scriptstyle\tilde{\zeta}} \\
EG^+ \wedge \Sigma^\infty_G \tilde{B}\rho^+ \cong \Sigma^\infty_G(EG\times\tilde{B}\rho)^+ & \xrightarrow{\Sigma^\infty_G(1\times\mu)} & \Sigma^\infty_G(EG\times B(G,\Pi))^+ & \xrightarrow{\Sigma^\infty_G(\varepsilon\times 1)} & & & \Sigma^\infty_G B(G,\Pi)^+
\end{array}$$

The parts of the diagram involving transfer maps have already been observed to commute and the parts involving classifying maps commute in view of the first diagram of the proof.

REMARKS 9. (i) Everything we have done so far is valid for G finite and Π any compact Lie group; compare Remark 2.
(ii) Unraveling notations, we see that

$$E\rho = EG\times_G[(G\times\Pi)\times_{N\rho}EW\rho]$$

and $B\rho = E\rho/\Pi$. There are alternative descriptions, such as

$$E\rho \cong (G\times\Pi/\Delta\rho)\times_{G\times W\rho}(EG\times EW\rho),$$

$W\rho$ being regarded as the group of automorphisms of the transitive $(G\times\Pi)$-set $G\times\Pi/\Delta\rho$.
(iii) To specialize to the situation of the ordinary Segal conjecture, let Π be the trivial group and write

$$CH = EG\times_G(G\times_{NH}EWH) \cong EG\times_{NH}EWH.$$

(Thus $CH = E\rho = B\rho$ for the trivial homomorphism ρ.) Here μ is just the trivial map $CH \to pt$, hence $\Sigma^\infty\mu^+ : \Sigma^\infty CH^+ \to S$ is the unit $1 \in \pi^0(CH)$ of the stable cohomotopy ring of CH. We conclude that $\varepsilon\circ\xi$ is the sum over (H) of the adjoints

$$\Sigma^\infty BWH^+ \to F(BG^+,S) = D(BG^+)$$

of the elements $\tau(1) \in \pi^0(BG\times BWH)$.
(iv) With the description in (iii), the work of Lin, Gunawardena, and Ravenel shows that $\varepsilon\circ\xi$ induces an equivalence upon completion at p when G is a cyclic p-group. See Gunawardena [2,1.5.3] and the appendix by Miller to Ravenel's paper [13]. When G is an elementary Abelian 2-group, unpublished work of Adams, Gunawardena, and Miller gives the same conclusion. Of course, by Theorem 8 and the results of [11], we are entitled to conclude the

equivariant form of the Segal conjecture for all finite groups all of whose p-Sylow subgroups are either cyclic or, if $p = 2$, elementary Abelian. Maunder's paper [10] is devoted to a different proof of this passage from the nonequivariant to the equivariant version of the Segal conjecture in the special case $G = Z_2$.

§2. THE PROOFS OF THEOREMS A AND B

By definition, $BW\rho$ is just $BW(\Delta\rho)$ for the subgroup $\Delta\rho$ of $G\times\Pi$. Thus $[\Sigma_G^\infty B(G,\Pi)^+]^G$ is a wedge summand of $(S_{G\times\Pi})^{G\times\Pi}$. It is simple group theoretical bookkeeping to identify the complementary wedge summand.

<u>PROPOSITION 10</u>. There is an identification

$$\bigvee_{(K)} \Sigma^\infty BWK^+ = \bigvee_{(\Lambda)} \bigvee_{(H)} \bigvee_{[(\rho)]} \Sigma^\infty BW\rho^+,$$

where the wedge on the left is taken over one K in each conjugacy class of subgroups of $G\times\Pi$ and the wedge on the right is taken over one Λ in each conjugacy class of subgroups of Π, one H in each conjugacy class of subgroups of G, and one ρ in each WH-orbit of $W\Lambda$-conjugacy classes of homomorphisms $H \to W\Lambda$. Therefore (for G finite)

$$(S_{G\times\Pi})^{G\times\Pi} \simeq \bigvee_{(\Lambda)} [\Sigma_G^\infty B(G,W\Lambda)^+]^G.$$

<u>PROOF</u>. The second statement will follow from the equivalences of (3). For the first statement, associate $\Lambda \subset \Pi$, $H \subset G$, and $\rho\colon H \to W\Lambda$ to a subgroup $K \subset G\times\Pi$ by

$$\Lambda = \{\lambda \mid (e,\lambda) \in K\}$$

$$H = \{h \mid (h,\sigma) \in K \quad \text{for some} \quad \sigma \in \Pi\},$$

and, with the observation that $\sigma \in N\Lambda$ if $(h,\sigma) \in K$,

$$\rho(h) = \sigma\Lambda, \quad \text{where} \quad (h,\sigma) \in K.$$

Conversely, associate a subgroup $K \subset G\times\Pi$ to a triple (Λ,H,ρ) by

$$K = \{(h,\sigma) \mid h \in H \quad \text{and} \quad \sigma \in \rho(h)\},$$

where the cosets $\rho(h)$ are regarded as subsets of Π. It is easily checked that these are inverse bijective correspondences such that K runs over a set of conjugacy class representatives as (Λ,H,ρ) runs over a set of representatives as in the statement. Another easy check shows that

$$NK = \{(g,\sigma) \mid g \in NH,\ \sigma \in N\Lambda,\ \sigma\rho(h)\sigma^{-1} = \rho(ghg^{-1})\}.$$

Visibly the quotient homomorphism $N\Lambda \to W\Lambda$ induces an isomorphism

$$WK = NK/K \to N\rho/\Delta\rho = W\rho.$$

REMARK 11. What really seems to be going on here is that there is an equivalence of G-spectra

$$(S_{G\times\Pi})^{\Pi} \simeq \bigvee_{(\Lambda)} \Sigma^{\infty}_{G} B(G,W\Lambda)^{+}.$$

The second author has an outline proof which, if correct, will appear in [3] since it depends on equivariant infinite loop space theory for the construction of a map.

We use the previous proposition to relate the maps ε of (1) and (2). Henceforward, G and Π are both to be finite.

PROPOSITION 12. The following diagram commutes:

$$\begin{array}{ccccc} (S_{G\times\Pi})^{G\times\Pi} & \cong & \bigvee_{(\Lambda)} [\Sigma^{\infty}_{G} B(G,W\Lambda)^{+}]^{G} & \xrightarrow{\vee\varepsilon} & \bigvee_{(\Lambda)} F(BG^{+},\Sigma^{\infty} BW\Lambda^{+}) \\ \downarrow{\scriptstyle\varepsilon} & & & & \wr\wr \\ F((BG\times B\Pi)^{+},S) & \cong & F(BG^{+},F(B\Pi^{+},S)) & \xleftarrow{F(1,\varepsilon)} & F(BG^{+},(S_{\Pi})^{\Pi}) \end{array}$$

PROOF. By Theorems 1 and 8, the following commutative diagram implies the result upon passage to adjoints. Let K correspond to (Λ,H,ρ) as in the previous proof.

$$\begin{array}{ccccccc} B\Pi^{+}\wedge\Sigma^{\infty}(BG\times BW\rho)^{+} & & & = & & & \Sigma^{\infty}(BG\times B\Pi\times BWK)^{+} \\ \downarrow{\scriptstyle\tau} & & & & & & \downarrow{\scriptstyle\tau} \\ B\Pi^{+}\wedge\Sigma^{\infty}B\rho^{+} & \cong & \Sigma^{\infty}(B\Pi\times B\rho)^{+} & \xrightarrow{\tau} & \Sigma^{\infty}(E\Pi\times_{N\Lambda}E\rho)^{+} & \cong & \Sigma^{\infty}CK^{+} \\ \downarrow{\scriptstyle 1\wedge\Sigma^{\infty}\mu} & & \downarrow{\scriptstyle\Sigma^{\infty}(1\times\mu)} & & \downarrow{\scriptstyle\Sigma^{\infty}(1\times\tilde{\mu})} & & \downarrow{\scriptstyle\Sigma^{\infty}\mu} \\ B\Pi^{+}\wedge\Sigma^{\infty}BW\Lambda^{+} & \cong & \Sigma^{\infty}(B\Pi\times BW\Lambda)^{+} & \xrightarrow{\tau} & \Sigma^{\infty}C\Lambda^{+} & \xrightarrow{\Sigma^{\infty}\mu} & S \end{array}$$

Here $C\Lambda = E\Pi\times_{N\Lambda}EW\Lambda$ and $\tilde{\mu}\colon E\rho \to EW\Lambda$ covers μ. The bottom middle square commutes by the naturality of transfer. We easily obtain a homeomorphism $E\Pi\times_{N\Lambda}E\rho \cong CK$ from

$$E\rho = EG\times_{G}[(G\times W\Lambda)\times_{N\rho}EW\rho],\quad CK = (EG\times E\Pi)\times_{NK}EWK,$$

and $W\rho \cong WK$. The bottom right square commutes trivially and the top rectangle commutes by the transitivity of transfer.

Of course, $W(e) = \Pi$ while all other $W\Lambda$ are proper subquotients of Π. To prove Theorem A when G and Π are p-groups, we proceed by induction on the order of Π. Upon completing the diagram of the previous proposition at p, we find in this case that the Segal conjecture for the groups Π and $G\times\Pi$ and the completion conjecture for the G-spectra $\Sigma_G^\infty B(G,W\Lambda)^+$, $\Lambda \neq e$, implies the completion conjecture for $\Sigma_G^\infty B(G,\Pi)^+$.

Because of the different topologies involved, this argument fails hopelessly when G and Π are not p-groups. We use the following observation to prove Theorem A when G is but Π is not a p-group.

LEMMA 13. The following diagram commutes for $i\colon \Lambda \subset \Pi$:

$$\begin{array}{ccccc}
[\Sigma_G^\infty B(G,\Pi)^+]^G & \xrightarrow{\tau^G} & [\Sigma_G^\infty B(G,\Lambda)^+]^G & \xrightarrow{(\Sigma_G^\infty Bi)^G} & [\Sigma_G^\infty B(G,\Pi)^+]^G \\
\downarrow \varepsilon & & \downarrow \varepsilon & & \downarrow \varepsilon \\
F(BG^+,\Sigma^\infty B\Pi^+) & \xrightarrow{F(1,\tau)} & F(BG^+,\Sigma^\infty B\Lambda^+) & \xrightarrow{F(1,\Sigma^\infty Bi)} & F(BG^+,\Sigma^\infty B\Pi^+)
\end{array}$$

PROOF. Clearly $E(G,\Pi)/\Lambda$ is a model for $B(G,\Lambda)$, hence we have a G-cover $Bi\colon B(G,\Lambda) \to B(G,\Pi)$ with fibre Π/Λ. The right square is an obvious naturality diagram. The left square relates the equivariant transfer of Bi to the nonequivariant transfer of $Bi\colon B\Lambda \to B\Pi$ and commutes by naturality and the third diagram in the proof of Theorem 8.

Now let G be a p-group, let Λ be a p-Sylow subgroup of Π, and complete the diagram of the lemma at p. The bottom composite is then clearly an equivalence, and Theorem B will imply that the top composite is also an equivalence. It follows that ε for Π is an equivalance if ε for Λ is an equivalence, and this completes the proof of Theorem A.

REMARK 14. Under the hypotheses of the previous paragraph, we may consider $(\Sigma_G^\infty Bi)^G$ as a map

$$\bigvee_{(H)} \bigvee_{[(\sigma)]} \Sigma^\infty BW\sigma^+ \to \bigvee_{(H)} \bigvee_{[(\rho)]} \Sigma^\infty BW\rho^+,$$

$\sigma\colon H \to \Lambda$ and $\rho\colon H \to \Pi$, and Theorem B implies that this map is a p-local retraction. Clearly the wedge summand indexed on $[(\sigma)]$ maps to the wedge summand indexed on $[(i\sigma)]$, but many $[(\sigma)]$

may map to the same $[(\rho)]$. Glauberman has proven that the number of $[(\sigma)]$ which map to a given $[(\rho)]$ and are such that $W(\sigma)$ is conjugate to a p-Sylow subgroup of $W(\rho)$ is prime to p. This provides a plausibility argument for the cited retraction property.

It remains to prove Theorem B. This is an easy exercise in the use of the ordinary RO(G)-graded cohomology theories on G-spectra which we introduced in [7] and will study in detail in [9].

A map $f: X \to Y$ of connective G-spectra induces an equivalence upon localization at p if and only if

$$f^*: H_G^*(Y;M) \to H_G^*(X;M)$$

is an isomorphism for all p-local Mackey functors M. Compare [12], where the analogous G-space level statement is proven.

Let $\pi: E \to B$ be a G-cover with fibre F and consider the composite

$$f: \Sigma_G^\infty B^+ \xrightarrow{\tau} \Sigma_G^\infty E^+ \xrightarrow{\Sigma_G^\infty \pi} \Sigma_G^\infty B^+.$$

If k_G^* is a ring valued RO(G)-graded cohomology theory and m_G^* is a k_G^*-module valued RO(G)-graded cohomology theory, then $f^*: m_G^*(B) \to m_G^*(B)$ is multiplication by $\tau(1) \in k_G^0(B)$, where $1 \in k_G^0(E)$ is the identity element.

For any Mackey functor M, $H_G^*(?;M)$ is module-valued over $H_G^*(?;\underline{A})$, where $\underline{A}$ is the Burnside ring Green functor, $\underline{A}(G/H) = A(H)$. The same holds for p-local M and the localization $\underline{A}_p$ of $\underline{A}$ at p. To prove Theorem B, it suffices to show that $\tau(1) \in H_G^0(B;\underline{A}_p)$ is a unit when G is a p-group and $|F|$ is prime to p.

By G-CW approximation and an easy colimit argument, we may assume without loss of generality that B is a G-CW complex with finite 0-skeleton B^0. The inclusion of B^0 in B induces a monomorphism $H_G^0(B;M) \to H_G^0(B^0;M)$ for any M. Our finiteness assumption ensures that $H_G^0(B^0;\underline{A}_p)$ is an integral extension of $H_G^0(B^0;\underline{A}_p)$. Thus it suffices to prove that $\tau(1) \in H_G^0(B^0;\underline{A}_p)$ is a unit, where τ is the transfer associated to the restriction $\pi: \pi^{-1}(B^0) \to B^0$.

Since B^0 is a (finite) disjoint union of orbits G/H, $\pi^{-1}(B^0) \to B^0$ breaks up into a disjoint union of G-covers $\pi^{-1}(G/H) \to G/H$ with fibre F. It suffices to show that

$$\tau(1) \in H^0_G(G/H;\underline{A}_p) = A(H)_p$$

is a unit for each orbit G/H. Now $\pi^{-1}(G/H) \to G/H$ has the form $G\times_H F \to G/H$ for some action of H on F. We may regard F as a finite H-set and thus as an element of $A(H)$ via this action. A check of definitions shows that $\tau(1) = F$.

Since H is a p-group, $|F^K| \equiv |F| \not\equiv 0 \mod p$ for $K \subset H$. It follows that the image of F under the natural embedding of rings $\chi\colon A(H)_p \to \underset{(K)}{\times} Z_{(p)}$, $\chi_K(S) = |S^K|$ for an H-set S, is a unit. Since χ is an integral extension, F is a unit in $A(H)_p$ and the proof is complete.

BIBLIOGRAPHY

1. T. tom Dieck. Orbittypen und aquivariante Homologie II. Arch. Math 26 (1975), 650-662.

2. J. H. Gunawardena. Cohomotopy of some classifying spaces. PhD thesis. Cambridge, 1981.

3. H. Hauschild, J. P. May, and S. Waner. Equivariant infinite loop space theory. In preparation.

4. C. Kosniowski. Equivariant cohomology and stable cohomotopy. Math. Ann. 210 (1974), 83-104.

5. R. K. Lashof. Equivariant bundles. Illinois J. Math. To appear.

6. R. K. Lashof, J.P. May, and G.B. Segal. Equivariant bundles with Abelian structure group. To appear.

7. L. G. Lewis, J. P. May, and J. E. McClure. Ordinary RO(G)-graded cohomology. Bull. Amer. Math. Soc. 4 (1981) 208-212.

8. L. G. Lewis, J. P. May, J. E. McClure, and M. Steinberger. Equivariant stable homotopy theory. Springer Lecture Notes in Mathematics. In preparation.

9. L. G. Lewis, J. P. May, J. E. McClure, and S. Waner. Equivariant cohomology theory. In preparation.

10. C. R. F. Maunder. On Segal's Burnside ring conjecture for the group Z_2. Preprint.

11. J. P. May and J. E. McClure. A reduction of the Segal conjecture. These proceedings.

12. J. P. May, J. E. McClure, and G. Triantafillou. Equivariant localization. Bull. London Math. Soc. To appear.

13. D. Ravenel. The Segal conjecture for cyclic groups and its consequences. Preprint.

14. G. Segal. Equivariant stable homotopy theory. Actes Congrès Intern. Math. 1970 Tome 2, 59-63.

15. K. Wirthmuller. Equivariant homology and duality. Manuscripta Math. 11 (1974), 373-390.

Canadian Mathematical Society
Conference Proceedings
Volume 2, Part 2 (1982)

On Generalization of Conner-Miller Classes to Z_p-actions I.

Tsau Y. Lin

ABSTRACT. Conner-Miller classes for free involutions are generalized to all cyclic group actions (of prime orders). Vanishing theorems for these classes are established. As applications some Borsuk-Ulam type theorems were obtained.

§ Introduction

This paper is motivated by a call from the joint work of P. E. Conner and E. Y. Miller [4].

In studying equivariant self-intersection, P. E. Conner and E. Y. Miller introduced a new type of characteristic classes for free involutions. These new characteristic classes (Conner-Miller classes) captured quite interesting properties on involutions [4] [6] [7]. A natural question raised in [4] is whether there are analogous theories for general prime p. This is one of the first papers to the positive answer of this question. In defining Conner-Miller classes, they rely on the Bredon operation [1] and the Gysin sequence of a 2-fold covering. There seems to be no Gysin type sequence of a p-fold covering suitable for us to define the characteristic classes. However, we have Proposition 2.2 to bypass the difficulty and define the new characteristic classes for any prime p. These new types of characteristic classes are closely related to Bredon-Hattori classes. (See [4] for p = 2 case). We should like to point out here that Hattori [5] had defined the Bredon-

---

1980 Mathematics Subject Classification. 57E99, 57D20.
Partially Supported by National Science Foundation.
© 1982 American Mathematical Society
0731-1036/82/0000-0466/$06.25

Hattori class for odd p from different angles. We will show in the next paper that Hattori's classes agree with our version. The main results of this paper are to establish some vanishing theorems for free Z_p-actions which preserve certain structure introduced by Browder [2]. As a simple application, by combining our new classes with Hattori classes, we obtain a generalization of Munkholm's Bosurk-Ulam type theorem [9]. Part of the present paper was announced in [8].

Most of the results presented here were developed while this author was visiting LSU (1979-80). We should like to express our sincere thanks to Professor P. E. Conner for his kind guidence and encouragement.

§2 The structure of $H_G^*(X^p)$ and Bredon operations.

2.1 The structure of $H_G^*(X^p)$.

In this section, we will briefly review Steenrod's external reduced power P [11, Ch VII] and the cohomological structure of extended p-th power, $H_G^*(X^p)$. Throughout the whole paper, all the cohomology shall use Z/p coefficients and will be suppressed from notations.

Let G be a cyclic group of order p. Let X^p be the p-fold Cartesian product on which G acts by permuting the factors cyclically. Let $EG = S^\infty$ be acted upon by G via complex multiplication on $S^\infty = \lim S^{2n-1}$. Then the orbit space EG/G = BG is the classifying space. The cohomology $H_G^*(BG)$ is the coefficient of the equivariant cohomology i.e.,

$$H^*(BG) = H_G^*(pt) = Z/p\,[b] \times \Lambda_p\,[a], \text{ if } p = \text{odd}$$

$$H^*(BG) = H_G^*(pt) = Z/2\,[c], \text{ if } p = 2$$

where deg a = deg c = 1, deg b = 2 and b = βa where β is the Bockstein operator.

We need to use many monomials in a, b (or in c). We shall denote the monomial of degree k, by w_k, e.g.,

$w_{2h} = b^h$, $w_{2h+1} = ab^h$ (or $w_i = c^i$)

Let X be a G-space. Then,

$$\nu : X \longrightarrow X/G$$

is the natural projection. If X is a free G-space, then there is a "umkehr homomorphism"

$$tr: H^*(X) \longrightarrow H^*(X/G).$$

We shall call it transfer and denote it by tr.

Since EG is contractible, the projection

$$\nu : X^p \times EG \longrightarrow X^p \times_G EG$$

gives rises to $\nu^*: H^*(X^p x_G EG) \longrightarrow H^*(X^p)$ and the transfer

$$tr: H^*(X^p) \longrightarrow H^*_G(X^p) = H^*(X^p x_G EG).$$

According to [11, Ch VII], there is an external cohomology operation

$$P: H^q(X) \longrightarrow H_G^{pq}(X^p)$$

satisfying the following properties.

(a) $P(f^*\alpha) = (f^p x_G 1)^* P(\alpha)$

(b) $\nu^* P(\alpha) = \underbrace{\alpha \times \alpha \times \ldots \times \alpha}_{\text{p-factors}}$

(c) $P(\alpha + \beta) = P(\alpha) + P(\beta) + tr\,(\sum_{k=1}^{p-1} \underbrace{\alpha x..x\alpha}_{\text{k-factors}} \times \overbrace{\beta x..x\beta}^{\text{p-k factors}})$

(d) $P(\alpha\beta) = P(\alpha)\, P(\beta)$

(e) $(dx_G 1)^* P(\alpha) = \sum_{i=0}^{k} Sq^i \alpha \times W_{k-i}$, if $p = 2$

$$(d\, x_G 1)^* P(\alpha) = \sum_{i=0}^{[\frac{q}{2}]} (-1)^s (r!)^{-q} p^i(\alpha) \times W_{(q-2i)(p-1)}$$

$$+\sum_{i=0}^{[\frac{q}{2}]} (-1)^s (r!)^{-q} \beta p^i(\alpha) \times W_{(q-2i)(p-1)-1}, \text{ if } p=\text{odd}$$

where $r = \frac{p-1}{2}$, $s = i + r\,(q^2 + q)/2$, and d is the diagonal map $d: X \longrightarrow X^p$ defined by $d(x) = \underbrace{(x, x, \ldots x)}_{\text{p-factor}}$

Next, we will give the structure of $H^*_G(X^p)$.

2.1 THEOREM. Let $\{x_i\}$ be a homogeneous basis for $H^*_G(X)$. Then,

$$H^*_G(X^p) = F + tr(H^*(X^p))$$

where F is the free $H^*_G(pt)$ - module with basis $\{P(X_i)\}$ and P is Steenrod's external cohomology operation [11, pp 100].

Note that $tr(H^*(X^p))$ is a Z/p- vector space with basis $tr(x_{i_1} \times x_{i_2} \times \dots \times x_{i_p})$, $i_1 \leq i_2 \leq \dots\ i_p$ and $i_1 < i_p$.

proof: See [5], [11].

It is clear that one can always regard the equivariant cohomology $H_G^*(X)$ as a module over $H_G^*(pt)$. The next proposition gives some information about the action of $H_G^*(pt)$ on the transfer.

2.2 PROPOSITION. The image $tr(H^*(X^p))$ of the transfer is annhilated by a, and b [or c if p = 2].

proof: Consider the following commutative diagram

$$\begin{array}{ccc} EG & \xleftarrow{\quad \pi \quad} & X^p \times EG \\ \Big\downarrow \nu_1 & & \Big\downarrow \nu_2 \\ BG & \xleftarrow{\quad \Pi \quad} & X^p \times_G EG \end{array}$$

where π is the projection on the second factor, ν_1 is the natural projection to the orbit space and Π is the induced map of π on the orbit spaces.

Now, let $x \in H^*(X^p)$ and $y \in H^+(BG)$, $+ > 0$.

then,

$$tr(x) \cup \Pi^* y = tr(x \cup \nu_2^* \Pi^*(y)) = tr(x \cup \pi^* \nu_1^*(y))$$

$= tr(x \cup o) = 0$, since $H^+(EG) = 0$, $+ > 0$. That is $tr(x)$ is annhilated by any element in $H^+(BG)$, in particular by a, b[or c].

Next, we shall quote a result from [10], which gives the relations among p^i, Steenrod's reduced p-th power and P, the external operation.

2.3 THEOREM. For odd prime p, let p^i be Steenrod's reduced pith power. Then

$$\text{(i) } p^k P\alpha = \sum_i \left\{ {(q-2i)r \atop k-pi} \right\} W_{2(k-pi)(p-1)} P(p^i\alpha)$$

$$-\mu(q) \sum_i \left\{ {(q-2i)r-1 \atop k-pi-1} \right\} W_{2(k-pi)(p-1)-p} P(\beta p^i \alpha)$$

$$+z$$

$$\text{(ii) } \beta P(\alpha) = z^1$$

where $r = \frac{p-1}{2}$, $\alpha(q) = (r!)^{-1}(-1)^{rq}$, $u \in H^q(X)$ and

z, z^1 are lying in the image of transfer, and w_h is a monomial of a and b with total degree = h.

Remark: If p = 2, then

$$Sq^k P\alpha = \sum_i \left\{ {q-i \atop k-2i} \right\} c^{k-2i} P(Sq^i \alpha) + z$$

2.2 The Bredon operations.

Let (G, X) be a free G-space, G is cyclic of order p. the diagonal action (G, X x EG) has an equivariant embedding

$$e: (G, X \times EG) \longrightarrow (G, \overbrace{X \times X \times \ldots \times X}^{p\text{-factors}} \times EG)$$

$$(x, s) \longrightarrow (x, Tx, T^2x, \ldots, T^{p-1}x, s)$$

where T is the generator of G. This induces an embedding

$$E: X \times_G EG \longrightarrow X^p \times_G EG$$

In addition the equivariant projection

$$X \times EG \longrightarrow X$$

will induce a fibration

$$X \times_G EG \xrightarrow{p} X/G$$

with fibre EG; that is, p is a homotopy equivalence.

2.4 DEFINITION. For $\alpha \in H^q(X)$, let $Q(\alpha) \in H^{pq}(X/G)$ be the unique cohomology class for which

$$p^* Q(\alpha) = E^* P(\alpha).$$

We shall name this proposition the Bredon operation $Q\colon H^q(X) \longrightarrow H^{pq}(X/G)$

Corresponding to the properties of P, we have

2.5 THEOREM. There is associated to a fixed point free periodic transformaton (G, X) an operation

$$Q\colon H^q(X) \longrightarrow H^{pq}(X/G)$$

with the following properties

1. The operation commutes with homomorphisms induced by equivariant maps.

2. $\nu^* Q(\alpha) = \alpha \cdot T^*\alpha \cdot T^{*2}\alpha \cdot \ldots \cdot T^{*(p-1)}\alpha$

3. $Q(\alpha+\beta) = Q(\alpha) + Q(\beta) + \mathrm{Tr}(\sum_{k=0}^{p-\hat{2}} \alpha \cdot T^*\alpha \cdot \ldots \cdot T^{*K}\alpha \cdot T^{*k+1}\beta \ldots \cdot T^{*p-1}\beta)$

4. $Q(\alpha\beta) = Q(\alpha)Q(\beta)$

5. For $\alpha \in H^q(X/G)$

For $p = 2$,

$$Q(\nu^*(\alpha)) = \sum_{j=0}^{q} w_{q-j} Sq^j(\alpha)$$

For p = odd,

$$Q(\nu^*(\alpha)) = \sum_{i=0}^{[\frac{q}{2}]} (-1)^s (r!)^{-q} w_{(q-2i)(p-1)} P^i \alpha$$

$$+\sum_{i=0}^{[\frac{q}{2}]} (-1)^s (r!)^{-q} w_{(q-2i)(p-1)-1} \beta P^i \alpha^{\sim}$$

where $r = \frac{p-1}{2}$, $s = i + \frac{(q+q^2)r}{2}$

2.6 THEOREM.

(i) $$p^k Q(\alpha) = \sum_i \binom{(q-2i)r}{k-pi} w_{2(k-pi)(p-i)} Q(p^i\alpha)$$

$$-\mu(q)\sum_i \binom{(q-2i)r-1}{k-pi-1} w_{2(k-pi)(p-1)-p} Q(\beta p^i\alpha) + z$$

(ii) $$\beta Q(\alpha) = z^1$$

where z, z^1 are in the image of transfer, $\mu(q) = (-1)^{qr}(r!)^{-1}$

proof: This is immediate from 2.3.

§3 Wu-like classes for periodic maps

Let M^n be an n-dimensional manifold. Let G be a cyclic group of order p, and act freely on M. In this section we will introduce some new types of characteristic classes (following Conner-Miller for p = 2) via Bredon operation in the spirit of tangential Wu classes. We shall establish some vanishing theorems about these classes. We shall also define the Hattori classes here and obtain some vanishing theorems by the above results.

Let us recall that

$$Q(\alpha+\beta) = Q(\alpha) + Q(\beta) + z$$

where z is in the image of transfer. Then, from 2.2, we see that the correspondence

$$\alpha \longrightarrow w_j Q(\alpha)$$

is linear if $j > 0$. In fact, we can have more, let $z \in H^s(M/G)$ be a fixed element, then the correspondence

$$\alpha \longrightarrow w_j \cdot z \cdot Q(\alpha)$$

is linear if $j > 0$. With this observation we can define the Wu-like classes.

3.1 DEFINITION. Let M^n be as above. If $\frac{p-1}{p} n < k \leq n$, then, for fixed $z \in H^s(M^n/G)$, the class $\theta_k(z)$ is the unique cohomology class for which

$$\langle \theta_k(z)\cdot\alpha,\ \sigma(M^n)\rangle = \langle w_{pk-(p-1)n-s}\cdot z\cdot Q(\alpha),\ \sigma(M^n/G)\rangle$$

for all $\alpha \in H^{n-k}(M^n)$.

if $n = pk$, then an exceptional class $e_k \in H^k(M^n/G)$ is defined to be the unique cohomology class for which

$$\langle e_k\alpha, \sigma(M^n/G)\rangle = \langle Q(\nu^*(\alpha)),\ \sigma(M^n/G)\rangle$$

for all $\alpha \in H^k(M^n/G)$.

For $p = 2$, this new $\theta_k(z)$ is exactly the same old $\theta_k(z)$

in Lin [7]. For $\theta_k(1)$, we shall, as in [7], simply write as θ_k. It is clear that $\theta_k(0) = 0$ and $\theta_k(w_j) = \theta_k$. In this section, we will establish some vanishing theorems for θ_k when G preserves a certain structure of the manifold. In Lin [7], we have shown that, for $p = 2$, all the generalized CM-classes $\theta_k(z)$ come from some θ_k of other manifolds. For odd p, there seems to be no such theorem.

Let us recall the right action of Steenrod's reduced p-th power operation p^i. According to [3, §6]

$$u \cdot a = \Sigma\, 1 \cdot a' \cdot \chi(a'')\, u, \text{ where } \psi(a) = \Sigma a' \times a''$$

Let $a = p^*$, where $p^* = p^\circ + p^1 + \ldots$, be the total Steenrod reduced p-th power operation, then $1 \cdot p^*$ is the total Wu-class of vector bundle ϕ in the sense [3]. For M^n, we will write $Wu_j(M)$ for the j-th Wu-class of the stable normal bundle

$$1 \cdot p^* = Wu_*(M^n)$$

(Note that $\deg(Wu_j(M^n)) = 2j(p-1)$)

Using Wu_j we can express the right action of p^i on θ_k.

3.2 PROPOSITION. Let $Wu_t = Wu_t(M^n/G)$ be the Wu-class of the orbit space. Then

for p = odd,

$$\sum_{i=0}^{[\frac{j}{p}]} \binom{(q-i)r}{j-pi} \theta_{k-2i(p-1)} p^i - \sum_{i=0}^{[\frac{j}{p}]} \mu(q) \binom{(q-2i)r-1}{j-pi-1} \theta_{k-2i(p-1)-1} \beta p^i$$

$$= \sum_{t=0}^{j} -\left(\frac{[pk-(p-1)n-2j(p-1)]}{2} \atop j-t \right) \theta_k(Wu_t) ,$$

for p = 2,

$$\sum_{i=0}^{[\frac{j}{2}]} \binom{q-i}{j-2i} \theta_{k-i} Sq^i = \sum_{t=0}^{j} \binom{2k-n-j}{j-t} \theta_k(Wu_t)$$

where $2k-n-j>0$ (for $p = 2$); and $q = n-k$, $pk-n-2j(p-1)>0$

$\mu(q) = (r!)^{-1}(-1)^{rq}$ (for p = odd)

Proof: From definition of θ_k, we have

$$(1) \quad <\theta_{k-2i(p-1)}P^i\alpha,\ \sigma(M^n)>$$
$$=<w_{p(k-2i(p-1))-(p-1)n}\ Q(P^i\alpha),\ \sigma(M^n/G)>$$

and

$$(2) \quad <\theta_{k-2i(p-1)-1}\beta P^i\alpha, \sigma(M^n)>=$$
$$=<w_{p(k-2i(p-1)-1)-(p-1)n}\ Q(\beta P^i\alpha).\ \sigma(M^n/G)>$$

multiply (1) by $\binom{(q-2i)r}{j-pi}$, (2) by $-\mu(q)\ \binom{(q-2i)r}{j-pi-1}$

and sum up, where $q = n-k$, $r = \frac{p-1}{2}$, then the right hand side can be reduced to, via 2.6,

$$<w_{pk-(p-1)n-2j(p-1)}P^jQ(\alpha),\ \sigma(M^n/G)>$$

$$=<\sum_{t=0}^{j} Wu_t(M^n/G)\ \chi(P^{j-t})\ w_{pk-(p-1)n-2j(p-1)}Q(\alpha),\ \sigma(M^n/G)>$$

$$=<\sum_{t=0}^{j} Wu_t(M^n/G)\{-\binom{[\frac{pk-(p-1)n-2j(p-1)}{2}]}{j-t}w_{pk-s}\}\ Q(\alpha),\ \sigma(M^n/G)>$$

where $s = (p-1)n-2t(p-1)$ and $[\text{—}]$ means the integral part of the argument in the bracket.

Observe the definition of $\theta_k(z)$ by taking $z=Wu_t$ we have the equality reduced to

$$=<\sum_{t=0}^{j} -\binom{[\frac{pk-n-2j(p-1)}{2}]}{j-t}\ \theta_k(Wu_t)\ \alpha,\ \sigma(M^n)>$$

where $Wu_t = Wu_t(M^n/G)$. Thus we have the proposition.

We quote some special cases

3.3 COROLLARY. Let $Wu_t = Wu_t(M/G)$. Then,

For p = odd

$$\{\frac{k-1}{2} -(p-1)\}\theta_k - \mu(q)\ \theta_{k-1}\beta = \theta_k(Wu_1),\ k>\frac{(p-1)n-2(p-1)}{p}$$

where $q = n-k$, $r = \frac{p-1}{2}$, $\mu(q) = (r!)^{-1}(-1)^{rq}$

For $p = 2$,

$$(k+1)\,\theta_k = \theta_k(Wu_1),\ k > \frac{n+1}{2}$$

With these we can set up vanishing theorems. The next theorem reflects the action of the Bockstein β (or Sq^1)

3.4 THEOREM. Let M be oriented and G preserve the orientation. For p = odd,

$$\theta_k = 0, \text{ for } k = \text{even and } k > \frac{p-1}{p}\,n$$

and for $p = 2$,

$$\theta_k = 0, \text{ for } k = \text{even and } k > \frac{n+1}{2}$$

Proof: For p = odd, $\langle\theta_k\alpha, \sigma(M^n)\rangle = \langle w_j Q(\alpha), \sigma(M^n/G)\rangle$ where $j = pk-(p-1)n$. Note that if k is even, then j is even. By assumption $j>0$, so $\beta w_{j-1} = w_j$. Thus

$$\langle w_j Q(\alpha), \sigma(M^n/G)\rangle = \langle(\beta w_{j-1})Q(\alpha), \sigma(M^n/G)\rangle$$

Note that from 2.6 and 2.2, we have $w_{j-1}\beta Q(\alpha) = 0$, so the equality reduces to

$$= \langle\beta(w_{j-1}Q(\alpha)), \sigma(M^n/G)\rangle,$$

$$= \langle w_{j-1}Q(\alpha), \beta\sigma(M^n/G)\rangle = 0,$$ since M^n/G is orientable. For $p = 2$, orientability implies $Wu_1 = 0$, so this follows immediately from 3.3.

3.5 THEOREM. Let $Wu_t = Wu_t(M^n/G)$. If $Wu_t = 0$, $1 \leq t \leq p^a$, then

$$\theta_k(M) = 0, \text{ if } k \not\equiv -1 \bmod p^{b+1} \text{ and } k > \frac{(p-1)n+2p^b(p-1)}{p}$$

for some b, $0 \leq b \leq a$.

Remark: If p = odd, then for all even k, $\theta_k = 0$, by 3.4.

If $p = 2$, this agree with [7, 3.6].

Proof: We will prove this theorem by induction on a: First, we shall establish the case $a = 0$.

From 3.3 (set j = 1), we have

$$\left(\frac{k-1}{2} - (p-1)\right)\theta_k - \mu(q)\theta_{k-1}\ \beta = \theta_k(Wu_1)$$

where $k > \frac{(p-1)n + 2(p-1)}{p}$ and k is odd (when p is odd)

By proposition 3.4 and assumption, the equation reduces to

$$\left(\frac{k-1}{2} - (p-1)\right)\theta_k = 0$$

Thus if $\frac{k-1}{2} - (p-1) \equiv \frac{k+1}{2} - p \not\equiv 0 \pmod{p}$, then

$$\theta_k = 0$$

that is, if $k \not\equiv -1 \pmod{p}$, then $\theta_k = 0$. This establishes case a = 0.

In general, we assume the theorem is true for a and will prove that it is true for a + 1. By induction assumption

$$\theta_k = 0, \text{ if } k \not\equiv -1 \pmod{p^{b+1}} \text{ and}$$

$$k > \frac{(p-1)n+2p^b(p-1)}{p}$$

for some b, $0 \le b \le a$.

Therefore we only need to consider the case b = a+1. That is, to show that

$$\theta_k(M) = 0 \text{ if } k \not\equiv -1 \pmod{p^{a+2}} \text{ and}$$

$$k > \frac{(p-1)n+2p^{a+1}(p-1)}{p}$$

However, from the case b = a, we can conclude that the only possible non-zero classes are

$$\theta_k(M), \quad k \equiv p^{a+1}-1 \pmod{p^{a+2}}$$

$$\theta_k(M), \quad k \equiv -1 \pmod{p^{a+2}}$$

To complete the induction we have to show that, for the case

$$k \equiv p^{a+1}-1 \pmod{p^{a+2}}$$

$$\theta_k(M) = 0, \text{ if} > k\ \frac{(p-1)n+2p^{a+1}(p-1)}{p}$$

Subtituting the conditions $Wu_t = 0$, $0 \le t \le p^{a+1}$, into the formula in proposition 3.3 (set $j = p^{a+1}$) and observing that $\theta_{even} = 0$, we have

$$\left\{\binom{qr}{p^{a+1}} + \binom{\left[\frac{(pk-(p-1)n-2p^{a+1}(p-1)}{2}\right]}{p^{a+1}}\right\} \theta_k(M)$$

$$+ \sum_{1\leq i\leq p^a} \binom{(q-i)r}{p^{a+1}-pi} \theta_{k-2i(p-1)p^i} = 0$$

where $k \equiv p^{a+1}-1 \pmod{p^{a+2}}$ and $k > \frac{(p-1)n-2p^{a+1}(p-1)}{p}$.

$$k - 2i(p-1) \equiv p^{a+1}-1 - 2i(p-1) \pmod{p^{a+2}}$$

and hence

$$k - 2i(p-1) \equiv p^{a+1}-1 - 2i(p-1) \pmod{p^{a+1}}$$

thus for $1\leq i\leq p^a$ we have

$$k - 2i(p-1) \equiv p^{a+1}-1 - 2i(p-1) \not\equiv -1 \pmod{p^{a+1}}$$

[Note $2i(p-1) \not\equiv 0 \pmod{p^{a+1}}$]

By induction assumption

$$\theta_{k-2i(p-1)} = 0, \quad 1\leq i\leq p^a$$

Therefore the summation term in the relation (3) vanishes. Next, note from Lemma 3.6 below, we see that the coefficient of θ_k in (3) is not zero (mod p), so the relation (3) gives

$$\theta_k(M) = 0 \text{ if } k \equiv p^{a+1}-1 \pmod{p^{a+2}} \text{ and}$$

$$k > \frac{(p-1)n+2p^{a+1}(p-1)}{p}$$

Thus from (1) and (2) we conclude that

$$\theta_k(M) = 0 \text{ if } k \not\equiv -1 \pmod{p^{a+2}}$$

$$\text{and } k > \frac{(p-1)n+2p^{a+1}(p-1)}{p}$$

This complete the induction.

3.6 LEMMA. Let k be an odd integer and let

$k > \frac{(p-1)n+p^{a+1}(p-1)}{p}$. If $k \equiv -1 \pmod{p^{a+1}}$ and $k \not\equiv -1 \pmod{p^{a+2}}$, then

$$\binom{qr}{p^{a+1}} + \binom{[\frac{pk-(p-1)n-2p^{a+1}(p-1)}{2}]}{p^{a+1}} \not\equiv 0 \pmod{p}$$

where $r = \frac{p-1}{2}$ and $q = n-k$

Proof: First let us compute the sum

$$S = qr + [\frac{pk-(p-1)n-2p^{a+1}(p-1)}{2}]$$

$$S = [\frac{(p-1)(n-k)}{2} + \frac{pk-(p-1)n-2p^{a+1}(p-1)}{2}]$$

$$S = [\tfrac{k}{2}] - p^{a+1}(p-1) .$$

By assumption on k, k can be written as

$$k = -1 + 2\alpha p^{a+1} + 2\beta p^{a+2}$$

for some integers α and β. Note that the condition $k \not\equiv -1 \pmod{p^{a+2}}$ implies that

$$0 < 2\alpha < p$$

Subtituting all these conditions into the computed sum S above we have

$$S = \alpha p^{a+1} + \beta p^{a+2} - 1 - p^{a+1}(p-1)$$

Note $\alpha \leq p-1$ and $S \geq 0$, so $\beta > 0$, thus we can write

$$S = \alpha p^{a+1} + (\beta-1)p^{a+2} + (p^{a+2}-1) - p^{a+1}(p-1)$$

$$= (p-1)\left(\sum_{t=0}^{a} p^t\right) + \alpha p^{a+1} + (\beta-1)p^{a+2}$$

Write $qr = \Sigma a_t p^t$, and

$$[\frac{pk-(p-1)n-2p^{a+1}(p-1)}{2}] = \Sigma\, b_t p^t$$

Then, $S - qr = \sum_{t=0}^{a} (p-1-a_t)p^t + (\alpha - a_{a+1})p^{a+1} + \dots$

Hence, $b_t = p-1-a_t$ for $0\leq t\leq a$

$$b_{a+1} = \alpha - a_{a+1}, \text{ or } p + (\alpha - a_{a+1})$$

By [11, Ch I. 2.6] the sum of binomial coefficients in the Lemma is equal to $a_{a+1} + b_{a+1}$. By the computation, we have above $a_{a+1} + b_{a+1} = \alpha$ or $p+\alpha$. In either case $\alpha\not\equiv 0 \bmod p$. So we proved the Lemma.

Let V be a Z_p-space (not necessary free). Let S^m, m odd, have Z_p-action with complex multiplication by p-th roots of unity. Then the diagonal action on $S^m \times V$ is a free action. By considering the θ_k on the diagonal action we can, following Conner-Miller, introduce the following characteristic classes.

In the next paper, we will show that what we introduced below are expressible by the data of fixed point set and they are the same characteristic classes defined by Hattori [5].

3.7 DEFINITION: Let V^n be a closed manifold with Z_p-action (not necessarily free). With $m > n$ we define the Bredon-Hattori characteristic classes to be the unique cohomology class for which

$$\langle B_k\alpha, \sigma(V^n)\rangle = \langle w_{pk-(p-1)n}Q(\alpha \times 1), \sigma(V^n \times_\tau S^m)\rangle$$

for all $\alpha \in H^{n-k}(V)$, where τ is the diagonal action.

From the definition, we immediately have the following vanishing theorem; this was the main theorem in Hattori [5].

Theorem $B_k = 0$ if $k > \frac{p-1}{p}\dim v$.

Applying our vanishing theorem on θ_k, we can obtain a vanishing theorem on B_k. Now let us recall Browder's notion of orientability [2]. Following him, we will call a manifold $B\langle Wu_1 \ldots Wu_p a\rangle$-oriented if the manifold has a lifting to the classifying space in which Wu_i, $1 \leq i \leq p^a$ are all killed. For convenience, we will say such a manifold has Wu_a-structure. With this notion the vanishing theorem 3.5 can be rephrased as follows:

3.8 THEOREM. If G preserves Wu_a-structure, then

$$\theta_k(M) = 0 \text{ if } k \not\equiv -1 \pmod{p^{b+1}} \text{ and}$$
$$k < \frac{(p-1)n+2p^b(p-1)}{p}$$

for some b, $0 \leq b \leq a$.

We say a G action on V preserves $B\langle Wu_1,\ldots Wu_{p^a}\rangle$-orientation, if for some m, the diagonal action on $S^m \times V$ preserves the structure. Under this notion we have

3.9 THEOREM. Let V be a manifold with Wu_a-structure. If G preserves the strucutre, then we have

$$B_k(V) = 0, k \not\equiv 0 \pmod{p^{a+1}}.$$

In particular the Wu-class $Wu_k(V) = 0$, $k \not\equiv 0 \pmod{p^{a+1}}$.

Remark: If G acts trivially on V, then $B_k(V) = Wu_k(V)$. This can be seen immediately by noting that Q acts on $\nu^*(\alpha) = \sigma \times 1$ via "tensor product" of p^i and w_j.

$$\langle B_k\alpha \quad \sigma(V)\rangle = \langle Q(\nu^*(\alpha)), \sigma(V \times S^m/G)\rangle$$
$$= \langle \Sigma\, p^i\alpha \otimes w_j, \sigma(V) \otimes \sigma(S^m/G)\rangle$$
$$= \langle p^k\alpha, \sigma(V)\rangle$$
$$= \langle Wu_k\alpha \quad \sigma(V)\rangle$$

§4 Some applications

We shall mention a few simple applications; they are analogous results of Lin [7].]. These applications are easy consequence of product formulas. We believe S. Kahn's formula for p = 2 should have analogous formulation in odd p, we refer the reader to his works and future works. We shall mention here the two simplest cases.

4.1 Theorem (i) Let Z_p acts on V^n trivially and on M^m freely.

$$\theta_k(V^n \times M^m) = \sum_h Wu_{k-h}(V^n) \otimes \theta_h(M^m)$$

(ii) Let Z_p acts on V (maybe with fixed points), Let S^m be a homology m-sphere with free Z_p-action. Then

$$\theta_k(V^n \times S^m) = B_{k-m}(V^n) \otimes \theta_m(S^m). \quad k > \frac{n+m}{2}$$

The proof of these two formulas are fairly routine and essentially the same as given in Lin [7].

Let $f : M^m \to V^n$ be any map and let $Z_p = \{I, T, T,^2 .., T^{p-1}\}$

and,

$$A(f) = \{x \mid f(T^i x) = T^i f(x),\ i = 1,2,\ldots,p-1\}$$

The set A(f) is closed and Z_p-invariant (a point in A(f) is called an equivariant point by M. Nakaoka).

Consider the diagonal action $T_2 = T \times T$ on $M^m \times V^n$ (dim M = m, dim V = n). Let

$$g : M^m \to M^m \times V^n$$

be a map defined by g(x) = (x,f(x)). Clearly, the image G of g is the graph of f. Note that $g|A(f)$ is an equivariant homemorphism of A(f) onto the "equivariant self-intersection $\bigcap_{i=0}^{p-1} T_2^i(G)$. By appealing to the geometric meaning of the Bredon operation we introduce the following "self-intersection number": Let $1 \in H^0(M^m)$. Then $g_!(1) \in H^n(M^m \times V^n)$ is the cohomology class dual to the homology class represented by G. Then the "self-intersection number" of $\bigcap_{i=0}^{p-1} T_2^i(G)$ is defined to be

$$C(f) = \langle w''_{m-(p-1)n} Q''(g_!(1)),\ \sigma(M X_{Z_p} V)\rangle$$

where w''_i and Q'' are class and Bredon operation on M x V. (For p = 2, this number is suggested by P. E. Conner. We shall call this number the Conner Number.

By appealing to the geometric meaning of Q we can see the following proposition.

4.2 PROPOSITION. If $C(f) \neq 0$, then $w_{m-(p-1)n} \in H^{m-n}(M/Z_p)$ restricts to a non-zero element in $H^{m-n}(A(f)/Z_p)$ and hence dim $A(f) \geq m-(p-1)n$

If m>n, then by the definition of θ_k, we have

$$C(f) = \langle \theta_m(M \times V) g_!(1),\ \sigma(M \times V)\rangle .$$

For the product formulas we established above, we can reexpress C(f) in the data of M and V. Namely,

4.3 PROPOSITION. If $\theta_m(M \times V) = \Sigma\, \theta_{m-j}(M) \times B_j(V)$, then

$$C(f) = \sum_{j=0}^{[\frac{n}{2}]} \langle \theta_{m-j}(M) f^*(B_j(V)),\ \sigma(M)\rangle$$

$$= \sum_{j=0}^{m} \langle\ B_j(V) f_!(\theta_{m-j}(M)),\ \sigma(V)\rangle .$$

Letting S^m be a homology m-sphere, we have

4.4 COROLLARY. $C(f) = \langle \theta_m(S^m) f^*(B_0(V)),\ \sigma(S^m)\rangle$

If V is a trivial Z_p-space, we have $1 = Wu_0(V) = B_0(V)$, so

$$C(f) = <\mu_m, \sigma(S^m)> = 1 .$$

That gives the following Borsuk-Ulam type theorem (see Munkholm [9]).

4.5 COROLLARY. Let $f: S^m \to V^n$ be a map $(m>n)$, then $\dim A(f) \geq m-(p-1)n$.

References

1. G. E. Bredon, Cohomological aspects of transformation groups, Proc. Conf. Transformation Groups (New Orleans, LA 1967) Springer-Verlag, 1968, 245-280.

2. W. Browder, The Kervaire invariant of framed Manifolds and its generalization, Ann. of Math, 1969.

3. H. Brown, Jr. and F. P. Peterson, Relations Among Characteristic Classes-I, Topology, Vol, 3, Suppl. 1, pp. 39-52.

4. P. E. Conner and E. Y. Miller, Equivariant self-intersection, Preprint, 1979.

5. A. Hattori, The fixed point set of an involution and theorems of the Borsuk-Ulam type, Proc., Japan Acad. 50(1974), 537-541.

6. S. Kahn, Conner-Miller classes of a product involution and Borsuk-Ulam type theorem. (Houston, J. 1981)

7. T. Y. Lin, Wu-like classes for involutions and generalized Peterson-Stein classes, Proc. fixed point theory, Univ. of Sherbrooke (1980), Springer-Verlag.

8. T. Y. Lin, Wu-like classes for periodic maps, Notice of AMS (Jan. 1981), #783-55-36.

9. H. J. Munkholm, "Borsuk-Ulam type theorems for proper Z_p-actions on (mod p homology) n-spheres," Math. Scand. 24 (1969), 167-185.

10. G. Nishida, Cohomology operations in iterated loop spaces, Proc. Japan Acad., 44(1968), 104-109.

11. N. Steenrod and D. B. A. Epstein, Cohomology operations, Ann. Math. studies No. 50, Princeton Univ. Press, 1962.

DEPARTMENT OF MATHEMATICS
UNIVERSITY OF SOUTH CAROLINA AT AIKEN
AIKEN, SOUTH CAROLINA 29801

Canadian Mathematical Society
Conference Proceedings
Volume 2, Part 2 (1982)

VECTOR BUNDLES ON HOMOGENEOUS SPACES

Arunas Liulevicius[1]

ABSTRACT. This paper gives a simple proof of a formula due to V. Snaith on equivariant K-theory of homogeneous spaces with a linear G-action.

1. STATEMENT OF RESULTS. Let U be a compact connected Lie group, H a closed subgroup. If $a:G \to U$ is a continuous homomorphism of a topological group into U, we let a^*U/H be the G-space U/H with $g \cdot uH = a(g)uH$.

THEOREM (Snaith's Formula). Let U be a compact connected Lie group with $\pi_1(U)$ free abelian, $H \subset U$ a closed connected subgroup of maximal rank, $a:G \to U$ a continuous homomorphism. Then

$$K_G(a^*U/G) \approx R(G) \otimes_{R(U)} R(H)$$

as algebras over $K_G(*) = R(G)$.

The aim of this paper is to present a simple proof of Snaith's Formula. Since the geometric constructions are very pleasant for the case U = U(n), the unitary group of C^n, we shall give the proof in that special case. Indeed we shall prove a more general result.

MAIN THEOREM. Let $p:P \to B$ be a principal U = U(n) bundle and a map of left G-spaces. If H is a closed subgroup of U define the map

$$A:K_G(B) \otimes_{R(U)} R(H) \to K_G(P/H)$$

by setting $A(V \otimes W) = \underline{V} \otimes (P \times_H W)$ for a vector bundle V over B and a representation W of H, where $\underline{V}$ is the pull-back of V to P/H. If H is connected and of maximal rank, then A is an isomorphism.

1980 Mathematics Subject Classification. 55N15, 55U25, 57S25.
[1]Supported by NSF grant MCS 80-02730.

© 1982 American Mathematical Society
0731-1036/82/0000-0467/$02.50

The action of $R(U)$ on $K_G(B)$ is obtained as follows: if V is a vector bundle over B and W is a representation of U, then $V \cdot W = \underline{V} \otimes (P \times_U W)$. The action of $R(U)$ on $R(H)$ is defined by the restriction homomorphism $i^*: R(U) \to R(H)$. In the special case of $B = *$, a point, $P = U$ and the action of G is given by a homomorphism $a: G \to U$, and the action of $R(U)$ on $K_G(*) = R(G)$ is given by the induced homomorphism of representation rings $a^*: R(U) \to R(G)$.

The key point in the argument is a result of Pittie [13] and Steinberg [18] that if U is a compact connected Lie group such that $\pi_1(U)$ is free abelian and H is a closed connected subgroup of maximal rank then $R(H)$ is a free $R(U)$ module. Indeed, if $T \subset H \subset U$ is a maximal torus and $WH \subset WU$ are the Weyl groups of H and U with respect to T, then $R(H)$ is a free $R(U)$ module of rank $n = [WU:WH]$. If $g_1, \ldots, g_n$ is an $R(U)$ basis of $R(H)$, then we have

$$g_i g_j = \sum_k u_{ij}^k g_k$$

for unique u_{ij}^k in $R(U)$. The module $K_G(a^*U/H)$ is free over $R(G)$ with basis $h_i = A(1 \otimes g_i)$, and its algebra structure is given by

$$h_i h_j = \sum_k a^*(u_{ij}^k) h_k \ .$$

Similarly, if ψ^i is the i-th Adams operation and

$$\psi^i g_j = \sum_k v_{ij}^k g_k$$

for v_{ij}^k in $R(U)$, then also

$$\psi^i h_j = \sum_k a^*(v_{ij}^k) h_k \ .$$

Thus Snaith's Formula gives also a functorial expression for the product in $K_G(a^*U/H)$ and the action of the Adams operations. The formulae are also very useful even in the special case of the inclusion of the identity $e: E \to G$ and give the structure of ordinary K-theory $K(U/H)$. The map $A: R(H) \to K(U/H)$ was first studied by Atiyah and Hirzebruch [5] and was shown to be onto in many cases. They conjectured that it was always onto if $\pi_1(U)$ is free abelian and H is a closed connected subgroup of maximal rank. This conjecture was shown to be true in papers [13], [15], and [17]. The result that $R(H)$ is a free $R(U)$ module was fed into the Künneth spectral sequence in equivariant K-theory (Hodgkin [8],[9]) and this sequence reduced to its edge term, hence collapsed.

Other important work in this area was done by Snaith [16], Seymour [15], and McLeod [12].

Snaith's Formula in the special case of an injection $a:G \to U$ can be proved by the use of the Künneth spectral sequence in K_U-theory. Notice that $K_G(a^*U/H) = K_U(U \times_G a^*U/H) = K_U(U/G \times U/H)$. the E_2-term of the Künneth spectral sequence is

$$E_2^{*,*} = \mathrm{Tor}_{*,*}^{R(U)}(R(G),R(H)),$$

and since $R(H)$ is $R(U)$-free, this reduces to the edge term $R(G) \otimes_{R(U)} R(H)$.

The advantage of the approach of this paper is that it is not necessary to assume that $a:G \to U$ is faithful, and also all spectral sequence arguments are avoided. The argument is based on the work of Atiyah [4] on Bott periodicity.

The author wishes to thank M.F. Atiyah, M. Karoubi, and R. Seymour for their comments, and V.P. Snaith for communicating his formula at the end of the Waterloo topology conference in 1978.

2. PROOF OF A SPECIAL CASE. In this section we will check the main theorem in the special case of $U = U(n)$, $H = U(1) \times U(n-1)$. If we let $E = P \times_U C^n$ be the total space of the G vector bundle associated to the principal U-bundle $p:P \to B$, then $P/(U(1) \times U(n-1)) = P \times_U CP^{n-1} = P(E)$ is the projective space bundle associated to E. We let $h:S(E) \to P(E)$ be the Hopf bundle which associates to a unit vector in E the projective line which it determines, and let $H = h^*$ be its dual line bundle. Let $r:P(E) \to B$ be the projection map. Notice that $\mathrm{Hom}(h,r^!E) = r^!E \otimes H$ has an obvious nowhere zero section, so its Koszul complex is acyclic and we have the relation (see Segal [14])

$$1 - \Lambda E \cdot H + \Lambda^2 E \cdot H^2 - \cdots + (-1)^n \Lambda^n E \cdot H^n = 0 ,$$

and since $H = h^{-1}$, multiplying both sides of the equation by h^n we obtain

$$h^n - \Lambda^1 E \cdot h^{n-1} + \Lambda^2 E \cdot h^{n-2} - \cdots + (-1)^n \Lambda^n E \cdot 1 = 0.$$

It is easy to see (Atiyah [3], Segal [14]) that $K_G(P(E))$ is free over $R(G)$ with $1,h,\dots,h^{n-1}$ as basis, so the above relation determines the $R(G)$-algebra structure of $K_G(P(E))$ completely.

We now examine the map

$$A:K_G(B) \otimes_{R(U(n))} R(U(1) \times U(n-1)) \to K_G(P(E)).$$

Notice that $R(U(n)) = Z[a_1,\dots,a_n,a_n^{-1}]$, where a_1 is the identity map of $U(n)$ and $a_i = \Lambda^i a_1$ (see for example Adams [1], p.167). Similarly, $R(U(1)) = Z[t,t^{-1}]$, $R(U(n-1)) = Z[b_1,\dots,b_{n-1}, b_{n-1}^{-1}]$, where t and b_1 are birth certificate representations. The inclusion $i:U(1) \times U(n-1) \to U(n)$ induces the restriction homomorphism

$$i^*:R(U(n)) \to R(U(1)) \otimes R(U(n-1))$$

with $i^*a_1 = b_1 + t$, $i^*a_2 = b_2 + b_1t, \ldots, i^*a_n = b_{n-1}t$, so $R(U(1)) \otimes R(U(n-1))$ is $R(U(n))$-free with basis $1, t, \ldots, t^{n-1}$ and the algebra structure is determined by the relation

$$t^n - a_1 \cdot t^{n-1} + a_2 \cdot t^{n-2} - \cdots + (-1)^n a_n \cdot 1 = 0.$$

We now notice that $A(1 \otimes t^i) = h^i$, so it is indeed true that the map A is an isomorphism in the special case of $H = U(1) \times U(n-1)$.

3. PROOF FOR THE CASE $H = T^n$. We will now use induction to show that A is an isomorphism for subgroups $H = T^k \times U(n-k) \subset U(n)$, where $T^k = U(1) \times \cdots \times U(1)$ k times. We have just done the case $k = 1$. Assume that the case of $T^k \times U(n-k)$ has been proved - we wish to show how it implies that A is an isomorphism for $T^{k+1} \times U(n-k-1)$. Notice that $P/T^k \times U(n-k) = F_k(E)$ is the bundle of k-flags associated with $E = P \times_U C^n$, where a k-flag

$$\varphi = : V_1 \subset V_2 \subset \cdots \subset V_k$$

is a chain of subspaces with $\dim_C V_i = i$. We let

$$E' = \{(\varphi, v) \mid \varphi \perp v\},$$

then $E' \to F_k(E)$ is an n-k plane bundle with $P(E') = F_{k+1}(E)$. The inductive hypothesis tells us that the map

$$A_1:K_G(B) \otimes_{R(U(n))} R(T^k \times U(n-k)) \to K_G(F_k(E))$$

is an isomorphism. The case $k = 1$ gives us that

$$A_2:K_G(F_k(E)) \otimes_{R(U(n-k))} R(T^1 \times U(n-k-1)) \to K_G(F_{k+1}(E))$$

is an isomorphism. We now notice that we have canonical isomorphisms

$$K_G(B) \otimes_{R(U(n))} R(T^k \times U(n-k)) \otimes_{R(U(n-k))} R(T^1 \times U(n-k-1))$$
$$\cong K_G(B) \otimes_{R(U(n))} R(T^k \times U(n-k)) \otimes_{R(T^k \times U(n-k))} R(T^{k+1} \times U(n-k-1))$$
$$\cong K_G(B) \otimes_{R(U(n))} R(T^{k+1} \times U(n-k-1)),$$

and under these identifications the composition $A_2(A_1 \otimes 1)$ corresponds to

$$A_3 : K_G(B) \otimes_{R(U(n))} R(T^{k+1} \times U(n-k-1)) \to K_G(F_{k+1}(E)) ,$$

so A_3 is an isomorphism. This does the inductive step and proves that A is an isomorphism for $k = 1,\ldots,n$. The case $k = n$ is that of the standard torus of $U = U(n)$.

4. THE GENERAL CASE. It is helpful to recast our notation a bit. Notice that U acts freely on P, so $K_{G \times U}(P) = K_B(P/U) = K_G(B)$ and our map is

$$A_H : K_{G \times U}(P) \otimes_{R(G \times U)} K_{G \times H}(*) \to K_{G \times H}(P) .$$

We have shown that if $H = T$, a maximal torus of U, then A_T is an isomorphism. We will now use the work of Atiyah [4] to reduce the general case to the case of the maximal torus. Let $i : T \to H$ be the inclusion of a maximal torus of U. Atiyah constructs a natural transformation of functors $i_* : K_{G\times T}(\ ) \to K_{G\times H}(\ )$ such that if $i^* : K_{G\times H}(\ ) \to K_{G\times T}(\ )$ is the restriction then $i_* i^* =$ identity and both transformations commute with A. That is, i^* is an injection onto a direct summand and the following diagram commutes:

$$\begin{array}{ccc}
K_{G\times U}(P) \otimes_{R(G\times U)} K_{G\times H}(*) & \xrightarrow{A_H} & K_{G\times H}(P) \\
\downarrow {\scriptstyle 1 \otimes i^*(*)} & & \downarrow {\scriptstyle i^*(P)} \\
K_{G\times U}(P) \otimes_{R(G\times U)} K_{G\times T}(*) & \xrightarrow{A_T} & K_{G\times T}(P) \\
\downarrow {\scriptstyle 1 \otimes i_*(*)} & & \downarrow {\scriptstyle i_*(P)} \\
K_{G\times U}(P) \otimes_{R(G\times U)} K_{G\times H}(*) & \xrightarrow{A_H} & K_{G\times H}(P)
\end{array}$$

and the composition of the vertical maps is the identity. Since $1 \otimes i^*(*)$ and $i^*(P)$ are injective, the injectivity of A_T implies that of A_H. Since $1 \otimes i_*(*)$ and $i_*(P)$ as well as A_T are onto, then A_H is also onto. This means that A_H is an isomorphism, as claimed, and the proof of the main theorem is complete.

The reader should compare this argument with the special case $H = U(k) \times U(n-k)$ in Hodgkin [9], p.59 which was used to show that there are enough Künneth spaces. The argument also shows the key role of the maximal torus - this is also the case in work on the convergence of the Künneth spectral sequence: Hodgkin [9], Snaith [15], McLeod [12].

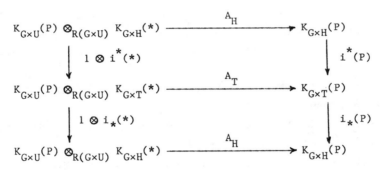

BIBLIOGRAPHY

1. J.F. Adams, Lectures on Lie groups, Benjamin, New York, 1967.

2. M.F. Atiyah, Vector bundles and the Künneth formula, Topology 1 (1962), 245-8.

3. M.F. Atiyah, (notes by D.W. Anderson) K-theory, Harvard University, Cambridge, 1964.

4. M.F. Atiyah, Bott periodicity and the index of elliptic operators, Quarterly Jour. Math. Oxford (2), 19(1968), 113-40.

5. M.F. Atiyah and F. Hirzebruch, Vector bundles and homogeneous spaces, Proceedings of Symposia in Pure Math. AMS 3(1961), 7-38.

6. M.R. Atiyah and G.B. Segal, Lectures in equivariant K-theory, Oxford, 1965.

7. J. Ewing and A. Liulevicius, Homotopy rigidity of linear actions on homogeneous spaces, Jour. of Pure and App. Algebra 18(1980), 259-267.

8. L. Hodgkin, An equivariant Künneth formula in K-theory, University of Warwick, 1968.

9. L. Hodgkin, The equivariant Künneth theorem in K-theory, Lecture Notes in Math. 496(1975), 1-100.

10. M. Karoubi, K-theory: an introduction, Springer-Verlag, Berlin-New York, 1978.

11. A. Liulevicius, Equivariant K-theory and homotopy rigidity, Lecture Notes in Math. 788(1980), 340-358.

12. J. McLeod, The Künneth formula in equivariant K-theory, Lecture Notes in Math. 741(1979), 316-333.

13. H.V. Pittie, Homogeneous vector bundles on homogeneous spaces, Topology 11(1972), 199-203.

14. G.B. Segal, Equivariant K-theory, Inst. Hautes Etudes Sci. Publ. Math. 34(1968),129-151.

15. R.M. Seymour, Lie groups and KR-theory, Thesis, University of Warwick, 1969.

16. V.P. Snaith, On the Künneth formula spectral sequence in equivariant K-theory, Proc. Camb. Phil. Soc. 72(1972), 167-77.

17. V.P. Snaith, On the K-theory of homogeneous spaces and conjugate bundles of Lie groups, Proc. London Math. Soc. (3)22(1971), 562-584.

18. R. Steinberg, On a theorem of Pittie, Topology 14(1975), 173-77.

DEPARTMENT OF MATHEMATICS
THE UNIVERSITY OF CHICAGO
CHICAGO, ILLINOIS 60637

Canadian Mathematical Society
Conference Proceedings
Volume 2, Part 2 (1982)

A REDUCTION OF THE SEGAL CONJECTURE

by J. P. MAY and J. E. McCLURE

In this note, we show that the Segal conjecture holds for a given finite group G if it holds for all subgroups of G which have prime power order. We also point out that the p-group case reduces to a question about p-adic completions.

First we must say what we mean by the Segal conjecture. Various forms are current, but our methods apply equally well to any of them. Let π_G^* denote equivariant stable cohomotopy and recall that $\pi_G^0(pt)$ is canonically isomorphic to the Burnside ring A(G) [7, 17]. In particular, $\pi_G^*(X)$ is a (Z-graded) module over A(G). Let I(G) denote the augmentation ideal of A(G) and let $\hat{\pi}_G^*(X)$ denote the I(G)-adic completion of $\pi_G^*(X)$. Let EG be a free contractible G-space and let ε denote the map $EG \to pt$. As we shall recall in section 2, $\varepsilon^*: \pi_G^*(pt) \to \pi_G^*(EG)$ factors to give a homomorphism $\hat{\varepsilon}^*: \hat{\pi}_G^*(pt) \to \pi_G^*(EG)$, and the target here is isomorphic to the ordinary stable cohomotopy $\pi^*(BG)$. The Segal conjecture asserts that $\hat{\varepsilon}^*$ is an isomorphism. As Segal understood and we shall explain in section 2, this is equivalent to a more elaborate statement involving RO(G)-graded cohomology.

We shall prove the following results.

<u>THEOREM A</u>. The Segal conjecture is true for G if it is true for all subgroups of G having prime power order.

<u>PROPOSITION B</u>. The Segal conjecture is true for a p-group G if and only if $\varepsilon^*: \pi_G^*(pt) \to \pi^*(BG)$ induces an isomorphism on passage to p-adic completion.

If, as seems not unlikely, the Segal conjecture turns out to be true for Abelian p-groups but false in full generality, then Theorem A will imply the Segal conjecture for groups with Abelian p-Sylow subgroups.

In fact, we shall see that these reductions apply not just to equivariant cohomotopy but to the analogous completion map for

© 1982 American Mathematical Society
0731-1036/82/0000-0468/$04.50

quite general RO(G)-graded equivariant cohomology theories. For example, these reductions shed some light on Atiyah's original calculation of $K^*(BG)$ [2]. We shall emphasize this generic aspect of our work throughout.

Theorem A is actually an easy consequence of the following purely algebraic fact. It will be proven (and its undefined terms explained) in section 1. The application to cohomology theories will be discussed in section 2.

THEOREM C. The completed Burnside ring functor $\underline{\hat{A}}$ is a Green functor which satisfies induction with respect to its set of subgroups of prime power order.

This result sheds considerable light on the structure of $\hat{A}(G)$ and should have other uses.

We wish to acknowledge the earlier work of Laitinen [11, §1], which gave enough information to deduce the monomorphism part of the reduction of Theorem A, and of Madsen [15, §1], which led us to the idea of deducing Theorem A from Theorem C. Quite recently, Segal gave a different proof of Theorem A in a letter to Adams.

§1 INDUCTION FOR THE COMPLETED BURNSIDE RING

Let G be a finite group. We recall some terminology from [7] or [9]. A Mackey functor M consists of a covariant and a contravariant functor, with the same object function, from the category of finite G-sets to the category of Abelian groups. For a G-map $f: S \to T$, we think of the contravariant map $f^*: M(T) \to M(S)$ as restriction and the covariant map $f_*: M(S) \to M(T)$ as induction (or transfer). These are to be related in a suitable way, and the axioms imply that the entire Mackey functor is determined by its restriction to the full subcategory of orbits G/H. The category of Mackey functors admits a tensor product, and there is a resulting notion of a ring object, or Green functor.

In particular, there is a Green functor $\underline{A}$ whose value on the orbit G/H is the Burnside ring $A(H)$ of finite H-sets. A G-map $f: G/J \to G/K$ is given by a subconjugacy relation $gJg^{-1} \subset K$, and we use the letter i generically for inclusions $J \subset K$ (or π for the corresponding projection $G/J \to G/K$ of G-sets when we prefer to emphasize that point of view). Restriction $i^*: A(K) \to A(J)$ assigns to a K-set T the same set regarded as a J-set. Induction $i_*: A(J) \to A(K)$ assigns to a J-set S the K-set $K\times_J S$; i^* is a morphism of rings and i_* is a morphism of

$A(J)$-modules (Frobenius reciprocity).

There is a notion of a module Mackey functor over a Green functor, and the Green functor $\underline{A}$ acts universally.

LEMMA 1. Any Mackey functor has a natural structure of $\underline{A}$-module Mackey functor.

Let $\hat{A}(H)$ be the completion of $A(H)$ in the $I(H)$-adic topology. The restriction maps are evidently continuous and it will follow from Lemma 6 below that the induction maps are also continuous. It follows easily that $\hat{\underline{A}}$ inherits a structure of Green functor from $\underline{A}$. Similarly, any Mackey functor gives rise to an $\hat{\underline{A}}$-module Mackey functor upon completion.

Choose a p-Sylow subgroup G_p for each prime dividing the order of G (to be denoted $|G|$). Theorem C can then be restated as follows.

THEOREM 2. The sum $\sum_p i_*: \sum_p \hat{A}(G_p) \to \hat{A}(G)$ is an epimorphism.

By a basic result in induction theory [7; 9, §6], we have the following immediate consequence.

COROLLARY 3. Let M be an $\hat{\underline{A}}$-module Mackey functor and let $S = \coprod_p G/G_p$. Then the following sequence is exact:

$$0 \longrightarrow M(pt) \xrightarrow{\pi^*} M(S) \xrightarrow{\pi_1^* - \pi_2^*} M(S \times S) .$$

Here $\pi: S \to pt$ and $\pi_i: S \times S \to S$ are the evident projections.

Any Mackey functor converts disjoint unions to direct sums, and $S \times S$ can be written as a disjoint union of orbits G/H where the H are p-groups for varying primes p. Thus, by naturality and a comparison of exact sequences, the previous corollary implies the following one.

COROLLARY 4. Let $\sigma: M \to N$ be a morphism of $\hat{\underline{A}}$-module Mackey functors, for example the completion of any morphism of Mackey functors. If $\sigma: M(G/H) \to N(G/H)$ is an isomorphism for all subgroups H of G of prime power order, then $\sigma: M(pt) \to N(pt)$ is an isomorphism.

Of course, this is the result we shall use to prove Theorem A.

We now turn to the proof of Theorem 2, and we need some preliminary recollections and observations. Embed Z in $A(G)$ by sending $n > 0$ to the trivial n-pointed G-set. For $H \subset G$, define a ring homomorphism $\chi_H: A(G) \to Z$ by sending a G-set S to the cardinality of S^H. Then $I(G) = \text{Ker}\, \chi_e$, where e is the trivial group. Let $\hat{I}(G)$ be the completion of $I(G)$ in the

$I(G)$-adic topology. We have an obvious induced splitting $\hat{A}(G) = Z \oplus \hat{I}(G)$, and the natural composite

$$Z \subset \hat{A}(H) \xrightarrow{i_*} \hat{A}(G) \xrightarrow{\chi_e} Z$$

is multiplication by $|G/H|$. Since the greatest common divisor of the numbers $|G/G_p|$ is one, we see that it suffices to prove Theorem 2 with $\hat{A}$ replaced by $\hat{I}$.

Let $C(G)$ be the ring direct product of one copy of Z, denoted Z_H, for each conjugacy class (H) of subgroups of G and let $IC(G)$ be the ideal of elements with $e^{\underline{th}}$ coordinate zero. Let $\chi\colon A(G) \to C(G)$ be the ring homomorphism with $H^{\underline{th}}$ coordinate χ_H. Then χ is a monomorphism with finite cokernel, $|G|\cdot C(G)$ being contained in the image of χ [7, §1]. The following observation is essentially well-known and will imply Proposition B; compare [7, 4.1.1] and [11, 1.12].

LEMMA 5. Let M be a Mackey functor and let $\pi^*\colon M(pt) \to M(G)$ be induced by the projection $\pi\colon G \to pt$. Then $M(pt)$ is an $A(G)$-module and π^* is a morphism of $A(G)$-modules, where $A(G)$ acts on $M(G)$ through χ_e. Further, $|G|\cdot \mathrm{Ker}\,\pi^*$ is contained in $I(G)\cdot \mathrm{Ker}\,\pi^*$. If G is a p-group, then the p-adic topology and $I(G)$-adic topology on $\mathrm{Ker}\,\pi^*$ coincide.

PROOF. The first statement is part of Lemma 1, multiplication by a G-set S being the composite

$$M(pt) \xrightarrow{\pi^*} M(S) \xrightarrow{\pi_*} M(pt).$$

Taking $S = G$, we see immediately that $G\cdot \mathrm{Ker}\,\pi^* = 0$, hence

$$|G|\cdot \mathrm{Ker}\,\pi^* = (|G| - G)\cdot \mathrm{Ker}\,\pi^* \subset I(G)\cdot \mathrm{Ker}\,\pi^*.$$

Now let $|G| = p^n$. We obviously have $|G|^m\cdot \mathrm{Ker}\,\pi^* \subset I(G)^m\cdot \mathrm{Ker}\pi^*$. We claim that $I(G)^{n+1} \subset pI(G)$. This will imply that

$$I(G)^{m(n+1)}\cdot \mathrm{Ker}\,\pi^* \subset p^m\cdot \mathrm{Ker}\,\pi^*$$

and so complete the proof of the last statement. For $H \subset G$, $H \neq e$, and $K \subset G$, $\chi_H(G/K - |G/K|)$ is congruent to zero mod p since $G/K - (G/K)^H$ is a disjoint union of non-trivial H-orbits and thus has cardinality divisible by p. Therefore $\chi_H I(G) \subset pZ_H$, hence $\chi I(G) \subset pIC(G)$ and

$$\chi I(G)^{n+1} \subset p^{n+1} IC(G) \subset p\chi I(G).$$

Of course, the last statement fails if we replace $\mathrm{Ker}\,\pi^*$ by $M(pt)$. For example, $\hat{A}(G) = Z \oplus \hat{I}(G)$, whereas the completion of $A(G)$ at p would have Z replaced by its p-adic completion.

We shall need to know the prime ideal spectrum of $A(G)$ [8; 7, §1]. Let $q(H,0)$ be the kernel of χ_H and let $q(H,p)$ be the kernel of the composite of χ_H and reduction mod p. These are all of the prime ideals of $A(G)$, and the lattice of prime ideals is determined by the relations

$$q(H,0) \subset q(H,p),$$

$$q(H,0) = q(K,0) \quad \text{if } H \text{ is conjugate to } K,$$

and

$$q(H,p) = q(K,p) \quad \text{if } H^p \text{ is conjugate to } K^p,$$

where H^p is the smallest normal subgroup of H such that H/H^p is a p-group. Note in particular that $q(e,p) = q(H,p)$ if and only if H is a p-group. Write $q(H,p) = q(H,p;G)$ when necessary for clarity.

We need three lemmas. In all of them, we focus attention on a fixed given subgroup H of G. Via $i^*: A(G) \to A(H)$, any $A(H)$-module is an $A(G)$-module. Via $\chi: A(H) \to C(H)$, any $C(H)$-module is an $A(H)$-module.

LEMMA 6. The following topologies on $A(H)$ and $I(H)$ coincide.

(1) The $I(G)$-adic topology.

(2) The $I(H)$-adic topology.

(3) The subspace topology induced from the $I(H)$-adic topology on $C(H)$.

PROOF. The agreement of the first two topologies is due to Laitinen [11, 1.14]. Since $i^*I(G) \subset I(H)$, it is enough to show that $I(H)^n \subset i^*(IG)$ for some n, and this holds provided that any prime ideal of $A(H)$ which contains $i^*I(G)$ also contains $I(H)$. Since $(i^*)^{-1}q(K,p;H) = q(K,p;G)$ for $K \subset H$ and any p (including 0), because $\chi_K i^* = \chi_K$, this is a simple check of cases from the facts just recorded. The agreement of the last two topologies is a standard consequence of the Artin-Rees lemma [3, 10.11].

For any $A(H)$-module N and any $n \geq 1$, define

$$P_n(N,H) = P_n N = N/I(H)^n N.$$

Observe that we have induced homomorphisms

$$i^*: P_n(A(G),G) \to P_n(A(H),G) \quad \text{and} \quad i_*: P_n(A(H),G) \to P_n(A(G),G).$$

By the lemma, $P_n(A(H),G)$ is a quotient of some $P_m(A(H),H)$. Define

$$J^n(H) = \chi^{-1}(I(H)^n C(H)) \subset A(H)$$

and

$$Q_n N = N/J^n(H)N.$$

Since $I(H)^n$ is contained in $J^n(H)$, we have a natural epimorphism $P_nN \to Q_nN$, and this is evidently the identity when $N = C(H)$. Further, χ induces a monomorphism

$$Q_nA(H) \to Q_nC(H) = P_nC(H) = \prod_{(K)} P_nZ_K,$$

where Z_K is Z regarded as an $A(H)$-module via χ_K.

<u>LEMMA 7</u>. (i) If $K = e$, then $P_nZ_K = Z$ for all n.
(ii) If K is not a p-group for any p, then $P_nZ_K = 0$ for all n.
(iii) If K is a p-group, then P_nZ_K is a p-group for all n.

<u>PROOF</u>. For any prime p, the kernel of the composite

$$I(H) = q(e,0) \subset A(H) \xrightarrow{\chi_K} Z \longrightarrow Z_p = Z/pZ$$

is $q(K,p) \cap q(e,0)$, and $q(K,p)$ contains $q(e,0)$ if and only if K is e or a p-group. If $K = e$, the composite is always zero and $I(H)Z_e = 0$. If K is not a p-group for any p, the composite is always non-zero and therefore $I(H)Z_K = Z_K$. If K is a p-group, the composite is non-zero for all primes other than p and is zero for p, hence $I(H)Z_K = p^rZ_K$ for some $r \geq 1$.

<u>LEMMA 8</u>. The group $P_nI(H)$ is finite for all $n \geq 1$.

<u>PROOF</u>. The agreement of the last two topologies in Lemma 6 implies that $P_nI(H)$ is a quotient of $Q_mI(H)$ for some m. Since $Q_mI(H)$ injects into $P_mIC(H) = \prod_{(K)\neq e} P_mZ_K$ and the latter is finite by the previous lemma, the conclusion follows.

In view of the agreement of the first two topologies in Lemma 6, we have the following commutative diagram:

$$\begin{array}{ccc} \sum_p \hat{I}(G_p) & \xrightarrow{\sum_p i_*} & \hat{I}(G) \\ \cong & & \cong \\ \lim_n \sum_p P_n(I(G_p),G) & \xrightarrow{\lim_n \sum_p i_*} & \lim_n P_nI(G) \end{array}$$

Thus Theorem 2 will hold if the bottom arrow is an epimorphism. Since the groups $P_n(I(G_p),G)$ are finite, the usual $\lim^1$ exact sequence shows that it suffices to prove that each map

$$\sum_p i_*: \sum_p P_n(I(G_p),G) \to P_nI(G)$$

is an epimorphism. Again by finiteness, this will hold provided the $p^{\underline{th}}$ map i_* is surjective on p-primary components. We claim that the composite

$$P_nI(G) \xrightarrow{i^*} P_n(I(G_p),G) \xrightarrow{i_*} P_nI(G)$$

becomes an isomorphism when localized at p. This composite is multiplication by the G-set G/G_p regarded as an element of A(G) or of its quotient ring $P_nA(G)$. By Lemma 6 again, the latter ring is a quotient of $Q_mA(G)$ for some m. Thus it suffices to check that G/G_p is a unit in $Q_mA(G)_{(p)}$. Since $Q_mC(G)/Q_mA(G)$ is finite, $Q_mC(G)_{(p)}$ is an integral extension of $Q_mA(G)_{(p)}$. It therefore suffices to check that G/G_p is a unit in $Q_mC(G)_{(p)}$; see e.g. [3, 5.10]. Lemma 7 shows that

$$Q_mC(G)_{(p)} = Z_{(p)} \times \prod_{(K)} Q_mZ_K,$$

where the product is restricted to the conjugacy classes of p-groups $K \subset G$. Since $q(K,p) = q(e,p)$ in A(G) and since $\chi_e(G/G_p) = |G/G_p|$ is prime to p, we must have that $\chi_K(G/G_p)$ is also prime to p. Thus G/G_p is a unit in $Q_mC(G)_{(p)}$. This completes the proof of our claim and thus the proof of Theorem 2.

§2. APPLICATIONS TO EQUIVARIANT COHOMOLOGY THEORIES

Let k_G^* be an RO(G)-graded cohomology theory on G-spaces Y. Thus we are given groups $k_G^\alpha Y$ for $\alpha \in RO(G)$ such that the $k_G^{\alpha+n}Y$ for fixed α and varying $n \in Z \subset RO(G)$ comprise a Z-graded cohomology theory and there are coherent natural isomorphisms

$$\tilde{k}_G^\alpha(X) \simeq \tilde{k}_G^{\alpha+V}(X \wedge S^V)$$

for based G-spaces X (with G-fixed basepoint), where S^V denotes the one-point compactification of a representation V. Such theories were introduced by Segal [17] and have been studied by Kosniowski [10] and others. Ordinary RO(G)-graded theories were introduced in [12], and a comprehensive treatment will appear in [14]. Some of our details here will be more intuitive than precise, and we will take for granted various facts from [13] and [14] about equivariant spectra and cohomology theories.

Our reason for considering RO(G)-graded theories is that the following result is false for general Z-graded theories. Let Y^+ denote the union of a G-space Y and a disjoint G-fixed basepoint.

PROPOSITION 9. Let k_G^* be an RO(G)-graded cohomology theory, let $\alpha \in RO(G)$, and let Y be a G-space. Then the correspondence $G/H \to k_G^\alpha(G/H \times Y)$ determines the object function of a Mackey functor, and this Mackey functor structure is functorial in Y.

PROOF. This is folklore; compare [5] and [10]. We give a sketch of our favorite argument. There is an equivariant stable category of G-spectra, constructed by Lewis and May [13], and there is a functor Σ^∞ from based G-spaces to G-spectra. The theory k_G^* is represented by a G-spectrum k_G. More precisely, k_G represents an RO(G)-graded cohomology theory on G-spectra such that

$$k_G^*(\Sigma^\infty X) = \tilde{k}_G^*(X) \quad \text{and} \quad \tilde{k}_G^*(Y^+) = k_G^*(Y).$$

Let $\mathcal{O}$ be the full subcategory of the stable category whose objects are the G-spectra $\Sigma^\infty(G/H)^+$ for $H \subset G$. As we observed in work with Lewis [12, 14], a Mackey functor determines and is determined by an additive contravariant functor $\mathcal{O} \to Ab$. With this homotopical interpretation of the algebraic notion of a Mackey functor, the conclusion becomes tautologically obvious. We have identifications of G-spectra

$$\Sigma^\infty((G/H^+ \wedge X) \simeq \Sigma^\infty(G/H)^+ \wedge X \simeq \Sigma^\infty(G/H)^+ \wedge \Sigma^\infty X.$$

For any G-spectrum E, such as $\Sigma^\infty Y^+$, the Abelian groups $k_G^\alpha(\Sigma^\infty(G/H)^+ \wedge E)$ and homomorphisms $(f \wedge 1)^*$ for morphisms $f \in \mathcal{O}$ specify an additive contravariant functor $\mathcal{O} \to Ab$.

For $H \subset G$ with inclusion i, a G-spectrum k_G determines an H-spectrum $k_H = i^* k_G$ and thus an RO(H)-graded cohomology theory k_H^*. The precise definition implies

$$k_H^{i^*\alpha}(Y) = k_G^\alpha(G \times_H Y)$$

for $\alpha \in RO(G)$ and an H-space Y. We abbreviate $k_e^* = k^*$, this being the underlying nonequivariant cohomology theory associated to k_G^*. In practice, as for K-theory and stable cohomotopy, we are given an RO(G)-graded cohomology theory k_G^* for every G and must check that $k_H^* = (i^* k_G)^*$. We shall need a bit of extra structure; compare [10, §2].

DEFINITION 10. A cohomology theory k_G^* is said to be split if there is a morphism of Z-graded nonequivariant cohomology theories $\zeta\colon k^*Y \to k_G^*Y$ (where spaces Y are given trivial G-action on the right) such that the composite

$$k^*Y \xrightarrow{\zeta} k_G^*Y \xrightarrow{\pi^*} k_G^*(G \times Y) = k^*Y,$$

$\pi\colon G \times Y \to Y$, is a natural isomorphism. Of course, since π^* is also a morphism of cohomology theories, this will hold provided that the composite is an isomorphism when Y is a point. We say that k_G^* is a split ring theory if k_G^* is ring-valued (from which it follows that all k_H^* for $H \subset G$ are ring-valued) and

ζ is a morphism of ring valued cohomology theories.

REMARK 11. In terms of spectra, a G-spectrum k_G determines a nonequivariant fixed point spectrum $(k_G)^G$ which represents the Z-graded theory k_G^*Y on spaces Y with trivial G-action. There is a natural map $(k_G)^G \to k$ which represents π^*, and the definition requires this map to be a retraction in the stable category. If k_G^* is ring-valued, then, ignoring $\lim^1$ questions, k_G is a ring G-spectrum and induces ring structures on the k_H and on $(k_G)^G$ such that $(k_G)^G \to k$ is a ring map. The last part of the definition requires a ring map $k \to (k_G)^G$ such that the composite $k \to (k_G)^G \to k$ is the identity.

Let S_G be the 0-sphere G-spectrum, so that $S_G^* = \pi_G^*$. Then S_G^* is a split ring theory, the unit $S \to (S_G)^G$ providing the required splitting map.

One reason for introducing split theories is the following observation, which is due to Kosniowski [10, p. 92].

LEMMA 12. If k_G^* is a split cohomology theory, then the composite

$$k^*(Y/G) \xrightarrow{\zeta} k_G^*(Y/G) \xrightarrow{\pi^*} k_G^*(Y),$$

$\pi\colon Y \to Y/G$, is an isomorphism for all free G-CW complexes Y.

PROOF. We have assumed the result for G, and it follows by suspension that the evident reduced analog holds for $G^+ \wedge S^n$ for all $n \geq 1$. By induction and the five lemma, the result holds for the skeleta of Y. By the $\lim^1$ exact sequence, it holds for Y.

With these preliminaries, we return to the study of completions. We say that the completion conjecture holds for the theory k_G^* if $\varepsilon^*\colon k_G^n(pt) \to k_G^n(EG)$ induces an isomorphism on passage to $I(G)$-adic completion for all integers n, $\varepsilon\colon EG \to pt$. There are theories for which this is false; it is true for real and complex K-theory by Atiyah and Segal [4]. Corollary 4 immediately implies the following reduction, which in fact applies separately to each grading n.

THEOREM 13. The completion conjecture holds for the theory k_G^* if it holds for the theories k_H^* for all subgroups H of prime power order.

Lemma 5 leads to the following further reduction.

PROPOSITION 14. If G is a p-group and k_G^* is split, then the completion conjecture holds for k_G^* if and only if $\varepsilon^*\colon k_G^n(pt) \to k_G^n(EG)$ induces an isomorphism upon passage to p-adic completion for all integers n.

PROOF. Consider the commutative diagram

$$\begin{array}{ccccccccc} 0 & \longrightarrow & K & \longrightarrow & k_G^*(pt) & \xrightarrow{\pi^*} & k^*(pt) & \longrightarrow & 0 \\ & & \downarrow{\varepsilon^*} & & \downarrow{\varepsilon^*} & & \cong\downarrow{\varepsilon^*} & & \\ 0 & \longrightarrow & L & \longrightarrow & k_G^*(EG) & \xrightarrow{\pi^*} & k^*(EG) & \longrightarrow & 0 \end{array}$$

where K and L denote the respective kernels. By assumption, the top row is split exact, and there results a compatible splitting of the bottom row. By Lemma 5, this is a diagram of $A(G)$-modules, where $A(G)$ acts trivially on $k^*(pt)$ and $k^*(EG)$, and the $I(G)$-adic and p-adic topologies on K and on L coincide. The conclusion follows.

The interest in the completion conjecture lies in the algebraic computability of $k_G^*(pt)$ and the homotopical interest of $k_G^*(EG)$. We quickly review the latter, following Atiyah and Segal [4]. We assume that k_G^* is a split ring theory. Then, by the proof of Proposition 9, $A(G)$ acts on k_G^*Y by pullback of the natural $k_G^0(pt)$-module structure along the unit

$$\eta_*: A(G) = \pi_0^G(S_G) \to \pi_0^G(k_G) = k_G^0(pt).$$

By Lemma 12, we have an isomorphism of Z-graded rings

$$k_G^*(EG) \simeq k^*(BG).$$

Let B^qG be the q-skeleton of BG. It is the union of $q + 1$ contractible subsets, hence all (q+1)-fold products are zero in $\tilde{k}^*(B^qG)$, hence the composite

$$k_G^*(pt) \to k_G^*(EG) \cong k^*(BG) \to k^*(B^qG)$$

factors through $k_G^*(pt)/I(G)^{q+1}k_G^*(pt)$. Passing to limits, we obtain $\hat{k}_G^*(pt) \to \lim_q k^*(B^qG)$. If $\lim_q{}^1 k^*(B^qG) = 0$, then the target here is $k^*(BG)$, and this is complete. Thus the completion conjecture asserts that

$$\hat{\varepsilon}^*: \hat{k}_G^*(pt) \to k^*(BG)$$

is an isomorphism. It is this form of the assertion that motivated our original Z-graded formulation. However, there is an easy generalization to an RO(G)-graded formulation. Recall that EG/H is a model for BH for any $H \subset G$.

PROPOSITION 15. Let k_G^* be an RO(G)-graded split ring theory such that each $k_G^\alpha(pt)$ is a finitely generated A(G)-module. Assume that $\lim_q{}^1 k^n(B^qH) = 0$ and the completion map

$$\hat{\varepsilon}^*: \hat{k}_H^n(pt) \to k_H^n(EG) \cong k^n(BH)$$

is an isomorphism for all integers n and subgroups H of G. Then, for any finite G-CW complex Y, the projection $\varepsilon: Y \times EG \to Y$ induces an isomorphism

$$\hat{\varepsilon}^*: \hat{k}_G^\alpha(Y) \to k_G^\alpha(Y \times EG)$$

for all $\alpha \in RO(G)$.

PROOF. Rather than consider $\lim^1$ terms, we observe that the following diagram commutes and define $\hat{\varepsilon}^*$ to be the displayed composite extension of ε^*:

$$\begin{array}{ccc} k_G^\alpha(Y) & \xrightarrow{\quad\varepsilon^*\quad} & k_G^\alpha(Y \times EG) \\ \downarrow & \dashrightarrow & \uparrow \phi \\ \hat{k}_G^\alpha(Y) \cong k_G^\alpha(Y)\otimes_{A(G)}\hat{A}(G) \xrightarrow{1\otimes\hat{\eta}} k_G^\alpha(Y)\otimes_{A(G)}\hat{k}_G^0(pt) & \xrightarrow{1\otimes\hat{\varepsilon}^*} & k_G^\alpha(Y)\otimes_{A(G)}k_G^0(EG) \end{array}$$

Here ϕ is the external product. Our finiteness assumptions ensure that $\hat{k}_G^\alpha(Y)$ is finitely $A(G)$-generated and so give the unlabeled isomorphism [3, 10.13], and of course $\hat{A}(G)$ is $A(G)$-flat [3, 10.14]. Thus $\hat{k}_G^*$ is an $RO(G)$-graded cohomology theory on finite G-CW complexes and is represented by a G-spectrum $\hat{k}_G$; $k_G^*(? \times EG)$ is also such a theory and is represented by the function G-spectrum $F(EG^+, k_G)$. Since $\hat{\varepsilon}^*$ is a morphism of cohomology theories, it is represented by a map of G-spectra $\hat{\varepsilon}: \hat{k}_G \to F(EG^+, k_G)$. By hypothesis (and Lemma 6), $\hat{\varepsilon}^*$ is an isomorphism when $Y = G/H$ and $\alpha = n$. This means that $\hat{\varepsilon}$ induces an isomorphism on equivariant homotopy groups where, for a G-spectrum k_G, $\pi_n^H(k_G) = k_H^{-n}(pt)$. By the Whitehead theorem in the equivariant stable category [13], it follows that $\hat{\varepsilon}$ is an isomorphism in that category and so induces an isomorphism of cohomology theories.

The cited Whitehead theorem asserts that a map $k_G \to k'_G$ of G-spectra is an isomorphism in the equivariant stable category if and only if its fixed point maps $(k_G)^H \to (k'_G)^H$ are isomorphisms in the nonequivariant stable category, and, as one would expect, $(k_G)^H = (k_H)^H$.

Returning to the situation of the proposition, the isomorphism $k_G^*(EG) \cong k^*(BG)$ of Lemma 12 implies an isomorphism

$$F(EG^+, k_G)^G \simeq F(BG^+, k)$$

in the stable category. Thus the completion conjecture asserts that

$$\hat{\varepsilon}: (\hat{k}_G)^G \to F(BG^+, k)$$

is an isomorphism in the stable category. If G is a p-group, we may replace I(G)-adic completion by p-adic completion provided we also p-adically complete the canonical sphere wedge summand of $F(BG^+,k)$, $F(BG,k)$ already being p-complete. Since passage to fixed point spectra commutes with p-adic completion, the completion conjecture here asserts that

$$\varepsilon\colon (k_G)^G \to F(BG^+,k)$$

becomes an isomorphism in the stable category upon completion at p.

In the case $k_G = S_G$, there is an equivalence of spectra

$$\bigvee_{(H)} \xi_H\colon \bigvee_{(H)} \Sigma^\infty BWH^+ \to (S_G)^G,$$

where $WH = N_GH/H$ and the wedge is taken over the conjugacy classes of subgroups H of G; see Segal [17], Kosniowski [10], and tom Dieck [6]. In a sequel, we shall verify that the composite $\varepsilon\cdot\xi_H\colon \Sigma^\infty BWH^+ \to F(BG^+,S)$ has adjoint

$$\tau(1)\colon \Sigma^\infty BWH^+ \wedge BG^+ = \Sigma^\infty(BWH \times BG)^+ \to S,$$

where τ is the transfer associated to the natural cover $E \to BWH \times BG$ with fibre G/H and $1 \in \pi^0E$ is the unit. This will recover the formulation of the Segal conjecture preferred by those engaged in its study by Adams spectral sequence techniques. We refer the reader to Adams [1] for a summary of work in that direction.

We close with the homology analog of Lemma 12, which will be needed in the sequel.

<u>LEMMA 16</u>. If k_G^* is a split cohomology theory, then the composite

$$k_*(Y/G) \xrightarrow{\zeta} k_*^G(Y/G) \xrightarrow{\tau} k_*^G(Y)$$

is an isomorphism for all free G-CW complexes Y, where τ is the equivariant transfer associated to $\pi\colon Y \to Y/G$.

<u>PROOF</u>. Here ζ is induced by $k \to (k_G)^G$; see Remark 11. The map π is a G-fibration, in fact a (G,A)-bundle where A is the group of automorphisms of G regarded as a discrete set. The transfer is induced by a map of G-spectra $\tau\colon \Sigma^\infty(Y/G)^+ \to \Sigma^\infty Y^+$. It must not be confused with the obvious nonequivariant transfer associated to π, which is induced by the map of nonequivariant spectra obtained from τ by neglect of G action. See [16, 18, 13]. There is a relative equivariant transfer compatible with connecting homomorphisms [13], and, as in Lemma 12, it suffices to prove the result for $Y = G$. Here the conclusion holds by hypothesis since we have commutative diagrams

$$\begin{array}{ccc} k_n^G(pt) & \xrightarrow{\tau} & k_n^G(G), \\ \| & * & \downarrow \delta \\ k_G^{-n}(pt) & \xrightarrow{\pi} & k_G^{-n}(G) \end{array}$$

where δ is an equivariant Spanier-Whitehead duality isomorphism [19, 13, 14].

BIBLIOGRAPHY

1. J. F. Adams. Graeme Segal's burnside ring conjecture. Bull. Amer. Math. Soc. To appear.

2. M. F. Atiyah. Characters and cohomology of finite groups. Inst. Hautes Etudes Sci. Publ. Math. No. 9 (1961), 23-64.

3. M. F. Atiyah and J. G. Macdonald. Introduction to commutative algebra. Addison-Wesley Publishing Co. Reading, Mass. 1969.

4. M. F. Atiyah and G. B. Segal. Equivariant K-theory and completion. J. Diff. Geometry 3 (1969), 1-18.

5. T. tom Dieck. Equivariant homology and Mackey functors. Math. Ann. 206 (1973), 67-78.

6. T. tom Dieck. Orbittypen und aquivariante Homologie II. Arch. Math 26 (1975), 650-662.

7. T. tom Dieck. Transformation groups and representation theory. Springer Lecture Notes in Mathematics Vol 766. 1979.

8. A. Dress. A characterization of solvable groups. Math. Z. 110 (1969), 213-217.

9. A. Dress. Contributions to the theory of induced representations. Springer Lecture Notes in Mathematics Vol 342, 1973, pp. 183-240.

10. C. Kosniowski. Equivariant cohomology and stable cohomotopy. Math. Ann. 210 (1974), 83-104.

11. E. Laitinen. On the Burnside ring and stable cohomotopy of a finite group. Math. Scand. 44 (1979), 37-72.

12. L. G. Lewis, J. P. May, and J. E. McClure. Ordinary RO(G)-graded cohomology. Bull. Amer. Math. Soc. 4 (1981), 208-212.

13. L. G. Lewis, J. P. May, J. E. McClure, and M. Steinberger. Equivariant stable homotopy theory. Springer Lecture Notes in Mathematics. To appear.

14. L. G. Lewis, J. P. May, J. E. McClure, and S. Waner. Equivariant cohomology theory. To appear.

15. I. Madsen. Smooth spherical space forms. Springer Lecture notes in Mathematics Vol. 657, 1978, 303-352.

16. G. Nishida. The transfer homomorphism in equivariant generalized cohomology theories. J. Math Kyoto 18(1978), 435-451.

17. G. Segal. Equivariant stable homotopy theory. Actes Congrès Intern. Math. 1970. Tome 2, 59-63.

18. S. Waner. Equivariant fibrations and transfer. Trans. Amer. Math. Soc. 258 (1980), 369-384.

19. K. Wirthmuller. Equivariant S-duality. Arch. Math. 26 (1975), 427-431.

DEPARTMENT OF MATHEMATICS
UNIVERSITY OF CHICAGO
CHICAGO, ILLINOIS
60637

DEPARTMENT OF MATHEMATICS
JOHNS HOPKINS UNIVERSITY
BALTIMORE, MARYLAND
21701

Canadian Mathematical Society
Conference Proceedings
Volume 2, Part 2 (1982)

THE EQUIVARIANT J HOMOMORPHISM AND SMITH EQUIVALENCE OF REPRESENTATIONS

Ted Petrie

ABSTRACT. In this paper G is a finite group and V is a representation of G. We treat equivariant cohomology theories $K_G^*(\,;V)$ and $\omega_G^*(\,;V)$ generalizing $K_G^*(\ )$ of [A] and $\omega_G^*(\ )$ of [Se]. In particular we study the equivariant J homomorphism $J_V\colon K_G^o(\,;V) \to J_G^o(\,;V)$, compare this with the ordinary equivariant J homomorphism $J\colon K_G^o(\ ) \to J_G^o(\ )$ and relate the latter to the Adams operations. In [P4] the results are applied to an old problem in transformation groups [Sm]. This application is discussed in sections 0 and 1 for motivation. The first use of the functors $K_G^*(\,;V)$ to transformation groups on manifolds is due to R. Schultz [Sc].

§0. INTRODUCTION

An old problem of Smith [Sm] states: Describe the representations (V,W) of G which occur as $(T_p\Sigma, T_q\Sigma)$ for Σ a smooth G sphere with $\Sigma^G = p \cup q$ (two points). Here $T_p\Sigma$ is the representation of G on the tangent space at p. Under these conditions we say V and W are Smith equivalent and write $V \sim W$. In [P1] and [P2] we considered the Smith problem for a family of abelian groups and exhibited an ideal in the representation ring of G which is realized as the set of differences of Smith equivalent representations. In particular this provided the first examples of inequivalent representations which are Smith equivalent. A stronger equivalence relation on representations is also interesting. We say V and W are s-Smith equivalent if $(V,W) = (T_p\Sigma, T_q\Sigma)$ and Σ^K is a homotopy sphere for all $K \subseteq G$ with $\Sigma^G = p \cup q$. For this we write $V \approx W$.

Using the methods of [P3] we give in [P4] a necessary and sufficient condition for $V \approx W$ provided one of the representations is stable. This result, stated in section 1 as Theorem A, involves the equivariant J homomorphism for an equivariant K theory $K_G^*(\cdot,V)$ depending upon a representation V of G. To implement Theorem A we need to compare $K_G^*(\cdot,V)$ and its J homomorphism J_V with ordinary equivariant K theory $K_G^*(\ )$ and its J homomorphism denoted by J. The point here is that Ker J can be studied via the Adams operations in $K_G^*(\ )$. This relationship (see e.g. 3.17 and 3.18) is

© 1982 American Mathematical Society
0731-1036/82/0000-0469/$03.75

essential in the proof of Theorem B from [P4]. See section 1. The main results of this paper 2.9, 3.15 and 3.18 are applied in [P4] (see Theorem A in section 1) to give a simple explicit sufficient condition for $V \approx W$. See Theorem C in section 1.

The author warmly thanks R. Schultz for several useful comments on this paper.

§1. MOTIVATION

In this section we set up notation and explain the application of the results here to the problem of s-Smith equivalence.

Though many of the results here and in [P4] hold more generally we restrict attention to the case G is cyclic of order $|G| = 2d$. The subgroup of index 2 is denoted by H and the two Sylow subgroup is denoted by G_2. We use various combinations of these properties of a representation V of G:

1.1) i) $V^{G_2} = 0$ and $V^{H_2} = V^H$.

ii) $H \in \mathrm{Iso}(V)$

iii) $\mathrm{Iso}(V/V^H - 0)$ is closed under subgroups.

Here $\mathrm{Iso}(X) = \{G_x | x \in X\}$ denotes the set of isotropy groups of the G space X. Restriction from G to a subgroup L is denoted by res_L. The representation V of G is stable if for each $L \in \mathrm{Iso}(V)$ and each non trivial real representation χ of K either the multiplicity of χ in V is zero or $m_\chi \dim \chi > \dim V^K$. Let $\mathrm{Vect}_G(Y;V)$ denote the additive semi-group of G vector bundles ξ over the G space Y such that $\mathrm{Iso}(\xi) \subset \mathrm{Iso}(V)$. The associated Grothendieck group $K_0(\mathrm{Vect}_G(Y;V))$ is denoted by $K_G^o(Y;V)$. When $\mathrm{Iso}(V)$ contains $\mathcal{S}(G)$ - the set of all subgroups of G - this is $K_G^o(Y)$. Like [A], the groups $K_G^i(Y,V)$ for $i \leq 0$ are defined. As usual $KO_G^o(\ ,V)$ will denote real G vector bundle theory and $K_G^o(\cdot,V)$ the complex case. A G map $f\colon X \to Y$ induces a homomorphism $f_V^*\colon K_G^*(Y;V) \to K_G^*(X;V)$ (f^* when $\mathrm{Iso}(V) = \mathcal{S}(G)$). The inclusion $\mathrm{Vect}_G(Y;V) \to \mathrm{Vect}_G(Y,V')$ induces a homomorphism $K_G^o(Y;V) \to K_G^o(Y;V')$ which is denoted by ρ when $\mathrm{Iso}(V')$ contains $\mathcal{S}(G)$. Here $V \subset V'$.

We now state two of the main results of [P4].

Theorem A [P4]. Let V and W be representations of G with $V^G = W^G = 0$ and V stable. A necessary and sufficient condition that $V \approx W$ is that there exist $\xi_+, \xi_- \in \mathrm{Vect}_G(Y,V)$ $(Y = S(V+R))$ - the unit sphere of $V+R$ where R is the trivial one dimensional representation) and a G fiber homotopy equivalence $\theta\colon \xi_- \to \xi_+$ such that

i) $i_V^*\xi = (0,V-W)$, $\xi = \xi_+ - \xi_- \in KO_G^o(Y;V)$ $i\colon Y^G \to Y$

ii) A sequence of obstructions $\{\sigma_K(\theta) \in L^h_{\dim V^K}(G/K,(-1)^{\dim V^K}) | K \in \mathrm{Iso}(V)\}$ vanish. Here $L_n^h(G,W)$ is the Wall group [P3].

Remark: Observe that $Y^G = p \cup q$ implies $KO_G^o(Y^G;V) = KO_G^o(p;V) \oplus KO_G^o(q;V)$ and $KO_G^o(q,V)$ is contained in the real representation ring $RO(G)$; so $(0,V-W) \in KO_G^o(Y^G;V)$ makes sense.

Let $\equiv$ be the relation on $\mathrm{Vect}_G(Y;V)$ (3.19) defined by G fiber homotopy equivalence. Set $\underline{\mathrm{Vect}}_G(Y;V) = \mathrm{Vect}_G(Y;V)/\equiv$ and $J_G^o(Y;V) = K_0(\underline{\mathrm{Vect}}_G(Y;V))$. Then $J_V\colon K_G^o(Y;V) \to J_G^o(Y;V)$ is defined in an obvious way. For the real case the symbol JO_V is used. Note that Theorem A implies $JO_V(\xi) = 0$. Since the Wall groups $L_{4n+1}^h(G,-1)$ vanish when G is cyclic, theorem A simplifies to

Theorem A'. Suppose V and W are as in Theorem A and $\dim V^K \equiv 1(4)$ whenever $K \in \mathrm{Iso}(V-0)$. A necessary and sufficient condition that $V \approx W$ is that there is a $\xi \in KO_G^o(Y;V)$ such that $JO_V(\xi) = 0$ and $i_V^*(\xi) = (0,V-W)$.

A simple argument shows that $V \approx W$ implies $V-W \in RO(G)$ lies in the image of $R(G) \to RO(G)$; so the following replacement in Theorem A' makes sense and gives a sufficient condition for s-Smith equivalence:

(*): There is a $\xi \in K_G^o(Y;V)$ with $i_V^*\xi = (0,V-W)$ and $J_V(\xi) = 0$.
(Working with complex G vector bundles simplifies a lot.)

In order to achieve *, we first treat the case where $\mathrm{Iso}(V) = \mathcal{S}(G)$ for then $\mathrm{Ker}\, J_V$ can be studied via Adams operations. Here is a result in that direction from [P4]. Let $L \subset G_2$ have index ≥ 4. Let $\mathrm{fix}_L\colon R(G) \to R(G/L)$ be the homomorphism induced by $V \to V^L$ for a representation V of G. Set $j(L) = 1$ if the index of L in G exceeds 8 and let $j(L) = 2$ otherwise.

Theorem B. Suppose V is a representation of G which satisfies 1.1 i-iii. If $\dim V^H$ is odd and for some $L \subset G_2$,
$z \in 2^{\nu(V)+j(L)} \cdot \mathrm{Ker}(R(G) \xrightarrow{\mathrm{res}_H \times \mathrm{fix}_L} R(H) \times R(G/L))$, then $(0,z) = i^*(\xi')$ with $J(\xi') = 0$ for some $\xi' \in K_G^o(Y)$. (Here $\nu(V)$ is a simple numerical function of V which does not concern us here. See [P4].)

Two of the main results here are theorems 2.9 and 3.15. They assert that i) $\xi' = \rho(\xi)$ for some $\xi \in K_G^o(Y;V)$ where $\rho\colon K_G^o(Y;V) \to K_G^o(Y)$ is the natural homomorphism and ii) $J_V(2^{b(V)}\xi) = 0$ where $b(V) = \dim V^H - 1$ when $\dim V^H$ is odd. Combining this with Theorem A' gives

Theorem C. Let V be a stable representation of G which satisfies 1.1 i-iii. Suppose $\dim V^K \equiv 1(4)$ for all $K \in \mathrm{Iso}(V-0)$. If $V-W \in 2^{\nu(V)+j(L)+b(V)} \cdot \mathrm{Ker}(R(G) \to R(H) \times R(G/L))$, for some $L \subset G_2$ of index at least 4, then $V \approx W$.

Cappell-Shaneson in "Fixed points of periodic differentiable maps" have announced some results on s-Smith equivalence. These results are special cases of theorems A and C. Alan Siegel in his Rutgers thesis has studied the rational Smith equivalence relation. This is defined like Smith equivalence except Σ is only required to be a rational homology sphere. He also has some results on Smith equivalence.

§2.

In this section we treat some aspects of equivariant K theory which are concerned with the image of

$$\rho: K_G^o(Y;V) \to K_G^o(Y)$$

when $Y = S(V+R)$. Here R is the one dimensional trivial representation of G and V is a representation of G which satisfies 1.1 i-iii). This assumption is retained throughout this paper. The unique non trivial one dimensional representation of G is denoted by R_-. Note that the subgroup H of index 2 acts trivially on R_- and $R_- \subset V$ because $H \in \mathrm{Iso}(V)$.

Lemma 2.1. If G acts on X with $X^H = X$, then $E \in \mathrm{Vect}_G(X)$ implies $\mathrm{Hom}_H(\chi,E) \in \mathrm{Vect}_{G/H}(X) \subset \mathrm{Vect}_G(X;V)$ for any complex representation χ of G.

Proof: First $\mathrm{Hom}_H(\chi,E)$ is a G/H vector bundle via $(gf)(x) = gf(g^{-1}x)$ for $f \in \mathrm{Hom}_H(\chi,E)$, $g \in G$ and $x \in \chi$. Since H acts trivially on $\mathrm{Hom}_H(\chi,E)$, this is in $\mathrm{Vect}_{G/H}(X) \subset \mathrm{Vect}_G(X;V)$. The last inclusion is an obvious consequence of the assumption that H (and G) are isotropy groups of V.

Lemma 2.2. For any G space X, $E \in \mathrm{Vect}_G(X;V)$ iff the restriction E_p of E to $p \in X$ is in $\mathrm{Vect}_{G_p}(p;V)$ for all $p \in X$.

Proof: This easy fact is left to the reader.

Definition: $\chi \in V^+$ means χ is a complex representation of G such that $\mathrm{Iso}(\chi) \subset \mathrm{Iso}(V)$.

Identify $R(G)$ with $Z[t]/(1-t^{2d})$ in the usual way.

Lemma 2.3. $\chi \in V^+$ iff $t^d \cdot \chi \in V^+$.

Proof: Note for any representations A and B:

i) $\mathrm{Iso}(A \oplus B) = \{H \cap K \mid H \in \mathrm{Iso}(A),\ K \in \mathrm{Iso}(B)\}$,

ii) $\mathrm{Iso}(A)$ is closed under intersections and

iii) $\mathrm{Iso}(nA) = \mathrm{Iso}(A)$ for any integer n.

This means $\chi \in V^+$ iff each irreducible constituent of χ is in V^+. Then $\chi = t^i$ and $\chi \in \mathrm{Iso}(V)$ implies either $|H_2|$ does not divide i or i is d (1.1i). If i is d, $t^{i+d} = t^0$ and $\mathrm{Iso}(t^0) = \{G\} \in \mathrm{Iso}(V)$; so $t^{i+d} \in V^+$. If $|H_2|$ does not divide i, $(|G|,i) = (|G|,i+d) = k$ and $\mathrm{Iso}(t^i) = \mathrm{Iso}(t^{i+d}) = \{G,Z_k\}$; so $t^i \in V^+$ implies $t^{i+d} \in V^+$.

Lemma 2.4. If G acts on X with $X^H = X$ and $E \in \mathrm{Vect}_G(X)$, then $E = E_V \oplus E_V^\perp$ where $E_V \in \mathrm{Vect}_G(X;V)$ and $E \in \mathrm{Vect}_G(X;V)$ iff $E_V^\perp = 0$.

Proof: For each irreducible complex representation ψ of H choose a representation $\tilde\psi$ of G such that $\mathrm{res}_H \tilde\psi = \psi$. (Here res_H denotes restriction to H.) Set

$$E_V = \bigoplus_{\psi \mid \tilde\psi \in V^+} \tilde\psi \otimes \mathrm{Hom}_H(\tilde\psi,E)$$

and define $E_V^\perp$ to be the complement of E_V in E. See [A2]. Then $E_V \in \mathrm{Vect}_G(X,V)$. To see this use lemmas 2.1-2.3. For $p \in X$

$$E_{Vp} = \bigoplus_{\psi|\tilde{\psi}\in V^+} \mathrm{res}_{G_p} \tilde{\psi} \otimes \mathrm{Hom}_H(\tilde{\psi}, E_p).$$

If $G_p = H$, this is $\bigoplus_{\psi|\tilde{\psi}\in V^+} \psi\cdot a_\psi$ with a_ψ a non negative integer. Now $\mathrm{Iso}(\psi) = \mathrm{Iso}(\tilde{\psi}) \cap H$. Since H is an isotropy group of V and $\mathrm{Iso}(\tilde{\psi}) \subset \mathrm{Iso}(V)$ and $\mathrm{Iso}(V)$ is closed under intersection (true for any representation V), it follows that $\mathrm{Iso}(\psi) \subset \mathrm{Iso}(V)$ for each ψ such that $\tilde{\psi} \in V^+$. Since $\mathrm{Iso}(E_{Vp})$ is the collection of intersections $\bigcap_{a_\psi>0} H_\psi$ $H_\psi \in \mathrm{Iso}(\psi)$, it follows that $E_{Vp} \in \mathrm{Vect}_{G_p}(p;V)$. If $G_p = G$, note $\mathrm{Hom}_H(\tilde{\psi}, E_p)$ is a representation of G on which H acts trivially; so is a sum of copies of 1 and t^d. Thus $\tilde{\psi} \otimes \mathrm{Hom}_H(\tilde{\psi}, E_p)$ is a sum of copies of $\tilde{\psi}$ and $\tilde{\psi}t^d$ and both are in V^+ by Lemma 2.3. Thus $E_p \in \mathrm{Vect}_{G_p}(p;V)$. Since $\mathrm{Iso}(X) \subset \{G,H\}$, this shows $E_V \in \mathrm{Vect}_G(X;V)$.

Now suppose $E \in \mathrm{Vect}_G(X;V)$; so $E_p \in \mathrm{Vect}_{G_p}(p;V)$ for $p \in X$. If $G_p = H$,

$$E_p = \bigoplus_{\psi \text{ irred rep of } H} \psi\cdot a_\psi \qquad a_\psi = \dim_{\mathbb{C}} \mathrm{Hom}_H(\psi, E_p).$$

If $\tilde{\psi} \notin V^+$, then $\mathrm{Iso}(\psi) = \mathrm{Iso}(\tilde{\psi}) \cap H \notin \mathrm{Iso}(V)$; so $a_\psi = 0$ and $E = E_V$. Similarly if $G_p = G$. This proves Lemma 2.4.

<u>Lemma 2.5</u>. $0 \to K_G^1(SV^H) \xrightarrow{\mathrm{res}_H} K_H^1(SV^H)$ is exact and $K_G^1(SV^H) = 0$ if $\dim V^H$ is odd.

Proof: By [A2] $K_G^1(S(V^H)) = R(G)\otimes_{R(G/H)} K^1_{G/H}(S(V^H))$. As $V^G = 0$, G/H acts freely on $S(V^H)$; so $K^1_{G/H}(S(V^H)) = K^1(PV^H)$ where PV^H is real projective space of dimension equal to $\dim V^H - 1$. If $\dim V^H$ is odd, this is zero. If $\dim V^H$ is even, we may view it as a complex representation of V. Then $K_G^1(S(V^H))$ is the ideal in $R(G) = Z[t]/(1-t^{2d})$ which annihilates $\lambda_{-1}(V^H) = (1-t^d)^{\dim_{\mathbb{C}} V^H}$. See [A]. Similarly $K_H^1(S(V^H))$ is the ideal in $R(H)$ which annihilates $\mathrm{res}_H \lambda_{-1}(V^H) = 0$ i.e. $K_H^1(S(V^H))$ is $R(H)$. One checks explicitly that res_H is injective in this case.

<u>Lemma 2.6</u>. The following diagram has exact rows. The left column is exact.

$$\begin{array}{ccccccc}
 & & 0 & & & & \\
 & & \downarrow & & & & \\
0 & \to & K_G^1(S(V^H)) & \to & K_G^0(S(V^H+R)) & \xrightarrow{j^*} & K_G^0(S(R)) \\
 & & \downarrow \mathrm{res}_H & & \downarrow \mathrm{res}_H & & \downarrow \mathrm{res}_H \\
0 & \to & K_H^1(S(V^H)) & \to & K_H^1(S(V^H+R)) & \to & K_H^0(S(R))
\end{array}$$

Proof: The rows arise from the Mayer-Vietoris sequence for $S(V^H+R)$. Apply Lemma 2.5 and use $K_G^1(S(R)) = 0$.

<u>Lemma 2.7</u>. Let X and Y be G c.w. complexes. Suppose $X \subset Y$ and $\mathrm{Iso}(Y-X)$ is closed under subgroups. Let W be a representation of G such

that $\mathrm{Iso}(W) \supset \mathrm{Iso}(Y-X)$. Then $\rho\colon K_G^*(Y,X;V) \to K_G^*(Y,X)$ is an isomorphism.

Proof: Suppose Y is obtained from X by adding one G cell of type $G/K\times D^i$; so $K \in \mathrm{Iso}(Y-X)$. Then $K_G^*(Y,X;W) \cong K_K^*(D^i,S^{i-1};\mathrm{res}_K W) = K_K^*(D^i,S^{i-1}) = K_G^*(Y,X)$. The first equality is implied by $\mathrm{res}_K W \supset S(K)$ because $\mathrm{Iso}(W) \supset \mathrm{Iso}(Y-X) \ni K$ and $\mathrm{Iso}(Y-X)$ is closed under subgroups.

Now consider the diagram 2.8 for $(Y,X) = (\Sigma Y',\Sigma X')$ where $\Sigma Y'$ is the suspension of Y'. Set $K_G^1(Y,X) = K_G^o(Y',X')$.

$$\text{2.8)}\qquad \begin{array}{ccccccc} K_G^o(Y,X;V) & \xrightarrow{j_V^*} & K_G^o(Y;V) & \xrightarrow{k_V^*} & K_G^o(X;V) & \xrightarrow{\delta_V} & K_G^1(Y,X;V) \\ \cong\downarrow\rho & & \rho\downarrow & & \rho\downarrow & & \cong\downarrow\rho \\ K_G^o(Y,X) & \xrightarrow{j^*} & K_G^o(Y) & \xrightarrow{k^*} & K_G^o(X) & \longrightarrow & K_G^1(Y,X) \end{array}$$

<u>Theorem 2.9</u>. Suppose $\xi' \in K_G^o(S(V+R))$ and $i^*\xi' \in K_G^o(S(R);V) \subset K_G^o(S(R))$ where $i\colon S(R) \to S(V+R)$ is the inclusion. If dimension V^H is even, suppose $\mathrm{res}_H \xi' = 0$. Then $\xi' = \rho(\xi)$ for some $\xi \in K_G^o(S(V+R);V)$.

Proof: Consider 2.8 with $(Y,X) = (S(V+R), S(V^H+R))$. By 2.4, $k^*(\xi') = \xi'_V + \xi'^{\perp}_V$ and $\xi'_V = \rho(\xi'')$ for some $\xi'' \in K_G(S(V^H+R);V)$. By hypothesis $i^*(\xi') = j^*k^*(\xi') = j^*\xi'_V + j^*\xi'^{\perp}_V \in K_G^o(S(R);V)$; so $j^*\xi'^{\perp}_V = 0$ by Lemma 2.4. By 2.5 $\xi'^{\perp}_V = 0$ if $\dim V^H$ is odd. If $\dim V^H$ is even, $\mathrm{res}_H \xi' = 0$; so $\mathrm{res}_H k^*(\xi') = 0$. This means $\mathrm{res}_H \xi'_V$ and $\mathrm{res}_H \xi'^{\perp}_V$ are both zero. This uses $K_G^*(X) = R(G)\otimes_{R(G/H)} K^*{}_{G/H}(X) = R(H)\otimes K^*_{G/H}(X)$ and $K_H^*(X) = R(H)\otimes K^*(X)$ [A2] when $X^H = X$. Since $\mathrm{res}_H \xi'^{\perp}_V$ and $j^*\xi'^{\perp}_V$ are both zero, 2.6 implies $\xi'^{\perp}_V$ is zero; so $k^*(\xi') = \rho(\xi'')$ for $\xi'' \in K_G^o(X;V)$. By 2.7, the ρ in the left and right hand columns of 2.8 is an isomorphism. From this it follows that $\delta_V(\xi'') = 0$; so $\xi'' = k_V^*(\xi_0)$ for some $\xi_0 \in K_G^o(Y;V)$. Then $k^*(\xi' - \rho(\xi_0)) = 0$; so $\xi' - \rho\xi_0 = j^*\rho(\mu) = \rho j_V^*(\mu)$; thus, $\xi' = \rho(\xi_0 + j_V^*(\mu))$.

§3. EQUIVARIANT HOMOTOPY

Let A be a representation of G. The identity component $M(A)$ of the space of self maps of $S(A)$ becomes a G space via

$$f \longmapsto gfg^{-1} \qquad f \in M(A), \quad g \in G.$$

For any representation W of G and subgroup K set

$$\text{3.0}\qquad F(W) = \varinjlim_{\mathrm{Iso}(A) \subset \mathrm{Iso}(W)} M(A) \qquad\qquad F(K) = \varinjlim_{\mathrm{Iso}(A-0) = K} M(A)$$

The identity map of $S(A)$ for each A gives rise to the basepoint of $F(W)$.

3.1 <u>Remark</u>: It is also useful to deal with $M_p(A)$ the space of proper self maps of A. There are natural G maps $M_p(A) \to M(A+R)$ and $M(A) \to M_p(A)$

obtained by the one point compactifications functor and radial extension. We get the same G space F(W) if M(A) is replaced in 3.0 by $M_p(A)$.

Composition of maps in M(A) gives rise to a multiplication in F(W) which is compatible with its G action. It has a base point, coming from the identity S(A). If $W \subset W'$ and $K \in \mathrm{Iso}(W')$, there are equivariant inclusions of F(W) and F(K) into F(W'). Using multiplication in F(W') these give a G map

3.2 $$T : F(W) \times \prod_{K \in \mathrm{Iso}(W') - \mathrm{Iso}(W)} F(K) \to B(W'), \text{ and}$$

3.3 T^G is a homotopy equivalence for G cyclic. Compare [Se].

Define a G cohomology theory by

3.4 $$\omega_G^{-i}(X;W) = [S^i \wedge X^+, F(W)]^G \quad i \geq 0.$$

This is abbreivated by $\omega_G^{-i}(X)$ when Iso(W) contains all subgroups of G. Here $[X,Y]^G$ denotes the set of G homotopy classes of base point preserving G maps from X to Y.

Lemma 3.5. Let A be a representation of G such that $\mathrm{Iso}(A) \subset \mathrm{Iso}(W)$. Let $i: S(R) \to S(A+R) = Y$ be the inclusion. Then there is a homomorphism

$$j_W: \mathrm{Ker}\{K_G^o(S(A+R);W) \xrightarrow{i_W^*} K_G^o(S(R);W) \xrightarrow{J_W} J_G^o(S(R);W)\} \to \omega_G^o(S(A);W)$$ such that if $\alpha \in K_G^o(S(A+R);W)$ and $J_W i_W^*(\alpha) = 0$; then $J_W(\alpha) = 0$ iff $j_W(\alpha) = 0$ (3.20).

Proof: Let $\alpha \in \mathrm{Ker}(J_W i_W^*)$. Then $\alpha = \xi - Y\times M$ for some representation M of G and there is a G fiber homotopy equivalence $\omega: i_W^*\xi \to i_W^* Y\times M$. Let C and B be the representations of G defined by $i_W^*\xi = (C,B) \in K_G^o(S(R))$; so if $S(R) = p \cup q$, C and B are the fibers of ξ over p and q. Let ω_p resp. ω_q be the restrictions of ω to these fibers. Then $\lambda = \omega_p\omega_q^{-1}: B \to C$ is a G fiber homotopy equivalence. Let $\theta: S(A)\times C \to S(A)\times B$ be the G vector bundle isomorphism which defines ξ through

$$\xi = D_+(A)\times C \cup_\theta D_-(A)\times B$$

where $D_+(A)$ and $D_-(A)$ are the upper and lower hemisphere of S(A+R). Then θ extends to a G fiber homotopy equivalence $\Theta: D_+(A)\times C \to D_+(A)\times B$ iff $1\times\lambda\cdot\theta: S(A)\times C \to S(A)\times C$ extends (stably) to $\psi: D_+(A)\times C \to D_+(A)\times C$ with ψ the identity on $o \times C$. Define

$$j_V(\alpha) = [1\times\lambda\cdot\theta] \in \omega_G^o(S(A);W).$$

(Here remark 3.1 is used.) If $j_V(\alpha) = 0$, then $\Theta: D_+(A)\times C \to D_+(A)\times B$ extends θ. Use Θ to construct a G fiber homotopy equivalence $\omega': \xi \to Y\times B$. Then $1\times\omega_q^{-1}\circ\omega': \xi \to Y\times M$ is a G fiber homotopy equivalence; so $J_V(\xi - Y\times M) = 0$.

In order to exploit 3.5 we need further information about F(W) and its associated cohomology theory $\omega_G^*(\cdot;W)$. We view the cyclic group G as a subgroup of $G_\infty = S^1$.

Lemma 3.6. The G action on $F(W)$ extends to a G_∞ action.

Proof: Since $F(W)$ only depends on $\mathrm{Iso}(W)$, we may suppose W is a complex representation of G. Then $W = \mathrm{res}_G W_\infty$ for some complex representation W_∞ of G_∞; so $\mathrm{res}_G F(W_\infty) = F(W)$.

Corollary 3.7. The G/H action on $F(W)^H$ extends to a $G_\infty/H = S^1$ action.

Corollary 3.8. G/H acts trivially on $\omega_H^*(pt)$.

Lemma 3.9. Let V be a representation of G which satisfies 1.1. Suppose $\dim V^H$ is odd. Then $2^{\dim V^H-1}$ annihilates $\omega_G^{-1}(S(V^H),S(R_-))$. (Note $R_- \subset V$ by 1.1 ii.)

Proof: There is a spectral sequence $H^p(S''/G, S'/G, \omega_H^q(pt)) \Longrightarrow \omega_G^{p+q}(S'',S')$ when G/H acts freely on S''. Apply this to $(S'',S') = (S(V^H), S(R_-))$. Then $(S''/G, S'/G) = (RP^n, \text{point})$ $n = \dim V^H-1$. By 3.8 the coefficients in the spectral sequence are not twisted; so for $S''/G = RP^n$ 2 annihilates each term in the spectral sequence because n is even. This implies 2^n annihilates $\omega_H^{-1}(S'',S')$ because it has a filtration of length n whose successive quotients are killed by 2.

Lemma 3.10. Let V satisfy 1.1. Then for $i \geq 1$, 2 annihilates the cokernel of $\omega_G^{-i}(S(V);W) \to \omega_G^{-i}(S(R_-);W)$ for any representation W of G. (Since $\mathrm{Iso}(V)$ contains H, $R_- \subset V$.)

Proof: Observe that $S(R_-)$ is G/H as a G space; so $\omega_G^{-i}(S(R_-)) = \omega_H^{-i}(pt)$. The composition of the induction homomorphism $\omega_H^{-i}(S(V)) \to \omega_G^{-i}(S(V))$ and the restriction homomorphism from G to H is multiplication by $|G/H| = 2$ (see 3.20); so $\omega_G^{-i}(S(V)) \to \omega_H^{-i}(S(V))$ has cokernel annihilated by 2. Since $\omega_H^{-i}(S(V)) \to \omega_H^{-i}(pt)$ is surjective, we are done.

Corollary 3.11. Let V satisfy 3.9. Then $\gamma\colon \omega_G^0(S(V),S(R_-)) \to \omega_G^0(S(V))$ has kernel annihilated by 2.

Proof: The kernel of the given homomorphism is the cokernel of $\omega_G^{-1}(S(V)) \to \omega_G^{-1}(S(R_-))$ which is zero by 3.10.

Lemma 3.12. Let X and Y be G c.w. complexes. Suppose $X \subset Y$ and $\mathrm{Iso}(Y-X)$ is closed under subgroups. Let W be a representation of G such that $\mathrm{Iso}(W) \supset \mathrm{Iso}(Y-X)$. Then $\omega_G^*(Y,X;W) \to \omega_G^*(Y,X)$ is an isomorphism.

Proof: The proof is the same as 2.7.

Lemma 3.13. Let $V^{G_2} = 0$ and $W \subset W'$. Then $\omega_G^0(S(V^H);W) \to \omega_G^0(S(V^H);W')$ is split injective.

Proof: Note $\omega_G^0(S(V^H);W) = [S(V^H)^+, F(W)^H]^{G/H}$ and similarly for W'. By 3.3 T^H is a G/H map which is a homotopy equivalence. Now use G/H acts freely on $S(V^H)$ to complete the proof.

Lemma 3.14. Let V be a representation of G which satisfies 1.1 and $\dim V^H$

is odd. Then 2^n annihilates $\mathrm{Ker}(\omega_G^o(S(V);V) \xrightarrow{\rho} \omega_G^o(S(V)))$ for $n = \dim V^H$.

Proof: Let $S = S(V)$, $S' = S(R_-)$. Consider the diagram:

$$\begin{array}{ccccc}
 & & \omega_G^{-1}(S,S') & & \\
 & & \downarrow\theta' & & \\
\omega_G^o(S,S^H;V) & \xrightarrow{\rho_1} & \omega_G^o(S,S^H) & \xrightarrow{\theta} & \omega_G^o(S,S') \\
\downarrow j_V^* & & j^*\downarrow & \swarrow\gamma & \\
\omega_G^o(S;V) & \xrightarrow{\rho_2} & \omega_G^o(S) & & \\
\downarrow k_V^* & & \downarrow k^* & & \\
\omega_G^o(S^H;V) & \xrightarrow{\rho_3} & \omega_G^o(S^H) & &
\end{array}$$

If $x \in \ker \rho_2$, then $x \in \mathrm{Ker}\, k_V^*$ by 3.13; so $x = j_V^*(x')$ and $\gamma\theta\rho_1(x') = 0$. By 3.11, $2\theta\rho_1(x') = 0$; so $2\rho_1(x') = \theta'(x'')$. By 3.9, $2^{n-1}\cdot x'' = 0$; so $\rho_1(2^n x') = 0$. But ρ_1 is an isomorphism by 3.12.

Theorem 3.15. Suppose V satisfies 1.1, $\dim V^H$ is odd and $\xi \in K_G^o(S(V+R);V)$ with $J(\rho(\xi)) = 0$. Then $J_V(2^n\xi) = 0$ $n = \dim V^H$.

Proof: Since $J_G^o(S(R);V) \to J_G^o(S(R))$ is a monomorphism (easy check for G cyclic), $J(\rho(\xi)) = 0$ implies $J_V(i_V^*\xi) = 0$ (i_V: $S(R) \to S(V+R)$); so $j_V(\xi)$ is defined (3.5). Since $J(\rho(\xi)) = 0$, $j(\rho(\xi)) = \rho j_V(\xi) = 0$. By 3.14, $j_V(2^n\xi) = 0$. By 3.5 $J_V(2^n\xi) = 0$.

The restriction $\dim V^H$ is odd can be removed using a trick which we now explain. Let $V' = V \oplus R_-$; so $\dim V'^H$ is odd if $\dim V^H$ is even.

Lemma 3.16. Let $\eta \in K_G^o(S(V+R))$ and suppose $\mathrm{res}_H\eta = 0$. Then $\eta = j^*\eta'$ for some $\eta' \in K_G^o(S(V'+R))$ $j{:}S = S(V+R) \to S(V'+R) = S'$.

Proof: The cokernel of j^* injects into $K_G^1(S',S) \cong K_G^o(S(R_-))$. This isomorphism is a consequence of the suspension axiom and the Thom isomorphism theorem for K_G^* theory. Now $S(R_-)$ as a G space is G/H; so $K_G^o(S(R_-)) \cong K_H^o(pt)$ where pt is a point. Put these isomorphisms together to see that coker j^* is the kernel of $(K_G^o(S) \to K_H^o(pt))$ obtained by the composition of res_H and restricting to pt. By hypothesis η is in this kernal.

The following theorem from [P4] may be used to produce $\eta \in K_G^o(Y)$ with $J(\eta) = 0$ for any G space Y.

Theorem 3.17 [P4]. Let a and b be units in $Z/|G|Z$ where G is any abelian group. For any $\xi \in K_G^o(Y)$ there are integers p and q which represent a and b in $Z/|G|Z$ such that $J((\psi^p-1)(\psi^q-1)\xi) = 0$. Here ψ^p is the Adams operation.

We combine 3.15 - 3.17 with 2.9

Theorem 3.18. Let V be a representation of G satisfying 1.1. Let

$b(V) = \dim V^H$ if this is odd and $\dim V^H+1$ if $\dim V^H$ is even. Let a and b be units in $Z/|G|Z$. Given any $\eta \in K_G^o(S(V+R))$ with $res_H \eta = 0$ if $\dim V^H$ is even, there is a $\xi \in K_G^o(S(V+R);V)$ with $J_V(\xi) = 0$ and $\rho(\xi) = 2^{b(V)}(\psi^p-1)(\psi^q-1)\eta$ for some integers p and q which represent a and b.

Proof: Suppose first $\dim V^H$ is odd. By 3.17 $J((\psi^p-1)(\psi^q-1)\eta) = 0$. By 2.9 $(\psi^p-1)(\psi^q-1)\eta = \rho(\xi_0)$ for some ξ_0 in $K_G^o(S(V+R);V)$. By 3.15 $J_V(2^{b(V)}\xi_0) = 0$. Set $2^{b(V)}\xi_0 = \xi$. Suppose next $\dim V^H$ is even. Let $V' = V+R$. Lift η to $\eta' \in K_G^o(S(V'+R))$ and apply the preceeding argument to it, to produce $\xi' \in K_G^o(S(V'+R);V)$ with $J_V(\xi') = 0$ and $\rho\xi' = 2^{b(V')}(\psi^p-1)(\psi^q-1)\eta'$. Let ξ be the restriction of ξ' to $S(V+R)$.

Remark 3.19. Two G vector bundles ξ and ξ^1 are stably G fiber homotopy equivalent if there is a G vector bundle ξ'' and a proper fiber preserving G map $\omega : \xi \oplus \xi'' \to \xi' \oplus \xi''$ such that the fiber degree of ω^K is 1 for all K. (When ξ is a complex G vector bundle, there is a natural orientation for ξ^K for all K. In the real case we must postulate the existence of an orientation for each ξ^K as part of its structure.) If $\xi,\xi' \in Vect_G(Y;V)$, then ξ'' is required to lie in $Vect_G(Y;V)$ as well. Stable G fiber homotopy equivalence gives the equivalence relation on $Vect_G(Y;V)$ mentioned earlier.

Remark 3.20. For $i \geq 1$ $\omega_G^{-i}(Y)$ defined here and in [Se] coincide. The zero of $\omega_G^o(Y;W)$ is the class of the map of Y to the base point of F(W) while addition is defined by multiplication in F(W).

REFERENCES

[A] Atiyah, M.F., K-theory, Benjamin (1967).

[A2] __________, Elliptic operators and compact groups, Springer Verlag, Lecture Series 401 (1974).

[P1] Petrie, T., Three theorems in transformation groups, Lecture series 763, Springer-Verlag (1978).

[P2] __________, Isotropy representations of actions on spheres, to appear.

[P3] __________, Pseudoequivalences of G manifolds, Proc. Symp. in Pure Math., XXXII AMS (1978) pp. 169-210.

[P4] __________, Smith equivalence of representations, to appear.

[Sc] Schultz, R., Spherelike G manifolds with exotic tangent bundles, Studies in Algebraic Topology (Adv Suppl. Studies Vol. 5) 1-39, Academic Press (1979).

[Se] Segal, G., Equivariant stable homotopy, Proc. ICM (1970) 59-63.

[Sm] Smith, P.A., Some new results and old problems in finite transformation groups, BAMS 66 (1960) 401-415.

DEPARTMENT OF MATHEMATICS
RUTGERS UNIVERSITY
NEW BRUNSWICK, NEW JERSEY 08903

Canadian Mathematical Society
Conference Proceedings
Volume 2, Part 2 (1982)

DIFFERENTIABILITY AND THE P.A. SMITH THEOREMS FOR SPHERES: I. ACTIONS OF PRIME ORDER GROUPS

Reinhard Schultz[1]

ABSTRACT. Let p be a prime. Given a mod p homology sphere A, the converse to the P.A. Smith theorem asks if A is the stationary set of a $\mathbb{Z}/p$ action on some sphere. We restrict attention to smooth objects here and consider the realizability of A for smooth actions on spheres of infinitely many different dimensions. Results of L. Jones and others show that every "reasonable" A is so realizable modulo taking connected sums with exotic spheres. Using the truth of the Strong Segal Conjecture, we clarify the role of exotic spheres and determine the proper subset of all A realizable for actions on spheres of infinitely many distinct dimensions.

One of the cornerstones of the theory of group actions in the following result, which is basically due to P.A. Smith. The version that we quote is actually a slight generalization that can be found in Bredon's book [12, Ch. III, Theorem 7.11]:

Let $G = Z/p$ (p prime) act on a compact T_2 space that has the Čech homology of a sphere or disk. Then the set X^G of fixed points of G also has the Čech homology of a sphere of disk respectively.

Of course, if X is a genuine sphere or disk and the action of G is an orthogonal linear action, then X^G is again a genuine sphere or disk. On the other hand, there are many well-understood ways of constructing smooth actions on spheres and disks so that the fixed point set is not an integral homology sphere or disk (compare [12], Ch. I). Thus one is led to ask how close the necessary conditions in the Smith Theorem come to being sufficient if X is a genuine disk or sphere. In other words, to what extend is there a converse to the P.A. Smith Theorem for $\mathbb{Z}/p$ actions on S^n or D^n?

To avoid purely topological problems - not for lack of interest, but for lack of time and resources - we assume henceforth that X is either a smooth disk or homotopy sphere and G acts smoothly. In this case, differentiability

AMS Subject Classifications 57S17, 57S25.

[1] Partially supported by National Science Foundation Grants MCS 78-02913 and MCS 81-04852 and Sonderforschungsbereich 40 "Theoretische Mathematik", Universität Bonn.

© 1982 American Mathematical Society
0731-1036/82/0000-0470/$10.75

yields additional necessary conditions on X^G:

(1) X^G is a compact smooth submanifold, and $(\partial X)^G = X^G \cap \partial X$. Furthermore, X^G and ∂X^G are finite or connected.

(2) If $p > 2$, then an orientation of X induces a canonical weak almost complex structure on X^G. If X is a sphere and $\dim X^G \neq 2$, then the rational Chern classes of the associated complex vector bundle are trivial.

The first assertion is true by local linearity at fixed points (see [12]) and Smith theory, and the first half of the second assertion follows from the equivariant splitting of the normal bundle of X^G in X as a sum of complex vector bundles (compare [23]). The statement on rational Chern classes is considerably less elementary, and the definitive reference is [19]. Although the rational Chern classes can be nonzero if $\dim X^G = 2$ (see [48]), this falls outside the main interests of this paper because X^G must be diffeomorphic to S^2 in this case by the classification theorems for surfaces.

NOTATION. We shall use the term unitary manifold to denote a manifold with a weak almost complex structure; if the rational Chern classes of the associated complex vector bundle are trivial, we shall use the term torsion unitary manifold. For such manifolds, the underlying complex vector bundle has finite order in reduced complex K-theory.

The necessary conditions stated as (1) and (2) above are very nearly sufficient; the initial breakthrough was due to L. Jones [23, 24], with subsequent contributions in [2, 4], [6, 7], and [17].

THEOREM 0. *If Y is a compact smooth $\mathbb{Z}/p$-homology disk — with a torsion unitary structure if $p > 2$ — then there is a smooth $\mathbb{Z}/p$-action on some disk D^N with fixed point set Y.* ■

THEOREM 00. *If Y is a closed smooth $\mathbb{Z}/p$-homology sphere — with a torsion unitary structure if $p > 2$ — then (at least) for certain congruence classes of $\dim Y \bmod 4$ there is a smooth $\mathbb{Z}/p$-action on some homotopy sphere Σ^N whose fixed point set is diffeomorphic to $Y \# E$ for some homotopy sphere E. (# means connected sum.)* ■

The purpose of this paper is to resolve the uncertainty in Theorem 00 regarding the exotic sphere E. To do this, we shall first formulate Theorem 00 using a generalized Pontrjagin-Thom invariant. Recall from [51:I, §1] that if A is an R-homology k-sphere for some subring $R \subseteq \mathbb{Q}$, then the canonical degree one map $f_A\colon A^k \to S^k$ is an R-homology equivalence and as such defines a class $q(f_A) = \mathfrak{P}(A)$ in $\pi_k(F/O) \otimes R$. For homotopy spheres this is just the image of the usual Pontrjagin-Thom construction as in [26]. Theorem 00 can then be reformulated and strengthened as follows:

THEOREM A. Let Y be a closed smooth $\mathbb{Z}/p$ (equivalently, $\mathbb{Z}_{(p)}$-)-homology k-sphere, with a torsion unitary structure if $p \neq 2$. Assume that $\mathfrak{p}(Y) = 0$ in $\pi_k(F/O)_{(p)}$. Let $M > k/2 + 1$ be given; if $p = 2$ and k is odd, assume $k + 2M \equiv 1 \bmod 4$. Then there is a smooth $\mathbb{Z}/p$-action on S^{k+2M} or the exotic Kervaire sphere with fixed point set Y. If $p \neq 2$, one can take the action on the standard sphere.

In a subsequent paper we shall deal with the anomalies that force extra assumptions if $p = 2$.

It is elementary to derive Theorem 00 from Theorem A. By surgery theory, one can always find a homotopy sphere Σ so that $\mathfrak{p}(\Sigma) = \mathfrak{p}(A)$ (compare [26, 38, 3]), and it is immediate that $\mathfrak{p}(A\#-\Sigma) = 0$. We shall prove Theorem A in Section 3.

Combining Theorem A with results of P. Löffler [33] and ideas from [51:IV], we obtain the following conclusion:

THEOREM B. Let $p > 2$, and let Y be a closed smooth $\mathbb{Z}/p$-homology sphere with a torsion unitary structure. Then for every even integer $2M$ such that $k < 2M < (p-1)k$ there is a smooth $\mathbb{Z}/p$-action on S^{k+2M} with fixed point set Y.

In Theorem 0, one can construct actions on all D^{N+2s}; $s \geq 0$, by crossing D^N with s copies of rotation by $2\pi/p$ on D^2 and rounding off corners. Therefore, in all the preceding theorems the dimension of the ambient sphere or disk was basically just "sufficiently large." Accordingly, it is natural to guess that the upper bound on M in Theorem B can be removed by additional work. However, this is not the case. In fact, if k is fixed, we shall prove that Theorem A is the best possible result for all but finitely many M.

THEOREM C. Let p be fixed. Then there is a sequence of positive integers N_k with the following property: If Y^k is the fixed point set of a smooth $\mathbb{Z}_p$-action on a $\mathbb{Z}_{(p)}$-homology sphere Σ^{k+N} and $N \geq N_k$, then $\mathfrak{p}(Y) = 0$.

We shall say that such actions lie in the super stable range.

Theorem C is essentially an application of the methods developed in [51: IV] and the solution of the Strong Segal Conjecture for $\mathbb{Z}/p$. Combining Theorems A and C, we see that Y^k is the fixed point set of $\mathbb{Z}/p$-actions on S^{k+N} for infinitely many N if and only if $\mathfrak{p}(Y) = 0$.

The main steps in the proof of Theorems A and C are in Sections 1-3 and Sections 4, 6, 7 and 8 respectively. Section 5 contains the applications of the Strong Segal Conjecture that we need. The Appendix gives examples of unitary homology spheres that do not admit torsion unitary structures.

Acknowledgments

I owe a great deal to my colleagues for their comments, questions, and willingness to communicate work in progress. Listeners to my lectures on this topic (particularly at Northwestern, Wuppertal, and Oxford) have forced me to rethink difficult parts of the argument and exposition, and the result has been a clearer understanding on my part. Those who communicated then unpublished work that fits into this paper include J. F. Adams, A. Assadi, K. H. Dovermann, J. Gunawardena, P. Löffler, M. Mahowald, and D. Ravenel. I would like to thank all these mathematicians (and others I did not mention) warmly.

1. Torsion unitary structures on homology spheres

By definition, a torsion unitary manifold M^n has trivial rational Chern classes. It is well-known (compare [18]) that the associated complex normal bundle ξ_M determines an element of finite order in $\widetilde{K}(M)$. Stated differently, if $x_M: M \to BU$ denotes the classifying map of ξ_M and $L: BU \to BU \otimes \mathbb{Q}$ is rational localization (as in [56], say), then L_{x_M} is nullhomotopic. Consequently, we may factor x_M through the homotopy fiber Λ of L. This makes it clear that the formal definition of a torsion unitary structure should be a (Λ, rj)-structure in the standard sense of [55], where $r: BU \to BO$ is realification and $j: \Lambda \to BU$ is inclusion of the homotopy fiber of L. Strictly speaking, we need to define a system of finite approximations (B_n, f_n) to (Λ, rj) in order to conform to [55]. This is easy; if $n = 2k$ we set B_n equal to the fiber of the localization $BU_k \to BU_k \otimes \mathbb{Q}$, and let f_{2k} be the composite $B_{2k} \to BU_k \to BO_{2k}$, while if $n = 2k+1$ we set $B_{2k+1} = B_{2k}$ and let f_{2k+1} be f_{2k} composed with the induced map from BO_{2k} to BO_{2k+1}. In this section we shall dispose of two technical points that arise naturally. First of all, when can a unitary homology sphere M be given a torsion unitary structure? Complete conditions can be given, and they imply that one can always do this for M if $\dim M \not\equiv 2 \bmod 8$; in the exceptional case, one can find a torsion unitary structure on $M \# M$ in any case. Examples where no torsion unitary structure exists are constructed in the Appendix. The second question is considerably more technical. Given a torsion unitary homology sphere (M,f) what can one say about the other torsion unitary structures on M, say (M,g), that give the same unitary structure? In other words, describe all (M,g) so that (M,jg) and (M,jf) are equivalent unitary structures. The simplest method of constructing such an (M,g) is to take the connected sum of (M,f) with (S^m, Φ), where Φ is a framing of S^m that defines the same unitary structure as the standard framing. We shall prove that such examples generate all the possibilities in a precise sense. Actually, we shall do this in a slightly more general setting in which Λ is replaced by the fiber Λ_ℓ of localization $L_\ell\colon BU \to BU_\ell$ at some set of primes ℓ (in the applications, we shall usually have ℓ = all primes except a given prime p).

We begin with a simple observation.

PROPOSITION 1.1. *Let* F^k *be a unitary rational homology sphere (closed, smooth), and let* $q\colon F^k \to S^k$ *have degree one. Then* q^* *is split injective in complex* K^**-theory.*

PROOF. Since F has a unitary structure, it is orientable in complex K-theory [8, 18]. Elementary calculations imply that the homology map $q_*\colon K_k(F) \to K_k(S^k)$ must send the K-theoretic fundamental class of F to the corresponding class for S^k. Given this, the proof in [14] goes through word for word. ∎

In particular, if ξ is a complex vector bundle over F^k, then the rational Chern classes $c_m(\xi;\mathbb{Q})$ equal the classes $c_m(q^*\xi;\mathbb{Q})$ for some pullback of a bundle over S^k. For the cokernel of q^* is finite by a simple calculation involving the Atiyah-Hirzebruch spectral sequence.

THEOREM 1.2. *Let* F^k *be a unitary rational homology sphere.*

(i) *If* $k \not\equiv 2 \bmod 8$, *then* F^k *admits a torsion unitary structure.*

(ii) *If* $k \equiv 2 \bmod 8$, *then* F^k *admits a torsion unitary structure if and only if it admits a unitary structure whose complex bundle maps to zero in* $(\widetilde{K}(F)/\text{Torsion}) \otimes \mathbb{Z}/2$ $(\cong \mathbb{Z}/2)$.

PROOF. If k is odd, there is nothing to prove. If $k = 4m$ and $c_m(\gamma,\mathbb{Q}) \neq 0$, then the underlying real vector bundle has nontrivial rational Pontrjagin classes. But this cannot happen for the tangent bundle of F by the Hirzebruch Signature Theorem. If $k = 4m + 2$, by Proposition 1.1 and the paragraph above we can find a bundle γ' over S^k so that $\gamma - q^*\gamma'$ is torsion in $K^*(F)$ and $C^*(\gamma;\mathbb{Q}) = C^*(q^*\gamma';\mathbb{Q})$. If the underlying real vector bundle of γ' is trivial, then there is an obvious unitary structure on F with complex bundle $\gamma - q^*\gamma'$; but the latter bundle has a trivial rational Chern class, and therefore the proof is finished.

If m is odd, then every element of $\widetilde{K}(S^{4m+2}) = \mathbb{Z}$ goes to zero in $\widetilde{KO}(S^{4m+2}) = 0$, and if m is even then an element of $\widetilde{K}(S^{4m+2})$ goes to zero in $\widetilde{KO}(S^{4m+2}) = \mathbb{Z}/2$ if and only if the element is divisible by two. These assertions can be read off from the strong form of Bott periodicity [11]. This gives us what we need if $k = 4m + 2 \equiv 6 \bmod 8$, and if $k \equiv 2 \bmod 8$ it completes the proof PROVIDED we know that γ goes to an even element of $\widetilde{K}(F)/\text{Torsion}$ (an infinite cyclic group). This completes the proof of (i) and the "if" direction of (ii). However, the "only if" direction of (ii) is trivial. ■

Since the rational Chern classes are additive for bundles over rational homology spheres, we have the following obvious consequence of Theorem 1.2:

COROLLARY 1.3. *If* $k \equiv 2 \bmod 8$ *and* F^k *is a unitary rational homology sphere, then* $F \# F$ *admits a torsion unitary structure.* ■

In the Appendix we shall show that a unitary rational homology (8s+2)-sphere need not have a torsion unitary structure. It is natural to guess that the connected sum of such a manifold with a torsion unitary homology sphere will also not admit a torsion unitary homology sphere; this would yield an additive $\mathbb{Z}/2$-valued obstruction. However, this is not true, and we shall explain why not in the appendix.

We now turn to the second problem of this section. As mentioned before, we really want to work with structures slightly more delicate than torsion unitary structures. The following elementary result can serve as a motivation:

PROPOSITION 1.4. *Let* M^k *be a torsion unitary rational homology sphere, and let* $x_M\colon M \to BU$ *classify the complex normal bundle. Suppose that* ℓ *is a set of primes such that the torsion of* $H_*(M^k; \mathbb{Z})$ *is prime to* ℓ. *Then the composite of* x_M *with the localization* $L_\ell\colon BU \to BU_\ell$ *(at* ℓ*) is null-homotopic.*

PROOF. The hypotheses imply that the kernel of localization map $\widetilde{K}(M) \to \widetilde{K}(M) \otimes \mathbb{Z}_\ell$ is the torsion subgroup of $\widetilde{K}(M)$. ■

If we let $j_\ell\colon \Lambda_\ell \to BU$ be the fiber of localization at ℓ, then it follows that a torsion unitary manifold can be viewed as a manifold with a (Λ_ℓ, rj_ℓ)-structure (idiomatically, an ℓ-*torsion unitary structure*). In the next section we shall have to compare two ℓ-torsion unitary structures with the same underlying unitary structure. The following results give us all the information we need:

THEOREM 1.5. *Let* (M^k,f) *be a* $\mathbb{Z}_\ell$*-homology sphere with an* ℓ*-torsion unitary structure. If* (M^k,g) *represents a second such structure such that* $(M^k, j_\ell f)$ *and* $(M^k, j_\ell g)$ *are equivalent unitary structures, then* (M^k,g) *is equivalent to a connected sum* $(M^k,f) \# (S^k,\Psi)$, *where* Ψ *represents an* ℓ*-torsion unitary structure whose underlying unitary structure is standard.*

The "standard" structure is the one which comes from the framing S^k inherits as the boundary of D^{k+1}.

THEOREM 1.6. Let (S^k,Ψ) represent an ℓ-torsion unitary structure on S^k whose underlying unitary structure is standard.

(i) There is a framing Φ of S^k and a positive integer T such that T is a monomial in the primes in ℓ and (S^k,Φ) is equivalent to a connected sum of T copies of (S^k,Ψ).

(ii) If q is a prime in ℓ, then (S^k,Ψ) may be written as a connected sum of q copies of a structure (S^k,Ψ') with standard underlying unitary structure.

PROOFS OF 1.5 AND 1.6. We use the definition of (B,f) structure as an isotopy class of liftings for the classifying map $\nu_M\colon M \to BO$. Let (M,a) and (M,a') be two (Λ_ℓ, rj_ℓ)-structures that define the same (BU,r)-structures that define the same (BU,r)-structure. Then $j_\ell a$ and $j_\ell a'$ are homotopic through liftings of ν_M by definition. An application of the covering homotopy property shows that (M,a') is equivalent to a structure (M,a'') with $j_\ell a'' = j_\ell a$.

The fiber of j_ℓ is the localization U_ℓ, and with proper models the fibration is principal. It follows that $a'' = \mu(c,a)$, where $c\colon M \to U_\ell$ is some map. If M^k is a $\mathbb{Z}_\ell$-homology sphere, then the degree one map $M^k \to S^k$ induces an isomorphism in $K^{-1}(\ ) \otimes \mathbb{Z}_\ell$, and hence (M,a'') is equivalent to the connected sum of (M,a) with an exotic structure (S^k,c^*). This proves 1.5.

Let $[c^*] \in \pi_k(U_\ell) \cong \pi_k(U)_\ell$ denote the corresponding homotopy class. Then there is a positive monomial T in the elements of ℓ such that $T[c^*] \in \text{Image } \pi_k(U)$. This proves 1.6(i); since multiplication by q is an isomorphism in $\pi_k(U)_\ell$, the validity of 1.6(ii) is also clear. ∎

2. Groups of unitary homology spheres

Although the fundamental work of Kervaire and Milnor [18] dealt with the classification of homotopy spheres, it was implicit in their work that one could also classify $\mathbb{Z}$-homology spheres up to $\mathbb{Z}$-homological h-cobordism in exactly the same way (compare [22]). Later research established the existence of parallel classification theories for A-homology spheres where A was an arbitrary subring of the rationals (compare [2,3], [9], and [34]; such results were surely also known to others). As we have already seen, the fixed point set of a smooth $\mathbb{Z}/p$ action on a homotopy sphere Σ^n has an almost complex structure, and to handle the technicalities efficiently it is useful to carry along this extra structure. The idea of defining groups of homology spheres with almost complex structures already occurs in the work of Jones [24].

One unavoidable feature of Jones' groups is the existence of elements having infinite order; such elements are detected by rational Chern classes. In this section we shall define variants of Jones' groups in which elements of infinite order do not arise. Of course, the vanishing of the rational Chern class for a fixed point set (noted in the introduction) is needed to justify our avoidance of the rational Chern classes, and the machinery of Section 1 was developed to handle some problems in dealing with such manifolds.

From the restriction in the introduction and Section 1, we know that the unitary structure for the fixed set of a $\mathbb{Z}/p$ action (p an odd prime) can be lifted back to a (Λ_ℓ, rj_ℓ)-structure (ℓ all primes except p). To simplify notation, we shall henceforth use (Λ, rj) to denote this pair (Λ_ℓ, rj_ℓ). Therefore it seems appropriate to consider a group of $\mathbb{Z}_{(p)}$-homology spheres with (Λ, rj)-structures. Once we understand this group we can try to use 1.5 and 1.6 to look at the same homology spheres with the underlying unitary structures. The general pattern is very similar to that employed by Alexander-Hamrick-Vick for $p = 2$ [3] and Barge-Lannes-Latour-Vogel more generally [9], in both cases without the complication of an almost complex structure. In the interests of brevity we shall suppress details that correspond naturally to those of [3] and [9]; and we shall concentrate on the places where significant changes are necessary.

We define $\Lambda U\Theta_k\langle p\rangle$ to be the group of (Λ, rj)-structured $\mathbb{Z}_{(p)}$-homology spheres modulo (Λ, rj)-structured $\mathbb{Z}_{(p)}$-homological h-cobordism. As usual, the binary operation is induced by connected sum, and inverses are given by reversing orientations.

There is a natural map from $\Lambda U\Theta_k\langle p\rangle$ to the bordism group of (Λ, rj)-structured k-manifolds, and it corresponds to taking the underlying bordism class. Of course, this bordism group is isomorphic to the stable homotopy group $\pi_k(M\Lambda)$, where $M\Lambda$ denotes the associated Thom spectrum. These homotopy groups may be calculated completely as follows:

THEOREM 2.1. (i) *The groups* $\pi_k(M\Lambda)$ *are torsion groups.*

(ii) *If* $i: S^0 \to M\Lambda$ *is the fiber inclusion, then* i_* *induces an isomorphism from* $\pi_*(S^0)_{(p)}$ *to* $\pi_*(M\Lambda)_{(p)}$.

(iii) *The groups* $\pi_*(M\Lambda) \otimes \mathbb{Z}[p^{-1}]$ *are canonically isomorphic to*

$$\pi_{*+1}(MU/S^0) \otimes \mathbb{Q}/\text{Image } \pi_{*+1}(MU) \otimes \mathbb{Z}[p^{-1}].$$

More precisely, the "quotient" map $MU/S^0 \to MU/M\Lambda$ *is a localization of* MU/S^0 *at* p.

REMARKS ON THE PROOF. In [3] and [9] a comparable result is proved for $M\Phi$, where Φ is the fiber of the localization $BSO \to BSO_{(p)}$. The proofs in our setting are entirely analogous. In order to complete the proof of (iii) one must notice that $\pi_*(M\Lambda) \to \pi_*(MU)$ is zero in positive dimensions. But this is obvious from (i) and the torsion freeness of $\pi_*(MU)$. ∎

COMPLEMENT 2.2. The analog of the image of J here is the subgroup ΛJ_k of all (Λ, rj)-structures on S^k that reduce to the standard unitary structure. By 1.5 and 1.6, this subgroup is the image of $\pi_k(U)_{(p)}$ in the following diagram:

$$\begin{array}{ccc} \pi_k(U) & \xrightarrow{J} & \pi_k(S^o) \\ \downarrow & & \downarrow \\ \pi_k(U)_{(p)} & \longrightarrow & \pi_k(M\Lambda) \end{array}$$

Obviously, the p-primary component of ΛJ_k is equal to the p-primary component of the image of J by 2.1(ii). Away from p, one can get a good hold on ΛJ_k by means of the "e-invariant lifting" of the J-image generator $\alpha\colon S^k \to S^o$ to a map $E\colon S^k \to S^{-1}(MU/S^o)$. ∎

The next step is to fit the map $\Lambda U\Theta_k\langle p\rangle \to \pi_k(M\Lambda)$ into an exact sequence by means of surgery theory; the main theorem is entirely analogous to the results of [3] and [9], at least in its statement. However, the proof is not quite the same because the classifying space Λ is not simply connected (as is Φ in [3] and [9]).

THEOREM 2.3. *Let* $L_m(\mathbb{Z}_{(p)})$ *be the usual homotopy Wall group. Then for* $k \geq 5$ *there is an exact sequence of abelian groups:*

$$L_{k+1}(\mathbb{Z}_{(p)}) \to \Lambda U\Theta_k\langle p\rangle \to \pi_k(M\ ) \to L_k(\mathbb{Z}_{(p)}). \tag{2.4}$$

In fact, as in [3] and [9] one can extend these sequences in both directions, but we shall not need this.

The structure of the groups $L_m(\mathbb{Z}_{(p)})$ is well-understood; a good reference is [40]. In the proof of 2.3 we shall need to know that $L_m(\mathbb{Z}_{(p)}) = 0$ if m is odd. Actually, one can settle this more or less directly without resorting to heavy machinery; in particular, for p odd the results of [43] give a complete proof.

PROOF OF 2.3. We begin by considering the construction of the map $\pi_k(M\Lambda) \to L_k(\mathbb{Z}_{(p)})$ and exactness at $\pi_k(M\Lambda)$. Suppose we are given a class in $\pi_k(M\Lambda)$; let (W^k, f) be a representative. Obviously, we wish to perform surgery on W to make it a homology sphere. There is no problem making W connected, and one can proceed to kill the kernel of $f_*\colon \pi_1(W) \to \pi_1(V) \cong \mathbb{Z}_{(p)}/\mathbb{Z}$. Since $\pi_1(W)$ is a finitely presented group, it follows that $\pi = \text{Image } f_*$ is a finite group; in fact, π is cyclic of order prime to p. We cannot expect π to be trivial in general, for this happens if and only if the first integral Chern class of W is trivial. Because of this, we cannot follow a treatment such as [9] word for word.

Since $\pi_1(W)$ is (now) finite, we know that W has finitely generated homotopy groups in every dimension. This allows us to perform surgeries that will make the maps $f_*: \pi_i(W) \to \pi_i(\Lambda)$ injective below the middle dimension. The standard arguments in [37] go through unchanged. It follows that $H_*(W)_{(p)}$ and $H_*(\widetilde{W})_{(p)}$ are zero except in the top, bottom, and (one or two) middle dimensions (here $\widetilde{W}$ denotes the universal covering of W).

Write $k = 2m + \varepsilon$ for $\varepsilon = 0$ or 1 (note that $m \geq 2$). The first thing we obviously need for surgery in dimension m is that the localized Hurewicz map $h_W: \pi_m(W)_{(p)} \to H_m(W)_{(p)}$ is onto. To prove this, consider the commutative square below:

$$\begin{array}{ccc} \pi_m(\widetilde{W})_{(p)} & \xrightarrow{\cong} & \pi_m(W)_{(p)} \\ h_{\widetilde{W}} \downarrow & & \downarrow h_W \\ H_m(\widetilde{W})_{(p)} & \xrightarrow{P_*} & H_m(W)_{(p)} \end{array}$$

By construction, the localized space $\widetilde{W}_{(p)}$ is (m-1)-connected, and therefore $h_{\widetilde{W}}$ is bijective. Since $\pi = \pi_1(W)$ is finite of order prime to p, a transfer argument shows that P_* is split surjective; therefore h_W is surjective. In fact, even more is true; since P_* is bijective on the π-invariant elements of $H_m(\widetilde{W})_{(p)}$, we can conclude that h_W maps the π-invariant elements of $\pi_m(W)_{(p)}$ bijectively to $H_m(W)_{(p)}$.

If k is odd, this gives us everything we need to generalize the standard arguments in Wall's book ([58]; compare [5] also for modifications to treat $\mathbb{Z}_{(p)}$-coefficients). This gives us a class in $L_k(\mathbb{Z}_{(p)})$ associated to (W,f). Using the standard techniques and the points we have mentioned, it is fairly straightforward to show that surgery can be completed if the class in $L_k(\mathbb{Z}_{(p)})$ vanishes. But we have already noted that $L_k(\mathbb{Z}_{(p)}) = 0$ for k odd, and therefore exactness of (2.4) at $\pi_k(M\Lambda)$ for k odd is established.

If k is even, proceed as follows: Say that a positive integer is p-free if it is not divisible by p. Given an element $\alpha \in \pi_m(W)_{(p)}$, some p-free multiple of α is representable by an immersion $A: S^m \times \mathbb{R}^m \to W^k$ such that the induced (Λ, rj)-structure on $S^m \times \mathbb{R}^m$ extends to $D^{m+1} \times \mathbb{R}^m$ (compare [58, Ch. 1]). These immersions have algebraic intersection and self-intersection numbers as in [58] and [5]; since $H_m(W)_{(p)}$ is torsion-free, one can use these numbers to define a $(-1)^m$-symmetric form over $\mathbb{Z}_{(p)}$ (one must divide these intersection integers by the p-free multiples that were used earlier). Thus we obtain an element of $L_k(\mathbb{Z}_{(p)})$. It remains to check that this assignment is well-defined and additive, and that surgery can be completed if and only if the Wall class is zero. This will yield exactness of (2.4) for $k = 2m$.

As in ordinary surgery, the key point is to show that the vanishing of homological intersection numbers implies that geometric intersections may be modified to be made empty. From Wall's book (and [5]) we see that the obstructions to pulling immersions apart and making them embeddings are refined intersection and self-intersection numbers in $\mathbb{Z}_{(p)}[\pi]$ or (for self-intersections if m is odd) $\mathbb{Z}_{(p)}[\pi]/(2)$. Of course, if p is odd the latter group vanishes, so we may ignore the latter group. The refined intersection numbers map to our intersection numbers under the augmentation map from $\mathbb{Z}_{(p)}[\pi]$ to $\mathbb{Z}_{(p)}$. But we are dealing with <u>invariant</u> homotopy classes (we only need to kill $H_m(W)_{(p)} \cong \pi_m(W)^{\pi}_{(p)}$), and the bilinearity properties of refined intersection numbers imply that π-invariant classes determine π-invariant intersection numbers. Since the augmentation map is bijective on π-invariant elements, we see that the vanishing of the ordinary intersection numbers implies the same for the refined intersection numbers.

With this information available, the proof that the map $\pi_{2m}(M\Lambda) \to L_{2m}(\mathbb{Z}_{(p)})$ is a well defined homomorphism and the exactness of (2.4) at $\pi_{2m}(M\Lambda)$ proceed in the standard fashion.

<u>Note</u>. If W is 1-connected, then the form we obtain is equivalent to the cup product form in the middle dimension. This is entirely analogous to the ordinary case, the key point being that one can alter an immersion to lower the self-intersection by $N \in \mathbb{Z}$, the price being that the normal bundle of the new immersion will be $N\tau_{S^m} \in \pi_m(BSO_m)$. (Compare [57]).

We now proceed to the proof of exactness at $\Lambda U\Theta_k\langle p\rangle$. First of all, we shall define the map from $L_{k+1}(\mathbb{Z}_{(p)})$. Since this group vanishes if k is even, we assume now that k is odd; set $k + 1 = 2m$. An element of the Wall group is represented by a $(-1)^m$-symmetric bilinear form over $\mathbb{Z}_{(p)}$. This form can be realized geometrically by a framed $(m-1)$-connected $2m$-manifold with boundary X (compare [9] or [34]), where ∂X^{2m-1} is a framed $\mathbb{Z}_{(p)}$-homology sphere. This gives us a map Δ of $L_{k+1}(\mathbb{Z}_{(p)})$ into $\Lambda U\Theta_k\langle p\rangle$, and it is fairly straightforward to check that Δ is well-defined and additive (see [34] for further details). Since ∂X is a framed boundary, it is clear that the composite map $L_{k+1}(\mathbb{Z}_{(p)}) \to \pi_k(M\Lambda)$ is zero.

We now drop the assumption on k, and we suppose that Σ^k represents an element in the kernel of the map

$$\Lambda U\Theta_k\langle p\rangle \to \pi_k(M\Lambda).$$

Then we know that $\Sigma^k = \partial W^{k+1}$ for some appropriate W, and we want to do surgery on W, holding Σ fixed, to make W $\mathbb{Z}_{(p)}$-homologically acyclic. We proceed exactly as before. Surgery below the middle dimension allows us to make the map $\pi_*(W) \to \pi_*(\Lambda)$ monic in this range, and from here one defines an element θ of $L_{k+1}(\mathbb{Z}_{(p)})$. If this element is zero, then surgery can be completed to make W a $\mathbb{Z}_{(p)}$-homology disk. On the other hand, if θ is nonzero, then we may construct a highly connected manifold X as above using the <u>negative</u> of θ. Consider the connected sum $W \# X$. By construction, it has Wall group invariant zero, and therefore one can complete the surgery process on $W \# X$. Therefore $\partial(W \# X) = \Sigma \# \partial X$ bounds a $\mathbb{Z}_{(p)}$-homology disk, and this tells us that the class of Σ in $\Lambda U\Theta_k\langle p\rangle$ is the negative of the class of ∂X. This implies exactness of (2.4) at $\Lambda U\Theta_k\langle p\rangle$. ■

<u>Note</u>. If $k = 4$ we can use the same methods to prove that the sequence

$$L_5(\mathbb{Z}_{(p)}) \to \Lambda U\Theta_4\langle p\rangle \to \pi_4(M\Lambda)$$

is exact. ■

DEFINITION. The <u>group</u> $U\Theta_k\langle p\rangle$ <u>of torsion unitary</u> p-<u>homology spheres</u> is the quotient of $\Lambda U\Theta_k\langle p\rangle$ by ΛJ_k (see 2.2 for the latter). The following gives several important properties of these groups.

THEOREM 2.5. <u>The groups</u> $U\Theta_k\langle p\rangle$ <u>are torsion groups if</u> $k \neq 3$, <u>and their</u> p-<u>primary components are isomorphic to</u> $\Theta_{k(p)}$ (<u>the localized Kervaire-Milnor group</u>) <u>if</u> p <u>is odd</u> ($k \neq 3$ <u>again</u>).

Before proving this result, we state the structure theorems for $L_m(\mathbb{Z}_{(p)})$ where m is even and p is odd; one reference is [40], but the result is actually a quite standard fact about Witt groups.

(2.6) <u>The group</u> $L_{4q+2}(\mathbb{Z}_{(p)})$ <u>is zero</u>. ■

(2.7) <u>The group</u> $L_{4q}(\mathbb{Z}_{(p)})$ <u>is the direct sum of an infinite cyclic group and a countable number of copies of</u> $\mathbb{Z}/2$ <u>and</u> $\mathbb{Z}/4$. <u>The torsion free part is detected by the signature homomorphism</u> $\tau\colon L_{4q}(\mathbb{Z}_{(p)}) \to \mathbb{Z}$, <u>and</u> $8\mathbb{Z}$ <u>lies in the image of</u> τ (<u>it comes from</u> $L_{4q}(\mathbb{Z})$, <u>in fact</u>). ■

PROOF OF 2.5. From the above results and 2.3, it is clear that $\Lambda U\Theta_k\langle p\rangle$/Torsion is zero if $k \not\equiv 3 \bmod 4$ and at most infinite cyclic in the latter case. We may define a nontrivial map from $\Lambda U\Theta_{4q-1}$ to $\mathbb{Q}$ as follows: Given a structured homology sphere (Σ, f), we know that some finite multiple of (Σ, f) goes to zero in $\pi_k(M\Lambda)$; therefore we may write $N(\Sigma, f) = (W^{4q}, g)$. Let $T(\Sigma, f) = \operatorname{sgn} W/N$. It is immediate that this number depends only on (Σ, f); for example, if $N(\Sigma, f)$ also bounds (X, h), then

$Y = Wu_{N\Sigma} - X$ is a unitary manifold with no rational Chern classes and thus $0 = \operatorname{sgn} Y = \operatorname{sgn} W - \operatorname{sgn} X$ by the Hirzebruch theorem and Novikov additivity. On the other hand, if Σ^{4q-1} is the Milnor exotic sphere which bounds a π-manifold of index 8, then $\tau(\Sigma, f) = 8$. Thus τ is nontrivial, and it follows that $\Lambda U\Theta_{4q-1}\langle p\rangle$/Torsion injects into the rationals. Also, the map

$$\mathbb{Z} = L_{4q}(\mathbb{Z}) \to L_{4q}(\mathbb{Z}_{(p)}) \to \Lambda U\Theta_{4q-1}\langle p\rangle/\text{Torsion}$$

has torsion cokernel.

The torsion assertion for $U\Theta_{4q-1}\langle p\rangle$ will follow if we can prove that some nonzero element of $L_{4q}(\mathbb{Z})$ maps into ΛJ_{4q-1}. But we know that some nonzero element of $L_{4q}(\mathbb{Z})$ corresponds to a framed manifold whose boundary is S^{4q-1} with some exotic framing. If this exotic framing determines the standard unitary structure, we are done. But a framing of S^{4q-1}, viewed as a class in $\pi_{4q-1}(O)$, determines the standard unitary structure if and only if its image in $\pi_{4q-1}(O/U)$ is zero. By Bott periodicity we know that twice every homotopy class does map trivially; since the exotic framing under consideration is divisible by the order of the image of J_{4q-1}[25] — a number divisible by 8 — this suffices.

To prove the statement on p-components, we first prove a comparable result for $\Lambda U\Theta_k\langle p\rangle$. Let $F\Theta_k$ denote the group of <u>framed homotopy</u> k-<u>spheres</u> (compare [27]). We then have the following commutative diagram with exact rows:

$$\begin{array}{ccccccc}
L_{k+1}(\mathbb{Z})_{(p)} & \xrightarrow{\partial} & F\Theta_{k(p)} & \longrightarrow & \pi_{k(p)} & \longrightarrow & L_k(\mathbb{Z})_{(p)} \\
\downarrow & & \downarrow{\scriptstyle\phi} & & \downarrow & & \downarrow \\
L_{k+1}(\mathbb{Z}_{(p)})_{(p)} & \xrightarrow{\partial'} & \Lambda U\Theta_k\langle p\rangle_{(p)} & \to & \pi_k(M\Lambda)_{(p)} & \to & L_k(\mathbb{Z}_{(p)})_{(p)}.
\end{array}$$

By the preceding results all the vertical arrows except perhaps ϕ denote isomorphisms. If $k \not\equiv 0 \bmod 3$, then the groups at the left are zero, and therefore ϕ is an isomorphism by the five lemma. On the other hand, if $k = 4q - 1$, then the five lemma tells us that ϕ is onto and a nonzero element of Kernel ϕ corresponds to a nonzero element of Kernel ∂'. But ∂' is monic by the first paragraph, for the composite $\tau\partial'$ is nontrivial. Therefore $F\Theta_{k(p)}$ is isomorphic to $\Lambda U\Theta_k\langle p\rangle$.

To obtain Θ_k from $F\Theta_k$, one factors out the framings of the standard sphere, a group isomorphic to $\pi_k(O)$. To obtain $U\Theta_k\langle p\rangle_{(p)}$ from $\Lambda U\Theta_k\langle p\rangle_{(p)}$, one factors out a group isomorphic to a quotient of $\pi_k(U)_{(p)}$. Furthermore, these groups are related by the following diagram:

$$\begin{array}{ccccc} & \nearrow & \pi_k(O) & \to & F\Theta_k \\ \pi_k(U) & & & & \downarrow \\ & \searrow & \pi_k(U)_{(p)} & \to & \Lambda U\Theta_k\langle p\rangle_{(p)}. \end{array}$$

If we localize at p, we see that the image of $\pi_k(U)_{(p)}$ in the lower right hand group is just the image of $\pi_k(O)_{(p)}$. Therefore the quotient groups $\Theta_{k(p)}$ and $U\Theta_k\langle p\rangle_{(p)}$ are isomorphic. ∎

For $p = 2$ we do not need unitary structures on our homology spheres. Hence everything we need is already in [3] and [9]; for the sake of completeness, we review what we need. In that case one has groups of oriented $\mathbb{Z}_{(2)}$-homology spheres we shall call $\Theta_k\langle 2\rangle$. If we let Φ be the fiber of the localization $BSO \to BSO_{(2)}$, then we obtain an exact sequence as in (2.4) involving $\Theta_k\langle 2\rangle$, $\pi_k(M\Phi)$, and the Wall groups of $\mathbb{Z}_{(2)}$. As before, $L_{odd}(\mathbb{Z}_{(2)}) = 0$, and in analogy with the integers $L_{4q+2}(\mathbb{Z}_{(2)}) \cong \mathbb{Z}/2$ is given by the Kervaire invariant. The group $L_{4q}(\mathbb{Z}_{(2)})$ has the same general sort of structures as does $L_{4q}(\mathbb{Z}_{(p)})$ for p odd. Finally, the group $\Theta_k\langle 2\rangle_{(2)}$ is isomorphic to $\Theta_{k(2)}$.

3. Partial converses to the P.A. Smith Theorem

Given an integer k and a free representation V of $\mathbb{Z}/p$ (p prime), we may form the group $\mathfrak{A}^{\mathbb{Z}/p}_{k+V}$ of $\mathbb{Z}/p$-oriented actions on homotopy spheres such that the fixed point set is k-dimensional and the normal representation at a fixed point is V (such a group may be constructed as in [12] or [24]). Passage to the fixed point set yields a homomorphism

$$I_p : \mathfrak{A}^{\mathbb{Z}/p}_{k+V} \to U\Theta_k\langle p\rangle \qquad (p \neq 2),$$

$$I_2 : \mathfrak{A}^{\mathbb{Z}/2}_{k+V} \to \Theta_k\langle 2\rangle .$$

Our first order of business is to add one term to the right so that we have exactness at the codomain of I_p; for this we must assume that V is suitably large (e.g., $\dim V \geq k + 2$) and unitypical. Specifically, we need that only one free irreducible representation of $\mathbb{Z}/p$ occurs in the irreducible decomposition of V.

The idea for the exact sequences we need goes back to Jones [24]. Following Rothenberg [44], we define a group

$$RS^{\mathbb{Z}/p,s}_{\ell+V}$$

consisting of s-cobordism classes of the following triples of objects:

(i) A $\mathbb{Z}/p$-homotopy $(\ell+V)$-sphere Σ such that the equivariant collapse $\Sigma \to S^V$ is simple in the sense of [45].

(ii) A diffeomorphism from $\sum^G$ to S^k.

(iii) An extension of (ii) to an equivariant tubular neighborhood $S^k \times V \to \sum$.

The difference between this group and Rothenberg's $RS^k(\mathbb{Z}/p, V)$ is the use of simple homotopy and s-cobordism rather than ordinary homotopy and h-cobordism. As in [48, §6], the forgetful map from our group to Rothenberg's group fits into a long "Rothenberg exact sequence". By Rothenberg's s-cobordism theorem [45], one sees that (nearly always) an element of $RS^{\mathbb{Z}/p,s}_{k+V}$ represents zero if and only if it is equivalent to the linear action on $S(k+1+V)$, the fixed point set identification $S^k \to S(k+1+V)^{\mathbb{Z}/p} \cong S^k$ is (isotopic to) the identity, and the tubular neighborhood is just the standard inclusion of $S^k \times V \cong S^k \times \operatorname{Int} D(V)$ in $S(k+1+V)$.

Define a map

$$C_p\colon U\Theta_k\langle p\rangle \to RS^{\mathbb{Z}/p,s}_{k-1+V}$$
($p \neq 2$, V stable with respect to k),

$$C_2\colon \Theta_k\langle 2\rangle \to RS^{\mathbb{Z}/2,s}_{k-1+V}$$

as follows: Given a homology sphere M^k with suitable extra structure, let $M_0 = M - \operatorname{Int} D^k$. By the work of Jones [23] and its refinements by Assadi [6,7] and Dovermann-Rothenberg [17], there is an $\mathbb{Z}/p$-manifold $N^{k+\dim V}$, unique up to equivariant diffeomorphism, such that N is nonequivariantly diffeomorphic to $D^{k+\dim V}$, the fixed point set is isomorphic to M_0 (with the correct unitary structure if $p \neq 2$), and the collapsing maps $(N,\partial N) \to (D(k+V), S(k+V))$, $N \to D(k+V)$, $\partial N \to S(k+V)$ are equivariant simple homotopy equivalences. By construction we have a canonical tubular neighborhood $S^{k-1} \times V \to \partial N$ of $(\partial N)^{\mathbb{Z}/p}$, and accordingly we obtain from ∂N an element of $RS^{\mathbb{Z}/p,s}_{k-1+V}$. Additivity follows by standard considerations involving connected sums and the uniqueness property; in particular the latter implies that the action associated to $M \# M'$ is $N \# N'$. One must also check that the mapping is well-defined. By additivity, it suffices to check that if M bounds a homology disk then ∂N goes to zero in the RS-group. But if $M = \partial L$, then the previously mentioned results of [6,7,17,23] imply that L is the fixed point set of some action on $P^{k+1+\dim V}$, where P is nonequivariantly diffeomorphic to $D^{k+1+\dim V}$. It follows by uniqueness that N is equivariantly diffeomorphic to $\partial P - \operatorname{Int} D(k+V)$. Therefore ∂N is just $S(k+V)$ and we are done.

In order to state the next result, we need simple-homotopy variants of the $\mathfrak{A}$ groups. One can define torsions for actions on homotopy spheres just as for disks; in fact, given $\sum$, its torsion equals that of the disk $\sum - \operatorname{Int} D(k+V)$. As usual, we denote such groups of s-cobordism classes by the symbol $\mathfrak{A}^{\mathbb{Z}/p,s}_{k+V}$.

THEOREM 3.1. The sequences

$$\mathfrak{A}^{\mathbb{Z}/p,s}_{k+V} \to U\Theta_k\langle p\rangle \to RS^{\mathbb{Z}/p,s}_{k-1+V} \qquad (p \neq 2),$$

$$\mathfrak{A}^{\mathbb{Z}/2,s}_{k+V} \to \Theta_k\langle 2\rangle \to RS^{\mathbb{Z}/2,s}_{k-1+V}$$

are exact.

REMARK. Jones has shown that this sequence fits into a long exact sequence in the following sense: One can embed our groups into larger groups $U^{\#}\Theta_k\langle p\rangle$ of $\mathbb{Z}_{(p)}$-homology spheres with arbitrary unitary structure; the results of Section 1 imply that $U^{\#}\Theta_k\langle p\rangle$ is generated by $U\Theta_k\langle p\rangle$ and $\pi_k(O/U)$ ($\cong$ unitary structures on S^k) if $k \not\equiv 2 \bmod 8$, and if $k \equiv 2 \bmod 8$ this subgroup has index at most (by the appendix, exactly if $k \neq 2$) two. One then has a long exact sequence

$$\to RS^{\mathbb{Z}/p,s}_{k+V} \to \mathfrak{A}^{\mathbb{Z}/p}_{k+V} \to U^{\#}\Theta_k\langle p\rangle \to RS^{\mathbb{Z}/p,s}_{k-1+V} \to$$

which terminates on the left when k becomes close to $\dim V$. The map $RS \to \mathfrak{A}$ is just a forgetful map. We shall not use this, however.

PROOF. We first show that the composites yield zero. Suppose that one has an action on the homotopy sphere Σ with fixed point set M and zero torsion. In the previous notation, the composite is a $\mathbb{Z}/p$-action on ∂N, where N is a disk and the fixed point set of the action is M_0. On the other hand, we also obtain such an action from Σ - Int $D(k+V)$. It follows from the uniqueness statement that $\partial N \cong S(k+V)$, and a more careful analysis shows that the other pieces of structure are also trivial.

On the other hand, suppose we are given a structured homology sphere (M^k,f) whose image in the RS-group is zero. This means that, if N is the action constructed before, then $\partial N \cong S(k+V)$, the diffeomorphism $S^k \to \partial M_0 \cong S(k+V)^{\mathbb{Z}/p} \cong S^k$ extends to a diffeomorphism of D^{k+1}, and the tubular neighborhood of S^k extends likewise. Let Σ be formed by gluing N and $D(k+V)$ together using the equivalence $\partial N \cong S(k+V)$. Since N is a disk, Σ is clearly a homotopy sphere. Since the diffeomorphism of S^k extends to D^{k+1}, the fixed point set is diffeomorphic to M; since the tubular neighborhood extends, one obtains the same unitary structure on M as one had originally if p is odd. ∎

We are almost ready to begin proving Theorem A. Before doing this, we shall fit the invariant $\mathfrak{z}$ into our setting more precisely. As we have already noted, a $\mathbb{Z}_{(p)}$-homology sphere M^k determines an element

$\mathcal{G}(M^k) \in \pi_k(F/O)_{(p)}$; specifically, this is the normal invariant of the degree 1 homology equivalence $M^k \to S^k$ in the sense of [51:I, §1]. A standard argument involving connected sums tells us that $\mathcal{G}(M_1 \# M_2) = \mathcal{G}(M_1) + \mathcal{G}(M_2)$. Furthermore, if M^k bounds a $\mathbb{Z}_{(p)}$-homology disk, then the results of [51:I §1] imply that $\mathcal{G}(M) = 0$. Consequently, $\mathcal{G}$ passes to a homomorphism from $U\Theta_k\langle p\rangle$ or $\Theta_k\langle 2\rangle$ to $\pi_k(F/O)_{(p)}$.

PROPOSITION 3.2. The map $\mathcal{G}$ is given by the composite

$$U\Theta_k\langle p\rangle_{(p)} \overset{\cong}{\to} \Theta_{k(p)} \xrightarrow[\text{Thom}]{\text{Pont.}} \pi_k(F/O)_{(p)}$$

if $p \neq 2$ and by the composite

$$F\Theta_k\langle 2\rangle_{(2)} \to \pi_k(M\Phi)_{(2)} \overset{\cong}{\to} \pi_{k(2)} \to \pi_k(F/O)_{(2)}$$

if $p = 2$($F\Theta_k\langle 2\rangle$ denotes mod 2 homology spheres with Φ-structures-such always exist [2,3]).

PROOF. The case $p \neq 2$ follows from 2.5 and the fact that $\Theta_k \to U\Theta_k\langle p\rangle \to \pi_k(F/O)_{(p)}$ is the p-localization of the usual Pontrjagin-Thom map. If $p = 2$, then the same sort of argument will apply, but we must replace 2.5 by an assertion that $\mathcal{G}$ is trivial on the image of $L_{k+1}(\mathbb{Z}_{(p)})$. To prove the latter take a representative homology k-sphere that is a framed boundary; let P be the coboundary. Let $f\colon (P,\partial P) \to (D^k, S^{k-1})$ be the degree one map; since P is parallelizable, f is a normal map. Thus we have Umkehr maps in S-theory as follows:

$$\begin{array}{ccc} (D^k)_+ & \xrightarrow{f^!} & P_+ \\ \uparrow i_+ & & \uparrow j_+ \\ S^{k-1}_+ & \xrightarrow{(\partial f)^!} & \partial P_+ \end{array} \qquad (Y_+ = Y \amalg \{\text{pt.}\}).$$

The map $(\partial f)^!$ is a $\mathbb{Z}_{(2)}$ homology (and hence a Z-local stable homotopy) equivalence, and the map $(i_+ \otimes \mathbb{Z}_{(2)}) \circ ((\partial f)^! \otimes \mathbb{Z}_{(2)})^{-1}|\partial P$ is the one from which the normal invariant in [51: I] is constructed. Since $f^!$ is a homotopy retract (collapse P_+ to $\{\text{pt.}\}_+$ and call this ε), the normal invariant will be zero if $(\varepsilon j_+) \otimes \mathbb{Z}_{(2)}|\partial P$ is zero. But clearly $\varepsilon j_+|\partial P$ is the constant map, and thus the argument is complete. ∎

We can now establish the first step in the proof of Theorem A.

THEOREM 3.3. The homomorphisms $C_p \otimes \mathbf{Z}[p^{-1}]$ are trivial. Therefore every representative of every element of $U\Theta_k\langle p\rangle$ or $\Theta_k\langle 2\rangle$ having order prime to p is the fixed point set of a smooth $\mathbb{Z}/p$-action on a homotopy $(k+\dim V)$-sphere.

Note. The result makes no assumption that the action is orientation preserving if $p = 2$.

PROOF. The groups $RS_{q+V}^{\mathbb{Z}/p,s}$ fit into long exact sequences as follows (compare [44]):

$$(3.4) \qquad \cdots \to hS_{q+1}^{s}(S(V)/G) \to RS_{q+V}^{G,s} \to \pi_q(F_G(V)) \to hS_q^s(S(V)/G) \to \cdots \quad (G = \mathbb{Z}/p)$$

Here hS^s denotes simple homotopy smoothings, and $F_G(V)$ is the space of G-equivariant self-maps of $S(V)$. The groups $hS_{q+1}^s(S(V)/G)$ are finitely generated (modulo low-dimensional troubles) by the long exact surgery sequence (compare [44]), and the groups $\pi_q(F_G(V))$ are well-known to be finitely generated (compare [46]). Hence $RS_{q+V}^{\mathbb{Z}/p,s}$ is finitely generated.

On the other hand, consider the groups $U\Theta_k\langle p\rangle \otimes \mathbb{Z}[p^{-1}]$ and $\Theta_k\langle 2\rangle \otimes \mathbb{Z}[2^{-1}]$. If $k \not\equiv 3 \bmod 4$, then these groups are isomorphic to quotients of $\pi_k(M\Lambda) \otimes \mathbb{Z}[p^{-1}]$ and $\pi_k(M\Phi) \otimes \mathbb{Z}[2^{-1}]$ respectively; this follows from (2.4) and the vanishing of $L_{k+1}(\mathbb{Z}_{(p)}) \otimes \mathbb{Z}[p^{-1}]$ and $L_k(\mathbb{Z}_{(p)}) \otimes \mathbb{Z}[p^{-1}]$. Since the localized bordism groups are divisible by 2.1(iii) if $p \neq 2$ and its analogs in [3] if $p = 2$, the groups $U\Theta_k\langle p\rangle \otimes \mathbb{Z}[p^{-1}]$ and $\Theta_k\langle 2\rangle \otimes \mathbb{Z}[2^{-1}]$ are also divisible.

Therefore, if $k \not\equiv 3 \bmod 4$ the map $C_p \otimes \mathbb{Z}[p^{-1}]$ maps a divisible group to a finitely generated $\mathbb{Z}[p^{-1}]$-module. Since the latter is a direct sum of cyclic modules, it is clear that 0 is the unique divisible subgroup, and therefore $C_p \otimes \mathbb{Z}[p^{-1}] = 0$ (the image of a divisible group is divisible).

If $k \equiv 3 \bmod 4$, let $\mathcal{D}$ be the unique maximal divisible subgroup of $U\Theta_k\langle p\rangle \otimes \mathbb{Z}[p^{-1}]$ or $\Theta_k\langle 2\rangle \otimes \mathbb{Z}[2^{-1}]$ (i.e., $\mathcal{D}(A) = \cap nA$, the intersection ranging over all positive integers n). As in the preceding paragraph, $C_p \otimes \mathbb{Z}[p^{-1}]$ restricted to $\mathcal{D}$ is trivial. On the other hand, by (2.4) we know that $L_{k+1}(\mathbb{Z}_{(p)}) \otimes \mathbb{Z}[p^{-1}]$ and $\mathcal{D}$ generate $U\Theta_k\langle p\rangle \otimes \mathbb{Z}[p^{-1}]$ or $\Theta_k\langle 2\rangle \otimes \mathbb{Z}[2^{-1}]$. Therefore we need only prove that the composite

$$(3.5) \qquad \psi\colon L_{k+1}(\mathbb{Z}_{(p)} \otimes \mathbb{Z}[p^{-1}] \to RS_{k-1+V}^{\mathbb{Z}/p,s} \otimes \mathbb{Z}[p^{-1}]$$

is trivial.

First of all, the composite of ψ with the induced map $L_{k+1}(\mathbb{Z}) \to L_{k+1}(\mathbb{Z}_{(p)})$ is trivial. Stated differently, every exotic sphere in bP_{k+1} is the fixed point set of some $\mathbb{Z}/p$-homotopy $(k+V)$-sphere. To prove this we use exact sequence (6.2) from [51: II]; by that sequence, such an action exists if the composite in the following diagram is zero:

$$\begin{array}{ccc} \mathbb{Z} \cong L_{k+1}(\mathbb{Z}) & \xrightarrow{xS(V)/G} & L^s_{k+\dim V}(\mathbb{Z}[G,w]) \\ \downarrow & & \downarrow \\ \Theta_k & \xrightarrow{xId[S(V)/G]} & hS^s_k(S(V)/G) \end{array}$$

($G = \mathbb{Z}/p$ here, $w(g)$ = determinant of g acting on V).

If p is odd, the composite is clearly zero because $L^s_{k+\dim V}(\mathbb{Z}[\mathbb{Z}/p]) = 0$ ($k + \dim V$ is odd; see [59] for the assertion about the Wall group). If $p = 2$ and $\dim V \equiv 2 \bmod 4$, the triviality of the composite (3.5) is immediate because $L_{4q+1}(\mathbb{Z}[\mathbb{Z}/2]) = 0$. If $p = 2$ and $\dim V \equiv 0 \bmod 4$, the group $L_{4q+3}(\mathbb{Z}[\mathbb{Z}/2])$ is detected by a codimension one Kervaire invariant [58]; therefore, if we let $V_0 \subseteq V$ have codimension 1, the map "$xS(V)/G$" is trivial if and only if the same is true for "$xS(V_0)/G$". Thus it suffices to prove that the map "$xS(V)/G$" is trivial if $\dim V$ is odd. But $L_{even}(\mathbb{Z}[\mathbb{Z}/2,-])$ is detected by a Kervaire invariant, and by standard product formulas the Kervaire invariant of $\sigma \times S(V)/G$ equals $(\text{Index } \sigma)\cdot\chi(S(V)/G)$. But Index σ is divisible by 8, so the map is zero.

We can now complete the proof; recall that $k \equiv 3 \bmod 4$ is the only case left. The case $p = 2$ is easiest; notice first that $L_{4q}(\mathbb{Z}_{(2)}) \otimes \mathbb{Z}[2^{-1}] \cong \mathbb{Z}[2^{-1}]$, the isomorphism being induced by the signature invariant. On the other hand, the map $L_{4q}(\mathbb{Z}) \xrightarrow{j^*} L_{4q}(\mathbb{Z}_{(2)}) \xrightarrow{sgn} \mathbb{Z}$ has image $8\mathbb{Z}$, and hence $j_* \otimes \mathbb{Z}[2^{-1}]$ is an isomorphism. Putting this together with the previous paragraph, we see that the map ψ in (3.5) must be zero. Therefore the case $p = 2$ is finished.

Assume now that p is odd. From the previous considerations we see that the image of ψ has exponent dividing 8, and thus we may as well localize (3.4) and (3.5) at 2. The study of (3.4) localized at two is simplified by two computational facts. First of all, $\pi_q(F_G(V)) \otimes \mathbb{Z}[p^{-1}]$ is isomorphic to $\pi_q(F(V)) \otimes \mathbb{Z}[p^{-1}]$ via the forgetful map [46], and this is just $\pi_q \otimes \mathbb{Z}[p^{-1}]$ because we are in the stable range. Secondly, in the surgery sequence for $S(V)/G$ the normal invariant group $[S^q[S(V)/G]_+, F/O]$ has the property that $\pi^*: [S^q[S(V)/G]_+, F/O] \otimes \mathbb{Z}[p^{-1}] \to [S^q(S(V)_+), F/O] \otimes \mathbb{Z}[p^{-1}]$ is bijective. These and other standard considerations (compare [44, 47]) give a complete description of $RS^{\mathbb{Z}/p,s}_{k-1+V} \otimes \mathbb{Z}_{(2)}$; we shall merely present the conclusions. If we surger an element of the RS group by cutting out the tubular neighborhood $S^{k-1} \times \text{Int } D(V)$ and replacing it with $D^k \times S(V)$, we get a free $\mathbb{Z}/p$-manifold which has an Atiyah-Singer invariant. Letting $\mathcal{R}$ denote the ring of complex valued functions on $\mathbb{Z}/p-\{1\}$, we obtain a homomorphism

$$A\colon RS^{\mathbb{Z}/p,s}_{k-1+V} \otimes \mathbb{Z}_{(2)} \to \mathcal{R} \otimes \mathbb{Z}_{(2)}.$$

Further details appear in [54]. The result we need is the following:

(3.7) The map A detects the torsion-free part of the domain, and the latter's torsion is detected by the forgetful map into $\Theta_{k+\dim V-1(2)}$. ∎ Since the classes in Image ψ are torsion, this means it suffices to show that the map

$$U\Theta_k\langle p\rangle \xrightarrow{C_p} RS^{\mathbb{Z}/p,s}_{k-1+V} \to \Theta_{k-1+\dim V}$$

is trivial. But this is immediate from construction, for an action representing some $C_p(x)$ was the boundary of some action on a disk. ∎

We may strengthen Theorem 3.3 as follows:

THEOREM 3.8. Let M^k be a $\mathbb{Z}_{(p)}$-homology sphere representing an element in $U\Theta_k\langle p\rangle$ or $\Theta_k\langle 2\rangle$ of order prime to p. Then M^k is the fixed point set of a smooth $\mathbb{Z}/p$-action on $S^{k+\dim V}$, where $\dim V > k+1$.

PROOF. By Theorem 3.3 we can realize M as the fixed point set of some $\mathbb{Z}/p$-action ϕ on some homotopy sphere $\Sigma^{k+\dim V}$. Let q be the order of M in the group of homology spheres, and let Σ have order $p^r s$, in $\Theta_{k+\dim V}$, where p and s are relatively prime. Choose $T > 0$ so that $T \equiv 0 \bmod p^r$ and $T \equiv 1 \bmod q$, and consider the connected sum of T copies of Σ with its $\mathbb{Z}/p$-action. The fixed point set is then a connected sum of T copies of M, and the ambient homotopy sphere Σ' has order prime to p. On the other hand, it is well-known that such a Σ' admits a smooth $\mathbb{Z}/p$-action with S^k as its fixed point set; one merely expresses $\Sigma' = p\Sigma''$ (possible since Σ' has order prime to p), takes an equivariant connected sum of the orbit space $S^{k+\dim V}$ — viewed as a differentiable variety with singularities — with Σ'' away from the singularities, and sees that $S^{k+\dim V}/(\mathbb{Z}/p) \# \Sigma''$ is the differentiable orbit space of an action on $p\Sigma'' = \Sigma'$ (compare [50]). Taking the connected sum of our original action on Σ' with the negative of this new action, we obtain a smooth action on $S^{k+\dim V}$ with $\#^T M$ as its fixed point set.

But by our assumptions $\#^T M$ is homologically h-cobordant to M (with the correct unitary structures if $p \neq 2$). Let N be this h-cobordism, and let E be its closed normal disk bundle (with its unitary structure if $p \neq 2$). We may then plumb $S^{k+\dim V} \times I$ and N together equivariantly at $E|\#^T M$ to obtain a cobordism P with $\partial P = \partial_1 P \cup -S^{k+\dim V}$ and $\partial P^G = M \cup -\#^T M$. Furthermore, P has the $\mathbb{Z}_{(p)}$-homology of $S^{k+\dim V}$ with $\mathbb{Z}$-homology torsion governed by the torsion in $H_*(N; \mathbb{Z})$. We would like to add handles to the top of P, equivariantly and away from the fixed point set, in order to make P an h-cobordism. However, one can do this exactly as in the original work of Jones [23] thanks to our assumptions on $\dim V$. ■

COROLLARY 3.9. *Theorem A is true if* $p \neq 2$.

PROOF. In view of 3.8 we need only check this for elements of $U\Theta_k\langle p\rangle$ having order a power of p and $\mathcal{G}$-invariant zero. However, by 2.3 such elements are given by $bP_{k+1(p)}$, and by the results stated in Section 2 this can be nonzero only if $k \equiv 3 \bmod 4$. Furthermore, in the proof of 3.3 we noted that the composite $\psi \cdot L_{k+1}(\mathbb{Z}) \to RS^{\mathbb{Z}/p,s}_{k-1+V}$ was trivial. Since $L_{k+1}(\mathbb{Z})$ maps onto bP_{k+1} in Θ_k, the desired conclusion follows. ■

We turn now to the case $p = 2$. It remains to consider $\mathbb{Z}_{(2)}$-homology spheres in the image of $L_{k+1}(\mathbb{Z}_{(2)})$. The argument splits into the two cases $k \equiv 1 \bmod 4$ and $k \equiv 3 \bmod 4$ (the L-group is zero is k is even). In the first case $L_{k+1}(\mathbb{Z}_{(2)})$ is detected by the Kervaire invariant, and therefore the following result completes the proof of Theorem A if $\dim V \equiv 0 \bmod 4$.

THEOREM 3.10. *If* Σ^{4q+1} *is the Kervaire sphere, then* Σ^{4q+1} *is the fixed point set of an involution on the Kervaire* $(4(q+r)+1)$*-sphere for every* $r > 0$. ■

This follows easily from the realization of Σ^{4s+1} as the intersection of the variety

$$\{z_0^3 + z_1^2 + \dots + z_{2s+1}^2 = 0\} \subseteq C^{4s+2}$$

with the unit sphere (see, for example, [15]). Explicitly, if $s = q+r$, then one takes the action induced by the linear involution that fixes the first $2q+2$ complex coordinates. ■

Of course, $L_{4q}(\mathbb{Z}_{(2)})$ is far more complicated as a group. However, we have the following result of Alexander, Hamrick, and Vick to help us (see [4] for details).

THEOREM 3.11. *Let* M^{4q-1} *be a* $\mathbb{Z}_{(2)}$*-homology sphere obtained from an element of* $L_{4q}(\mathbb{Z}_{(2)})$ *by some appropriate plumbing. Then for every* V *with* $\dim V \equiv 2 \bmod 4$ *and* $\dim V > 4q$, *one can realize* M *as the fixed point set of a smooth involution on a homotopy* $(4q-1 + \dim V)$*-sphere.* ■

The proof of Theorem A will be completed by the following result:

(3.12) COMPLEMENT TO (3.11). *One can construct a smooth involution on* $S^{4q-1+\dim V}$ *or the corresponding Kervaire sphere.*

PROOF. The results of [4] do not state that the homotopy sphere bounds a parallelizable manifold, but this is implicit in the proof. In the notation of [4], one constructs a manifold M^{4r+1} that bounds a parallelizable manifold and performs surgery on M in dimensions $\leq 2r$ to get a homotopy sphere M''. Clearly it suffices to notice that the surgeries may be chosen so that the surgery cobordism has a framing extending that of M. But it is well-known that twisting the embedding $S^k \times D^{4r-k+1} \to M$ by an element of $\pi_k(SO_{4r-k+1})$ will accomplish this if $k < 2r$ (compare [14]). ∎

4. Equivariant Pontrjagin-Thom diagrams

In this section we describe some of the machinery needed to prove Theorem C. A complete account of the results presented will appear in [51: IV].

Let G be a finite group, and let α be a representation of G; a special α-framing of a smooth G-manifold M is a stable class of isomorphisms from $\tau_M \oplus k$ to $\alpha \oplus k$, where k denotes the product of M with a trivial representation. One can define a bordism group $\Omega_\alpha^{G,fr,sp}$ of specially α-framed smooth G manifolds, and one can prove that the Pontrjagin-Thom construction induces an isomorphism from $\Omega_\alpha^{G,fr,sp}$ to the equivariant stable homotopy group π_α^G (compare [21] or [39]).

Likewise, one can also define groups of specially framed G-homotopy U-spheres -- denoted by $F\Theta_U^{G,sp}$. Its elements are h-cobordism classes of pairs (Σ,Φ), where (i) Σ is a G-homotopy U-sphere in the sense of [51: II], (ii) Φ is a special U-framing of Σ. As noted in [51: II, §6], one can study the groups Θ_U^G of G-homotopy U-spheres by means of a long exact sequence. The following exact sequence is an analog of [51 : II, 6.2] for studying $F\Theta_U^{G,sp}$ if G acts semifreely on U:

$$(4.1) \qquad \to hS_{k+1}^T(L(V)) \to F\Theta_{k+V}^{G,sp} \to \begin{matrix}\pi_k(F_G(V))\\ \oplus \\ F\Theta_k\end{matrix} \to hS_k^T(L(V)) \to$$

$$(U = k+V, \quad G \text{ acts freely on } V)$$

Here is a description of the notation in (4.1) that is not introduced above or in [51: II, §6]. The symbol hS^T denotes tangential homotopy smoothings as defined in (say) [35]; specifically, one takes homotopy equivalences $F: M \to X$ together with specific bundle maps $F: \nu_M \to \nu_X$ lefting f. The symbol $F\Theta_k$ denotes the group of homotopy spheres endowed with specific framings (compare [27]). Each of the maps in (4.1) has an obvious analog in the earlier sequence [51: II, (6.2)], and the derivation of (4.1) proceeds in the expected fashion. Further comments will appear in [51: IV].

Assume now that $G = \mathbb{Z}/p$, where p is prime.

Of course, the group $F\Theta^{G,sp}_{k+V}$ admits a natural map into $\Omega^{G,fr,sp}_{k+V} \cong \pi^G_{k+V}$ by passage to the underlying (specially) framed bordism class. Furthermore, the coexact sequence $S^k \to S^{k+V} \to S^{k+V}/S^k$ yields a long exact sequence as follows:

$$(4.2) \to \{S^{k+V}/S^k, S^o\}^G \to \pi^G_{k+V} \to \pi^G_k \to \{S^{k-1+V}/S^{k-1}, S^o\}^G \to$$

Using some standard isomorphisms, we may rewrite (4.2) as follows:

$$(4.3) \to \{S^{k+1}(L(V)_+), S^o\} \to \pi^G_{k+V} \to \begin{matrix}\pi^S_k(BG_+)\\ \oplus \\ \pi_k\end{matrix} \overset{\delta}{\to} \{S^k(L(V)_+), S^o\}$$

A reader with some experience in surgery theory and equivariant homotopy will notice that each group in (4.1) admits a natural map to a corresponding group in (4.3). The main result of [51: IV] states that these sequences and maps fit together in the best possible fashion:

THEOREM 4.4. <u>Let</u> $P_\alpha: F\Theta^{H,sp}_\alpha \to \pi^H_\alpha$ <u>be the Pontrjagin-Thom map for the</u> H-<u>representation</u> α (H = 1 <u>or</u> G <u>later on</u>), <u>let</u>

$$q': hS^T_m(X) \to \{S^m(X_+), S^o\}$$

<u>be the normal invariant as defined in</u> [35], <u>and let</u> $\gamma_V = -\lambda s_*$, <u>where</u> $s: F_G(V) \to F_G$ <u>is stabilization and</u> λ <u>is the homotopy equivalence from</u> F_G <u>to</u> $\Omega^\infty S^\infty(BG_+)$ <u>defined in</u> [10]. <u>Then the following diagram commutes</u>:

$$\begin{array}{ccccccccc} \to & hS^T_{k+1}(L(V)) & \to & F\Theta^{G,sp}_{k+V} & \to & \pi_k(F_G(V)) \oplus F\Theta_k & \to & hS^T_k(L(V)) & \to \\ (4.5) & \downarrow q' & & \downarrow P_{k+V} & & \downarrow \gamma_V \oplus P_k & & \downarrow q' & \\ \to & \{S^{k+1}(L(V)_+), S^o\} & \to & \pi^G_{k+V} & \to & \{S^k, BG_+\} \oplus \pi_k & \to & \{S^k(L(V)_+), S^o\} & \to \end{array}$$

In this diagram the rows are given by exact sequences (4.1) and (4.3).

Diagram (4.5) is central to the results of [51: 4] and [52] concerning group actions on exotic spheres. In this paper it is the means by which the solution of the Strong Segal Conjecture is applied to prove Theorem C.

5. <u>Consequences of the Strong Segal Conjecture for</u> $\mathbb{Z}/p$

In a prepublication draft of [53], G. Segal made a bold and remarkable conjecture about the stable cohomotopy of BG for G a finite group. For groups of prime order the truth of this conjecture has been established [20, 28-29], and in fact its truth is understood in a much broader context for these groups. This result has been called the <u>Strong Segal Conjecture for</u>

$\mathbb{Z}/p$ (one has analogous conjectures for any finite group). We shall state one version of this result below.

First of all, we need some notation. If α is a virtual representation of G and X is a finite complex approximating BG, then there is no problem in constructing a Thom complex $X^{\phi(\alpha)}$, where ϕ is the obvious map from $RO(G)$ to $\widetilde{KO}(G)$. By taking limits we can form a Thom spectrum BG^{α}; note that the bottom cell of this spectrum lies in dimension $= \dim \alpha$ (which may be negative). If V is a free representation of G, then one knows that the (-1)-skeleton of BG^{-V} is just the Thom complex $L(V)^{\nu}$, where ν is the normal bundle of $L(V)$, and consequently one can split the (-1)-skeleton as a wedge of the (-2)-skeleton and the finite spectrum S^{-1}. Combining the map of S^{-1} into the (-1)-skeleton with inclusion into all of BG^{-V}, we obtain a map of spectra

$$f_V: S^{-1} \to BG^{-V}$$

Let $\omega_V: BG^{-V} \to BG_+$ be the map which collapses the (-1)-skeleton to a point. With this notation, we may state the facet of the Strong Segal Conjecture that we need as follows:

(5.1) _Let_ p _be a prime, and let_ $k > 0$ _be given. Then there is a positive integer_ P_k _with the following property: If_ V _is a free_ $\mathbb{Z}/p$_-module and_ $\dim V \geq P_k$, _then the map_

$$f_{V*}: \pi_i(S^{-1})_{(p)} \to \pi_i(BG^{-V})_{(p)}$$

is injective and the map

$$\omega_{V*}: \pi_i(BG^{-V})_{(p)} \to \pi_i(BG_+)_{(p)}$$

is zero for all i _such that_ $0 < i \leq k$. ■

In Section 8 we shall need the following related result:

PROPOSITION 5.2. _Let_ k, P_k, _and_ V _be defined as in_ (5.1). _Then the localized map_

$$\delta_{(p)}: \pi_k(BG_+)_{(p)} \oplus \pi_{k(p)} \to \{S^k(L(V)_+), S^0\}_p,$$

where δ _is defined as in_ (4.3), _is injective_.

PROOF. As in [16] one can form a stable group $\pi_{k+V}^{G,\text{free}}$ by stabilizing with with only free G-modules. A computation as in [51: II] shows that $\pi_{k+V}^{G,\text{free}}$ is isomorphic to $\pi_k(BG^{-V})$. Furthermore, the map $\phi: \pi_{k+V}^{G} \to \pi_k$ given by passage to the fixed point set fits into a long exact sequence

$$(5.3) \qquad \to \pi_{k+1} \overset{\phi'}{\to} \pi_{k+V}^{G,\mathrm{free}} \overset{\phi''}{\to} \pi_{k+V}^{G} \overset{\phi}{\to} \pi_k \to,$$

and under the isomorphism of the previous sentence ϕ' corresponds to the map f_{V*} defined before the statement of (5.1). Therefore $\phi_{(p)}$ is zero by (5.1), so that $\phi''_{(p)}$ is onto.

Consider now the map

$$\pi_{k+V}^{G,\mathrm{free}} \overset{\phi''}{\to} \pi_{k+V}^{G} \overset{\Gamma}{\to} \pi_k \oplus \pi_k(BG_+),$$

where Γ is given as in (4.3). The first coordinate of Γ is given by ϕ, and the second coordinate of $\Gamma\phi''$ is given by the map ω_V defined before (5.1). Applying (5.1) once again, we see that $(\Gamma\phi'')_{(p)}$ is zero on the second coordinate; of course, from the previous paragraph we know that the first coordinate is zero too. Finally, we also know that $\phi''_{(p)}$ is onto, and from this it is immediate that $\Gamma_{(p)}$ is zero. By the exactness of (4.3) localized at p, it follows that $\delta_{(p)}$ must be injective. ■

6. Pseudotangential homology equivalences

In [51; I, §1] it is shown that an R-homology equivalence $f: M \to X$ ($\mathbb{Z} \subseteq R \subseteq \mathbb{Q}$) has a naturally defined normal invariant $q(f)$ in $[X, F/O] \otimes R$. The image of $q(f)$ in $\widetilde{KO}(x) \otimes R$ measures the difference between ν_X and ν_M, and therefore if $\nu_M - f^*\nu_X \in \widetilde{KO}(M)$ is annihilated by an invertible integer in R the normal invariant lifts back to $[X,F] \otimes R \cong \{X,S^0\} \otimes R$. If $R = \mathbb{Z}$, then f is tangential and the lifting of the normal invariant can be done in a fairly canonical fashion. We shall prove in this section that canonical liftings exist for arbitrary choices of R.

DEFINITION. An R-homology equivalence (R($\mathbb{Z}/2$)-twisted homology equivalence in the unoriented case; compare [51: I, §1]) $f: A \to B$ is called R-<u>pseudotangential</u> if $\tau_A - f^*\tau_B$ maps to zero in $\widetilde{KO}(A) \otimes R$. An R-<u>pseudoframed homology equivalence</u> is a triple (f,g,λ) where

(i) $f: M \to X$ is an R-homology equivalence,

(ii) $g: M \to \Phi_p = \mathrm{Fiber}(BO \to BO_{(p)})$ is a map,

(iii) $\lambda: E(\nu_M) \to E(\nu_X) \times E(\gamma_p)$ is a stable bundle map covering (f,g) — the bundle γ_p represents the universal vector bundle pulled back to Φ_p.

To be more specific in (iii), one should consider for each N the N-plane bundle γ_p^N over the pullback $\Phi_{p,N} = \Phi_p \times_{BO} BO_N$ and use factorizations of g through $\Phi_{p,N}$. However, such points are fairly routine (compare [55]) and therefore will be left to the reader.

Henceforth we shall restrict ourselves to the case $R = \mathbf{Z}_{(p)}$ for the sake of simplicity. In this case we usually use the term p-*pseudotangential* as a synonym for R-*pseudotangential*.

The space Φ_p is path connected with homotopy of finite order prime to p. These facts are easy consequences of basic localization theorems. We shall need one more fact about Φ_p that follows from the above discussion and the realizability of 1-connected free abelian chain complexes as cellular chain groups of 1-connected complexes:

PROPOSITION 6.1. *There is a directed system of finite complexes* $\{K_\alpha\}$ *with the following properties*:

(i) Φ_p *has the homotopy type of* $\text{inj lim}_\alpha K_\alpha$.
(ii) *If* $p = 2$, *then* K_α *is* 1-*connected with all homology and homotopy finite of order prime to* p. *If* p *is odd, then* K_α *is the product of* $RP^{2n(\alpha)}$ *with such a space*.
(iii) *Given any* α *and* $q > 0$, *there is a* $\beta \geq \alpha$ *such that* $\gamma \geq \beta$ *implies* (Φ_p, K_α) *is at least* q-*connected*. ■

This proposition allows us to factor the extra structure map g for a pseudotangential homology equivalence through a manageable finite complex with much the same homology and homotopy properties as Φ_p. In addition, it gives us a family of choices for this factorization for which the specific choice is relatively unimportant. It will be particularly convenient to assume K_α is a compact (bounded) codimension 0 submanifold of some large Euclidean space. Denote the universal bundle over K_α (with order prime to p) by γ.

Let (f,g,λ) be a p-pseudotangential homology equivalence, and assume that g factors through the finite subcomplex K_0 (given as in 6.1. Construct a smooth embedding of M in $X \times D^\ell$ that is homotopic to f, and let $f^!: X_+ \to M^\xi$ be the associated collapse map. As in [51: I, §1] the localized map $f^!_{(p)}$ is a stable homotopy equivalence and as such yields an S-map $R(f): M^\xi \to S^0_{(p)}$ that corresponds to a localized fiber retraction.

Recall that by standard methods one can classify localized fiber retractions on vector bundles, the classifying space being the fiber of $BO \to BF_{(p)}$. As in [51: I; §1] we shall call this fiber $F_{(p)}/O$. Since $F_{(p)}$ and $BF_{(p)}$ are H-spaces, it is immediate that $[K_\alpha, F_{(p)}] = [K_\alpha, BF_{(p)}] = 0$ for all K_α in 6.1 and therefore the map $\Delta: [K_\alpha, F_{(p)}/O] \to [K_\alpha, BO]$ is bijective. Thus given a vector bundle ζ over some K_α, there is a (basically) unique localized fiber retraction $K^\zeta \to S^0_{(p)}$ that we shall call $\Delta\#(\zeta)$.

We can now define the normal invariant of (f,g,λ) as an element of $\{M,S^o\}_{(p)}$. To do this, we first note that the bundle $\xi - g^*\gamma$ over M has a canonical trivialization. For λ gives an isomorphism from ν_M to $f^*\nu_X \oplus g^*\gamma$, and f itself gives a canonical isomorphism $K(f)$ from $f^*\tau_X$ to $\tau_M \oplus \xi$. Combining these isomorphisms yields the desired trivialization; call it Γ_λ. With this notation, the normal invariant of (f,g,λ) is the composite

$$M \subseteq M_+ \xrightarrow{Th(\Gamma_\lambda)} M^{\xi - g^*\gamma} \xrightarrow{Th(1,g)} M^{\xi} \wedge K^{-\gamma} \xrightarrow{R(f)\wedge\Delta\#(-\gamma)} S^o_{(p)} \wedge S^o_{(p)} \simeq S^o_{(p)}.$$

We call this composite $\overline{q}(f,g,\lambda)$. As in [51: I], the map $f^*\colon \{M,S^o\}_{(p)} \to \{X,S^o\}_{(p)}$ is bijective, and we take $q'(f,g,\lambda) = -(f^*)^{-1}\overline{q}(f,g,\lambda)$; the reasons for the minus sign were mentioned in [51: I, §1].

THEOREM 6.2. <u>Let</u> $j\colon Q_1S^o \simeq SF \to F/O$ <u>be the usual map, and let</u> (f,g,λ) <u>be as above. Then</u> $j_*q'(f,g,\lambda) = q(f)$, <u>where</u> $q(f)$ <u>is the normal invariant of</u> [51: I, §1].

PROOF. Let $R_o(f) \in [M,F_{(p)}/O]$ be the element classifying $R(f)$, and let $j'\colon Q_1S^o_{(p)} \simeq SF_{(p)} \to F_{(p)}/O$ be the standard map. Then by construction one has the following identity in $[M,F_{(p)}/O]$:

$$j'_*\overline{q}(f,g,\lambda) = R_o(f) \oplus \Delta^{\#}(-\gamma)$$

(Note that $F_{(p)}/O$ has a good notion of direct sum). If we localize $F_{(p)}/O$ at p, we obtain $(F/O)_{(p)} \simeq F_{(p)}/O_{(p)}$. Localization kills $\Delta^{\#}(-\gamma)$, and consequently we have

$$(*) \quad j_*\overline{q}(f,g,\lambda) = R_o(f)_{(p)} \quad \text{in} \quad [M,F/O]_{(p)}.$$

By construction, $q(f) = -(f^*)^{-1}R_o(f)_{(p)}$. Hence the theorem follows by applying $(-f^*)^{-1}$ to both sides of $(*)$. ■

There is a natural relationship between the ideas of this section and the first two of this paper. Let Φ_p denote the fiber of the localization map $BO \to BO_{(p)}$. Then every mod p homology sphere Σ^k admits a Φ_p structure (also compare [9]), and a pair $(\Sigma, \Phi_p$ structure) determines a bordism class in $\pi_k(M\Phi_p)$. On the other hand, one can use $(\Sigma, \Phi_p$ structure) to form a p-pseudotangential homology smoothing of S^k. This yields a normal invariant $q'(\Sigma, \Phi_p \text{ str.})$ in $\pi_{k(p)}$. These are related in obvious fashion.

PROPOSITION 6.3. <u>Under the isomorphism from</u> $\pi_{k(p)}$ <u>to</u> $\pi_k(M\Phi_p)_{(p)}$, <u>the</u> class $q'(\Sigma, \text{str.})$ corresponds to the image of the bordism class of $(\Sigma, \text{str.})$. ■

7. Modifications of the knot invariant

This section contains the remaining technical material needed to prove Theorem C.

Results in the work of P. Löffler state that, for any representation α of $\mathbb{Z}/p$ with $\dim \alpha^{\mathbb{Z}/p} \neq 2$, a $\mathbb{Z}/p$-homotopy α-sphere has a stably trivial equivariant tangent bundle [30, 31]. The previous results on realizing homology spheres as fixed point sets show that one cannot expect this always. If the fixed point set is not a π-manifold, the remarks of [48, Appendix] show that the equivariant tangent bundle cannot be trivial. However, something close to Löffler's result is true:

PROPOSITION 7.1. Let $\mathbb{Z}/p$ act smoothly on a homology sphere Σ, and assume $\dim \Sigma^{\mathbb{Z}/p} \neq 2$. Then the tangent bundle of Σ maps to zero in $\widetilde{KO}_{\mathbb{Z}/p}(\Sigma)_{(p)}$.

PROOF. If $p \neq 2$, this may be proved almost exactly as in [30]. If $p = 2$, the main points that might seem different from [31] are (i) the problem of showing that the equivariant normal bundle of $\Sigma^{\mathbb{Z}/2}$ has odd order in $J(\Sigma^{\mathbb{Z}/2})$, (ii) the problem of showing that the equivariant normal bundle of Σ in some linear representation has odd $J_{\mathbb{Z}/2}$-order. Since the tangent bundle of $\Sigma^{\mathbb{Z}/2}$ has odd J-order, it is clear that (ii) implies (i) by naturality. But as in [31] one can show that ν admits a 2-local equivariant fiber retraction, and from this one can proceed as in [31]. ■

We may now proceed to define torsion structures as in [3,9] and Section 1. To do this, it is useful to know that equivariant KO-theory has a representable localization at p and there is a representable theory $KO_G(\ ;\mathbb{Z}_{(p)}/\mathbb{Z})$ that fits into a long exact sequence with the localization map. One can in fact realize this via the (equivariant) homotopy fiber $\Phi_{(p)}(G)$ of the equivariant localization map $BO(G) \to BO(G)_{(p)}$ in the sense of [36] ($BO(G)$ is supposed to represent $\widetilde{KO}_G$). We may define a p-equivariant pseudoframing of M to be an equivariant $\Phi_{(p)}(G)$-structure in the obvious appropriate sense.

The principal aim of this section is to modify the machinery of [51: I] to give results for actions on equivariantly pseudoframed homotopy (and homology) spheres.

Our first aim is to define a natural analog of the knot invariant. We shall begin by fixing our notation. Let $G = \mathbb{Z}/p$ act smoothly on the $\mathbb{Z}_{(p)}$-homology sphere Σ, let K^k be the fixed point set of the action, let V be the normal representation of G at a fixed point, and let Ψ be a G-equivariant p-pseudoframing (recall that one exists if $k \neq 2$). In analogy with [48, Appendix], the equivariant pseudoframing of Σ yields an ordinary pseudoframing of K upon restriction to the fixed point set.

In [51: I, §2] the unlocalized knot invariant of the action is defined by a class ω_0 in $[K, F_G(V)_{(p)}/C_G(V)]$. One can recover the equivariant normal bundle of K by taking the image $\nu(\omega)$ of ω in $[K, BC_G(V)]$.

Let C_G be the limit of the centralizers $C_G(V')$ over all free G-modules V'. The p-pseudoframing Ψ yields a nullhomotopy for the image of $\nu(\omega)$ in $[K, BC_G]_{(p)} \cong [K, BC_{G(p)}]$, and from this one obtains a lifting of ω to the fiber of the composite

$$F_G(V)_{(p)}/C_G(V) \to BC_G(V) \to BC_{G(p)}.$$

It is an elementary exercise to verify that this fiber equals the fiber of the map

$$BC_G(V) \to BF_G(V)_{(p)} \times BC_{G(p)},$$

and accordingly we denote the common fiber by the symbol $F_G(V)_{(p)} \times C_{G(p)}/C_G(V)$. To summarize, Ψ induces a canonical lifting of ω_0 to a class ω_0' in $[K, F_G(V)_{(p)} \times C_{G(p)}/C_G(V)]$.

This invariant has many of the standard properties described in [51: I, §2]. For example, it is additive in the sense of [51: I, 2.1]. As in [51: I] we may localize ω_0' at p to obtain a localized pseudoframed knot invariant

$$\omega'(\Sigma,\Psi) \in \pi_k(F_G(V) \times C_G/C_G(V))_{(p)}.$$

In order to use this effectively, we need some results parallel to Section 2 of [51: I]. One can define a map

$$\Theta': \pi_k(F_G(V) \times C_G/C_G(V)) \to hS^T(S^k \times L(V))$$

analogous to Θ in [51: I, 2.3]; the projection of the domain to $\pi_k(F_G(V)/C_G(V))$ yields a homotopy smoothing by the construction for Θ, and the projection of the domain to $\pi_k(C_G/C_G(V))$ gives an explicit tangential structure for the homotopy smoothing. We can now formulate the analog of [51: I, Prop. 2.2].

PROPOSITION 7.2. _Let K be a p-pseudoframed smooth manifold, let V be a free G-module, let $L(V) = S(V)/G$, and let Λ' be the map of homotopy fibers induced by the following commutative square_:

$$\begin{array}{ccc} BC_G(V) & \to & BF_G(V) \times BC_G \\ \downarrow & & \downarrow \\ BC_G(V) & \to & BF_G(V)_{(p)} \times BC_{G(p)} \end{array}$$

Then there is a map ν_K': $[K, SF_G(V))_{(p)} \times SC_{G(p)}/SC_{G(p)}(V)] \to \{K \times L, S^o\}_{(p)}$ *such that the following diagram commutes*:

$$\begin{array}{ccc} [K, SF_G(V) \times SC_G/SC_G(V)] \overset{\Theta'}{\to} hS^T(K \times L) & \overset{q'}{\to} & \{K \times L, S^o\} \\ \Lambda_*' \downarrow & & \text{loc.} \downarrow \\ [K, SF_G(V)_{(p)} \times SC_{G(p)}/SC_G(V)] & \xrightarrow{\nu_K'} & \{K \times L, S^o\}_{(p)} \end{array}$$

PROOF (*Sketch*). The main difference with [51: I, Prop. 2.2] involves self-maps of $L(V)$. Namely, one must check that a self-map f with degree $\ell|G| + 1$ is a tangential homology equivalence. It suffices to note that the induced map on the universal coverings is equivariant and the tangent bundle of $L(V)$ is stably equivalent to $S(V) \times_G V$, for then $\tilde{f} \times_G \text{id}$ will give an explicit tangential structure. ∎

We also have an analog of [51: I, 2.4].

PROPOSITION 7.3. *Under the hypotheses of 7.2, the map* ν_K' *factors through the set* $[K,(SF_G(V) \times SC_G/SC_G)_{(p)}]$. *Furthermore, both* ν_K' *and its factorization are natural in* K. ∎

This leads to an analog of the simplification principle in [51: II, Prop. 4.7]:

PROPOSITION 7.4. *In the notation of this section, let* $c: K^k \to S^k$ *have degree* 1. *Then there is a class* $\bar{\omega} \in \pi_k(SF_G(V) \times SC_G/SC_G(V))$ *such that the image of* $c^*\bar{\omega}$ *in*

$$\{K \times L, S^o\}_{(p)} \cong \{S^k \times L, S^o\}_{(p)}$$

corresponds to that of ω_o'. ∎

Finally, we have that $\nu_K(\omega_o')$ *is invariant under stabilization* via the canonical maps

$$\oplus\, W: SF_G(V) \times SC_G/SC_G(V) \to SF_G(V \oplus W) \times SC_G/SC_G(W).$$

The proof proceeds as in [51: I, 2.6-2.7]. ∎

Of course, there are natural maps from the spaces $SF_G(V)$ to $SF_G(V) \times SC_G/SC_G(V)$, and these are consistent with stabilization; in fact, the induced map $SF_G \to \lim_V SF_G(V) \times SC_G/SC_G(V)$ is a homotopy equivalence. The following corollary of this will be extremely important in the proof of Theorem C.

PROPOSITION 7.5. Given the notation of 7.4, *assume also that* $k < \dim V$. *Then one can choose the class* $\bar{\omega}$ *to lie in the image of* $\pi_k(SF_G(V))$.

PROOF. Let α be any choice for $\bar{\omega}$ consistent with 7.4. By the previous paragraph we know that the stabilization of α equals the stabilization of some element β in the image $\pi_k(SF_G(V \oplus W))$ for suitable W. By the dimension restriction and the results of [46, §3], we can even say that β lies in the image of $\pi_k(SF_G(V))$. But $c^*\alpha$ and $c^*(\text{Image } \beta)$ go to the same element of $\{S^k \times L, S^0\}$ by the principle of invariance under stabilization that was noted above. Hence Image β is also a perfectly good choice for the class $\bar{\omega}$. ■

REMARK. Proposition 7.5 motivates one to ask if the pseudoframed knot invariant always lies in the image of $\pi_k(F_G(V))_{(p)}$. This is clear if $p = 2$ or 3 because the normal bundle of K in Σ turns out to be a stably trivial (at p, at least) real or complex vector bundle. For $p > 3$, Petrie [41] and Dovermann and Rothenberg [17] have considered an important case in which this is automatic. Namely, it holds if V is stable with respect to k (each irreducible free representation has multiplicity 0 or greater than $k/2$). At the end of this section we shall show that for V unstable (but $\dim V > k$) the knot invariant can lie outside the image of $\pi_k(F_G(V))_{(p)}$ whenever the map

$$\pi_k(SF_G(V) \times SC_G)_{(p)} \to \pi_k(SF_G \times SC_G/SC_G(V))_{(p)}$$

has a nonzero cokernel. In particular, the unstable normal bundle components are essentially arbitrary if $p > 3$.

Returning to our central theme, we are ready to restate the main results of [41: I, §3] in a form suitable for actions with pseudoframings:

THEOREM 7.6. *Suppose we are given a p-pseudoframed G-action* (Σ, Φ, Ψ) *with the previous notation. Let* γ' *be the pseudoframed normal invariant of* K *in* $\pi_{k(p)}$, *and identify* $\pi_{k(p)}$ *with its image in* $\{S^k \times L(V), S^0\}_{(p)}$. *Then* $q'\Theta'\omega'(\Sigma, \Phi, \Psi) + \gamma'$ *is zero in* $\{S^k \times L(V)), S^0\}_{(p)}$. ■

THEOREM 7.7. *Let* M *be a free 2-dimensional G-module, and let* c *be the collapse map*

$$S^k \times L(M \oplus V) \to S^k \times L(M \oplus V)/S^k \times L(V) \simeq S^n \vee S^{n+1}.$$

Then the element $q'\Theta'[\omega' \oplus M] + \gamma'$ *is equal to* $-c^*(\mathfrak{p}'(\Sigma) \oplus \xi)$, *where* $\mathfrak{p}'(\Sigma) \in \pi_{n(p)}$ *is the normal invariant of* Σ *with its underlying nonequivariant pseudoframing and* ξ *is undetermined.* ■

One can relate ξ to the isotopy representation of the action as in [49]; however, we shall not do so here.

This gives us all the machinery we need to prove Theorem C.

ADDENDUM. *$\mathbb{Z}/p$-actions on spheres that satisfy the Gap Hypothesis but have twisted fixed point sets* (in the sense of [13]). We shall now provide the examples discussed in the remark following the proof of Proposition 7.5. The examples are all generated by the following result.

THEOREM 7.8. *Let $p \geq 5$, and choose stably trivial complex n-plane bundles $\zeta_1, \ldots, \zeta_{(p-1)/2}$ over S^k with $\dim_{\mathbb{C}} \zeta_j = q_j$. Assume that $\Sigma 2q_j > k+1$. Then there is a smooth $\mathbb{Z}/p$-action on S^{k+2n} with the following properties:*

(i) *The fixed point set is S^k.*

(ii) *If ν is the equivariant normal bundle of S^k, then ν corresponds to the free G-vector bundle $\Sigma \zeta_j \otimes \chi_j$, where $\{\chi_j\}$ is the set of real irreducible representations of G.*

PROOF. Construct a free G-vector bundle ν over S^k as in (ii), and let V be its linear fiber. We must first check that $S(\nu)$ is G-fiber homotopically trivial; i.e., we must check that ν goes to zero in $\pi_k(F_G(V))$. By assumption, ν goes to zero in $\pi_k(C_G)$, so at least it must go to zero in $\pi_k(F_G)$. On the other hand, in the given range the map $\pi_k(F_G(V)) \to \pi_k(F_G)$ is bijective. Let $z \in \pi_k(F_G(V)/C_G(V))$ project to ν (the preceding argument argument shows that such a class exists). By exact sequence (6.2) in [51: II], we shall get an action on a homotopy sphere with normal bundles ν if the image of z in $hS_k^s(L(V))$ is zero. The first step is to show that z may be chosen to have zero normal invariant. But this follows easily from the earlier ideas in this section; the stabilization of z in $\pi_k(F_G/C_G)$ lifts back to $\pi_k(F_G)$ and therefore also back to $\pi_k(F_G(V))$. Thus we may change z by an element of $\pi_k(F_G(V))$ so that the new element $z' \in \pi_k(F_G(V)/C_G(V))$ is stably trivial. By invariance under stabilization, we know the normal invariant associated to z' is zero. Therefore the image of z' in $hS_k^s(L(V))$ comes from the action of the Wall group $L_{k+n}^s(\mathbb{Z}/p)$ on the latter set. By construction there is no Atiyah-Singer invariant, and therefore by the computations of $L_*^s(\mathbb{Z}/p)$ we know the image of z' comes from the image of $L_*^s(1)$. In other words, $S(\nu)/G$ is diffeomorphic to $S^k \times L(V)_{\#}\Sigma'$, where Σ' is a homotopy sphere bounding a π-manifold. However, we can now proceed as in [49] to show that Σ' must be the standard sphere.

To show that such an action exists on a standard sphere, notice that $z' \oplus M$ also has trivial normal invariant by invariance under stabilization. By a refinement of [51: I, 3.2] to actions with homotopy spheres as fixed point sets (specifically, in that case one can drop the need to localize at p), we see that in any case the ambient homotopy sphere bounds a π-manifold.

To conclude, it suffices to show that every π-manifold boundary has an action with S^k as its untwisted fixed point set (use such actions to modify the existing action via connected sum). But this follows from the commutative diagram below, for the transfer map τ is well-known to be onto (compare [59]):

$$\begin{array}{ccccc} L^s_{N+1}(\mathbb{Z}/p) & \xrightarrow{\tau} & L^s_{N+1}(1) & \longrightarrow & bP_{N+1} \\ \downarrow & & & & \downarrow \\ hS_{k+1}(L(V)) & \xrightarrow{\Gamma} & \Theta^G_{k+V} & \xrightarrow[\text{map}]{\text{forgetful}} & \Theta_{N=k+\dim V} \end{array}$$

The map Γ is just the map in [51: II, 6.2]. ∎

8. Proof of Theorem C

With the machinery now at our disposal, the proof of Theorem C becomes straightforward. Suppose that we are given an action of $\mathbb{Z}/p$ on Σ^{k+M} with k-dimensional fixed point set, where $M > P_k$ (the latter is defined in Section 5). By 7.1 we may choose a pseudoframing Ψ on Σ that will be fixed henceforth. Let $\omega'(\Sigma,\Phi,\Psi)$ be the pseudoframed knot invariant, and let γ' be the pseudoframing of the fixed point set K. By 7.6 we know that the image of (ω',γ') in $\{S^k \times L,S^o\}_{(p)}$ is zero. On the other hand, by 7.5 we know that there is an $\omega'' \in \pi_k(F_G(V))$ such that ω'' and ω' have the same image in $\{S^k \times L,S^o\}_{(p)}$. But the map $q'\Theta' \oplus$ inclusion equals the map δ from Section 5 by construction. Therefore we conclude that $\delta_{(p)}(\omega'' \oplus \gamma') = 0$.

But according to Proposition 5.2, in the range $M > P_k$ the map $\delta_{(p)}$ is injective. Therefore $\gamma' = 0$ in $\pi_{k(p)}$. Since $\wp(F)$ = image γ' in $\pi_k(F/O)_{(p)}$, this tells us that $\wp(F)$ must be zero. ∎

The results of this paper lead to many further questions that will be treated in subsequent papers. One problem is to deal with anomalies in this paper and [32, 33] for involutions. Another is to investigate similar questions for actions of larger groups such as $\mathbb{Z}/p^r$ and S^1. To illustrate the complications that arise, we state without proof a result that contrasts sharply with Theorem C:

(8.1) Let $p > 2$, $r \geq 2$, and let K^k be a homotopy sphere. Then there exists a smooth $\mathbb{Z}/p^r$-action on some homotopy sphere Σ with three orbit types and fixed point set F. Given $M > 0$, there is an $r > 0$ so that one can take $\dim \Sigma \geq M$. ∎

Despite this, one can still ask if there is a superstable range as in Theorem C for p and r fixed. This would require more detailed generalizations of the Strong Segal Conjecture; it appears that the methods of Ravenel [42] might allow one to study this effectively.

9. Appendix. Unitary homology spheres admitting no torsion unitary structures

The following result fills the obvious gap in the statement of Theorem 1.2.

THEOREM 9.1. *Let* $k > 1$ *be given. Then there is a closed smooth* $\mathbb{Z}[2^{-1}]$*-homology sphere* M^{8k+2} *with the following properties:*

(i) M *admits a unitary structure but does not admit a torsion unitary structure. In fact, the same is true for a connected sum* $\#^{2s+1}M$ *of an arbitrary odd number of copies of* M.

(ii) *If* N *is a 2-connected* $\mathbb{Z}_{(p)}$*-homology sphere with a torsion unitary structure, then* $M\#N$ *does not admit a torsion unitary structure (although it admits a unitary structure).*

REMARK. Any homology (8k+2)-sphere satisfying (i) must have 2-torsion in its integral homology. Suppose it did not. Since the inclusion map $\pi_{8k+2}(BU) \to \pi_{8k+2}(BO)$ is onto by Bott periodicity, such an M^{8k+2} would be a $\mathbb{Z}_{(2)}$-homology sphere whose tangent bundle mapped nontrivially into $\widetilde{KO}(M^{8k+2})_{(2)} \cong \widetilde{KO}(S^{8k+2})_{(2)} \cong \mathbb{Z}/2$. But this cannot happen [3].

PROOF OF 9.1. There are two main steps. We must first construct a unitary $\mathbb{Z}[\frac{1}{2}]$ homology (8k+2)-sphere whose complex stable normal bundle is an indivisible element of $\tilde{K}(M^{8k+2})/\text{Torsion} = \mathbb{Z}$. Then we must show that one cannot find another unitary structure on M whose complex stable normal bundle is an even element of $\tilde{K}(M^{8k+2})/\text{Torsion}$. (After completing the proof we shall show that the first condition does not imply the second).

The construction of M resembles the construction of almost parallelizable manifolds with nonzero Pontrjagin classes (compare [25]). Given an element $\alpha \in \pi_q(SO)$, let (S^q,α) be S^q with the framing obtained by twisting the standard framing through α. Following standard notation, let $\rho_k \in \pi_{8k-1}(SO) = \mathbb{Z}$ be the generator; if η generically denotes the Hopf map in $\pi_{m+1}(S^m)$, then $\rho_k\eta^2$ generates $\pi_{8k+1}(SO) = \mathbb{Z}/2$. The first step is to prove that *the framed manifolds* $(S^{8k+1},\rho_k\eta^2)$ *bounds an SU-manifold* W^{8k+2} *whose normal bundle map* $W \to BSU$ *is a 2-primary element of* $[W, BSU]$.

This translates directly into a statement in bordism theory. Specifically, let $\Gamma(2)$ be the homotopy fiber of $BSU \to BSU[\frac{1}{2}]$, and consider the appropriate bordism theory for $\Gamma(2)$-structured manifolds. The assertion above is true if and only if the image of $\rho_k\eta^2$ under the composite

$$\pi_*(SO) \xrightarrow{J} \pi_*(S^0) \xrightarrow[\text{inclusion}]{\text{fiber}} \pi_*(M\Gamma(2))$$

is zero. To prove this, consider first the image $\phi(\rho_k)$ of ρ_k in $\pi_{8k-1}(M\Gamma(2))$. Since $\pi_{8k-1}(MSU) = 0$ (compare [15]), we know that $\phi(\rho_k)$ lifts to a class $x_k \in \pi_{8k}(MSU/M\Gamma(2))$. By an analog of 2.1(iii), we know that $MSU/M\Gamma(2)$ is equivalent to the localized spectrum $MSU/S^0[\frac{1}{2}]$. In dimension 8k the latter has 2-divisible stable homotopy ([15] again), so we may write $x_k = 2y_k$. If we let $\partial: \pi_{*+1}(MSU/M\Gamma(2)) \to \pi_*(M\Gamma(2))$ be the boundary map, then by naturality of composition we may write

$$\phi(\rho_k\eta^2) = \phi(\rho_k)\eta^2 = \partial(x_k)\eta^2 = 2\partial(y_k)\eta^2.$$

Since $2\eta = 0$ and composition is stably bilinear, it is immediate that that $\phi(\rho_k\eta^2) = 0$.

We now do surgery on W and try to make it $\mathbb{Z}[\frac{1}{2}]$ acyclic. Below the middle dimension there is no problem in killing homotopy to make $\pi_*(W) \to \pi_*(\Gamma(2))$ injective; hence we may assume W is 2-connected and has 2-primary homology below the middle dimension. One possibly encounters a Kervaire invariant in the middle dimension; however, this does not stop us from killing the middle homology group modulo 2-primary elements. Thus we can make W into a $\mathbb{Z}[\frac{1}{2}]$ acyclic SU-manifold.

Form M by adding D^{8k+2} along $\partial W = S^{8k+1}$. Then M admits a unitary structure because the class $\rho_k\eta^2 \in \pi_{8k+1}(SO)$ goes to zero in $\pi_{8k+1}(SO/U)$. Furthermore, the same considerations as in [25] show us that the complex normal bundle of M must go to an odd element of $\tilde{K}(M)/\text{Torsion} = \mathbb{Z}$. As in the proof of 1.2 we may modify the unitary structure to get a new structure corresponding to a generator of this infinite cyclic group. This completes the first step.

We must now check that M admits no unitary structure whose complex normal bundle gives an even element of $\tilde{K}(M)/\text{Torsion}$. The unitary structures of M are given by elements of [M, SO/U], and it suffices to check that the map

$$f: [M, SO/U] \to [M, BU] \cong \tilde{K}(M) \to \tilde{K}(M)/\text{Torsion} = \mathbb{Z}$$

has $2\mathbb{Z}$ as its image; for a change in the unitary structure by $\gamma \in [M, SO/U]$ will change the complex normal bundle by the image of γ in $\tilde{K}(M)$. It will suffice to check that the degree 1 map $h: M \to S^{8k+2}$ induces a split injection

$$(**)\ h^*: \pi_{8k+2}(SO/U) \to [M, SO/U].$$

For the left hand side is $\mathbb{Z}$ and the right hand side is $\mathbb{Z}$ plus a finite abelian group, and therefore the assumption on h^* tells us that the image of f is just the image of the usual map $\mathbb{Z} = \pi_{8k+2}(SO/U) \to \pi_{8k+2}(BU = \mathbb{Z}$.

But this image is $2\mathbb{Z}$ by Bott periodicity.

By Bott periodicity we also know that SO/U is homotopy equivalent to $\Omega^2 BO$. Therefore we may rewrite (**) in terms of the map $h^*: KO^{-2}(S^{8k+2}) \to KO^{-2}(M)$. In the spirit of 1.1, we see that this holds if M is KO-orientable. But M is 2-connected by construction, and therefore it satisfies the familiar Spin criterion for KO-orientability. Therefore h^* is split surjective as was to be shown. This proves (i).

The statements regarding $\#^t M$ and $M\#N$ are proved similarly. In each case there is an obvious unitary structure with the right sort of complex normal bundle, and in each case the argument above goes through to show that one cannot find a torsion unitary structure. ■

As noted before, if one takes the connected sum of two unitary homology (8k+2)-spheres that admit no torsion unitary structures, the resulting manifold will admit a torsion unitary structure. On the other hand if A admits a torsion unitary structure but B does not, then $A\#B$ admits an obvious unitary structure with an "odd" complex normal bundle, BUT the manifold $A\#B$ may also admit a torsion unitary structure. We proceed formally as follows:

(9.2) <u>Let M be constructed as in Theorem 9.1. Then there is a torsion unitary $\mathbb{Z}[\frac{1}{2}]$-homology sphere N such that $M\#N$ admits a torsion unitary structure.</u>

PROOF. Let $W = (\mathbb{R}P^5 - \text{Int } D^5) \times D^{8k-2}$ with corners rounded, and let $N = \partial W$. then the $\mathbb{Z}[\frac{1}{2}]$-homology disk W has a natural torsion unitary structure that induces one on N.

It will suffice to show that the degree one map $h^*: KO^{-2}(S^{8k+2}) \to KO^{-2}(M\#N)$ sends the generator of the (infinite cyclic) domain into an element y that is divisible by 2. It will then follow that the map from $KO^{-2}(M\#N)$ to $\tilde{K}(M\#N)/\text{Torsion}$ is onto (since the image of y is twice a generator), and from this it is clear that one can change the unitary structure on $M\#N$ to make the latter torsion unitary. To prove the assertion on h^*, first notice that $A\#B$ modulo its (8k-1)-skeleton is homotopic to $S^{8k-3}(\mathbb{R}P^5/\mathbb{R}P^2)$. It is convenient to write the latter as an (8k-2)-fold suspension of $P \cup L$, where $P = \mathbb{C}P^2 = S^2 \cup_n e^4$ and $L = S^2 \cup_2 e^3$. Of course, it is well-known that $\widetilde{KO}(P) = \mathbb{Z}$ and the collapsing map $P \to S^4$ is multiplication by 2 in $\widetilde{KO}$ [1]. If we can show that the restriction map $j^*: \widetilde{KO}(P \cup L) \to \widetilde{KO}(P)$ is bijective, then it will be immediate that the image of $\widetilde{KO}(S^4)$ in $\widetilde{KO}(P \cup L)$ is divisible by 2, and therefore the composite

$$\mathbb{Z} = KO^{-2}(S^{8k+2}) \to KO^{-2}(A\#B/(8k-1)\text{-skel.}) \to KO^{-2}(A\#B),$$

which corresponds to h^*, will also have image divisible by 2. But consider the associated Mayer-Vietoris sequence:

$$0 = \widetilde{KO}(S^3) \to \widetilde{KO}(P \cup L) \to \begin{matrix} \widetilde{KO}(P) \\ \oplus \\ \widetilde{KO}(L) \end{matrix} \to \widetilde{KO}(S^2) = \mathbb{Z}/2.$$

We know that $\widetilde{KO}(P) \to \widetilde{KO}(S^2)$ is surjective [1] and $\widetilde{KO}(L)$ is bijective (by coexactness of $S^2 \overset{2}{\to} S^2 \to L$). Therefore j^* is bijective as claimed. ■

Purdue University
West Lafayette, Indiana 47907

REFERENCES

1. J. F. Adams and G. Walker, On complex Stiefel manifolds, Proc. Camb. Philos. Soc. 61(1965), 81-103.
2. J. P. Alexander, G. C. Hamrick, and J. W. Vick, Involutions on homotopy spheres, Invent. Math. 24(1974), 35-50.
3. ————, ————, and ————, Cobordism of manifolds with odd order normal bundle, Invent. Math. 24(1974), 83-94.
4. ————, ————, and ————, A construction for involutions on homotopy spheres, Topology Conference (Virginia Polytech., 1973), Lecture Notes in Mathematics Vol. 375, 1-5. Springer, New York, 1974.
5. G. A. Anderson, Surgery with Coefficients, Lecture Notes in Mathematics Vol. 591. Springer, New York, 1977.
6. A. Assadi, Extensions of group actions from submanifolds of disks and spheres, preprint, University of Virginia, 1981.
7. ————, Finite group actions on simply connected manifolds and CW complexes, preprint, University of Virginia, 1981.
8. M. F. Atiyah, K-Theory. Benjamin, New York, 1967.
9. J. Barge, J. Lannes, C. Latour, and P. Vogel, Λ-sphères, Annales Sci. de ℓ'É.N.S.(4) 7(1974), 463-505.
10. J. C. Becker and R. E. Schultz, Equivariant function spaces and stable homotopy theory I, Comment. Math. Helv. 49(1974), 1-34.
11. R. Bott, Lectures on K(X). Benjamin, New York, 1968.
12. G. Bredon, Introduction to Compact Transformation Groups, Pure and Applied Mathematics Vol. 46. Academic Press, New York, 1972.
13. W. Browder, Surgery and the theory of differentiable transformation groups, Proc. Conf. on Transformation Groups (New Orleans, 1967), 1-46. Springer, New York, 1968.
14. ————, Surgery on simply connected manifolds, Ergebrisse der Math. Bd. 65. Springer, New York, 1972.
15. P. E. Conner and E. E. Floyd, Torsion in SU-bordism, Memoirs Amer. Math. Soc. 60(1966).
16. T. tomDieck, The Burnside ring and equivariant stable homotopy theory (notes by M. Bix), Lecture Notes Series, University of Chicago, 1975.

17. K. H. Dovermann and M. Rothenberg, An equivariant surgery sequence and equivariant diffeomorphism and homeomorphism classification, Topology Symposium Siegen 1979, Lecture Notes in Mathematics Vol. 788, 257-280. Springer, New York, 1980.

18. E. Dyer, Cohomology Theories. Benjamin, New York, 1967.

19. J. Ewing, Spheres as fixed point sets, Quant. J. Math. Oxford (2) 27(1976), 445-455.

20. J. Gunawardena, Segal's conjecture for groups of (odd) prime order, J. T. Knight prize essay, Cambridge University, 1980.

21. H. Hauschild, Allgemeine Lage and Äquivariante Homotopie, Math. Z 143 (1975), 155-164.

22. W.-C. Hsiang and W.-Y. Hsiang, Differentiable actions of compact connected classical groups I, Amer. J. Math. 89(1967), 705-786.

23. L. Jones, The converse to the fixed point theorem of P.A. Smith-I, Ann. of Math. 94(1971), 52-68.

24. ————, Ibid.: II, Indiana Univ. Math. J. 22(1972), 309-325; correction, ibid. 24(1975), 1001-1003.

25. M. Kervaire and J. Milnor, Bernoulli numbers, homotopy groups, and a theorem of Rohlin, Proc. Int. Congr. Math. (Edinburgh, 1958), 454-458. Cambridge University Press, New York, 1960.

26. ———— and ————, Groups of homotopy spheres, Ann. of Math. 78(1963), 514-537.

27. J. Levine, A classification of differentiable knots, Ann. of Math. 82 (1965), 15-50.

28. W. H. Lin, On conjectures of Mahowald, Segal, and Sullivan, Math. Proc. Camb. Philos. Soc. 87(1980), 449-458.

29. ————, D. M. Davis, M. E. Mahowald, and J. F. Adams, Calculation of Lin's Ext groups, ibid., 459-469.

30. P. Löffler, Über die G-Rahmbarkeit von G-Homotopiesphären Arch. Math. (Basel) 29(1977), 628-634.

31. ————, Equivariant framability of involutions on homotopy spheres, Manuscripta Math. 23(1978), 161-171.

32. ————, Homotopielineare Involutionen auf Sphären, Topology Symposium Siegen 1979, Lecture Notes in Math. Vol. 788, 359-363. Springer, New York, 1980.

33. ————, Homotopielineare $\mathbb{Z}/p$-Operationen auf Sphären, Topology 20 (1981), 291-312.

34. P. Lynch, Framed R-homology spheres, Ph.D. Thesis, Brandeis University, 1971.

35. I. Madsen, L. Taylor, and B. Williams, Tangential homotopy equivalences, Comment. Math. Helv. 55(1980), 445-484.

36. J. P. May, J. McClure, and G. Triantafillou, The construction of equivariant localizations, preprint, University of Chicago, 1981.

37. J. Milnor, A procedure for killing the homotopy groups of differentiable manifolds, Proc. A.M.S. Sympos. Pure Math. 3(1960), 39-55.

38. S. P. Novikov, Homotopically equivalent smooth manifolds (Russian), Izv. Akad. Nauk. SSSR Ser. Mat. 28(1964), 365-474; English transl., Amer. Math. Soc. Transl. (2) 48(1965), 271-396.

39. J. J. O'Connor, Equivariant stable homotopy theory, D. Phil. Thesis, Oxford, 1975.

40. W. Pardon, Local surgery and the exact sequence of a localization for Wall groups, Memoirs A.M.S. Vol. 12, Issue 2, No. 196 (1977).

41. T. Petrie, Pseudoequivalences of G-manifolds, Proc. A.M.S. Sympos. Pure Math. 32(1978) Pt. 1, 169-210.

42. D. Ravenel, The Segal conjecture for cyclic groups, Bull. London Math. Soc. 13(1981), 42-44.

43. N. S. Rege, On certain classical groups over Hasse domains, Math. Z. 102(1967), 120-157.

44. M. Rothenberg, Differentiable group actions on spheres, Proc. Adv. Study Inst. on Algebraic Topology (Aarhus, 1970), 455-475. Matematisk Institut, Aarhus Universitet, 1970.

45. ————, Torsion invariants and finite transformation groups, Proc. A.M.S. Sympos. Pure Math. 32(1978) Pt. 1, 267-311.

46. R. Schultz, Homotopy decompositions of equivariant function spaces I. Spaces of principal bundle maps, Math. Z. 131(1973), 49-75.

47. R. Schultz, Homotopy sphere pairs admitting semifree differentiable actions, Amer. J. Math. 96(1974), 308-323.

48. ————, Spherelike G-manifolds with exotic equivariant tangent bundles, Studies in Algebraic Topology (Adv. in Math. Suppl. Studies Vol. 5), 1-39. Academic Press, New York, 1979.

49. ————, Isotopy classes of periodic diffeomorphisms of spheres, Algebraic Topology Waterloo 1978 (Proc.) Lecture Notes in Mathematics Vol. 741, 334-354. Springer, New York, 1978.

50. ————, $\mathbb{Z}_2$-torus actions on homotopy spheres, Proc. Second Conf. on Compact Transformation Groups (Univ. of Mass., Amherst, 1971), Lecture Notes in Mathematics 298, 117-118. Springer, New York, 1972.

51. ————, Differentiable group actions on homotopy spheres:
I. Differential structure and the knot invariant, Invent. Math. 31(1975), 105-128.
II. Ultrasemifree actions, Trans. Amer. Math. Soc., to appear.
IV. Normal invariant formulas and applications, in preparation.

52. ————, Smooth actions of small groups on exotic spheres, Proc. A.M.S. Sympos. Pure Math. 32, Pt. 1 (1978), 155-160.

53. G. B. Segal, Equivariant stable homotopy theory, Actes, Congres internat. de math. (Nice, 1970) T. 2, 59-63, Gauthier-Villars, Paris, 1971.

54. D. Smith, Invariants for group actions on spheres. Ph.D. Thesis, University of Chicago, 1980.

55. R. Stong, Notes on Cobordism Theory. Princeton University Press, Princeton, 1968.

56. D. Sullivan, Genetics of homotopy theory and the Adams conjecture, Ann. of Math. 100(1974), 1-79.

57. C. T. C. Wall, Classification of (n-1)-connected 2n-manifolds, Ann. of Math. 75(1962), 163-189.

58. ————, Surgery on Compact Manifolds, L. M. S. Mathematical Monographs Vol. 1, Academic Press, New York, 1970.

59. ————, Classification of Hermitian forms: VI. Group rings, Ann. of Math. 103(1976), 1-80.

Canadian Mathematical Society
Conference Proceedings
Volume 2, Part 2 (1982)

EQUIVARIANT MAPS OF SPHERES WITH CONJUGATE ORTHOGONAL ACTIONS

J. Tornehave

§0. INTRODUCTION.

Let G be a finite group acting orthogonally on Euclidean spaces V and W, such that the fixed point sets of V^H, W^H have equal dimensions for all subgroups $H \subset G$. We get induced actions on the unit spheres S(V) and S(W). Our aim is firstly to classify (at least stably) the equivariant homotopy classes of G-maps $f : S(V) \to S(W)$ in terms of the simplest invariants imaginable: Degrees at fixed point sets

$$d_H(f) = \deg f^H, \quad f^H : S(V^H) \to S(W^H). \tag{0.1}$$

In particular we describe the equivariant stable homotopy group ω_G^α of Segal [16] for the case $\dim \alpha^H = 0$ $(H \subset G)$ in terms of congruences analogous to those found by Ted Petrie [7] for complex representations of compact Lie groups. Secondly we construct particular examples of G-maps with sufficiently good naturality properties so as to yield a precise "solution to the Adams conjecture" $A : BO \to G/O$*) along with a formula for A on a sum of two vectorbundles. This depends on the methods of Quillen [14], whereas the G-maps arise from ideas closely related to Sullivans work [18].

In the real case the degrees (0.1) can a priori only be defined up to sign. This difficulty is present even in the basic case, where W is conjugate to V by a field automorphism ψ of $\mathbb{C}$. The remedy is to choose a specific way in which W (with its quadratic form) is conjugate to V - hence the notion af a ψ-conjugation $\phi : V \otimes \mathbb{C} \to W \otimes \mathbb{C}$ (def. 1.1). In general we choose orthogonal decompositions

$$V = V_1 \perp V_2 \perp \dots \perp V_r, \quad W = W_1 \perp W_2 \perp \dots \perp W_r, \tag{0.2}$$

where W_j is conjugate to V_j by $\psi_j \in \mathrm{Aut}(\mathbb{C}/\mathbb{Q}(i))$, and ψ_j-conjugations

---

1980 Mathematics Subject Classifications. 55P10, 57S25.

*) G in the constellations G/O, SG will always refer to the well known infinite loop spaces.

© 1982 American Mathematical Society
0731-1036/82/0000-0471/$07.75

$\phi_j : V_j \otimes \mathbb{C} \to W_j \otimes \mathbb{C}$. We explain in §1 how to get well defined degrees $d_H(f) = d_H(f;\ \phi_1,\dots,\phi_r)$.

For subgroups $H \lhd K \subset G$ with $C = K/H$ cyclic let C^o denote the set of generators of C. The eigenvalues distinct from ± 1 of the map $V_j^H \to V_j^H$ induced by $g \in C^o$, come in complex conjugate pairs. From each pair choose a representative $\lambda_{j\ell}(g)$ $(\ell=1,\dots,s_j)$, and choose a square root $\mu_{j\ell}(g)$ of $\lambda_{j\ell}(g)$. An integer is defined by the formula

$$(0.3)\qquad n_C = \sum_{g\in C^o} \prod_{j=1}^{r} \prod_{\ell=1}^{s_j} \frac{\psi_j(\mu_{j\ell}(g)) - \psi_j(\mu_{j\ell}(g)^{-1})}{\mu_{j\ell}(g) - \mu_{j\ell}(g)^{-1}} .$$

Let $C = C_G(V_j,\psi_j)$ be the abelian group of families of integers $(d_H)_{H\subset G}$ indexed by subgroups $H \subset G$, constant on conjugacy classes of subgroups, and satisfying for each $H \subset G$ the congruence

$$(0.4)\qquad \sum_K n_{K/H}\, d_K \equiv 0 \pmod{|NH/H|}$$

where K runs through subgroups with $H \lhd K$ and K/H cyclic.

THEOREM A. For each subgroup $H \subset G$ let d_H be an integer. There exists a G-map $f : S(V) \to S(W)$ with $d_H(f;\ \phi_1,\dots,\phi_r) = d_H$ $(H \subset G)$ if and only if the following conditions hold:

i) The family $(d_H)_{H\subset G}$ belongs to C.
ii) If $\dim V^H = 0$, $d_H = 1$.
iii) If $\dim V^H = 1$, d_H equals 1,0 or -1.
iv) If $V^H = V^K$, $d_H = d_K$.

Two such G-maps are stably G-homotopic.

The necessity of i) is proved in §2 using real equivariant K-theory. In outline this is similar to the complex case treated in [7], but it involves the construction (by algebraic means) of suitable spin reductions. The remaining part of the proof (§3) uses equivariant obstruction theory as in [7].

Let $\Delta_G \subset \text{Units}\ \omega_G^o$ be the subgroup of <u>orthogonal units</u> in the Burnside ring ω_G^o. It consists of the elements represented by orthogonal equivariant maps $S(V) \to S(V)$. The ω_G^o-module structure on ω_G^α $(\alpha = [W] - [V])$ defines an action of Δ_G on ω_G^α. As a corollary to thm.A we show in §3:

THEOREM B. Fix V_j, W_j and ψ_j as above. For a choice of ψ_j-conjugations $\phi_j : V_j \otimes \mathbb{C} \to W_j \otimes \mathbb{C}$ an isomorphism

$$\omega_G^\alpha \to C$$

is given by $f \mapsto (d_H(f;\ \phi_1,\dots,\phi_r))_{H\subset G}$. Two such isomorphisms differ by the action of an orthogonal unit on ω_G^α.

Thm.B shows that an element in the orbit set ω_G^α/Δ_G may be specified simply be giving a list of degrees d_H.

The profinite completion $\hat{C} = C \otimes \hat{\mathbb{Z}}$ may be identified with the families $(d_H)_{H\subset G}$ of adic integers, that are constant on conjugacy classes and satisfy the congruences (0.4). For the next result we assume $r = 1$, i.e. V and W are conjugate by $\psi \in \mathrm{Aut}(\mathbb{C}/\mathbb{Q}(i))$. The action of ψ on the roots of unity is given by $\psi(\zeta) = \zeta^e$, where e is an adic unit congruent to 1 mod 4. With this notation we show in §4 (the square bracket indicates integer part):

THEOREM C. There is an element of $\hat{C}$ given by

$$d_H = e^{[\frac{1}{2}\dim V^H]}. \tag{0.5}$$

The degrees (0.5) are suggested by Sullivan's work on the Adams conjecture ([18] p. 70). In fact d_H equals the degree at H-fixed points of a ψ-conjugation $\phi : V \otimes \mathbb{C} \to W \otimes \mathbb{C}$ defined using etale cohomology of the unit quadrics $S(V^H \otimes \mathbb{C})$, $S(W^H \otimes \mathbb{C})$ and the orientation conventions of §1.

Combining theorems B and C we can associate to V an element

$$A(V,\psi) \in \hat{\omega}_G^{(\psi-1)V}/\Delta_G \tag{0.6}$$

given by the list of degrees (0.5). Recall from [7] the homomorphism

$$\Delta : RO(G) \to \Delta_G \subset \text{Units } \omega_G^o \tag{0.7}$$

assigning to V the antipodal map on S(V), and the subgroup $RO_o(G) \subset RO(G)$ consisting of elements α with $\dim \alpha^H = 0$ for $H \subset G$. We define a commutative monoid graded over $RO_o(G)$ by

$$E_G = \coprod_{\alpha\in RO_o(G)} \hat{\omega}_G^\alpha/\Delta_G \tag{0.8}$$

The product is given by composition of maps.

THEOREM D. The map $V \to A(V,\psi)$ extends to a map natural on the category of finite groups

$$A : RO(G) \to E_G$$

satisfying the sum formula

$$A(\alpha+\beta) = D(\alpha,\beta)A(\alpha)A(\beta) \tag{0.9}$$

where $D : RO(G) \times RO(G) \to \hat{\omega}_G^o$ is defined by

(0.10) $$D(\alpha,\beta) = 1 + \frac{e-1}{4}(\Delta(\alpha)-1)(\Delta(\beta)-1).$$

The map A may be thought of as a "solution to the equivariant Adams conjecture over a point".

Let p be a prime and $k \in \mathbb{Z}$ prime to p. The affirmed Adams conjecture ([5],[14],[18]) is equivalent to the existence of a map $A_p : BO_{(p)} \to G/O_{(p)}$, whose composite with the inclusion $G/O_{(p)} \to BO_{(p)}$ is $\psi^k - \psi^1$. By the recent results of Friedlander and Seymour [10], A_p may be chosen as an infinite loop map, when p is odd. When p = 2 and $k \equiv \pm 3 (\mathrm{mod}\ 8)$ A_2 cannot be chosen as an H-map (I. Madsen [12]).

Let $\Delta : BO \to SG$ be the composite of $\eta : BO \to SO$, representing multiplication by the generator of $KO^{-1}(pt) \cong \mathbb{Z}/2$, and $J : SO \to SG$. Interpreting Δ as a element of the stable cohomotopy ring $\pi_s^o(BO)$ of augmentation 1 the expression (with the sign chosen, so that $\pm k \equiv 1 (\mathrm{mod}\ 4)$)

(0.11) $$D = 1 + \frac{\pm k - 1}{4}(\Delta pr_1 - 1)(\Delta pr_2 - 1),$$

defines an element in the ring $\pi_s^o(BO \times BO)$ of augmentation 1. It is represented by a map

$$D : BO \times BO \to SG.$$

From thm.D we deduce (§5) using Brauer lifting as in Quillen [14]:

THEOREM E. The lifting $A_2 : BO_{(2)} \to G/O_{(2)}$ may be chosen to satisfy the formula

(0.12) $$A_2(x+y) = D(x,y)\ A_2(x)\ A_2(y).$$

Multiplying the map A_2 of thm.E by maps $BO_{(2)} \to SG_{(2)}$ of the form $1 + t(\Delta - 1)$ with $t \in \hat{\mathbb{Z}}_2$, the constant $\frac{\pm k - 1}{4}$ appearing in (0.11) can be changed arbitrarily within its congruence class mod 2. In particular we find somewhat unexpectedly:

THEOREM F. When $k \equiv \pm 1 (\mathrm{mod}\ 8)$ the lifting A_2 may be chosen as an H-map.

The author wishes to thank T. Petrie for some very helpful discussions.

§1. CONJUGATIONS AND ORIENTATIONS.

Let G be a finite group. Our $\mathbb{R}G$-modules V will unless otherwise specified be equipped with G-invariant inner products, i.e. we consider positive definite quadratic $\mathbb{R}G$-modules (V,Q). It is well known, that the isomorphism type of (V,Q) is determined by the character χ_V.

DEFINITION 1.1. Let $\psi \in \mathrm{Aut}(\mathbb{C})$ be a field automorphism. A ψ-conjugate pair (of positive definite quadratic $\mathbb{R}G$-modules) (V,W,ϕ) is a pair of $\mathbb{R}G$-modules (V,Q) and (W,Q') equipped with a ψ-conjugation ϕ, i.e. a map

$$\phi : V \otimes_{\mathbb{R}} \mathbb{C} \to W \otimes_{\mathbb{R}} \mathbb{C}$$

satisfying the following conditions:

i) ϕ is a ψ-linear isomorphism (i.e. an isomorphism of abelian groups, such that $\phi(av) = \psi(a)\phi(v)$ for $a \in \mathbb{C}$, $v \in V \otimes \mathbb{C}$).

ii) ϕ is G-equivariant.

iii) ϕ is compatible with the scalar extended forms (i.e. $Q'(\phi v) = \psi Q(v)$ for $v \in V \otimes \mathbb{C}$).

Observe that i) and ii) imply that W is conjugate to V by ψ in the ordinary sense of representation theory, that is $\chi_W = \psi\chi_V$. Note that ϕ generally does not map V into W.

PROPOSITION 1.2. Let (V,Q) and (W,Q') be $\mathbb{R}G$-modules such that W is conjugate to V by $\psi \in \mathrm{Aut}(\mathbb{C})$. There exists a ψ-conjugation $\phi : V \otimes \mathbb{C} \to W \otimes \mathbb{C}$.

The construction given in the proof will be needed in §2.

PROOF. Let n be the exponent of G, $\zeta \in \mathbb{C}$ a primitive n'th root of unity, and $L = \mathbb{Q}(\zeta + \zeta^{-1})$. The underlying $\mathbb{R}G$-module V can be obtained by scalar extensions from an LG-module V_o (Dress [9], Prop. 5.6). By choosing on V_o the form Σx_i^2 with respect to some basis and adding its conjugates by G, we obtain on V_o a G-invariant totally positive definite quadratic form Q_o. This means that Q_o is positive definite with respect to every ordering of L, or equivalently that Q_o maps V_o-0 into the set P(L) of totally positive elements in L ([13], Ch. III). Since (V,Q) is determined up to isomorphism by V, we can assume Q to be the scalar extension of Q_o. Let the quadratic LG-module (W_o,Q_o') be obtained from (V_o,Q_o) by scalar extension through $\psi|_L \in \mathrm{Aut}(L)$. Then Q_o' is totally positive definite, and forgetting forms W_o scalar extends to W. As we did for V, we may assume Q' to be the scalar extension of Q_o'. Now ϕ may be obtained by scalar extension from L.

The following constructions may be performed on ψ-conjugate pairs:

(1.3) Orthogonal direct sum:

$$(V_1,W_1,\phi_1) \perp (V_2,W_2,\phi_2) = (V_1 \perp V_2,\ W_1 \perp W_2,\ \phi_1 \oplus \phi_2).$$

(1.4) Fixed points at $H \subset G$: This construction associates to (V,W,ϕ) the ψ-conjugate pair of NH/H-modules (V^H, W^H, ϕ^H).

(1.5) Exterior power:

$$\Lambda^j(V,W,\phi) = (\Lambda^j V, \Lambda^j W, \Lambda^j \phi)$$

Here $\Lambda^j\phi$ is well defined since Λ^j commutes with scalar extension, and the form

$\Lambda^j Q$, on $\Lambda^j V$ is given by $\Lambda^j Q(e_1 \wedge \ldots \wedge e_j) = \det B(e_i, e_j)$, where B is the symmetric bilinear form with $Q(x) = B(x,x)$.

(1.6) Determinant: This is the special case $j = \dim V = \dim W$ of (1.5). We use the notation

$$\det(V,W,\phi) = (\det V, \det W, \det \phi).$$

The unit sphere in (V,Q) is denoted $S(V)$. Similarly we define the unit quadric in $V \otimes \mathbb{C}$ by

$$S(V,\mathbb{C}) = \{v \in V \otimes \mathbb{C} \mid Qv = 1\}.$$

A ψ-conjugation induces a (discontinuous) bijective map

$$S(\phi) : S(V,\mathbb{C}) \to S(W,\mathbb{C}).$$

The two orientations on V are in 1-1 correspondence with the two points of $S(\det V) = S(\det V,\mathbb{C})$. Hence

$$S(\det \phi) : S(\det V,\mathbb{C}) \to S(\det W,\mathbb{C})$$

defines a bijection between the set of orientations on V and the set of orientations on W. For a subgroup $H \subset G$ we find using (1.4), that $S(\det \phi^H)$ gives a 1-1 correspondence between orientations on V^H and orientations on W^H. Any orientations of V^H can be "transported along ϕ to an orientation of W^H". Therefore we can define:

DEFINITION 1.7. Let (V,W,ϕ) be a ψ-conjugate pair of $\mathbb{R}G$-modules and $f : S(V) \to S(W)$ a continuous G-map. For a subgroup $H \subset G$ <u>the degree at H-fixed points of f with respect to ϕ</u> $d_H(f;\phi)$ is the degree of $f^H : S(V^H) \to S(W^H)$ defined using any orientation on V^H and the orientation on W^H obtained by transportation along ϕ.

In the following lemma we consider fixed quadratic $\mathbb{R}G$-modules (V,Q), (W,Q') as above and a fixed $\psi \in \mathrm{Aut}(\mathbb{C})$. We determine how $d_H(f,\phi)$ depends on ϕ.

LEMMA 1.8. Let ϕ and $\bar{\phi}$ be ψ-conjugations $V \otimes \mathbb{C} \to W \otimes \mathbb{C}$. There exists a G-equivariant orthogonal map $A : S(V) \to S(V)$ with the following property: For every subgroup $H \subset G$ the bijections $S(\det \phi^H)$ and $S(\det \bar{\phi}^H)$ from $S(\det V^H)$ to $S(\det W^H)$ are equal if $A^H : V^H \to V^H$ preserves orientation, and distinct if A^H reverses orientation.

PROOF. There is an isomorphism of quadratic $\mathbb{C}G$-modules $A_1 : V \otimes \mathbb{C} \to V \otimes \mathbb{C}$ determined by $\phi A_1 = \bar{\phi}$. Hence A_1 is an element of the complex Lie group $O(V,\mathbb{C}) = O(V \otimes \mathbb{C})$ centralizing the image of G in $O(V) \subseteq O(V,\mathbb{C})$. We can write A_1 uniquely in the form $A_1 = A \exp(iX)$, where $A \in O(V)$ and X belongs to the Lie algebra of $O(V)$. It follows that A centralizes G, and that X is invariant under the adjoint action of G. The path of ortogonal G-maps

$A_t : V \otimes \mathbb{C} \to V \otimes \mathbb{C}$ joining $A_o = A \otimes 1$ to A_1 given by $A_t = A\exp(itX)$ shows, that $\det(A^H) = \det(A_1^H)$ for every subgroup H of G. The lemma follows, since

$$S(\det \phi^H) \cdot S(\det A_1^H) = S(\det \bar{\phi}^H).$$

COROLLARY 1.9. In the notation of 1.8 we have

$$d_H(f,\bar{\phi}) = d_H(fA,\phi)$$

for every continuous G-map $f : S(V) \to S(W)$.

In order to deal with the general case, where $\dim V^H = \dim W^H$ for every $H \subset G$ we need the following result in which quadratic forms play no role:

PROPOSITION 1.10. For $\mathbb{R}G$-modules V and W the following conditions are equivalent.

1) $\dim V^H = \dim W^H$ for every subgroup $H \subset G$.
2) $\dim V^C = \dim W^C$ for every cyclic subgroup $C \subset G$.
3) There are direct sum decompositions

$$V = V_1 \oplus \ldots \oplus V_r, \quad W = W_1 \oplus \ldots \oplus W_r$$

such that W_j is conjugate to V_j by a field automorphism ψ_j of $\mathbb{C}$.

PROOF. Obviously 3) implies 1) and 1) implies 2). In order to prove that 2) implies 3) we may as well prove the analogue for LG-modules, where $L = \mathbb{Q}(\zeta + \zeta^{-1})$ as in the proof of 1.2. From Serre [17], §13.1 follows, that V and W are isomorphic as $\mathbb{Q}G$-modules. Since

$$\coprod_\psi \psi V \cong V \otimes_Q L \cong W \otimes_Q L \cong \coprod_\psi \psi W$$

where ψ runs through Aut(L), each irreducible component of V has a conjugate appearing in W. Splitting these off, the proof is completed by induction on dim V.

If we have quadratic forms on V and W, the decompositions in 3) may be improved to orthogonal direct sum decompositions. By composing ψ_j with complex conjugation if necessary, we may assume, that ψ_j fixes $i \in \mathbb{C}$ (otherwise hideous signs crop up in §2). Combining 1.10 with 1.2 we obtain:

PROPOSITION 1.11. If V and W are positive definite $\mathbb{R}G$-modules satisfying the conditions of 1.10, there exist orthogonal direct sum decompositions

$$V = V_1 \perp \ldots \perp V_r, \quad W = W_1 \perp \ldots \perp W_r$$

and ψ_j-conjugate pairs (V_j, W_j, ϕ_j) where $\psi_j(i) = i$.

In the notation of 1.11 we can define the degree at H-fixed points of a continuous G-map $f : S(V) \to S(W)$ with respect to $\phi_1, \ldots, \phi_r$

(1.12) $$d_H(f) = d_H(f;\ \phi_1,\dots,\phi_r)$$

as follows: Choose orientations of V_j^H for $H \subset G$, $j=1,\dots,r$ and transport along ϕ_j to orientations of W_j^H. Define $d_H(f)$ using the product orientations on

$$V^H = V_1^H \oplus \dots \oplus V_r^H, \quad W^H = W_1^H \oplus \dots \oplus W_r^H.$$

An analogue to cor. 1.9 for this case is easily derived.

In §3 the following will be needed:

LEMMA 1.13. Let V and W be $\mathbb{R}G$-modules satisfying the conditions of 1.10. The maps

$$g : V^H \to V^H, \quad g : W^H \to W^H$$

induced by any $g \in NH/H$ both preserve or both reverse orientation.

PROOF. It suffices to treat the case $H = 1$. Using prop. 1.11 we can reduce to the case, where W is conjugate to V. Since det V and det W are then given by the same homomorphism $G \to \{\pm 1\}$, the lemma follows.

§2. THE CONGRUENCES.

We consider ψ_j-conjugate pairs (V_j, W_j, ϕ_j), $1 \leq j \leq r$ where $\psi_j(i) = i$, and form the quadratic $\mathbb{R}G$-modules

(2.1) $$V = V_1 \perp \dots \perp V_r, \quad W = W_1 \perp \dots \perp W_r.$$

The quadratic forms in V and V_j will be denoted Q, those on W and W_j will be denoted Q'.

For $H \triangleleft K \subset G$ we define $n_{K/H}$ by the formula (0.3). It is easily seen to be well defined. Abbreviating ψ_j, $\lambda_{j\ell}(g)$ and $\mu_{j\ell}(g)$ to ψ, μ and λ we have

(2.2) $$(\psi\mu - \psi\mu^{-1})(\mu - \mu^{-1})^{-1} = \sum_{\nu=-m}^{m} \lambda^{\nu} \quad (m = \tfrac{1}{2}(k-1)),$$

where $\psi(\zeta) = \zeta^k$ for a primitive root of unity ζ of order $2|K/H|$. The expression (2.2) is an algebraic integer in $\mathbb{Q}(\zeta^2)$. In the formula (0.3) the sum over generators of K/H has the effect of applying the trace Tr: $\mathbb{Q}(\zeta^2) \to \mathbb{Q}$ to the product corresponding to a particular generator g. Hence $n_C \in \mathbb{Z}$.

This section is devoted to the proof of the following real analogue of the congruences discovered by T. Petrie [7] for the complex case.

THEOREM 2.3. Let $f : S(V) \to S(W)$ be a G-equivariant map with fixed point degrees $d_H(f) = d_H(f;\ \phi_1,\dots,\phi_r)$. (cf. 1.12). For each subgroup $H \subset G$ we have

(2.4) $$\sum_K n_{K/H}\, d_K(f) \equiv 0 \pmod{|NH/H|}$$

where we sum over subgroups K of G with $H \triangleleft K$ and K/H cyclic.

Since $n_{H/H} = 1$ and $n_{K/H}$ as well as $d_K(f)$ only depends on the conjugacy class of K/H in NH/H, we can rewrite (2.4) in the form adopted in [7]:

(2.5) $$d_H(f) = -\Sigma(NH: NH \cap NK)\, n_{K/H}\, d_K(f) \pmod{|NH/H|},$$

where we sum over a set of representatives of the conjugacy classes of non-trivial cyclic subgroups K/H of NH/H.

The congruence (2.4) for the group H follows from the case H = 1 of (2.4) applied to the NH/H-modules V^H and W^H. Thus we need only prove the congruence

(2.6) $$\sum_C n_C\, d_C(f) \equiv 0 \pmod{|G|},$$

where C runs through the cyclic subgroups of G.

A change in the choice of ψ_j-conjugations ϕ_j leaves the integers n_C unchanged, whereas 1.8 implies the existence of an orthogonal G-map $A : S(V) \to S(V)$ so that $d_C(f)$ with respect to the old ϕ_j's equals $d_C(fA)$ with respect to the new ϕ_j's. Hence we are free to choose ψ_j-conjugations ϕ_j with particularly good properties. By adding further 1-conjugate pairs of the form $(U,U,1_U)$ and suspending f we may assume in the decomposition (2.1) of V, that each component appears 8 times, i.e. $r = 8s$ and $V_{8k-\ell} = V_{8k}$ for $k = 1,\dots,s$ and $\ell = 0,\dots,7$.

Let L be the field considered in the proof of 1.2. As shown there V_j has a G-invariant L-subspace $V_j(L)$, such that $V_j = V_j(L) \otimes_L \mathbb{R}$ and Q maps $V_j(L) - 0$ into P(L). We can choose $W_j(L) \subset W_j$ with similar properties and assume ϕ_j chosen so that $\phi_j V_j(L) = W_j(L)$. We arrange, that $V_{8k-\ell}(L) = V_{8k}(L)$ $(0 \leqq \ell \leqq 7)$.

The Clifford algebra of (E,-Q), where (E,Q) is a $\mathbb{R}$-vectorspace with a positive definite quadratic form Q, will be denoted C(E). Its complexification $C(E,\mathbb{C})$ may be identified with the Clifford algebra of -Q scalar extended to $E \otimes \mathbb{C}$. The unit sphere S(E) generates the subgroup Pin(E) of the invertible elements in C(E). Similarly the unit quadric $S(E,\mathbb{C})$ generates a subgroup $Pin(E,\mathbb{C})$ of the invertible elements in $C(E,\mathbb{C})$. There are double covering homomorphisms $Pin(E) \to O(E)$, $Pin(E,\mathbb{C}) \to O(E,\mathbb{C})$, which send $x \in S(E)$ (resp. $x \in S(E,\mathbb{C})$) into the orthogonal reflection in the hyperplane perpendicular to x. We have the usual $\mathbb{Z}/2$-gradings of C(E) and $C(E,\mathbb{C})$. Intersecting the above groups with $C^o(E)$ and $C^o(E,\mathbb{C})$ we obtain the double coverings $Spin(E) \to SO(E)$, $Spin(E,\mathbb{C}) \to SO(E,\mathbb{C})$.

In preparation for the construction of suitable Spin reductions, we consider a subgroup $H \subset G$, a totally positive definite LH-module E(L) with scalar extension E to $\mathbb{R}$, and 8 ψ_j-conjugate pairs (E,F_j,ϕ_j), $j = 1,\dots,8$, such that $F_j = \mathbb{R}F_j(L)$, where $F_j(L) = \phi_j(E(L))$.

The diagonal map $Pin(E) \to Pin(8E)$, where nE denotes the orthogonal direct sum of n copies of E, factors to give a homomorphism

$$O(E) \to Spin(8E) \tag{2.7}$$

Composing (2.7) with $H \to O(E)$ we obtain a canonical Spin reduction $H \to Spin(8E)$ of 8E.

In order to Spin reduce $F = F_1 \perp \ldots \perp F_8$ we first observe, that ϕ_j induces a (discontinuous) homomorphism

$$Pin(\phi_j) : Pin(E,\mathbb{C}) \to Pin(F_j,\mathbb{C})$$

extending the map $S(\phi_j)$ of unit quadrics. The composite

$$Pin(E) \xrightarrow{(Pin\phi_j)} \prod_{j=1}^{8} Pin(F_j,\mathbb{C}) \to Pin(F,\mathbb{C})$$

factors to give a homomorphism

$$O(E) \to Spin(F,\mathbb{C}) \tag{2.8}$$

Composing (2.8) with $H \to O(E)$ we obtain a homomorphism $H \to Spin(F,\mathbb{C})$ lifting $H \to O(F,\mathbb{C})$. Since $F(L) = F(L_1) \oplus \ldots \oplus F(L_2)$ is H-invariant, it follows, that (2.8) maps H into the "real points" $Spin(F) \subset Spin(F,\mathbb{C})$. This defines our canonical Spin reduction of F.

We also need canonical orientations on 8E and F. On 8E take the product orientation of 8 copies of a fixed orientation of E. On F we take the product orientation of orientation on F_j obtained from a fixed orientation on E by transportation along ϕ_j (§1).

From the orientations and Spin reductions constructed above we obtain real Spin representations Δ^+ and Δ^- on Spin 8E and Spin F. By pulling these back along the Spin reduction we get representations of H.

$$\Delta^+(8E),\ \Delta^-(8E),\ \Delta^+(F) \text{ and } \Delta^-(F).$$

We define "Euler classes" in RO(H) by

$$\begin{aligned} e(8E) &= \Delta^+(8E) - \Delta^-(8E) \\ e(F) &= \Delta^+(F) - \Delta^-(F). \end{aligned} \tag{2.9}$$

We shall calculate the characters of these classes in two basic cases.

LEMMA 2.10. Assume that H is cyclic generated by g, $\psi_j(i) = i$ $(j=1,\ldots,8)$, $\dim E = 2$, E is oriented, and that g acts on E by rotation through an angle θ (the orientation makes θ well defined modulo $2\pi\mathbb{Z}$). Let λ be the root of unity determined by

$$\cos\theta = \frac{1}{2}(\lambda + \lambda^{-1}), \quad \sin\theta = \frac{1}{2i}(\lambda - \lambda^{-1})$$

and let μ be a square root of λ. Then we have

$$\chi_{e(F)}(g) = \prod_{j=1}^{8} (\psi_j(\mu) - \psi_j(\mu)^{-1}) \tag{2.11}$$

$$\chi_{e(8E)}(g) = (\mu - \mu^{-1})^8. \tag{2.12}$$

PROOF. It suffices to prove (2.11). Since the expression in (2.11) is independent of the choice of square root, we can assume

$$\cos\frac{1}{2}\theta = \frac{1}{2}(\mu + \mu^{-1}), \ \sin\frac{1}{2}\theta = \frac{1}{2i}(\mu - \mu^{-1}).$$

Let ε_1, ε_2 be an orthogonal positively oriented basis for $E(L)$ over L. Let c_1 and c_2 be the positive real numbers, such that $c_\nu^2 = Q(\varepsilon_\nu)$, where Q is the quadratic form on E, and put $e_\nu = c_\nu^{-1}\varepsilon_\nu$. Now e_1, e_2 is a positively oriented orthonormal basis for E, and the action of g is given by

$$g\,e_1 = \cos\theta\, e_1 + \sin\theta\, e_2$$

$$g\,e_2 = -\sin\theta\, e_1 + \cos\theta\, e_2.$$

Put $e_{j\nu} = \phi_j(e_\nu) = \psi_j(c_\nu)^{-1}\,\phi_j(\varepsilon_\nu)$ for $j = 1,\dots,8$ and $\nu = 1,2$. Since c_ν^2 is a totally positive number $\psi_j(c_\nu)$ is real, so that e_{j1}, e_{j2} is a positively oriented orthonormal basis of F_j (with respect to the orientation obtained by transportation along ϕ_j). The rotation g lifts to the element

$$\bar{g} = (\sin\frac{\theta}{2}\,e_1 - \cos\frac{\theta}{2}\,e_2)\cdot e_2 = \cos\frac{\theta}{2} + \sin\frac{\theta}{2}\,e_1e_2$$

of $\mathrm{Pin}(E) \subset C(E)$. If we choose $\theta_j \in \mathbb{R}$ so that

$$\cos\frac{\theta_j}{2} = \frac{1}{2}(\psi_j(\mu) + \psi_j(\mu)^{-1}), \ \sin\frac{\theta_j}{2} = \frac{1}{2i}(\psi_j(\mu) - \psi_j(\mu)^{-1}),$$

we have in $\mathrm{Pin}(F_j)$

$$\mathrm{Pin}(\phi_j)\bar{g} = \cos\frac{\theta_j}{2} + \sin\frac{\theta_j}{2}\,e_{j1}\,e_{j2},$$

(if $\psi_j(i) = -i$ the last term appears with a minus).

The canonical Spin reduction of F maps g into the point

$$\sum_{j=1}^{8} (\cos\frac{\theta_j}{2} + \sin\frac{\theta_j}{2}\,e_{j1}\,e_{j2})$$

of the maximal torus T in Spin(F) associated to the orthonormal basis $e_{11}\ e_{12}\ e_{21}\ \ldots\ e_{82}$ of F. From the well known formulas for the characters of Δ^+ and Δ^- on T (Husemoller [11], Ch. 13), we can easily obtain (2.11).

LEMMA 2.13. Assume that H is cyclic of even order with a generator g acting on E by multiplication by -1, $\dim E = 1$ and $\psi_j(i) = i$ $(j = 1,\ldots,8)$. Then

$$\chi_{e(F)}(g) = \chi_{e(8E)}(g) = 16.$$

The details are similar to those of the previous proof. We leave them to the reader.

We return to the proof of (2.6). For $H \subset G$ we have canonical orientations and Spin reductions of the NH/H-modules $8V_{8k}^H$ and $W(k)^H$, where $W(k) = W_{8k-7} \perp \ldots \perp W_{8k}$. Taking direct sums we get canonical orientations and Spin reductions of V^H and W^H (in particular of V and W). Similarly we obtain canonical orientations and Spin-reductions of the orthogonal complements V_H and W_H to V^H and W^H in V and W respectively (V_H and W_H are NH-modules).

By the construction of [3] we get specific Thom classes (Bott classes in the terminology of Atiyah [2]) using the spinor representations Δ^+ and Δ^-

(2.14) $$U \in KO_G(V),\ U' \in KO_G(W),$$

where we use equivariant real K-theory with compact supports. For each cyclic subgroup $C \subset G$ we have specific Thom classes

(2.15) $$U_C \in KO_C(V^C),\ U'_C \in KO_C(W^C).$$

According to Theorem 6.1 of [2] the groups in (2.14) and (2.15) are free as modules over the respective representation rings with the indicated classes as generators.

We extend $f : S(V) \to S(W)$ radially to a proper G-equivariant map $V \to W$ again denoted f. Following [7] we consider the commutative diagram

$$\begin{array}{ccc}
KO_G(W) & \xrightarrow{f^*} & KO_G(V) \\
\downarrow{\scriptstyle Res} & & \downarrow{\scriptstyle Res} \\
KO_C(W) & \xrightarrow{f^*} & KO_C(V) \\
\downarrow{\scriptstyle i^*} & & \downarrow{\scriptstyle j^*} \\
KO_C(W^C) & \xrightarrow{f^*} & KO_C(V^C)
\end{array}$$

where $i : W^C \to W$ and $j : V^C \to V$ are the inclusions and Res indicates restriction from G to C. Define $a \in RO(G)$ by

$$f^*(U') = a\ U.$$

Since C acts trivially on W^C and V^C, we have

$$f^*(U'_C) = d_C(f)\ U_C.$$

Morever

$$j^*\mathrm{Res}\ U = e(V_C)U_C, \quad i^*\mathrm{Res}\ U' = e(W_C)U'_C.$$

These formulas combine to show

(2.16) $$(\mathrm{Res}\ a)\ e(V_C) = d_C(f)\ e(W_C).$$

As in [7] the next step is to evaluate characters at a generator $g \in C^o$. First we have

(2.17) $$e(V_C) = \prod_{k=1}^{s} e(8V_{8k,C}), \quad e(W_C) = \prod_{k=1}^{s} e(W(k)_C)$$

As an $\mathbb{R}C$-module without the quadratic form V_{jC} breaks up into a direct sum of 1- and 2-dimensional irreducible components of the types occuring in 2.13 and 2.10. This splitting can be realized over L, since L is a real splitting field for C. Taking successive orthogonal complements, we can write

(2.18) $$V_{jC}(L) = E_{j1}(L) \perp \ldots \perp E_{js_j}(L) \perp \ldots \perp E_{jt_j}(L),$$

where the scalar extension $E_{j\ell} = \mathbb{R}E_{j\ell}(L) \subset V_{jC}$ is of the type described in 2.10 for $1 \leq \ell \leq s_j$, and the type described in 2.13 for $s_j < \ell \leq t_j$. Furthermore we may assume, that $E_{8k,\ell}(L) = E_{8k-m,\ell}(L)$ for $1 \leq k \leq s$, $0 \leq m \leq 7$. Defining $F_{j\ell}(L) = \phi_j\ E_{j\ell}(L)$ and $F_{j\ell} = \mathbb{R}F_{j\ell}(L) \subset W_{jC}$ we obtain decompositions

(2.19) $$W_{jC}(L) = F_{j1}(L) \perp \ldots \perp F_{js_j}(L) \perp \ldots \perp F_{jt_j}(L).$$

From (2.18) and (2.19) follows

$$e(8V_{8k,C}) = \prod_{\ell=1}^{t_j} e(8E_{8k,\ell})$$

(2.20)

$$e(W(k)_C) = \prod_{\ell=1}^{t_j} e(F(k)_\ell),$$

where $F(k)_\ell = F_{8k-7,\ell} \perp \ldots \perp F_{8k,\ell}$. Let $\lambda_{8k,\ell}(g)$ be the root of unity appearing in 2.10 applied to $E_{8k,\ell}$ and let $\mu_{8k,\ell}(g)$ be a square root of $\lambda_{8k,\ell}(g)$. Put $\mu_{8k-m,\ell}(g) = \mu_{8k,\ell}(g)$ for $0 \leq m \leq 7$. Evaluating characters in (2.16) at g using (2.17), (2.20) and lemmas 2.10 and 2.13 we obtain

$$\chi_a(g) = d_C(f) \prod_{j=1}^{r} \prod_{\ell=1}^{s_j} \frac{\psi_j\mu_{j\ell}(g) - \psi_j\mu_{j\ell}(g)^{-1}}{\mu_{j\ell}(g) - \mu_{j\ell}(g)^{-1}}$$

Summing over the generators C^o of C we find

$$\sum_{g\in C^o} \chi_a(g) = d_C(f)\, n_C$$

where n_C is the integer given by the formula (0.3) of the introduction. The multiplicity of 1_G in $a \in RO(G)$ is the integer given by

$$\langle a,1_G\rangle = \frac{1}{|G|}\sum_{g\in G} \chi_a(g) = \frac{1}{|G|}\sum_{C}\sum_{g\in C^o} \chi_a(g) = \frac{1}{|G|}\sum_{C} n_C d_C(f)$$

where we sum over cyclic subgroups C of G. This proves the congruence (2.6).

§3. CLASSIFICATION OF G-MAPS.

In order to make sense of the following result no particular care with orientations is needed: Just choose orientations of all fixed point sets.

PROPOSITION 3.1. Let V and W be $\mathbb{R}G$-modules such that $\dim V^H = \dim W^H$ for all subgroups $H \subset G$. Two G-maps $f_0, f_1 \colon S(V) \to S(W)$ such that $d_H(f_0) = d_H(f_1)$ are stably *) G-homotopic. Moreover f_0 and f_1 are themselves G-homotopic, if one of the following conditions hold.

a) $\dim V^H$ is even for every $H \subset G$.

b) G is nilpotent.

PROOF. We assume a) to hold, and prove that f_0 and f_1 are G-homotopic. The first statement will follow by replacing V and W by $V \perp V$ and $W \perp V$ respectively. The method is to construct the G-homotopy inductively over the orbit types in $S(V)$ as in [7] starting at the G-fixed points. For $H \subset G$ we let

$$S(V^H)_s \subset S(V^H) \tag{3.2}$$

be the points in $S(V^H)$ with an isotropy group (for the G-action on all of $S(V)$) strictly larger than H. In the inductive step one extends a G-homotopy on $GS(V^H)_s$ to a G-homotopy on $GS(V^H)$. Equivalently one extends a NH-homotopy on $S(V^H)_s$ to an NH-homotopy on $S(V^H)$.

*) The word stably refers to suspension by arbitrary $\mathbb{R}G$-modules.

The obstructions to doing this are elements

(3.3) $$o_n \in H^n(S(V^H)/NH,\ S(V^H)_s/NH;\ \pi_n(S(W^H))).$$

For dimensional reasons these groups vanish unless $n+1 = \dim V^H = \dim W^H$. In that case $\pi_n(SW)) \cong \mathbb{Z}$, and the coefficients in (3.3) are twisted by the action of NH on $S(W^H)$. By lemma 1.13 the restriction to $M(H) = (S(V^H) - S(V^H)_s)/NH$ of the coefficient system in (3.3) is isomorphic to the orientation bundle of the manifold M(H). Because of a) $S(V^H)_s$ has codimension at least 2 in $S(V^H)$, so M(H) is connected. Hence the group in (3.3) for $n+1 = \dim V^H$ is infinite cyclic, and o_n may be identified as $\frac{1}{|NH/H|}(d_H(f_1) - d_H(f_0))$ times a generator. The assumption on degrees assures us that o_n vanishes.

When we assuem b) but not a) to hold, we run into difficulties in case $S(V^H)_s$ has codimension 1 in $S(V^H)$. Then there are subgroups $K \subset G$ strictly containing H such that

$$\dim V^H = 1 + \dim V^H \cap V^K.$$

The nilpotency of G implies $NH \cap K \neq H$. An element $g_K \in NH \cap K - H$ act non-trivially on V^H and by the identity on $V^H \cap V^K$. Therefore g_K acts on V^H by orthogonal reflection in $V^H \cap V^K$. Using all reflection g_K of this type we see that M(H) is connected.

REMARK 3.4. The simplest example where f_0 and f_1 are not G-homotopic occurs for two copies of the symmetric group Σ_3 acting on $\mathbb{R}^3$ by permutation of coordinates (Rubinsztein [15]). The author is indepted to H. Hauschild for pointing this out.

We proceed to derive a complete set of relations among the fixed point degrees $d_H(f)$. For this we consider ψ_j-conjugate pairs (V_j, W_j, ϕ_j), $1 \leq j \leq r$, where $\psi_j(i) = i$ and we consider the $\mathbb{R}G$-modules

$$V = V_1 \perp \dots \perp V_r, \quad W = W_1 \perp \dots \perp W_r.$$

Let $C = C_G(V_j, \psi_j)$ be the abelian group of families $(d_H)_{H\subset G}$ of integers constant on conjugacy classes of subgroups and satisfying for each $H \subset G$ the congruence $\sum_K n_{K/H}\, d_H \equiv 0 \pmod{NH/H}$.

THEOREM 3.5. For each subgroup $H \subset G$ let d_H be an integer. There exists a G-map $f : S(V) \to S(W)$ with $d_H(f;\phi_1,\dots,\phi_r) = d_H (H \subset G)$, if and only if the following conditions hold.

i) The family $(d_H)_{H\subset G}$ belongs to C.

ii) If $\dim V^H = 0$, $d_H = 1$.

iii) If $\dim V^H = 1$, d_H equals 1,0 or -1.

iv) If $V^H = V^K$, $d_H = d_K$.

PROOF. The degrees of a G-map f obviously satisfy ii) - iv), whereas i) follows from theorem 2.3. Assume conditions i) - iv) to hold. We construct f inductively over orbit types in $S(V)$. In the inductive step we have to extend a G-map $GS(V^H)_s \to S(W)$ to $GS(V^H)$. Equivalently we extend an NH-map $S(V^H)_s \to S(W^H)$ to $S(V^H)$. The obstructions to doing this are elements in the groups

$$H^n(S(V^H)/NH,\ S(V^H)_s/NH;\ \pi_{n-1}(SW^H))),$$

which are zero for dimensional reasons. Thus the only problem is, at each stage to obtain an extension of the correct degree. Consider a NH/H-equivariant map $f : S(V^H) \to S(W^H)$, such that $d_K(f) = d_K$ for every subgroup K strictly containing H. We are going to modify f away from $S(V^H)_s$ in order to adjust $d_H(f)$. If $S(V^H)_s = S(V^H)$ we have $d_H(f) = d_H$ because of iv), so we can assume $S(V^H)_s \neq S(V^H)$. If $\dim V^H \leq 1$ we are finished by ii) or iii), so we can further assume, that $\dim V^H \geq 2$. By thm. 2.3 applied to V^H and W^H we have

$$\sum_K n_{K/H}\, d_K(f) \equiv 0 \qquad (\mathrm{mod}\ |NH/H|).$$

From i) we have

$$\sum_K n_{K/H}\, d_K \equiv 0 \qquad (\mathrm{mod}\ |NH/H|).$$

Since $d_K(f) = d_K$ for $K \neq H$ and $n_{H/H} = 1$ it follows that $d_H(f) \equiv d_H$ (mod $|NH/H|$). Let $\{gx \mid g \in NH/H\}$ be the NH/H-free orbit of $x \in S(V^H) - S(V^H)_s$, and choose a disc D around x, so that the discs gD ($g \in NH/H$) are disjoint and contained in $S(V^H) - S(V^H)_s$. We can deform f radially in discs slightly larger than gD to a new map f with $f(D)$ equal to the point $f(x)$. Any map $h : (D,\partial D) \to (S(W^H), f(x))$ of degree d, and its conjugates by elements $g \in NH/H$, can be added at the discs gD to give an equivariant map $f_d : S(V^H) \to S(W^H)$. Using lemma 1.13 we find

$$d_H(f_d) = d_H(f) + d|NH/H|.$$

Hence d can be chosen, so that f_d is the correctly modified map.

COROLLARY 3.6. In the notation of 3.5 define $\alpha \in RO(G)$ by $\alpha = [W] - [V]$. There is an isomorphism of abelian groups

$$I = I(\phi_1,\dots,\phi_r) : \omega_G^\alpha \to C_G(V_j,\psi_j) \tag{3.7}$$

mapping $f : S(V \perp U) \to S(W \perp U)$ into

$$(d_H(f;\phi_1,\dots,\phi_r,1_U))_{H\subset G}.$$

If $\bar\phi_j : V_j \otimes \mathbb{C} \to W_j \otimes \mathbb{C}$, $j = 1,\dots,r$ is another collection of ψ_j-conjugations, there exists an orthogonal unit $u \in \Delta_G$ such that

$$I(\bar\phi_1,\dots,\bar\phi_r) = I(\phi_1,\dots,\phi_r) \circ u.$$

PROOF. The isomorphism (3.7) follows from 3.1 and 3.5. The last statement follows from 1.8.

Note that the isomorphism (3.7) can be applied for any $\alpha \in RO(G)$ with $\dim \alpha^H = 0$ $(H \subset G)$ because of 1.10.

§4. G-MAPS APPROXIMATING A CONJUGATION.

In this section we apply the results of §3 to a ψ-conjugate pair (V,W,ϕ) of positive definite quadratic $\mathbb{R}G$-module. In order to avoid cumbersome signs we assume $\psi(i) = i$. We are going to "approximate" ϕ by G-maps $S(V) \to S(W)$. The analogous step in Sullivan's treatment of the Adams conjecture is the use of etale homotopy in order to approximate the Galois action on an algebraic variety defined over $\mathbb{Q}$ by actual maps.

Let n be the least common multiple of 4 and twice the exponent of G, and assume ψ acts on the n'th roots of unity ζ by $\psi(\zeta) = \zeta^k$ $(k \in \mathbb{Z})$. Note that $(k,|G|) = 1$ and $k \equiv 1 \pmod 4$.

THEOREM 4.1. There exists a G-equivariant map $f : S(V) \to S(W)$ with fixed point degrees

$$d_H(f,\phi) = k^{[\frac{1}{2}\dim V^H]}.$$

The square bracket indicates integer part.

It suffices to verify the conditions of thm. 3.5. The only problem is to verify the congruences (0.4). This is done by induction on $|G|$. In particular the induction hypothesis implies (0.4) for $H \neq 1$, so we need only prove

$$\sum_C n_C\, k^{[\frac{1}{2}\dim V^C]} \equiv 0 \pmod{|G|} \tag{4.2}$$

where C runs through the cyclic subgroups of G. We may assume $k > 0$, and that $\dim V^H \geqq 2$ for every $H \subset G$ (replace V by $V \perp 2$).

For an even dimensional representation E of a compact Lie group K with det E trivial, Adams ([1]) has defined an element $\rho_{\mathbb{R}}^{k}E \in RO(K)$ for each odd natural number k. The eigenvalues of $g \in K$ acting on E occur in complex conjugate pairs $(\lambda_j, \lambda_j^{-1})$, $j = 1,\dots,r$. Adams proved that the character of $\rho_{\mathbb{R}}^{k}E$ at g is given by

$$\prod_{j=1}^{s} \Big(\sum_{\nu=-m}^{m} \lambda_j^{\nu} \Big) \qquad (m = \tfrac{1}{2}(k-1)). \tag{4.3}$$

In particular the eigenvalues equal to -1 contribute a factor 1, and the eigenvalues equal to 1 contribute k^d, where $d = \frac{1}{2}\dim E^g$ is an integer.

For a cyclic group $C \subset G$ we let V_C' denote the orthogonal complement in V to the sum of the (± 1)-eigenspaces of a generator $g \in C^o$. We put

$$a(C) = \rho_{\mathbb{R}}^{k} V_C' \in RO(C).$$

Comparing formulas (4.3) and (2.2) we find from (0.3)

$$n_C = \sum_{g \in C^o} \chi_{a(C)}(g).$$

We are now ready to prove (4.2) in the case where dim V is even and det V trivial: Put $a = \rho_{\mathbb{R}}^{k} V \in RO(G)$ and evaluate

$$|G|\langle \mathbf{a}, 1_G\rangle = \sum_{g \in G} \chi_a(g) = \sum_C \sum_{g \in C^o} \chi_a(g)$$

$$= \sum_C k^{\frac{1}{2}\dim V^C} \Big(\sum_g \chi_{a(C)}(g)\Big) = \sum_C n_C\, k^{\frac{1}{2}\dim V^C}.$$

Next we consider the case, where dim V is odd. Applying the previous case to the orthogonal direct sum of (V,W,ϕ) and its determinant (see 1.3 and 1.6), we find

$$\sum_C n_C\, k^{\{\frac{1}{2}\dim V^C\}} \equiv 0 \pmod{|G|},$$

where $\{x\}$ means the least integer, such that $\{x\} \geq x$. Let T be the antipodal map on $S(V \perp 1)$. Using the trivial component to add and subtract, we can realize the expressions

$$1 + \frac{k-1}{2}(1 \pm T)$$

as equivariant maps $f^{\pm} : S(V \perp 1) \to S(V \perp 1)$ with fixed point degrees

$$d_H(f^+) = \begin{cases} k & \text{for } \dim V^H \text{ even} \\ 1 & \text{for } \dim V^H \text{ odd} \end{cases}$$

$$d_H(f^-) = \begin{cases} 1 & \text{for } \dim V^H \text{ even} \\ k & \text{for } \dim V^H \text{ odd.} \end{cases}$$

By our induction hypothesis the obstruction arguments of 3.5 produce a G-map $f : S(V \perp 1) \to S(W \perp 1)$ with $d_H(f;\phi,1) = k^{[\frac{1}{2}\dim V^H]}$ for $H \neq 1$. Hence $d_H(ff^-;\phi,1) = k^{\{\frac{1}{2}\dim V^H\}}$ for $H \neq 1$. Theorem 2.3 applied to H-fixed points for $H \neq 1$ shows

$$\sum_K n_{K/H}\, k^{\{\frac{1}{2}\dim V^H\}} \equiv 0 \pmod{|NH/H|}.$$

The case $H = 1$ of this was proved above, so we conclude from thm 3.5, that there is a G-map $f' : S(V \perp 1) \to S(W \perp 1)$ with $d_H(f';\phi,1) = k^{\{\frac{1}{2}\dim V^H\}}$ for all subgroups H. Hence $d_H(f'f^+;\phi,1) = k^{[\frac{1}{2}\dim V^H]+1}$.

Now we apply the case $H = 1$ of thm. 2.3 and divide by k in the resulting congruence to obtain (4.2).

Finally, we have the general case with dim V even. We apply the odd dimensional case to the direct sum of the ψ-conjugate pairs (V,W,ϕ) and $(1,1,\psi)$. This gives a G-map $f : S(V \perp 1) \to S(W \perp 1)$ with $d_H(f;\phi,\psi) = k^{\{\frac{1}{2}\dim V^H\}}$. One proceeds as in the last part of the odd dimensional case.

By applying thm. 4.1 in the case $V = W$ with a non-trivial ϕ one can produce interesting elements in the Burnside ring ω_G^0.

EXAMPLE 4.4. Let G be cyclic of odd prime order p, and let $V = W$ be the regular representation of G over $\mathbb{R}$ with the group elements as an orthonormal basis. For an integer $k \equiv 1 \pmod 4$ prime to p we can find $\psi \in \mathrm{Aut}(\mathbb{C})$ acting on the 4p'th roots of unity ζ by $\psi(\zeta) = \zeta^k$. Let $\phi : V \otimes \mathbb{C} \to W \otimes \mathbb{C}$ be defined by $\phi(\sum_{g\in G} a_g g) = \sum_{g\in G} \psi(a_g)g$. If we let the group elements in some fixed order be a positive basis for V, the group elements taken in the same order will be a positive basis for W with respect to the transported orientation. On the other hand $S(V^G) = \{\pm e\}$, where $e = \frac{1}{\sqrt{p}} \sum_{g \in G} g$. If e is chosen as a positive basis for V^G, we find that $(\psi(\sqrt{p})/\sqrt{p})e$ is a positive basis for W^G for the transported orientation. For the degrees of the map f of thm. 4.1 with respect to identical orientations of fixed point sets V and W we find

$$d_1(f) = d_1(f;\phi) = k^{[\frac{1}{2}p]} = k^{\frac{1}{2}(p-1)}$$

$$d_G(f) = (\psi(\sqrt{p})/\sqrt{p})d_G(f;\phi) = \psi(\sqrt{p})/\sqrt{p}.$$

The congruences describing ω_G^o tell us that $d_1(f)$ and $d_G(f)$ are congruent modulo p. Modulo p $k^{\frac{1}{2}(p-1)}$ becomes the Legendre symbol $(\frac{k}{p})$. Hence

$$\psi(\sqrt{p})/\sqrt{p} = (\frac{k}{p}). \tag{4.5}$$

In the case $k = (-1)^{\frac{1}{2}(q-1)} g$, where q is an odd prime distinct from p, ψ acts the same way on $\sqrt{p}$ as the Frobenius automorphism at q. Hence (4.5) is equivalent to the classical quadratic reciprocity law

$$(\frac{q}{p})(\frac{p}{q}) = (-1)^{\frac{1}{4}(p-1)(q-1)}.$$

The action of $\psi \in \mathrm{Aut}(\mathbb{C}/\mathbb{Q}(i))$ on the roots of unity in $\mathbb{C}$ is given by an adic unit $e \in \hat{\mathbb{Z}}^*$. As an immediate corollary to thm. 4.1 we have

COROLLARY 4.6. For any $\mathbb{R}G$-module V there is an element in $\hat{C}_G(V,\psi)$ given by

$$d_H = e^{[\frac{1}{2}\dim V^H]}. \tag{4.7}$$

Combining this with 3.6 we can to V associate a well defined element

$$A(V,\psi) \in \hat{\omega}_G^{(\psi-1)V}/\Delta_G$$

For any choice of a ψ-conjugation $\phi : V \otimes \mathbb{C} \to W \otimes \mathbb{C}$, where $W = \psi V$, $A(V,\psi)$ is represented by the inverse image by $\hat{I(\phi)} : \hat{\omega}_G^{\alpha} \to \hat{C}$ of the family (4.7). We can now prove thm. D of the introduction (which also contains an explanation of the notation).

THEOREM 4.8. The map $V \mapsto A(V,\psi)$ extends (uniquely) to a map natural on the category of finite groups

$$A : RO(G) \to E_G = \coprod_{\alpha \in RO_o(G)} \hat{\omega}_G^{\alpha}/\Delta_G$$

satisfying the sum formula

$$A(\alpha + \beta) = D(\alpha,\beta)\, A(\alpha)\, A(\beta) \tag{4.9}$$

where $D : RO(G) \times RO(G) \to \omega_G^o$ is given by

$$D(\alpha,\beta) = 1 + \frac{e-1}{4}(\Delta(\alpha) - 1)(\Delta(\beta) - 1) \tag{4.10}$$

PROOF. The degrees of $D(\alpha,\beta)$ at fixed point sets are given by

$$d_H(D(\alpha,\beta)) = \begin{cases} 1 & \text{for } \dim \alpha^H \text{ or } \dim \beta^H \text{ even} \\ e & \text{for } \dim \alpha^H \text{ and } \dim \beta^H \text{ odd.} \end{cases}$$

From this (4.9) follows in case α and β are actual representations by calculating degrees.

Consider a ψ-conjugate pair (V,W,ϕ). Applying 4.6 to the ψ^{-1}-conjugate pair (W,V,ϕ^{-1}) we get an element $y \in \hat{\omega}_G^{V-W}$ with

$$d_H(y,\phi^{-1}) = e^{-[\frac{1}{2}\dim V^H]} = e^{\{\frac{1}{2}\dim(-V)^H\}}$$

As in the proof of 4.1 this can be modified to an element $A(-V,\psi) \in \hat{\omega}_G^{(\psi-1)(-V)}$ with

$$d_H(A(-V,\psi);\phi^{-1}) = e^{[\frac{1}{2}\dim(-V)^H]}.$$

Now consider for $j = 1,2$ ψ-conjugate pairs (V_j,W_j,ϕ_j). The expression

$$x = D(V_1,-V_2)A(V_1,\psi)A(-V_2,\psi)$$

defines an element $x \in \hat{\omega}_G^{(\psi-1)}$, where $\alpha = [V_1] - [V_2]$, such that

$$d_H(x;\phi_1,\phi_2^{-1}) = e^{[\frac{1}{2}\dim \alpha^H]}$$

If we in the above replace the two ψ-conjugate pairs with their orthogonal direct sums with a third ψ-conjugate pair (V_3,W_3,ϕ_3) x is replaced with an element $x' \in \hat{\omega}_G^{(\psi-1)\alpha}$, such that

$$d_H(x'; \phi_1, \phi_3, \phi_2^{-1}, \phi_3^{-1}) = e^{[\frac{1}{2}\dim \alpha^H]} .$$

If we orient the fixed point sets in V_1, V_3, W_2 and W_3, such that those in W_3 are transported from V_3 along ϕ_3, and transport by respectively $\phi_1, \phi_3, \phi_2^{-1}$ and ϕ_3^{-1} to W_1, W_3, V_2 and V_3, the resulting orientations in W_3 and V_3 agree with the original ones. This shows that x and x' are equivalent modulo Δ_G, so that the element

$$A(V_1 - V_2, \psi) \in \hat{\omega}_G^{(\psi-1)\alpha}/\Delta_G$$

given by

$$d_H(A(V_1 - V_2, \psi); \phi_1, \phi_2^{-1}) = e^{[\frac{1}{2}\dim \alpha^H]} \tag{4.11}$$

only depends on $\alpha = [V_1] - [V_2] \in RO(G)$. The formula (4.9) follows in the general case by a degree calculation. The naturality in G is obvious in view of (4.11).

§5. APPLICATION TO THE ADAMS CONJECTURE.

In this section we give a treatment of the Adams conjecture for real vectorbundles at the prime 2, which includes a formula for its behavior on a sum of two vectorbundles. The method is that of Quillen [14] replacing his treatment of vectorbundles with finite structural group by an application of the results of §4.

We denote finite groups with small greek letters in order to avoid confusion with the infinite loop space G. Let $IO(\pi) \subset RO(\pi)$ be the augmentation ideal. For $\alpha \in IO(\pi)$ let $\theta_\pi^\alpha \subset \omega_\pi^\alpha$ be the subset represented by π-maps f with $d_1(f) = \pm 1$.

We consider a principal π-bundle $P \to X$ over a finite connected complex X. For a π-map $f : S(V) \to S(W)$ representing an element in θ_π^α we can form $1_P \times_\pi f : P \times_\pi S(V) \to P \times_\pi S(W)$. This is a fiberhomotopy equivalence (Dold [8]) between the vectorbundles associated to V and W, and it is classified by a map $h(f) : X \to G/O$. This defines a map

$$h : \theta_\pi^\alpha \to [X, G/O].$$

Let $\hat{\theta}_\pi^\alpha$ be the closure of θ_π^α in $\hat{\omega}_G^\alpha$.

LEMMA 5.1. The map h extends to a continuous map

$$h : \hat{\theta}^{\alpha}_{\pi} \to [X,G/O]$$

where $[X,G/O]$ is given the discrete topology.

PROOF. Let the finite group $\tilde{\pi}^{o}_{s}(X)$ be annihilated by the integer $n \neq 0$. Consider f_1 and f_2 in θ^{α}_{π} such that $f_1 - f_2 = nf$ for some $f \in \omega^{\alpha}_{\pi}$ and $d_1(f_1) = d_1(f_2)$. It suffices to show that $h(f_1) = h(f_2)$. If $a \in \widetilde{KO}^{o}(X)$ is associated to α, the associated bundle construction defines a homomorphism $\phi : \omega^{\alpha}_{\pi} \to \omega^{a}(X)$ (Segal [15]). We have $\phi(f_1) - \phi(f_2) = n\phi(f)$, where $\phi(f)$ has fiber degree 0, and $\phi(f_2)$ is represented by a fiberhomotopy equivalence. Multiplying by a fiberhomotopy inverse in $\omega^{-a}(X)$, we get $\phi(f_1) = \phi(f_2)$. This implies $h(f_1) = h(f_2)$.

The map A of 4.8 sends $\alpha \in IO(\pi)$ into an element of $\hat{\theta}^{(\psi-1)\alpha}_{\pi}/\Delta_{\pi}$ by (4.11), and the map h factors through this quotient. Hence we can compose

$$hA : IO(\pi) \to [X,G/O].$$

If we apply this to the universal π-bundle over skeletons in $B\pi$, we get a welldefined natural map

$$a(\psi) : IO(\pi) \to [B\pi,G/O], \tag{5.2}$$

since $[B\pi,G/O]$ is given as an inverse limit over skeletons. Moreover we have the commutative diagram

$$\begin{array}{ccc} IO(\pi) & \xrightarrow{a(\psi)} & [B\pi,G/O] \\ \downarrow{\scriptstyle \psi-1} & & \downarrow{\scriptstyle i_*} \\ IO(\pi) & \longrightarrow & [B\pi,BO] \end{array} \tag{5.3}$$

where i_* is induced by the standard infinite loop map $i : G/O \to BO$. A good choice of a lifting $A : BO \to G/O\hat{}$ of $\psi^{e} - \psi^{1} : BO \to BO\hat{}$ (consult Sullivan [18] for Adams operations indexed by adic numbers) is given by the following result (e is the adic unit from thm. 4.8).

THEOREM 5.4. There is a map $A(\psi) : BO \to G/O\hat{}$ unique up to homotopy with the property that for every finite group π and $\alpha \in IO(\pi)$, the composite

$$B\pi \xrightarrow{\alpha} BO \xrightarrow{A(\psi)} G/O\hat{}$$

represents $a(\psi)\alpha$. This map satisfies

$$i \cdot A(\psi) \simeq \psi^{e} - \psi^{1}.$$

PROOF. We treat the various p-adic completions $G/O_p^{\wedge}$ separately. Fixing p we choose an odd prime $q \neq p$ and consider an algebraic closure k of the prime field $\mathbb{F}_q$. The infinite orthogonal group $O(k)$ is a union of an increasing sequence of finite orthogonal groups $\pi = O_n(\mathbb{F}_{q^r})$. Applying $a(\psi)$ to the Brauer lift of the standard representation of π on $\mathbb{F}_{q^r}^{\oplus n}$ reduced by its dimension, and passing to the limit, we obtain a certain homotopy class $BO(k) \to G/O_p^{\wedge}$. In [14] Quillen constructed by means of Brauer lifting a map $BO(k) \to BO$ inducing an isomorphism in mod p cohomology. By obstruction theory we can find a map $A(\psi)$ unique up to homotopy making the following diagram homotopy commutative

$$\begin{array}{ccc} BO(k) & \searrow & \\ \big\downarrow & & G/O_p^{\wedge} \\ BO & \nearrow_{A(\psi)} & \end{array}$$

The stated property holds by construction, when $\pi = O_n(\mathbb{F}_{q^r})$ and α is the reduced Brauer lift employed above. In order to verify it in general, we can reduce to a Sylow p-subgroup by a transfer argument. For a p-group the statement follows, since the decomposition homomorphism in that case becomes an inverse to Brauer lifting (Quillen [14], Appendix), so that we may factor through one of the Brauer lifts. The last statement of the theorem follows from (5.3).

The map $\Delta : RO(\pi) \to \omega_\pi^0$ defined in [7] has an immediate geometric analogue. To a vectorbundle E one associates the fibrewise antipodal map $S(E) \to S(E)$ on a unit sphere bundle. This construction is classified by a certain map $\Delta : BO \times \mathbb{Z} \to G$.

PROPOSITION 5.5. The map Δ is the composite of the map $\eta : BO \times \mathbb{Z} \to O$ representing multiplication by the generator of $KO^{-1}(pt) \cong \mathbb{Z}/2$, and the J-map $O \to G$. For a finite group π there is a natural commutative diagram with associated bundle constructions horizontally

(5.6)
$$\begin{array}{ccc} RO(\pi) & \longrightarrow & KO^0(B\pi) \\ \Delta\big\downarrow & & \Delta\big\downarrow \\ \text{Units}(\omega_\pi^0) & \longrightarrow & [B\pi, G] \end{array}$$

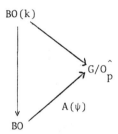

PROOF. The fibrewise antipodal map $E \to E$ is a vectorbundle automorphism, so Δ factors naturally through the J-map to give some map $\eta : BO \times \mathbb{Z} \to 0$. The product map $(BO \times \mathbb{Z}) \times 0 \to 0$ corresponds geometrically to the construction, which to a vectorbundle E over X and a vectorbundle $T : E' \to E'$ over Y associates the automorphism $1_E \otimes T$ of the exterior tensorproduct $E \otimes E'$ over $X \times Y$. If T is the antipodal map, so is $1_E \otimes T$. This shows that η represents a KO^0-module homomorphism $KO^0 \to KO^{-1}$. Therefore η is multiplication by the image of 1 under the homomorphism $KO^0(pt) \xrightarrow{\eta} KO^{-1}(pt)$. Since the trivial line bundle over the unit interval, glued at the ends by the antipodal map, becomes the Möbius bundle over S^1, the first statement of the proposition follows. The second statement is trivially verified.

EXAMPLE 5.7. Let us apply prop. 5.5 to $\pi = \mathbb{Z}/2$. In (5.6) we evaluate at $[L] - 1 \in RO(\mathbb{Z}/2)$, where L is the non-trivial 1-dimensional representation. In $KO^0(B\pi) = KO^0(RP^\infty)$ we get the reduced Hopf bundle. The image by η becomes the unique non-trivial map $f : RP^\infty \to SO$. The image of $[L] - 1$ in ω_π^0 is the unit with fixed point degrees $d_1 = 1$, $d_{\mathbb{Z}/2} = -1$. Under the identification of ω_π^0 with the Grothendieck group of finite π-sets, this unit corresponds to the virtual $\mathbb{Z}/2$-set $[\mathbb{Z}/2] - 1$. Hence the composite

$$a : RP^\infty \xrightarrow{f} SO \xrightarrow{J} SG \xrightarrow{\text{incl.}} \mathbb{Q}S^0$$

is related to the Dyer-Lashof map $b : RP^\infty = B\Sigma_2 \to \mathbb{Q}S^0$ by the formula $a = b - 1$ in $\pi_S^0(RP^\infty)$. There is apparently no proof of this fact in the literature. It was proved (differently) by the author in 1973 and independently by R. Schultz.

We proceed to describe $A(\psi)$ on reduced line bundles. For this we apply thm. 5.4 to $\pi = \mathbb{Z}/2$, so we must calculate $a(\psi)([L] - 1)$ using (4.11). We see that the composite $RP^\infty \to BO \to G/O^\wedge$ lifts to $G^\wedge = G$, and that this lifting arises from the element in $\hat{\omega}_{\mathbb{Z}/2}^0$ with $d_1 = 1$, $d_{\mathbb{Z}/2} = e^{-1}$. Hence we have a commutative diagram

$$\begin{array}{ccc} RP^\infty & \longrightarrow & BO \\ \downarrow{\scriptstyle h} & & \downarrow{\scriptstyle A(\psi)} \\ G & \longrightarrow & G/O^\wedge \end{array}$$

where h is given as an element of $\pi_S^0(RP^\infty)$ by

$$h = 1 + \frac{e^{-1} - 1}{2}(1 - a).$$

We define the map for a finite connected complex X

$$D = D_e : \widetilde{KO}^o(X) \times \widetilde{KO}^o(X) \to [X, SG]$$

for $e \in \hat{\mathbb{Z}}$, $e \equiv 1 \pmod 4$ by the following formula in $\pi_S^o(X)$

(5.8) $$D(x_1,x_2) = 1 + \frac{e-1}{4}(\Delta(x_1) - 1)(\Delta(x_2) - 1).$$

This is well defined, since $\tilde{\pi}_S^o(X)$ is finite.

THEOREM 5.9. For a finite complex X the map

$$A = A(\psi) : \widetilde{KO}^o(X) \to [X, G/O\hat{\ }]$$

satisfies the sum formula

(5.10) $$A(x_1 + x_2) = D(x_1,x_2)A(x_1)A(x_2).$$

PROOF. The formula (5.10) means that a certain pair of maps $BO \times BO \to G/O\hat{\ }$ are homotopic. By an application of Brauer lifting (Quillen [14]) if suffices to prove this for the composites with maps $B\pi \times B\pi \to BO \times BO$ arising from $\alpha \in IO(\pi)$. We may further restrict to finite subcomplexes of $B\pi$. Now (5.10) follows from (4.9) by associated bundle constructions using prop. 5.6 to relate the deviation maps (4.10) and (5.8).

It is standard procedure to pass from thm. 5.9 to the corresponding 2-local statement, i.e. theorem E of the introduction. Observe that, since $\psi^k = \psi^{-k}$ in real K-theory, it suffices to deal with the case $k \equiv 1$ (4).

From 5.5 follows that Δ and D are trivial at odd primes, so that $A(\psi) : BO \to G/\hat{O}_p$ is an H-map for p odd. We conclude with a proof, that $A(\psi)$ for $p = 2$ can be modified to an H-map, when $e \equiv 1 \pmod 8$, i.e. precisely the cases of least interest in geometric topology.

THEOREM 5.11. When $e \in \hat{\mathbb{Z}}_2^*$ satisfies $e \equiv 1 \pmod 8$ the exists an H-map $\hat{A} : BO \to G/\hat{O}_2$, such that $i \cdot A \simeq \psi^e - \psi^1$.

PROOF. Let $c \in \mathbb{Z}_2^*$ be a square root of e, and define A by multiplying $A(\psi)$ with the map $R : BO \to SG$ given by the expression

$$R = 1 + \frac{c-1}{2}(1 - \Delta).$$

To check that A is additive we need the formula

$$R(x + y)D(x,y) = R(x)R(y).$$

This may be verified by interpreting R as a map $IO(\pi) \to \hat{\omega}_\pi^o$ and calculating degrees at fixed point sets.

REFERENCES

1. J.F. Adams, On the groups J(X) - II, Topology 3(1965), 137-171.

2. M.F. Atiyah, Bott periodicity and the index of elliptic operators, Quart. J. Math. 19(1968), 113-140.

3. M.F. Atiyah, R. Bott and A. Shapiro, Clifford modules, Topology 3, Suppl. 1(1964), 3-38.

4. M.F. Atiyah and D.O. Tall, Group representations, λ-rings and the J-homomorphism, Topology 8(1969), 253-297.

5. J. Becker and D. Gottlieb, The transfer map and fiber bundles, Topology 14(1975), 1-13.

6. G. Brumfiel and I. Madsen, Evaluation of the transfer and the universal surgery classes, Invent. Math. 32(1976), 133-169.

7. T. tom Dieck and T. Petrie, Geometric modules over the Burnside ring, (Preprint).

8. A. Dold, Über fasernweise Homotopie äquivalenz von Faserräumen, Math. Z. 62(1955), 111-136.

9. A. Dress, Induction and structure theorems for orthogonal representations of finite groups, Annals of Math. 102(1975), 291-325.

10. E.M. Friedlander and R. Seymour Two proofs of the stable Adams conjecture, Bull. Amer. Math. Soc. 83(1977), 1300-1302.

11. D. Husemoller, Fibre bundles, McGraw-Hill, New York (1966).

12. I. Madsen, On the action of the Dyer-Lashof algebra in $H_*(G)$, Pacific J. Math. 69(1975), 235-275.

13. J. Milnor and D. Husemoller, Symmetric bilinear forms, Ergebnisse d. Math. u.i. Gretzgebiete 73(1973), Springer-Verlag.

14. D. Quillen, The Adams conjecture, Topology 10(1971), 67-80.

15. R.L. Rubinsztein, On the equivariant homotopy of spheres, Dissert. Mat. (Roszprawy Mat.) 134(1976).

16. G. Segal, Equivariant stable homotopy theory, Proceedings I.C.M., Nice(1970).

17. J.-P. Serre, Linear representation of finite groups, Graduate Texts in Math. 42(1977), Springer-Verlag.

18. D. Sullivan, Genetics of homotopy theory and the Adams conjecture, Annals of Math. 100(1974), 1-79.

DEPARTMENT OF MATHEMATICS
AARHUS UNIVERSITY
AARHUS, DENMARK

Canadian Mathematical Society
Conference Proceedings
Volume 2, Part 2 (1982)

Some Regularity Theorems for Smooth Actions on Complex Stiefel Manifolds

McKenzie Y. Wang

In [5], W. C. Hsiang and W. Y. Hsiang classified the principal isotropy types of arbitrary smooth actions of compact connected classical Lie groups on acyclic cohomology manifolds. In unpublished work, W. Y. Hsiang obtained analogous results for smooth actions on homotopy spheres and projective spaces. Briefly stated, the above results assert that for arbitrary smooth actions on acyclic manifolds, homotopy spheres, and projective spaces, one may define the notion of a geometric weight system, which generalizes weights of a linear representation. Moreover, the principal isotropy type and the geometrical weight system of such an arbitrary smooth action are the same as that of a linear model of the same group on the same manifold. In other words, every smooth action has a linear model, with which it shares such characteristics as orbit structure and fixed point variety structure.

We announce here (with brief indications of proof) similar regularity results for smooth actions of $SU(m)$, $Sp(m)$, $SO(m)$, and $Spin(m)$ on $W_{n,2}$, the complex Stiefel manifold of orthonormal 2-frames in complex n-space, under the following assumptions:

(i) the fixed point set of some maximal torus T of the Lie group G is non-empty,

and (ii) m and n satisfy certain mild dimension restrictions.

We are interested in the manifolds $W_{n,2}$ with n odd because while they have the same integral cohomology type as $S^{2n-3} \times S^{2n-1}$, they are distinguishable from the product of odd spheres by cohomology operations (Sq^2). Our results show that this difference in homotopy type is reflected by

© 1982 American Mathematical Society
0731-1036/82/0000-0472/$03.25

differences in the behaviour of transformation groups.

Furthermore, we chose to study $W_{n,2}$ because they form one of the simplest yet important families of compact homogeneous spaces. Aside from acyclic manifolds, homotopy spheres, and projective spaces, the compact homogeneous spaces are probably the nicest test-spaces for studying transformation groups as they are rich in both topological types and group actions.

1. THE MAIN THEOREM.

In stating the main theorem, it is convenient to consider "large" groups separately from "small" groups.

THEOREM 1 ("Large" groups). Let $G = SU(m)$, $Sp(m)$, $SO(m)$, or $Spin(m)$ act smoothly on $W_{n,2}$, n odd, with $F(T,W_{n,2}) \neq \emptyset$ and assume that the connected component of the principal isotropy type (H^0) is non-trivial. Let T_2 be a maximal 2-torus of $SO(m)$. In the cases $G = SO(m)$ or $Spin(m)$ we assume in addition that $F(T_2,W_{n,2}) \neq \emptyset$.(1)

Let $\Omega'_T(X)$ be the reduced geometrical weight system (see section IV-3 in [6]) of the action restricted to T. Then we have the following table:

G	(H^0)	$\Omega'_T(X)$
$SU(m), m \geq 5$	$(SU(r_0))$ with $m-r_0$ even if $r_0 \geq 6$	$(m-r_0)\{\pm\theta_i\}$
$Sp(m), m \geq 5$	$(Sp(r_0))$ with $m-r_0$ even if $r_0 \geq 4$	$(m-r_0)\{\pm\theta_i\}$
$SO(m), m \geq 7$	$(Spin(r_0))$ with $m-r_0$ even if $r_0 \geq 11$	$(m-r_0)\{\pm\theta_i\}$
$Spin(m), m \geq 14$	$(Spin(r_0))$ with $m-r_0$ even if $r_0 \geq 11$	$(m-r_0)\{\pm\theta_i\}$

(1) Under the hypotheses and dimension assumptions on m it follows that the $Spin(m)$ actions all factor through $SO(m)$.

THEOREM 1' ("Small" groups). Let G be one of the following groups acting smoothly on $W_{n,2}$, n odd, with $F(T,W_{n,2}) \neq \emptyset$ and assume that the connected component of the principal isotropy type (H^0) is non-trivial. Then the possibilities for (H^0) and $\Omega'_T(X)$ are given as follows:

G	(H^0)	$\Omega'_T(X)$	Remarks
SU(3)	T	$\Delta(SU(3))$	
SU(4) = Spin(6)	T	$\Delta(SU(4))$	(a)
	k=1 Sp(2) k=2 Sp(1)×Sp(1) k=3 Sp(1) k=4 S^1	$k\{\pm(\theta_1+\theta_2),\pm(\theta_1+\theta_3),$ $\pm(\theta_1+\theta_4)\}$ k=1,2,3,4	The k=4 case is realized by the linear model $\Lambda^2\mu_4$ + trivial
	SU(2)	$\{\pm\theta_i\}\cup\{\pm(\theta_1+\theta_2),$ $\pm(\theta_1+\theta_3),\pm(\theta_1+\theta_4\}$	
Sp(2) = Spin(5)	T	$\Delta(Sp(2))$	(b)
	Sp(1)	$2\{\pm(\theta_i+\theta_j)\}$	realized by $\Lambda^2[c\nu_2]$ + trivial
Sp(3)	T	$\Delta(Sp(3))$	
	S^1	$2\{\pm(\theta_i+\theta_j)\}$	realized by $\Lambda^2[c\nu_3]$ + trivial
Sp(4)	T	$\Delta(Sp(4))$	
Spin(7)	SU(3)	$2\{\frac{1}{2}(\pm\theta_1\pm\theta_2\pm\theta_3)\}$	
	SU(2)	$2\{\frac{1}{2}(\pm\theta_1\pm\theta_2\pm\theta_3)\}\cup\{\pm\theta_i\}$	

Spin(8)	SU(3)	$\Delta_8^+ + \Delta_8^- + \rho_8$	
	SU(2)	$\Delta_8^+ + 3\rho_8$ $2(\Delta_8^+ + \rho_8)$ $\Delta_8^+ + \Delta_8^- + 2\rho_8$ or the conjugate of any of these under an outer automorphism of Spin(8)	
Spin(9)	SU(3)	$2\{\frac{1}{2}(\pm\theta_1\pm\ldots\pm\theta_4)\}$	
	SU(2)	$2\{\frac{1}{2}(\pm\theta_1\pm\ldots\pm\theta_4),\pm\theta_i\}$	
Spin(12)	SU(2)×SU(2)×SU(2)	$2\{\frac{1}{2}(\varepsilon_1\theta_1+\ldots+\varepsilon_6\theta_6)\}$ $\varepsilon_i = \pm 1$, where either an even or an odd number of the ε_i's are -1	
SU(3), SU(4) = Spin(6)	SU(r_0)	$(m-r_0)\{\pm\theta_i\}$	$m-r_0$ even is realized by actions of regular type
Sp(2) = Spin(5), Sp(3), Sp(4)	Sp(r_0)	$(m-r_0)\{\pm\theta_i\}$	$m-r_0$ even is realized by actions of regular type
SO(m), m = 5,6	SO(r_0)	$(m-r_0)\{\pm\theta_i\}$	$m-r_0$ divisible by 4 is realized by actions of regular type
Spin(m), $7 \le m \le 14$	Spin(r_0)	$(m-r_0)\{\pm\theta_i\}$	$m-r_0$ divisible by 4 is realized by actions of regular type

2. IMPLICATIONS OF THE MAIN THEOREM.

For smooth actions of the "large" groups the connected principal isotropy type is either trivial or of regular type (see [4] for the terminology). In the latter case, the weight system is also of regular type. Indeed we have the following more precise theorem concerning orbit and fixed point set structure.

THEOREM 2. Let $G(m) = Sp(m)$, $SU(m)$, $SO(m)$, or $Spin(m)$ act smoothly on $W_{n,2}$, n odd, such that $(H^0) = (G(r_0))$ with $r_0 \geq 4,6,11,11$ respectively. In the first two cases the actions have the same orbit and fixed point variety structures as the linear models (<u>of regular type</u>) $\frac{1}{2}(m-r_0)\ c[\nu_m] + (\text{trivial})$ and $\frac{1}{2}(m-r_0)\mu_m + (\text{trivial})$ respectively. In the other cases, many aspects of the action resemble the regular linear models, but there is a Sq^1 obstruction to complete resemblance.

The proof of theorem 2 is quite long and can be found in [9] or [10]. (In the above, ν_m and μ_m denote respectively the usual actions of $Sp(m)$ and $SU(m)$ on $\mathbb{H}^m$ and $\mathbb{C}^m$. c denotes complexification.)

The results in theorem 1 also show that transformation group behaviour is different for $S^{2n-3} \times S^{2n-1}$ and $W_{n,2}$, n odd. In particular, we have

THEOREM 3. There are no smooth actions of $SU(m)$, $Sp(m)$, $SO(m)$, or $Spin(m)$, $m \geq 5$, on $W_{n,2}$, n odd, with connected principal isotropy type (T), where T is a maximal torus of the group. For $G = SO(m)$ or $Spin(m)$, we need the additional hypothesis that $F(T_2, W_{n,2}) \neq \emptyset$, T_2 being a maximal 2-torus of $SO(m)$.(2)

The proof of theorem 3 can be found in [9] or [11]. We point out that the groups in theorem 3 can act with principal isotropy type (T) on $S^{2n-3} \times S^{2n-1}$ for example by the adjoint representation on S^{2n-1} and by the trivial action on S^{2n-3}. Also, (a) and (b) in the second table of

(2) see footnote (1).

section 1 cannot occur if we assume further that $F(T_2, W_{n,2}) \neq \emptyset$, in view of the isomorphisms $SU(4) = Spin(6)$ and $Sp(2) = Spin\ 5$.

Our results also imply that $SU(m)$, $m \geq 5$, cannot act smoothly on $W_{n,2}$, n odd, with $(H^0) = (SU(2)\times\ldots\times SU(2))$ ($[\frac{m}{2}]$ times) and $Sp(m)$ cannot act smoothly on $W_{n,2}$, n odd, with $(H^0) = ([Sp(1)]^m)$. Again, such principal isotropy types are possible on $S^{2n-3} \times S^{2n-1}$.

3. ROUGH SKETCH OF THE PROOF OF THE MAIN THEOREM.

In proving theorem 1, we first observe that since $F(T, W_{n,2}) \neq \emptyset$ by assumption, the rational geometrical weight system as defined in section IV-3 of [6] exists, and properties of the weight system described in chapter V of [6] carry over when the space acted upon is a rational cohomology product of odd spheres. (See theorem IV.6 in [6] or [2].) By spectral sequence arguments, it is easy to show that $F(T,X)$ is a rational cohomology product of two odd spheres if X is. Using arguments in [5,6], we obtain a list of possibilities for the connected principal isotropy type and the geometrical weight system. This list is the same as that for acyclic cohomology manifolds.

We next consider each possibility in the list. For the purpose of illustration, we assume $G = SU(m)$. Theorems 2 and 3 help us to rule out the cases $(H^0) = (T)$ and $(H^0) = (SU(m-k))$ where k is <u>odd</u>. The case $(H^0) = (SU(2)\times\ldots\times SU(2))$ ($[\frac{m}{2}]$ times) is ruled out by the proof of theorem 3 (which uses $\mathbb{Z}/2$-coefficients so that $\theta_i - \theta_j = \theta_i + \theta_j$). There are also the following possibilities, which are ruled out by arguments similar to those in the proofs of theorems 2 and 3, or by other special arguments:

$$r\{\pm(\theta_1+\theta_2+\theta_3)\} \cup s\{\pm\theta_i\}, \quad m = 6$$

$$(r,s) = (1,0) \qquad (H^0) = (SU(3) \times SU(3))$$

$$(r,s) = (2,0) \qquad (H^0) = (S^1 \times S^1)$$

$$(r,s) = (1,1) \qquad (H^0) = (Sp(1) \times Sp(1))$$

The cases which cannot be ruled out by any of the above arguments appear in tables 1 and 2.

A typical argument for ruling out a particular possibility consists of studying the homomorphism

$$j^*: H_T^*(X;\mathbb{Z}/2) \to H_T^*(F;\mathbb{Z}/2) \quad ,$$

where $j: F \subset X$ is the inclusion map, $X = W_{n,2}$, and $F = F(T,X)$ or $F(T_2,X)$. As far as possible, the structures of $H_T^*(X;\mathbb{Z}/2)$ and $H_T^*(F;\mathbb{Z}/2)$ over the mod 2 Steenrod algebra are determined. For this we use localization theorems of the Atiyah-Segal type (see chapter III, [6]), spectral sequence arguments, theorem IV.1 in [6], the theorem of J. C. Su on $F(T_2,X)$ (see [1,p.410],[8]), and properties of Steenrod operations.

For example, when $G = SU(m)$, $m \geq 5$, and $(H^0) = (T)$, then it turns out that $H_T^*(X;\mathbb{Z}) \approx \Lambda_R(\tilde{x},\tilde{y})$, where $R = H^*(B_T;\mathbb{Z})$ and $\tilde{x},\tilde{y}$ are lifts of generators of $H^*(X;\mathbb{Z})$. In mod 2 coefficients, we may also assume $Sq^2\tilde{x} = \tilde{y}$. Moreover, $F(T,X) = F(T_p,X)$ for any prime p , and is an integral cohomology product of two odd spheres. The homomorphism j^* can be shown to be injective and $H_T^*(F;\mathbb{Z}/2) \approx \Lambda_R(f_1,f_2)$, where $H^*(F;\mathbb{Z}/2) = \Lambda_{\mathbb{Z}/2}(f_1,f_2)$ and R denotes also $H^*(B_T;\mathbb{Z}/2)$.

Let $j^*(x) = a\otimes f_1 + b\otimes f_2$. We apply the topological splitting principle of Chang and Skjelbred [3] to obtain $a^2 + aSq^2b + bSq^2a = \prod_{i<j}(\theta_i-\theta_j)$ if $Sq^2f_1 = f_2$ or $aSq^2b + bSq^2a = \prod_{i<j}(\theta_i-\theta_j)$ if $Sq^2f_1 = 0$. By studying how the Weyl group acts on $H_T^*(X;\mathbb{Z}/2)$ and $H_T^*(F;\mathbb{Z}/2)$ we conclude that a and b are polynomials in the θ_i invariant under the alternating subgroup of $W(G)$.

Let $(\theta_1-\theta_2) = T'$ be the corank 1 subtorus corresponding to the root $\theta_1 - \theta_2$. The action of T on $F(T',X)$ is then studied as above. Similar equations are obtained when we apply the topological splitting principle to the homomorphism $j_1^*: H_T^*(F(T',X);\mathbb{Z}/2) \to H_T^*(F;\mathbb{Z}/2)$. (Put differentily, this is just localization.) These equations are easy to solve, and using the action of $W(G)$, the solutions of these equations can be pieced together to determine a and b . However, one then checks that these a's and b's do not solve the original equation(s). Hence we obtain a contradiction, showing that the possibility (T) cannot be a reality.

The proofs of theorem 1 for SU(m), Sp(m), SO(m), and Spin(m) vary in complexity. The case of SU(m) is the most straight-forward, and the cases of SO(m) and Spin(m) are the most complicated. This is because a maximal 2-torus of SO(m) does not lie within any maximal torus, and the structure of $F(T_2,X)$ can a priori be any one of the various possibilities given in J. C. Su's theorem. However, in order to use Steenrod operations, we have to work with $\mathbb{Z}/2$ coefficients and so have to work without the help of a strong Smith-type fixed point theorem.

The complete classification of connected principal isotropy types assuming $F(T,X) \neq \emptyset$ and occasionally $F(T_2,X) \neq \emptyset$ is given in [9]. In forth-coming papers of N. Ercolani and the author, the classification of connected principal isotropy types <u>without</u> the assumption $F(T,X) \neq \emptyset$ will be undertaken.

BIBLIOGRAPHY

1. G. Bredon, Introduction to Compact Transformation Groups, Academic Press, N. Y. (1972).

2. G. Bredon, Homotopical Properties of Fixed Point Sets of Circle Group Actions, I, Amer. Jour. Math. 91 (1969), 874-888.

3. T. Chang & T. Skjelbred, The Topological Schur Lemma and Related Results, Ann. Math. 100 (1974), 307-321.

4. M. Davis, W. C. Hsiang & W. Y. Hsiang, Differentiable Actions of Compact Simple Lie Groups on Homotopy Spheres and Euclidean Spaces, Proc. Symp. Pure & Applied Math. Vol. 32 (1977), 99-109.

5. W. C. Hsiang & W. Y. Hsiang, Differentiable Actions of Compact Connected Classical Groups I, Amer. Jour. Math. 89 (1967), 705-786; II, Ann. Math. 92 (1970), 189-223; III. Ann. of Math. 99 (1974), 220-256.

6. W. Y. Hsiang, Cohomology Theory of Topological Transformation Groups, Ergebnisse der Mathematik und ihrer Grenzgebiete, Band 85, Springer-Verlag, N. Y. (1975).

7. W. Y. Hsiang, On Characteristic Classes of Compact Homogeneous Spaces and Their Application in Compact Transformation Groups I (to appear).

8. J. C. Su, Periodic Transformation on the Product of Two Spheres, Transac. Amer. Math. Soc. 112 (1964), 369-380.

9. M. Wang, Dissertation, Stanford University, 1980.

10. M. Wang, On Actions of Regular Type on Complex Stiefel Manifolds (to appear).

11. M. Wang, On Actions of Adjoint Type on Complex Stiefel Manifolds, (to appear).

DEPARTMENT OF MATHEMATICS
UNIVERSITY OF PENNSYLVANIA
PHILADELPHIA, PENNSYLVANIA 19104

Canadian Mathematical Society
Conference Proceedings
Volume 2, Part 2 (1982)

ON GEORGE COOKE'S THEORY OF HOMOTOPY AND TOPOLOGICAL ACTIONS

A. Zabrodsky

We are following here the work of George Cooke with the main reference being his 1978 paper ([Cooke]). In that paper an obstruction theory is established for the following problem:

Given a homotopy action of a group G on a space X, i.e., a morphism $\eta: G \to [X,X]$ = the pointed homotopy classes of maps $X \to X$. Is G conjugate to a group of homeomorphisms? i.e., Is there a homotopy equivalence $h: X \to \hat{X}$ so that $c_h \circ \eta: G \to [\hat{X},\hat{X}]$ is represented by a topological action $G \to \mathrm{Homeo}(\hat{X}) \subset [\hat{X},\hat{X}]$ where $(c_h \circ \eta)(g) = [h]\, \eta(g) [h]^{-1}$?

The main theorem of [Cooke] is the following

0.1 THEOREM: $\eta: G \to \mathrm{Aut}_\circ X \subset [X,X]$ is conjugate to a group of homeomorphisms if and only if the following lifting problem has a solution:

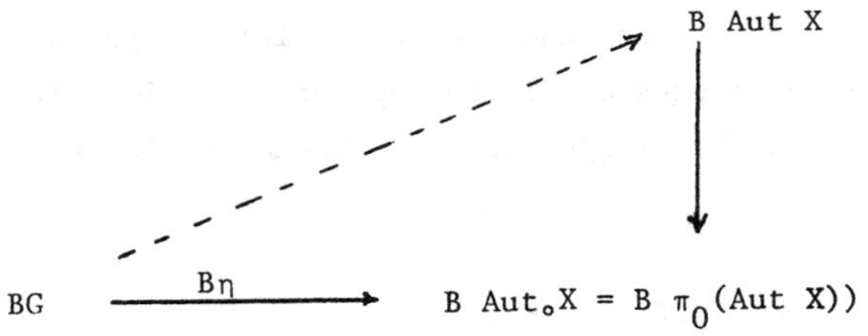

B Aut X - the Stasheff's classifying space for the monoid $\mathrm{Aut}\, X \subset X^X$.

This theorem establishes an obstruction theory with obstructions in

$$H^n(G, \pi_{n-2}(X^X, 1)) \quad n \geq 3 .$$

One of the consequences of the above is that any homotopy action of a free group is conjugate to a group of homeomorphisms.

Hence:

0.2 COROLLARY TO COOKE'S THEOREM: Given a CW complex X. There exists a CW complex $\hat{X}$, homotopy equivalent to X, so that all the self homotopy equivalences of $\hat{X}$ are homotopic to homeomorphisms.

PROOF: Let G be a free group with a surjection $G \to \mathrm{Aut}_\circ X$. By the above one has

© 1982 American Mathematical Society
0731-1036/82/0000-0473/$02.25

$$\begin{array}{ccc} G & \longrightarrow & \text{Homeo}\,\hat{X} \\ \eta\downarrow & \overset{c_h}{\approx} & \downarrow u \\ \text{Aut}_\circ X & \longrightarrow & \text{Aut}_\circ \hat{X} \end{array}$$

As the left vertical is a surjection so is the diagonal map $c_h\eta$. Consequently u is a surjection.

In the same paper ([Cooke]) an example is given to show that in general relations in $\text{Aut}_\circ X \subset [X,X]$ cannot be realized as relations among homeomorphisms even if X is replaced by its homotopy equivalent:

0.3 EXAMPLE (SECTION 3 OF [COOKE]): There exists a space X and a self homotopy equivalence $T:X \to X$ so that $T^2 \sim 1$ but there is no homotopy equivalence $h:X \to \hat{X}$ so that $[h][T][h]^{-1}$ is homotopic to a homoemorphism of order 2.

It follows that the difficulty is not in representing self homotopy equivalences as homeomorphisms but to realize relations in $\text{Aut}_\circ X$ as relations in the function space of some homotopy equivalent $\hat{X}$.

In this note we shall describe how relations in $[X,X]$ induce relations in $\text{Homeo}(\hat{X})$, $\hat{X} \approx X$.

We shall bring here only the major statements and a very general outline of proofs. The detailed proofs will appear elsewhere.

We always assume that spaces and maps are pointed, spaces are of the homotopy type of CW complexes of finite type or their localizations. Hence, a group action of G on X always will have the basepoint as a G-fixed point.

1. HOMOTOPY ACTIONS OF FINITE GROUPS.

1.1. PROPOSITION: Let X be a CW complex with a homotopy action $G \to [X,X]$, G - a finite group of order, say, t. Then:

(a). If $\pi_n(X) = 0$ for $n > N$ there exists a finite group $\hat{G}$ with a surjection $\hat{G} \to G$, so that:

(a1). $|\hat{G}|$ divides a power of t.

(a2). $\hat{G} \to G \to \text{Aut}\,X$ is conjugate to a group of homeomorphisms.

(b). If X is arbitrary then there exists a profinite group $\hat{G}$ with a surjection $\hat{G} \to G$ so that:

(b1). $\hat{G}$ is an inverse limit of groups of order dividing a power of t.

(b2). $\hat{G} \to G \to \text{Aut}\,X$ is conjugate to a group of homeomorphisms.

1.1. follows from the following:

1.2. LEMMA: Let G be a finite group of order, say, t. Let M be a $Z(G)$ module, finitely generated as an abelian group. Given $u \in H^n(G,M)$, $n \geq 3$, there exists a group $\hat{G}$ and a surjection $\hat{G} \to G$ so that $|\hat{G}|$

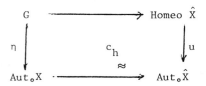

divides a power of t, and $u \in \ker\ H^n(G,M) \to H^n(\hat{G},M)$.

PROOF OF 1.1: The two parts are proved the same way using 0.1 and 1.2. One constructs inductively groups with surjections

$$G_n \xrightarrow{i_n} G_{n-1} \to \cdots \to G_0 = G, \quad |G_n| \text{ divides a power of } t, \text{ so that partial}$$

solutions of the lifting problem are obtained:

$$\begin{array}{ccccccccc}
 & & & & B\ \mathrm{Aut}\ X & & \\
 & & & & \downarrow & & \\
 & & & \nearrow & B_n & \xrightarrow{u} & K(\pi_{n+1}(B(\mathrm{Aut}\ X)_1),n+2) \\
 & & f_n & & \downarrow & & \\
 & & & & B_{n-1} & & \\
 & & & & \downarrow & & \\
BG_{n+1} & \xrightarrow{Bi_{n+1}} & BG_n \to \cdots \to BG_1 \to BG_0 & \to & B_0 = B\ \mathrm{Aut}_\circ X & &
\end{array}$$

$G_{n+1} \to G_n \to 1$ is a group (exists by 1.2) satisfying

$$[u \circ f_n] \in \ker\ (H^{n+2}(G_n,\ \pi_n(X^X,1)) \to H^{n+2}(G_{n+1},\ \pi_n(X^X,1))$$

Thus $f_n \circ Bi_{n+1}$ lifts to $f_{n+1}: BG_{n+1} \to B_{n+1}$.

If $\pi_n(X) = 0$ for $n > N$ then $\pi_n(X^X) = 0$ for $n > N$, $B\ \mathrm{Aut}\ X \approx B_{N+1}$ and $\hat{G} = G_{N+2}$ is the desired group in (a). For (b) one has $\hat{G} = \varprojlim G_n$ as the desired group.

1.3. EXAMPLES: If in 0.3 one replaces X by a finite Postnikov approximation then $T: X \to X$, $T^2 \sim 1$, could be realized by a homeomorphism of order a power of 2 in some homotopy equivalent $\hat{X}$ of X.

2. RELATIVE COOKE OBSTRUCTIONS. We outline here a relative version of Cooke's obstruction theory.

Consider the following problem:

Let G act homotopically (abr. "hoact") on X and act topologically on a space Y (abr. "Y is a G space"). Let $f: X \to Y$ be homotopically equivariant G-map (abr. G-hoequivariant) i.e., For every $g \in G$ $f \circ T_g \sim S_g \circ f$ where $T: G \to X^X$ represents $G \to [X,X]$, $S: G \to \mathrm{Homeo}\ Y \subset Y^Y$ is the action. Is there a G-space $\hat{X}$, a hoequivariant homotopy equivalence $h: X \to \hat{X}$ and a G-map $\hat{f}: \hat{X} \to Y$ so that $\hat{f} \circ h \sim f$? (If these exist we shall say that f is equivalent to a G-map.) Obviously if f is equivalent to a G-map then $G \to [X,X]$ is a conjugate to a group of homeomorphisms.

Assume first that G acts topologically on X and Y. Let $f: X \to Y$ be a G-hoequivariant map. G acts on the function space Y^X and on $(Y^X)_f$ -

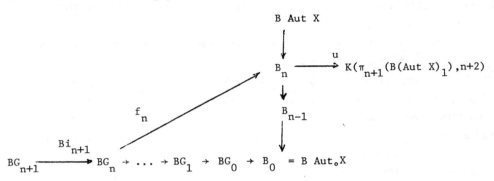

the path component of f. The latter, however may be without a fixed point. The action is given by: $g,h \to S_g \circ h \circ T_{g^{-1}}$, $g \in G$, $h \in (Y^X)_f$.

2.1. LEMMA: There exists a G- map $\hat{f}:\hat{X} \to Y$ and a G equivariant homotopy equivalence $h:\hat{X} \overset{\approx}{\to} X$ so that $f \circ h \sim \hat{f}$ if and only if there exists a G-map $E(G) \to (Y^X)_f$ where $E(G)$ is a contractible free G-space. (Both actions on $E(G)$ and $(Y^X)_f$, in this particular instance are (possibly) fixed point free.)

2.2. PROPOSITION: Given a G hoequivariant map $f:X \to Y$ where G hoacts on X and acts topologically on Y.

The obstructions for f to be equivalent to a G-map are elements of $H^2(G,A_n)$, $H^2(G,B_n)$ and $H^i(G,H^{n-i+2}(X,\pi_n(F(f)))$ $i \geq 3$, where A_n, B_n are quotients of G submodules of $H^n(X,\pi_n(F(f)))$, $F(f)$ - the homotopy fiber of f.

3. HOMOTOPY ACTIONS OF POLYCYCLIC GROUPS. A polycyclic group is a group G admitting a finite sequence of surjections.

(P) $$G = G_n \overset{\phi_n}{\to} G_{n-1} \to \dots \to G_0 = 1$$

where ker ϕ_i is a finitely generated abelian group.

The following cohomological property of polycyclic groups is proved in [Zabrodsky] (theorem 1.1).

3.1. PROPOSITION: Let G be a polycyclic group, M - a finite abelian group of order, say, t. Suppose G acts on M. Then there exists a subgroup $\tilde{G}$ of G of index dividing a power of t so that for every $i > 0$ $H^i(G,M) \to H^i(\tilde{G},M)$ is the zero homomorphism.

Using 2.2 and 3.1 one can prove:

3.2. THEOREM: Let X be a $\mathbb{P}_1$ local H_0 space. ($\mathbb{P}_1$ is a set of primes. The set of all primes and the empty set are not excluded.) Suppose either $\pi_n(X) = 0$ $n > N$ or $H_n(X,Z_{\mathbb{P}_1}) = 0$, $n > N$. Given a homotopy action of a polycyclic group G, $G \to [X,X]$ so that $H^*(X,Q) \to QH^*(X,Q)$ splits as a $Z(G)$ module. Then there exists a subgroup $\tilde{G}$ of G, $[G:\tilde{G}] = t < \infty$, t - a product of primes in $\mathbb{P}_1$, so that $\tilde{G}$ is conjugate to a group of homeomorphisms.

3.3. EXAMPLE: Let X be a $\mathbb{P}_1$ local finite dimensional H-space, $H^*(X,Q)$ - primitively generated. Given H-self equivalences $T_1,T_2,\dots,T_r$, $T_i:X \to X$, $\{T_i\}$ commute up to homotopy. Then for some product of primes in $\mathbb{P}_1$, say, t, T_i^t are conjugate to commuting homeomorphisms: There exists $h:X \to \hat{X}$, $[h'] = [h]^{-1}$, $h \circ T_i^t \circ h' \sim \hat{T}_i$, $\hat{T}_i$ - commuting homeomorphisms.

REFERENCES

1. G. Cooke, "Replacing homotopy actions by topological actions", Trans. AMS 237 (1978), pp. 391-406.

2. A. Zabrodsky, "Homotopy actions II. Polycyclic groups and the lifting theorem". (To appear.)

Manifolds and Structures on Manifolds

Canadian Mathematical Society
Conference Proceedings
Volume 2, Part 2 (1982)

INERTIAL PROPERTIES OF STABLY DIFFEOMORPHIC MANIFOLDS

Michael Frame[1]

ABSTRACT. Two n-manifolds M and N are called stably diffeomorphic if $M\#(S^p\times S^{n-p})\#\dots\#(S^r\times S^{n-r})$ is diffeomorphic to $N\#(S^p\times S^{n-p})\#\dots\#(S^r\times S^{n-r})$. For $n\geq 9$ it is shown that stably diffeomorphic manifolds have the same special inertia group I_h. In addition, $\pi_i(N) = 0$ for $i\leq \max(p,\dots,r)$, then M and N are diffeomorphic.

1. INTRODUCTION. An exotic n-sphere Σ is a smooth manifold homeomorphic to the standard n-sphere S^n. It is a well-known theorem of Milnor [9] that Σ need not be diffeomorphic to S^n. Under connected sum the exotic n-spheres form an abelian group denoted Θ_n. The inertia group of a smooth n-manifold M is the subgroup of exotic spheres preserving the differentiable structure of M under connected sum - that is, $I(M) = \{\Sigma \in \Theta_n : \Sigma \# M \cong M\}$. Many inertia groups have been computed ([1,7,13,14,15,16], for example); in general the results can be quite complex. In this paper a special inertia group will be considered. Using the isomorphism $\pi_0\mathrm{Diff}(S^{n-1}) \to \Theta_n$, $n \geq 5$, let $\Sigma(f) = D^n \cup_f D^n$ denote the exotic sphere corresponding to the diffeomorphism f. Then $M \# \Sigma(f)$ can be written $(M\setminus \mathrm{int}\, D^n)\cup_f D^n$, and for any M and any $\Sigma(f)$ there is a homeomorphism $\iota:(M\setminus\mathrm{int}\, D^n)\cup_f D^n \to (M\setminus\mathrm{int}\, D^n)\cup_{id} D^n$ defined by the identity on $M\setminus\mathrm{int}\, D^n$ and on D^n by the cone map on f. The special inertia group $I_h(M)$ is given by $I_h(M) = \{\Sigma \in \Theta_n:$ there is a diffeomorphism $g\colon \Sigma \# M \to M$ with $g \simeq \iota\}$. It was shown by Brumfiel (II.12 of [2]) that $I_h(M)$ is not a homotopy type invariant of M, while in [3] it was shown that $I_h(M)$ is an h-cobordism invariant of M. This paper is an extension of [3] since here it is shown that $I_h(M)$ is an invariant of the stable diffeomorphism type of M. Also, it is shown that highly-connected stably diffeomorphic manifolds are diffeomorphic.

2. INERTIA GROUPS. Let M and N be smooth n-manifolds. For any positive integer $p < n$ let $S(p) = S^p \times S^{n-p}$ and let $D(p) = D^{p+1} \times S^{n-p-1}$. Given

1980 Mathematics Subject Classification 57R50, 57R55, 57R65

[1]Supported by the grant number A7579 of the National Research Council of Canada and by the University of Wisconsin-Parkside.

© 1982 American Mathematical Society
0731-1036/82/0000-0474/$02.50

a set of integers $p_1,\ldots,p_m$, $1 \le p_i \le n - p_i$, M and N are called $\{p_1,\ldots,p_m\}$-stably diffeomorphic (or stably diffeomorphic when the index set is immaterial) if there is a diffeomorphism $M\#_i S(p_i) \cong N\#_i S(p_i)$. From the proof of the h-cobordism theorem (see the proof of the Proposition of [4]) it follows that if M and N are h-cobordant then they are $\{p_1,\ldots,p_m\}$-stably diffeomorphic for sufficiently large m , and moreover all the p_i can be taken equal to one another and can be any integer k , $2 \le k \le n - 2$. The existence of "non-inertial" h-cobordisms (Corollary 12.13 of [10], Theorem 6.1 of [5]) shows there are stably diffeomorphic manifolds which are not diffeomorphic. A step in comparing stably diffeomorphic manifolds can be obtained by a direct generalization of Theorem 2 of [3].

PROPOSITION. If M^n and N^n are $\{p_1,\ldots,p_m\}$-stably diffeomorphic and $n \ge 9$, and if $\max(p_i) \le [n/2]-2$, then $I_h(M) = I_h(N)$. Here $[x]$ denotes the integer part of x .

Proof. Let $g\colon M\#_i S(p_i) \to N\#_i S(p_i)$ be a diffeomorphism and suppose $\Sigma \in I_h(N)$ so there is a diffeomorphism $f\colon N\#\Sigma \to N$ homotopic to ι . The composition

$$h = g^{-1}(f\#1)(1\#g)\colon \Sigma\#M\#_i S(p_i) \to M\#S(p_i)$$

shows $I(N) \subset I(M\#_i S(p_i))$. (Since $f \simeq \iota$ has not been used yet, Σ can be taken to be any element of $I(N)$.) The homotopy between f and ι gives rise to a homotopy between $f \# 1$ and $\iota \# 1$, so h is homotopic to ι (see lemma 4 of [3]). Thus $h|+_i S^{p_i} \simeq$ inclusion and by the restrictions on the p_i these homotopies can be general positioned to imbeddings, hence pseudo-isotopies. By Theorem 2.1 of [6], $h|+_i S^{p_i}$ is isotopic to the inclusion and so by the isotopy extension theorem it may be assumed that $h|+_i S^{p_i}$ = inclusion. The surgery indicated in the diagram modifies each connected summand $S(p_i)$ to S^n and so the horizontal maps give a diffeomorphism $\Sigma \# M \to M$, hence $I_h(N) \subset I(M)$.

$$\begin{array}{ccc}
+_i D(p_i) & \xrightarrow{\text{identity}} & +_i D(p_i) \\
\cup & & \cup \\
+_i \partial D(p_i) & & +_i D(p_i) \\
\Big\downarrow \text{identity framing} & & \Big\downarrow h|\partial\nu(+_i S^{p_i}) \\
(\Sigma\#M\#_i S(p_i))\setminus \operatorname{int} \nu(+_i S^{p_i}) & \xrightarrow{h|\ldots} & (M\#_i S(p_i))\setminus \operatorname{int} \nu(+_i S^{p_i})
\end{array}$$

To show $I_h(N) \subset I_h(M)$, observe that the homotopy between $f \# 1$ and ι can be taken to be the identity on $+_i S^{p_i} \times I$ and so after modifying h by an isotopy, h and ι are connected by a homotopy G satisfying

$G|((+_i \nu S^{p_i})\times I)$ = inclusion (this requires making certain maps $S^{p_i} \times I^2 \to (M\#_i S(p_i))\times I$ into imbeddings - see lemma 4 of [3]). Replacing each space in the diagram by its product with I and replacing h by G, the resulting surgery gives a homotopy G' between ι and a diffeomorphism. The left side of the diagram is $(\Sigma \# M \# mS^n)\times I$ and the right side is $(M \# mS^n)\times I$, so $\Sigma \in I_h(M)$. This completes the proof.

COROLLARY. If M^n and N^n are h-cobordant and $n \geq 9$, then $I_h(M) = I_h(N)$.

Proof. As already observed, if M and N are h-cobordant then for some integer m there is a diffeomorphism $g: M \# m(S^2 \times S^{n-2}) \to N \# m(S^2 \times S^{n-2})$. The result follows from the theorem.

The reason for the restriction $\max(p_i) \leq [n/2] - 2$ in the proposition is that certain homotopies $S^p \times I \to M\#_i S(p_i)$ and $S^p \times I^2 \to (M\#_i S(p_i))\times I$ need to be pseudo-isotopies and the dimensional restrictions give this by transversality. By using a form of the Norman trick [11] these restrictions can be removed. (From the proof of the proposition it follows that making all the homotopies $S^p \times I$ into imbeddings gives $I_h(M) \subset I(N)$; making the homotopies $S^p \times I^2$ into imbeddings is needed to show $I_h(M) \subset I_h(N)$.)

THEOREM 1. If M^n and N^n are stably diffeomorphic and $n \geq 9$, then $I_h(M) = I_h(N)$.

Proof. For $n = 2k$ or $n = 2k + 1$ the cases not covered by the proposition are $\max(p_i) = k-1$ or k. Let M_0 denote $M\#_i S(p_i)$. To destabilize the diffeomorphism h and the homotopy G, the proof of the proposition works after certain homotopies $S^p \times I \to M_0$ and $S^p \times I^2 \to M_0 \times I$ have been modified to imbeddings. The case $n = 2k + 1$ and $\max(p_i) = k - 1$ will be considered first.

The arguments of the proposition can be used to remove all the $S(p_i)$, $p_i < k - 1$. By transversality, $S^{k-1} \times I \to M_0$ is already an imbedding and $S^{k-1} \times I^2 \to M_0 \times I$ is an immersion with self-intersection only at q isolated interior points $\{z_i\}$. Select $D^{2k+2} \subset M_0 \times I$ away from $(\#\Sigma) \times I$ and $(\#_i S(p_i))\times I$ and modify G (as in the proof of the proposition) so $G|D^{2k+2}$ = inclusion. Within D^{2k+2} further stabilize G by the identity on an additional $\#_q(S^{k+1} \times S^{k+1})$. Using disjoint imbedded cylinders, form the connected sum $(S^{k-1} \times I^2)\#_i S_i^{k+1} \equiv A$, where $S_i^{k+1} = S^{k+1} \times\{\text{point}\} \subset (S^{k+1} \times S^{k+1})_i$. Observe that each S_i^{k+1} intersects transversely in one point a_i a sphere $\bar{S}_i^{k+1} = \{\text{point}\} \times S^{k+1} \subset (S^{k+1} \times S^{k+1})_i$. Let γ_i be pairwise disjoint arcs in A from z_i to a_i. Then $\partial\nu(A)|\gamma_i$ intersects $S^{k-1} \times I^2$ in ∂D_i^{k+1}, $z_i \in D_i^{k+1}$, and intersects $\bar{S}_i^{k+1}$ in $\partial\bar{D}_i^{k+1}$, $a_i \in \bar{D}_i^{k+1}$. Modify A to $A' = (A \setminus +_i \operatorname{int} D_i^{k+1}) \cup (+_i \partial\nu(A)|\gamma_i) \cup (+_i \bar{S}_i^{k+1} \setminus \operatorname{int} \bar{D}_i^{k+1})$. Then $A' \cong (S^{k-1}\times I^2)\#_q S^{k+1}\#_q S^{k+1} \cong S^{k-1}\times I^2$ is imbedded as desired. Notice that G

is the identity along the added copies of $S^{k+1} \times S^{k+1}$ and so these can be removed immediately by surgery on $S^{k+1} \times \{point\}$. Of course, this argument is just Norman's trick in higher dimensions. For the other three cases, an extension of Norman's trick is necessary.

Consider $n = 2k$ and $\max(p_i) = k - 1$. By transversality $S^{k-1} \times I$ immersed in M_0^{2k} intersects itself only at isolated points, and these can be removed by the standard Norman trick adding copies of $S^k \times S^k \times I$ trivially to $M_0 \times I$. For $S^{k-1} \times I^2$ immersed in $M_0 \times I$ with $S^{k-1} \times I \times \partial I$ imbedded, the self-intersections are interior circles and line segments. The circles will be removed first. Let q denote the number of these circles S_i^1 and further stabilize G by the identity over $\#_q(S^{k-1} \times S^k)$. Select an $S_i^{k+1} = S^{k+1} \times \{point\} \subset (S^{k+1} \times S^k)_i$ in each of these connected summands and observe that each $(S^{k+1} \times S^k)_i$ contains an $\bar{S}_i^{k+1}$ intersecting S_i^{k+1} transversely in $S^1 \times \{point\} \equiv \bar{S}^1$.

Using small imbedded cylinders $S^k \times I$, form $(S^{k-1} \times I^2)\#_i S_i^{k+1} \equiv A$. Since $k \geq 5$ each S_i^1 is homotopic, hence pseudo-isotopic, to $\bar{S}_i^1$. Let $\gamma_i \cong S^1 \times D^1$ denote disjoint imbedded cylinders with $\partial\gamma_i = S_i^1 + \bar{S}_i^1$. The normal bundle of A restricted to γ_i is a (trivial) D^k-bundle ξ_i intersecting $S^{k-1} \times I^2$ in $S_i^1 \times D^k$ and intersecting $\bar{S}_i^{k+1}$ in $\bar{S}_i^1 \times D^k$. Modify A by removing $+_i S_i^1 \times \text{int } D^k$ and glueing in $+_i(\partial\xi_i \cup (\bar{S}_i^{k+1} \setminus \bar{S}_i^1 \times \text{int } D^k))$, obtaining $A\#_q(S^2 \times S^{k-1})$ with only the previous set of line segments for self-intersections. To remove the $\#_q(S^{k+1} \times S^k)$ from the homotopy, observe that each $(S^{k+1} \times S^k)$ contains an $S^{k+1} \times \{point\}$ missing $A\#_q(S^2 \times S^{k-1})$ and so since G is the identity on each $S^{k+1} \times S^k$ surgery on $S^{k+1} \times \{point\}$ replaces $\#(S^{k+1} \times S^k)$ with $\#S^{2k+1}$. To remove $\#_q(S^2 \times S^{k-1})$ from $A\#_q(S^2 \times S^{k-1})$, observe that each S^2 bounds a D^3 in $M_0 \times I$ with $\text{int } D^3$ missing $A\#_q(S^2 \times S^{k-1})$. Inside $M_0 \times I$ remove $+_i S_i^2 \times \text{int } D^{k-1}$ from $A\#_q(S^2 \times S^{k-1})$ and replace it by $+_i D_i^3 \times \partial D^{k-1}$ obtaining $A\#_q S^{k+1} \cong S^{k-1} \times I^2$ with only the previous set of line segments for self-intersections.

To remove line segments of self-intersection, let $p: S^{k-1} \times I^2 \to I^2$ denote the projection and let s be one of the line segments. Observe that it is always possible to find a disc D in I^2 containing $p(s)$ and intersecting ∂I^2 in a single line segment. Then $S^{k-1} \times cl(I^2 \setminus D)$ has the same self-intersection as before, except the segment s is missing. In this way all self-intersections can be removed from $S^{k-1} \times I^2$, as desired.

The remaining two cases are similar, though more complicated, and will not be included here.

3. DESTABILIZATION. For highly connected manifolds the method of the last section can be applied to show that stably diffeomorphic manifolds are

diffeomorphic. Information about the inertia groups of certain connected sums is obtained as corollaries.

THEOREM 2. If M^n and N^n are $\{p_1,\dots,p_m\}$-stably diffeomorphic, $n \ge 9$, $2\max(p_i) < n$, and if $\pi_j N = 0$ for $j \le \max(p_i)$, then M is diffeomorphic to N.

Proof. Let $g\colon M\#_i S(p_i) \to N\#_i S(p_i)$ be a diffeomorphism. By the argument of Theorem 1, in order to do surgery to remove a connected summand $S(p)$ from both sides it suffices to show $g|S^p \simeq$ inclusion. Suppose $p_1 = \min(p_i)$; also suppose $p_1 = \dots = p_k$ and $p_i > p_1$ for $i > k$. If $p_1 = 1$ then $\pi_1(N\#_i S(p_i)) \cong F_k = \mathbb{Z}*\dots*\mathbb{Z}$ (k copies) and $\mathrm{Aut}(F_k)$ is generated by the set of Whitehead automorphisms of F_k (see [8]). It is easy to see that each of these automorphisms can be realized by a self-diffeomorphism of $\#_i S(p_i)$ and so there is $h \in \mathrm{Diff}(N\#_i S(p_i))$ satisfying $hg|+_i S^{p_i} \simeq$ inclusion. If $p_1 = r > 1$, then $\pi_r(N\#_i S(p_i)) \cong \mathbb{Z}^k$ and since any element of $\mathrm{Aut}(\mathbb{Z}^k) = GL(k,\mathbb{Z})$ is induced by a diffeomorphism of $\#_i S(p_i)$, there is again $h \in \mathrm{Diff}(N\#_i S(p_i))$ with $hg|+_i S^{p_i} \simeq$ inclusion. Applying the argument of Theorem 1 gives a diffeomorphism $M\#S(p_{k+1})\#\dots\#S(p_m) \cong N\#S(p_{k+1})\#\dots\#S(p_m)$. Continuing in this way, all $\#S(p_i)$ are removed and so a diffeomorphism $M \cong N$ is obtained.

The contrapositive of Theorem 2 limits which manifolds can be stably diffeomorphic. For example, if $n \ge 5$ the only manifold which is $\{1,1,\dots,1\}$-stably diffeomorphic to $\mathbb{C}P(n)$ is $\mathbb{C}P(n)$ itself.

COROLLARY 1. If $\pi_i(M^n) = 0$ for $i \le \max(p_i)$, $n \ge 9$, and if $2\max(p_i) < n$, then $I(M\#_i S(p_i)) = I(M)$.

Proof. The Disc Theorem of Cerf and Palais implies $I(M) \subset I(M\#_i S(p_i))$. (See section 6 of [16].) For the reverse inclusion, if there is a diffeomorphism $\Sigma\#M\#_i S(p_i) \to M\#_i S(p_i)$ then Theorem 2 implies there is a diffeomorphism $\Sigma\#M \to M$.

COROLLARY 2. If $n \ge 9$ and $2\max(p_i) < n$, then $I(\#_i S(p_i)) = 0$.

Proof. If $\Sigma \in I(\#_i S(p_i))$ then Σ is $\{p_1,\dots,p_m\}$-stably diffeomorphic to S^n. Since S^n satisfies the connectivity hypothesis of Theorem 2, it follows that $\Sigma \cong S^n$.

It is well-known that $I(S^p \times S^q) = 0$ (see [13]). In fact, Schultz has shown (Theorem A of [12]) that any product of standard spheres has trivial inertia group. It does not follow immediately that $I(\#_i S(p_i)) = 0$ since although $I(M\#N) \supset I(M) + I(N)$, the reverse inclusion fails in general due to an example of Wilkens (section 6 of [16]).

BIBLIOGRAPHY

1. W. Browder, "On the action of $\Theta^n(\partial\pi)$", pages 23-36 in Differential and Combinatorial Topology, Princeton Mathematical series, number 27, Princeton

University Press, 1965.

2. G. Brumfiel, "Homotopy equivalences of almost smooth manifolds", Comment. Math. Helv., 46 (1971), 381-407.

3. M. Frame, "On the inertia groups of h-cobordant manifolds", preprint.

4. A. Hatcher, T. Lawson, "Stability theorems for 'concordance implies isotopy' and 'h-cobordism implies diffeomorphism'", Duke Math. J., 43 (1976), 555-560.

5. J-C Hausmann, "Open books and h-cobordisms", Comment. Math. Helv., 55 (1980), 330-346.

6. J. Hudson, "Concordance, isotopy, and diffeotopy", Ann. of Math. (2), 91 (1970), 425-448.

7. A. Kosinski, "Inertia groups of π-manifolds", Amer. J. Math., 89 (1967), 227-248.

8. J. McCool, "A presentation for the automorphism group of a free group of finite rank", J. London Math. Soc. 8 (1974), 259-266.

9. J. Milnor, "On manifolds homeomorphic to the 7-sphere", Ann. of Math. (2), 64 (1956), 399-405.

10. J. Milnor, "Whitehead torsion", Bull. Amer. Math. Soc. 72 (1966), 358-426.

11. R. Norman, "Dehn's lemma for certain 4-manifolds", Invent. Math., 7 (1969), 143-147.

12. R. Schultz, "On the inertia group of a product of spheres", Trans. Amer. Math. Soc., 156 (1971), 137-153.

13. R. Schultz, "Smooth structures on $S^p \times S^q$", Ann. of Math. (2), 90 (1969), 187-198.

14. R. Schultz, "Smoothings of sphere bundles over spheres in the stable range", Invent. Math., 9 (1969), 81-88.

15. I. Tamura, "Sur les sommes connexes de certaines variétés différentiable", Compt. Rendu Acad. Sci. Paris, 255 (1962), 3104-3106.

16. D. Wilkens, "On the inertia groups of certain manifolds", J. London Math. Soc. (2), 9 (1975), 537-548.

DEPARTMENT OF MATHEMATICS
UNIVERSITY OF WISCONSIN-PARKSIDE
KENOSHA, WISCONSIN 53141
U.S.A.

Canadian Mathematical Society
Conference Proceedings
Volume 2, Part 2 (1982)

FUNDAMENTAL GROUP PROBLEMS RELATED TO POINCARÉ DUALITY

Jean-Claude Hausmann

1. GEOMORPHISMS

(1.1) Let $\varphi : G \longrightarrow Q$ be a homomorphism between two finitely presented groups. Let $N = \ker\varphi$ and $I = \operatorname{Im}\varphi$. The extension

$$1 \longrightarrow N/[N,N] \longrightarrow G/[N,N] \xrightarrow{\varphi} I \longrightarrow 1, \qquad N/[N,N] = H_1(N)$$

is determined by a (twisted) cohomology class $\theta_\varphi \in H^2(I;H_1(N))$. One denotes by $|\theta_\varphi| \in \mathbf{N} \cup \{\infty\}$ the order of θ_φ in $H^2(I;H_1(N))$. If Λ denotes a subring of the rational numbers $\mathbf{Q}$, one denotes by θ_φ^Λ the image of θ_φ under the homomorphism $H^2(I;H_1(N)) \longrightarrow H^2(I;H_1(N) \otimes \Lambda) = H_2(I;H_1(N;\Lambda))$.

(1.2) Let d and n be integers. A homomorphism φ as in (1.1) is called a geomorphism of degree d in dimension n if there exists a continuous map $f : V^n \longrightarrow W^n$ of degree d between two closed oriented (smooth) manifolds V^n and W^n of dimension n, such that $\pi_1 f = \varphi$ (i.e., there are identifications $\pi_1(V) \cong G$ and $\pi_1(W) \cong Q$ under which $\pi_1 f = \varphi$.)

(1.3) Conjecture: For any $n \geq 4$, a homomorphism φ as in (1.1) is a geomorphism of degree d in dimension n if and only if the product $[Q:I]\cdot|\theta_\varphi|$ is finite and divides d(*).

The justification for this conjecture comes from results (1.4) to (1.7) below, in which the following definition is used: an oriented Λ-Poincaré complex of formal dimension n is a CW-complex X equipped with a fundamental class $[X] \in H_n(X;\Lambda)$ such that the homomorphism

$$\Delta_X = -\cap[X] \quad : \quad H^i(X;B) \longrightarrow H_{n-i}(X;B)$$

is an isomorphism for any $\Lambda\pi_1(X)$-module B. Observe that we do not ask that X has the homotopy type of a finite complex.

1980 AMS Subject Classification: 57N65,57P10, 20F99

(*) The problem of an algebraic characterisation of geomorphisms has been suggested to me by Sylvain Cappell.

© 1982 American Mathematical Society
0731-1036/82/0000-0475/$03.50

(1.4) Lemma Suppose that there exists a map $f : X \longrightarrow Y$ of degree d between two oriented $\mathbb{Z}$-Poincaré complexes of the same formal dimension, with $\pi_1 f = \varphi$. Then, the product $[Q:I]\cdot|\theta_\varphi|$ is finite and divides d.

Proof The proof that $[Q:I]$ divides d is classical (see [Ho, §6] for manifolds, [Br, proof of Proposition (1.2) for Poincaré complexes). We recall the idea, to fix some notation: the map f admits a lifting $\tilde{f} : X \longrightarrow \tilde{Y}_I$ where $\tilde{Y}_I$ is the covering of Y with $\pi_1(\tilde{Y}_I) = I$). Therefore $\tilde{Y}_I$ is a $\mathbb{Z}$-Poincaré complex and the covering projection $p : \tilde{Y}_I \longrightarrow Y$ has degree $[Q:I]$. Thus, one has $d = \tilde{d}\cdot[Q:I]$, where $\tilde{d}$ is the degree of $\tilde{f}$. It remains to prove that $\tilde{d}\cdot\theta_\varphi = 0$.

Let B be a $\mathbb{Z}I$-module and let $u \in H^*(\tilde{Y}_i;B)$. The formula

$$\Delta_{\tilde{Y}_I}^{-1} \circ \tilde{f}_* \circ \Delta_X \circ \tilde{f}^*(u) = \Delta_{\tilde{Y}_I}^{-1}(\tilde{f}_*(\tilde{f}^*(u) \cap [X])) = \Delta_{\tilde{Y}_I}^{-1}(u \cap \tilde{f}_*([X])) =$$

$$= \Delta_{\tilde{Y}_I}^{-1}(u \cap \tilde{d}[Y]) = \tilde{d}u$$

shows that $\tilde{d}\cdot\ker \tilde{f}^* = 0$. On the other hand, the homomorphism $H^2(\pi_1(T);A) = H^2(K(\pi_1(T),1);A) \longrightarrow H^2(T;A)$ is injective for any space T and any $\mathbb{Z}\pi_1(T)$-module A, since the Eilenberg-McLane space $K(\pi_1(T),1)$ can be obtained by attaching to T cells of dimensions ≥ 3. Thus $\tilde{d}\cdot\theta_\varphi = 0$ since $\theta_\varphi \in \ker\varphi^*$ (see [St, §II.3 and II.4]).

(1.5) Theorem Let Λ be a subring of $\mathbf{Q}$. Then, with the notations of (1.1), the following statements are equivalent:

(i) For any $n \geq 4$, φ is a geomorphism of degree d in dimension n with d invertible in Λ.

(ii) There exists a map $f : X \longrightarrow Y$ of degree 1 between two oriented Λ-Poincaré complexes X and Y of the same formal dimension.

(iii) a) $[Q:I]$ is finite and invertible in Λ and
b) the homomorphism $\varphi^* : H^2(I;B) \longrightarrow H^2(G;B)$ is injective for any ΛI-module B.

(iv) a) $[Q:I]$ is finite and invertible in Λ and
b) the homomorphism $\varphi^* : H^2(I;H_1(N,\Lambda)) \longrightarrow H^2(G;H_1(N,\Lambda))$ is injective.

(v) $[Q:I]$ is finite and invertible in Λ and $\theta_\varphi^\Lambda = 0$.

Proof The implications (i)$\Longrightarrow$ (ii)$\Longrightarrow$(iii)$\Longrightarrow$ (iv)$\Longrightarrow$ (v) are either obvious or come from Lemma (1.4) and its proof. It remains thus to prove that (v)$\Longrightarrow$ (i) .

Let K be a 2-dimensional finite complex with $\pi_1(K) = Q$. Let Σ_0^n be the boundary of a regular neighborhood of K in $\mathbb{R}^{n+1}$, with $n \geq 6$. Thus $\pi_1(\Sigma_0) = Q$. Let $\Sigma \longrightarrow \Sigma_0$ be the covering projection with $\pi_1(\Sigma) = I$. Since $[Q,I]$ is finite, Σ is a closed manifold.

Let L be a 2-dimensional complex with $\pi_1(L) = G$. The epimorphism $\varphi : G \twoheadrightarrow I$ gives rise to a map $p : L \longrightarrow BI$, where BI denotes an Eilenberg-McLane space $K(I,1)$. Let us consider the map p as a Serre fibration with homotopy fiber $\widetilde{L}_N$, the covering of L with $\pi_1(\widetilde{L}_N) = N$. Let us form the pull-back diagram of Serre fibrations:

$$\begin{array}{ccccc} \widetilde{L}_N & \longrightarrow & X & \xrightarrow{q} & \Sigma \\ \| & & \downarrow & & \downarrow \\ \widetilde{L}_N & \longrightarrow & L & \xrightarrow{p} & BI \end{array}$$

As the map $\Sigma \longrightarrow BI$ induces an isomorphism on the fundamental groups, the exact homotopy sequences of these fibrations imply that $\pi_1(X) = G$ and $\pi_1 q = \varphi$. We will prove the following

Claim: The homomorphism $q_* : H_n(X;\Lambda) \longrightarrow H_n(\Sigma;\Lambda)$ is surjective.

This assertion is equivalent to the nullity of all the successive differentials $d_k : E^k_{n,0} \longrightarrow E^k_{n-k,k-1}$ of the Serre spectral sequence for the homology with coefficients Λ of the fibration $\widetilde{L}_N \longrightarrow X \longrightarrow \Sigma$. But $\widetilde{L}_N$ is 2-dimensional and Σ has a handle decomposition with handles of index $0,1,2,n-2,n-1,n$. This makes the term $E^2_{u,v} = H_u(\Sigma;H_v(\widetilde{L}_N;\Lambda))$ look as follows:

	0		0		0		
*	*	*		*	*	*	
*	*	*	0	*	*	*	
*	*	*		*	*	*	
0	1	2		n-2	n-1	n	

Thus, if $n \geq 6$, the only differential of which we have to check the nullity is $d_2 : H_n(\Sigma;\Lambda) \longrightarrow H_{n-2}(\Sigma;H_1(\widetilde{L}_N;\Lambda))$.

Let E be the semi-direct product of I with the $\mathbb{Z}I$-module $H_1(I,H_1(N;\Lambda))$. The corresponding fibration of Eilenberg-McLane spaces admits a section, so the same is true for the induced fibration:

$$\begin{array}{ccccc} BH_1(N;\Lambda) & \longrightarrow & Y & \rightleftarrows & \Sigma \\ \| & & \downarrow & & \downarrow \\ BH_1(N;\Lambda) & \longrightarrow & BE & \rightleftarrows & BI \end{array}$$

Since $\theta_\varphi^\Lambda = 0$, one has a morphism of fibration:

$$\begin{array}{ccccc} \widetilde{L}_N & \longrightarrow & X & \longrightarrow & \Sigma \\ \downarrow & & | & & \downarrow \\ BH_1(N;\Lambda) & \longrightarrow & Y & \longrightarrow & \Sigma \end{array}$$

inducing a morphism on the corresponding Serre spectral sequences. For the term E^2 , one gets homomorphisms:

$$H_u(\Sigma;H_v(\widetilde{L}_N;\Lambda)) \longrightarrow H_u(I;H_v(H_1(N;\Lambda);\Lambda) = H_u(I;H_v(N;\Lambda))$$

which are isomorphisms for all u if $v = 0$ or 1 (since $H_1(\widetilde{L}_N;\Lambda) \cong$ $\cong H_1(N;\Lambda))$. The fibration $Y \longrightarrow \Sigma$ having a section, one has $H_n(Y;\Lambda) \longrightarrow H_n(\Sigma;\Lambda)$ surjective and thus the differential $d_2 : H_n(\Sigma;\Lambda) \longrightarrow$ $\longrightarrow H_{n-2}(\Sigma;H_1(N;\Lambda))$ is zero. Therefore, the differential $d_2 : H_n(\Sigma;\Lambda) \longrightarrow$ $\longrightarrow H_{n-2}(\Sigma;H_1(\widetilde{L}_N;\Lambda)$ is also identically zero.

Now, specialize at $n = 6$. By a classical result of Thom [Th, Corollary III.7] the map:

$$\Omega_6^{SO}(X) \otimes \Lambda \longrightarrow H_6(X;\Lambda)$$

is surjective, where $\Omega_*^{SO}(X)$ denotes the oriented bordism group of X . Thus one can find a closed oriented manifold M^6 together with a map $V^6 \xrightarrow{h} X \xrightarrow{q} \Sigma^6$ such that the degree of $q \circ h$ is invertible in Λ . By surgery in dimension 0 and 1 , one can obtain that $\pi_1 h$ is an isomorphism. Therefore the composed map $f : V^6 \longrightarrow X \longrightarrow \Sigma^6 \longrightarrow \Sigma_0^6$ satisfies statement (i) and $f \times id : V^6 \times S^{n-6} \longrightarrow \Sigma_0^6 \times S^{n-6}$ provides such examples in dimension $n \geq 8$.

To complete the proof of (v) $\Longrightarrow$ (i), it remains to establish that if one has a map $f : V^n \longrightarrow W^n$ satisfying the requirements of statement (i), then, if $n \geq 5$, there exists a map $f_1 : V_1^{n-1} \longrightarrow W_1^{n-1}$ which also satisfies (i). For that, embed a 2-dimensional polyedron K with $\pi_1(K) = Q$ in W (such that the embedding induces the identity on the fundamental groups) and take a regular neighborhood U of K in W . Let W_1 be the boundary of U . Make f transverse to W_1 and let $V_1 = f^{-1}(W_1)$. An easy homology argument shows that $f_1 = f|V_1$ has the same degree as f . Using the diagram:

$$\begin{array}{ccccc} & & \pi_1(V) = G & & \\ & \nearrow & & \searrow & \\ \pi_1(V_1) & \longrightarrow & \pi_1(W_1) & \longrightarrow & \pi_1(W) = Q \end{array}$$

it is possible by surgeries in dimension 0 and 1 to obtain $\pi_1(V_1) = G$ and $\pi_1 f_1 = \varphi$, as required in statement (i) of (1.5).

Putting $\Lambda = \mathbb{Z}$, one obtains:

<u>(1.6) Corollary</u> Conjecture (1.3) is true for $d = \pm 1$ (i.e., φ is a geomorphism of degree ± 1 in any dimension ≥ 4 if and only if φ is surjective and $\theta_\varphi = 0$).

<u>(1.7) Proposition</u> Conjecture (1.3) is true if I acts trivially on $H_1(N)$.

<u>Proof</u> Follow the proof of (1.5) (with $\Lambda = \mathbb{Z}$) till we get to the group E . Replace E by the extension E_φ $(1 \longrightarrow H_1(N) \longrightarrow E_\varphi \longrightarrow I \longrightarrow 1)$ with characteristic class θ_φ . Take the pull-back diagram of fibrations:

$$\begin{array}{ccccc} BH_1(N) & \longrightarrow & Y_\varphi & \longrightarrow & \Sigma \\ \| & & \downarrow & & \downarrow \\ BH_1(N) & \longrightarrow & BE_\varphi & \longrightarrow & BI \end{array}$$

and consider the morphism of fibrations:

$$\begin{array}{ccccc} \widetilde{L}_N & \longrightarrow & X & \longrightarrow & \Sigma \\ \downarrow & & \downarrow & & \| \\ BH_1(N) & \longrightarrow & Y_\varphi & \longrightarrow & \Sigma \end{array}$$

Now we prove that $|\theta_\varphi| \cdot H_n(\Sigma) \subset (\ker d_2 : H_n(\Sigma) \longrightarrow H_{n-2}(\Sigma : H_1(\widetilde{L}_N))) = (\ker d_2 : H_n(\Sigma) \longrightarrow H_{n-2}(\Sigma : H_1(N)))$. To do that, observe that the restriction of the fibration $BH_1(N) \longrightarrow Y_\varphi \longrightarrow \Sigma$ to the 1-skeleton of Σ is a product of fibration (since I acts trivially on $H_1(N)$). Thus the differential $d_2 : H_n(\Sigma) \longrightarrow H_{n-2}(\Sigma; H_1(N))$ can be identified with a cap-product with θ_φ [Sh, Chapter 5]. The rest of the argument goes like in the proof of (1.5).

<u>(1.8) Examples</u> To shorten the text, let us call a homomorphism $\varphi : G \longrightarrow Q$ as in (1.1) a Λ-geomorphism if, for some n, φ is a geomorphism in dimension n of degree invertible in Λ . Conditions (iii) - (v) of (1.5) make the following statements or examples obvious:

a) The composition of two Λ-geomorphisms is a Λ-geomorphism.

b) If the composition of two homomorphisms is a Λ-geomorphism, then the second one is a Λ-geomorphism.

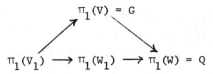

c) A split-epimorphism is a $\mathbb{Z}$-geomorphism.

d) If Q is finite of order q, then φ is a $\mathbb{Z}[\frac{1}{q}]$-geomorphism.

Here is a very strong property of $\mathbb{Z}$-geomorphisms:

(1.9) Theorem Let $\varphi : G \longrightarrow Q$ be a $\mathbb{Z}$-geomorphism and let S be a subgroup of Q. If $\varphi^{-1}(S)$ is free, then S is free.

Proof For any $\mathbb{Z}S$-module B, let us consider the following commutative diagram:

$$\begin{array}{ccccc} H^2(S;B) & & \longrightarrow & & H^2(\varphi^{-1}(S);B) \\ \uparrow\simeq & & & & \uparrow\simeq \\ H^2(Q;\mathrm{Hom}_{\mathbb{Z}S}(\mathbb{Z}Q;B)) & \overset{\varphi^*}{\rightarrowtail} & H^2(G;\mathrm{Hom}_{\mathbb{Z}S}(\mathbb{Z}Q;B)) & \overset{\simeq}{\longrightarrow} & H^2(G;\mathrm{Hom}_{\mathbb{Z}\varphi^{-1}(S)}(\mathbb{Z}G;B)) \end{array}$$

in which the vertical isomorphisms are given by the Shapiro lemma. As $\varphi^{-1}(S)$ is free, one has $H^2(\varphi^{-1}(S);B) = 0$. One deduces from the above diagram that $H^2(S;B) = 0$ for any $\mathbb{Z}S$-module B (φ^* being injective by (1.5)). This implies that S is free by the well-known Swann-Stallings Theorem.

2. OTHER PROBLEMS CONCERNING GEOMORPHISMS

(2.1) Let $\varphi : G \longrightarrow Q$ be a geomorphism of degree d in dimension n. Let V_0^n be an oriented closed manifold with $\pi_1(V_0) = G$. Does there exist a closed oriented manifold W^n and a map $f : V_0 \longrightarrow W$ of degree d with $\pi_1 f = \varphi$?

(2.2) Proposition Let $n \geq 5$ and $d = 1$. Then the answer to Question(2.1) is "yes" if $N = \ker\varphi$ is perfect (i.e., $N = [N,N]$).

Proof: The Quillen plus construction with respect to N can be performed on V_0 by adding to $V_0 \times I$ handles of index 2 and 3. This method, exposed in [Ha, p. 115], gives a cobordism (U^{n+1}, V_0, W) with $\pi_1(W) = \pi_1(U) = Q$ which retracts onto W.

(2.3) Let $\varphi : G \longrightarrow Q$ be a geomorphism of degree d in dimension n. Let W_0^n be a closed oriented manifold with $\pi_1(W_0) = Q$. Does there exist a closed oriented manifold V^n and a degree d map $f : V \longrightarrow W_0$ with $\pi_1(f) = \varphi$?

(2.4) Proposition The answer to Question (2.3) is "yes" if $d = 1$ and d is a split epimorphism.

Proof: There are sections in the pull-back diagram:

$$\begin{array}{ccc} X & \rightleftarrows & W_0 \\ \downarrow & & \downarrow \\ BG & \rightleftarrows & BQ \end{array}$$

and thus the map $\Omega_n^{SO}(X) \longrightarrow \Omega_n^{SO}(W_0)$ is surjective. The argument goes then as in the proof of (1.5).

(2.5) By (1.5), an epimorphism $\varphi : G \longrightarrow Q$ with N abelian is a $\mathbb{Z}$-geomorphism if and only if φ splits. Is that true with N solvable?

(2.6) Can a $\mathbb{Z}$-geomorphism $\varphi : G \longrightarrow Q$ with $\ker \varphi$ finite always be written as a composition $\varphi = \varphi_2 \circ \varphi_1$ with φ_2 split surjective and φ_1 an epimorphism with perfect kernel?

<u>(2.7) The Hopf problem</u>: Fifty years ago, H. Hopf asked the following problem: suppose that there exist degree one maps $f : V^n \longrightarrow W^n$ and $g : W^n \longrightarrow V^n$ between closed manifolds V and W. Is $\pi_1(V)$ then isomorphic to $\pi_1(W)$ via $\pi_1 f$? The answer "yes" to this problem is sometimes called "the Hopf conjecture."

In 1952, B. H. Neumann [Ne] found the first example of two non-isomorphic finitely presented groups G and Q admitting epimorphisms $\varphi : G \twoheadrightarrow Q$ and $\psi : Q \twoheadrightarrow G$. These groups have the following presentations:

$$G = \{a,b,c \mid a^{-1}ba=b^2, bc = cb\}$$

$$G = \{a,b,c \mid a^{-1}ba=b^2, bc=cb, [aba^{-1},c]^2 = 1\}$$

The epimorphism φ is induced by the identity on the generators and ψ is determined by $\psi(a) = a$, $\psi(b) = b^2$, $\psi(c) = c$.

Theorem (1.9) shows that this example cannot be used to construct a contrexample to the Hopf conjecture. Indeed, it is shown in [Ne] that $\ker(\psi \circ d) = \varphi^{-1}(\ker \psi)$ is free while $\ker \psi$ contains the element $[aba^{-1},c]$ of order 2 and thus $\ker \psi$ is not free. Therefore, φ is not a $\mathbb{Z}$-geomorphism.

Many other examples of pairs of finitely presented groups which are homomorphic images of each other but are not isomorphic have been constructed. All these known by the author can be ruled out by the above procedure.

However, the Hopf problem remains wide open. The Hopf conjecture is known to be true for $n = 2$ [Ho2] and is likely to be obtainable for $n = 3$ after the recent progress in the theory of 3-dimensional manifolds. Here are two questions related to the Hopf problem:

(2.8) Can one find a $\mathbb{Z}$-geomorphism $\varphi : G \longrightarrow Q$ with φ non-injective and Q isomorphic to G ? (see (3.3)). Using Mayer-Vietoris sequences, one can check that the example of [Hi] provides a non-injective epimorphism $\psi : G \longrightarrow G$ which is a $\mathbb{Z}[\frac{1}{2}]$-geomorphism.

(2.9) Is a degree one map $f : V \longrightarrow V$ a homotopy equivalence? (this is true if $\pi_1(V)$ is hopfian and $\mathbb{Z}\pi_1(V)$ is noetherian; for example: $\pi_1(V)$ finite, or finitely generated abelian).

3. THE FUNDAMENTAL GROUP OF A Λ-POINCARÉ COMPLEX(*)

It is known that the fundamental group of a Λ-Poincaré complex is finitely generated (see [Br]). On the other hand, any finitely presented group is the fundamental group of a closed manifold. The class of fundamental groups of Λ-Poincaré complexes is thus intermediate between the classes of finitely generated and of finitely presented groups. Such an intermediate class is introduced algebraically in [B, p. 20] or [B-S]:

Definition A group G is called almost finitely presented over Λ if there exists an exact sequence $1 \longrightarrow R \longrightarrow F \longrightarrow G \longrightarrow 1$ where F is free of finite rank and $H_1(R) \otimes \Lambda$ is finitely generated as a ΛG-module.

The aim of this section is to prove the following:

(3.1) Theorem Let G be a group and $n \geq 4$. There exists a Λ-Poincaré-complex X of formal dimension n with $\pi_1(X) \cong G$ if and only if G is almost finitely presented over Λ.

Proof: Let X be a Λ-Poincaré-complex. To prove that $\pi_1(X)$ is almost finitely presented over Λ, we just follow the argument of [Bn, §3] replacing $\mathbb{Z}$ by Λ : the cellular chain complex $C_*(\widetilde{X}) \otimes \Lambda$ (where $\widetilde{X}$ is the universal cover of X) is equivalent to a complex A_* of finitely generated $\Lambda\pi_1(X)$-free modules. This gives us a short exact sequence $A_2 \longrightarrow A_1 \longrightarrow A_0 \longrightarrow \Lambda \longrightarrow 0$ with A_1 a finitely generated $\Lambda\pi_1(X)$-free module which is equivalent to $\pi_1(X)$ being almost finitely presented over Λ (see [B, p. 20]).

Conversely, let G be almost finitely presented over Λ. By [B-S, Lemma (1.3)] there exists an exact sequence of groups $1 \longrightarrow P \longrightarrow H \longrightarrow G \longrightarrow 1$ where H is finitely presented and $H_1(P) \otimes \Lambda = 0$. Take a closed orientable manifold V^n $(n \geq 4)$ with $\pi_1(V) = H$. As $H_1(P;\Lambda) = H_1(P) \otimes \Lambda = 0$, one can perform to V the "Quillen plus construction with coefficients Λ" [Me, § 1] which gives a map $q : V \longrightarrow V^+$

(*) Conversations with K. Brown were useful for the reduction of this section.

where:

1) $\ker \pi_1 q = P$, $\pi_1(V^+) = H/P = G$

2) $q_* : H_*(V;B) \longrightarrow H_*(V^+;B)$ is an isomorphism for any ΛG-module B . Thus V^+ is a Λ-Poincaré complex (of formal dimension n) and $\pi_1(V^*) = G$.

Theorem (3.1) leaves open the following questions:

(3.2) If a group G is almost finitely presented over Λ , is it finitely presented? Positive answers in special cases are given in [B-S].

(3.3) Suppose that there exists an epimorphism $\varphi : G \twoheadrightarrow G$ with G finitely presented and $P = \ker\varphi$ non-trivial and Λ-perfect (i.e., $H_1(P;\Lambda) = 0$). Then $\Pi = \cup_{n\in\mathbb{N}} \ker\varphi^n$ (φ^n = n-th iterate of φ) is a Λ-perfect subgroup of G which is not the normal closure of finitely many elements (since $\ker\varphi \neq 1$). Thus G/Π is almost finitely presented over Λ but not finitely presented, and we would get a negative answer to Question (3.2). Also, if $\Lambda = \mathbb{Z}$, the homomorphism φ would be a $\mathbb{Z}$-geomorphism and give an example for Question (2.8).

(3.4) Theorem (3.1) does not tell us which group G arise as the fundamental group of a Λ-Poincaré complex of formal dimension $n = 2$ or 3 . If $n = 2$ and $\Lambda = \mathbb{Z}$, G is known to be a surface group unless $H_1(G;Q) = 0$, in which case some finiteness question remains open (see [E-M]). For $n = 3$, an algebraic description of fundamental groups of $\mathbb{Z}$-Poincaré complexes has been obtained by V. G. Tuarev [Tu].

REFERENCES

[B] Bieri, R., Homological dimension of discrete groups, Queen Mary College Math Notes (Univ. of London), 1976.

[B-S] Bieri, R. and Strebel, R., Almost finitely presented soluble groups, Comm. Math. Helv. 53 (1978) 258-278.

[Br] Browder, W., Poincaré spaces, their normal fibrations and surgery, Invent. Math. 17 (1972) 191-202.

[Bn] Brown, K., Homology criteria for finiteness, Comm. Math. Helv. 50 (1975) 129-135.

[E-M] Eckmann, B. and Muller, H., Poincaré duality groups of dimension 2, Comm. Math. Helv 55 (1980) 510-520.

[Ha] Hausmann, J-Cl., Homological surgery, Annals of Math. 104 (1976) 573-586.

[Hi] Higman, G., A finitely related group with an isomorphic proper factor subgroup, J. London Math. Soc. 26 (1951) 59-61.

[Ho] Hopf, H., Zür Algebra der Abbildungen von Mannigfaltigkeiten, Jal für die reine und angew. Math. 163 (1930) 71-88.

[Ho2] Hopf, H., Beiträge zur Klassifizierung der Flächenabbildungen Jal für die reine und angew. Math. 165 (1931) 225-236.

[Me] Meier, W., Acyclic maps and knot complements, Math. Annalen 243 (1979) 247-259.

[Ne] Neumann, B. H., On a problem of Hopf, J. of London Math. Soc. 28(1953) 351-353.

[Sh] Shih Weishu, Homologie des espaces fibrés, Publ. Math. IHES 13 (1962).

[St] Stambach, U., Homology in group theory, Springer Lect. Notes 359.

[Th] Thom, R., Quelques proprietes globales des varietes differentiables, Comm. Math. Helv. 28 (1954) 17-86.

[Tu] Tuarev, V. G., Fundamental groups of manifolds and Poincaré complexes, Math USSR-Sbornik (Translations) 38 (1981) 255-270.

UNIVERSITY OF GENEVA
GENEVA, SWITZERLAND

The Institute for Advances Study
(Summer 1981)
Princeton, New Jersey, USA

Canadian Mathematical Society
Conference Proceedings
Volume 2, Part 2 (1982)

THE RIGID HANDLEBODY THEORY I

W.-C. Hsiang[1] and B. Jahren

ABSTRACT. Based on the "manifold model", we rework $Wh^{Comb}(\ )$ of [10], which shall be called the "rigid handlebody space." We shall also derive a "manifold version" of the fibration $\Omega^{fr}(\ ) \to A(\ ) \to Wh^{Comb}(\ )$ of [10]. More importantly, we can put involutions on each term of the fibration such that they are compatible with the involution on $Wh^{Comb}(\ )$ essentially by turning $M \times I$ upside down.

I. INTRODUCTION. Let M be a compact manifold and let $C(M)$ be the space of concordance of M, i.e. automorphisms of $M \times I$ fixed on $M \times 0$. According to the category, these will be diffeomorphisms, and we write $C^{Diff}(M)$, $C^{PL}(M)$ etc. when we wish to specify the category. Sometimes we replace $C(M)$ by the subspace $C(M,\partial M)$ consisting of concordances fixed on $\partial M \times I$, since $(M\times I, M\times 0)$ is isomorphic to $(M\times I, M\times 0 \cup \partial M\times I)$ by "bending around the corners". (This also involves the usual smoothing of the corners in Diff.) We have the suspension map

$$(1) \qquad \Sigma: C(M) \longrightarrow C(M\times I)$$

sending f to $f \times id_I$. Define $C(M) = \cup_n C(M\times I^n)$ which becomes a homotopy functor and extend $C(M)$ to $C(X)$ for X a CW-complex. It turns out that the functors $X \longrightarrow C^{Diff}(X)$, $C^{PL}(X)$ are from spaces to infinite loop spaces. Let $Wh^{Diff}(X)$, $Wh^{PL}(X)$ be the double deloopings of $C^{Diff}(X)$, $C^{PL}(X)$ respectively [3] [10]. In his outline of the study of $Wh^{Diff}(X)$ and $Wh^{PL}(X)$, Waldhausen [10] introduced the fundamental functor $X \longrightarrow A(X)$ from spaces to infinite loop spaces, the algebraic K-theory of the space X, and homotopy fibration sequences of infinite loop spaces

$$(2) \qquad \begin{array}{l} Wh^{Diff}(X) \longrightarrow A(X) \longrightarrow h(X;A^S(*)) \\ h(X;A(*)) \longrightarrow A(X) \longrightarrow Wh^{PL}(X) \end{array}$$

AMS (MOS) subject classifications (1970). Primary 55-xx. 57-xx.

[1]Partially supported by NSF Grant GP 34324X1.

© 1982 American Mathematical Society
0731-1036/82/0000-0476/$13.25

Relating $A(X)$ to $Wh^{Diff}(X)$ and $Wh^{PL}(X)$, where $h(X;A^s(*))$, $h(X;A(*))$ are the infinite loop spaces associated to the homology theory with coefficients in the spectra $A^s(*)$, $A(*)$ respectively. (For the definition of $A^s(*)$, see [10, pp. 45-48].) In order to study $Wh^{Diff}(X)$ and $Wh^{PL}(X)$, Waldhausen also introduced an intermediate category called "expansion space" and its delooping $Wh^{Comb}(X)$, and then proved that there is a homotopy fibration

$$\Omega^{fr}(X) \longrightarrow A(X) \longrightarrow Wh^{Comb}(X) \tag{3}$$

which is mapped into the fibrations of (2). In many ways, the expansion space and $Wh^{Comb}(X)$ are more tractible than $Wh^{Diff}(X)$ and $Wh^{PL}(X)$. In fact, we may anticipate from Igusa's recent work [4] that $Wh^{Comb}(X)$ and $Wh^{Diff}(X)$ are probably homotopically equivalent.

Lately, some soft spots in [2] were found. Since [10] is based on [2] and it only uses simplicial complexes instead of manifolds, it is desirable to rework [10], based on manifold models. There is another reason why we prefer to have a manifold model. We have an involution on $C(M)$ essentially coming from turning $M \times I$ upside down. By stabilizing, we have an involution on $C(M)$ (and hence on $Wh^{PL}(M)$). It is important to define an involution on $A(X)$ compatible with that on $C(M)$.

In this paper, we shall use the manifold model of the "expansion space" of [10] and call it the "rigid handelbody theory". We shall prove that Quillen's Q-construction [6] actually works for rigid handlebody theory, and we also have an unstable version of the fibration of (3). We then put involutions on each term of (the manifold version of) the fibration (3) such that they are compatible with the involution defined on $Wh^{Comb}(M)$ (essentially by turning upside down of $M \times I$). In particular, we have an involution on $A(X)$. In a later paper, we shall identify this involution algebraically.

As the readers can easily see from the arguments, we are very much indebted to [6] and [10]. In fact, our main contribution is really in Chapter V.

II. THE RIGID HANDLEBODY SPACE.

2.1. The Model Category E_k^n .

Let ∂_0 be an $(n+k-1)$-dim manifold and $\pi_0 : \partial_0 \longrightarrow \Delta^k$ a differentiable bundle map. Hence the fibers are $(n-1)$-dim manifolds, possibly with boundary. Suppose that ∂_0 is a codim 0 submanifold of ∂Y of a manifold Y, and $\pi : Y \longrightarrow \Delta^k$ is a bundle projection extending π_0. We say that Y is a k-parameter family of rigid n-dim handlebodies on ∂_0 if there is a filtration

(4) $$Y^{(0)} = \partial_0 \subset Y^{(1)} \subset \ldots \subset Y^{(\ell)} = Y$$

satisfying the following conditions:

(a) For each $i > 0$, there is an embedding

$$f_i : S^{j_i - 1} \times D^{n-j_i} \times \Delta^k \longrightarrow Y^{(i-1)}$$

and a homeomorphism

$$Y^{(i)} \xrightarrow[\simeq]{d_i} Y^{(i-1)} \cup_{f_i} D^{j_i} \times D^{n-j_i} \times \Delta^k$$

rel $Y^{(i-1)}$ such that f_i and d_i preserve the projections onto Δ^k . For technical reasons we shall also assume that $2 < j_i < n-2$.

(b) Let $M^{(i)} = Y^{(i)} \cup_{\partial_0 = \partial_0 \times \{1\}} \partial_0 \times I$. This is a manifold, even though $Y^{(i)}$ may not be. ($M^{(i)}$ is the mapping cylinder of the inclusion $\partial_0 \subset Y^{(i)}$.)

(c) Let

$$\partial_+ Y^{(i)} = C\ell(\partial M^{(i)} - (\partial_0 \times \{0\} \cup \partial M^{(i)} | \partial \Delta^k))$$

where $\partial M^{(i)} | \partial \Delta^k$ is the part that projects to $\partial \Delta^k$. Then we assume that

$$f_i(S^{j_i - 1} \times D^{n-j_i} \times \Delta^k) \subset \partial_+ Y^{(i-1)}$$

and that f_i is a differentiable embedding into $\partial_+ Y^{(i-1)}$ (after we smooth the corners). d_i has an obvious extension to

$$M^{(i)} \simeq M^{(i-1)} \cup D^{j_i} \times D^{n-j_i} \times \Delta^k ,$$

and we also assume that this is a diffeomorphism (again we may have to smooth the corners).

Example:

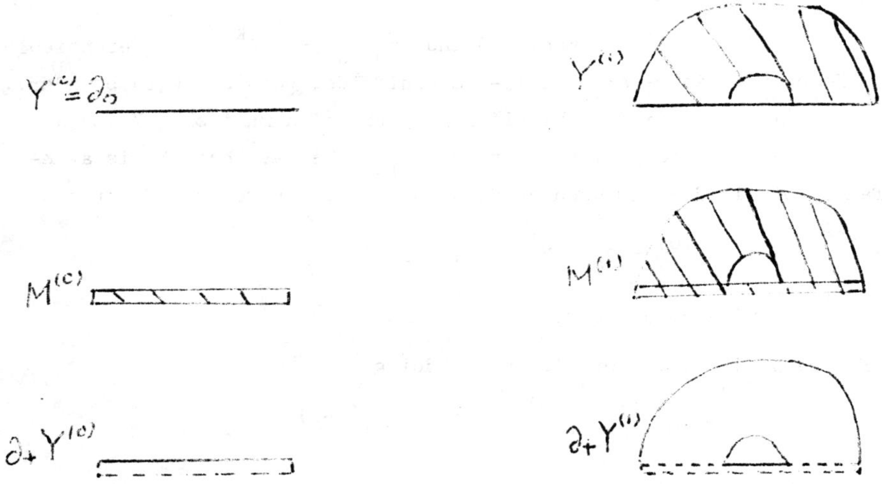

The attaching data (the f_i's and d_i's) are part of the structure, but independent handles may be attached in any order. The isomorphisms of a k-parameter family of rigid n-dim handlebodies on ∂_0 are the handle-preserving homeomorphisms (becoming diffeomorphisms after we add $\partial_0 \times I$ and smooth the corners) which commute with all the attaching data.

We shall construct a category E_k^n which has the k-parameter families of rigid n-dim handlebodies as objects. The morphisms correspond to cancelling trivial pairs of handles and isomorphisms. By a "trivial pair of handles" we mean a union of a k-handle and a (k+1)-handle

$$J = D^k \times D^{n-k} \cup D^{k+1} \times D^{n-k-1} \tag{5}$$
$$\text{along } D^k \times H_1 = H_2 \times D^{n-k-1} ,$$

where $H_1 \subset \partial D^{n-k}$ and $H_2 \subset \partial D^{k+1}$ are codim 0 discs. Hence J is a connected sum of n-dim discs along the boundaries -- hence $J \approx D^n$. A family of trivial handle-pairs is obtained by letting H_1 and H_2 vary by isotopies. Suppose now that Y' is an object of E_k^n with a family of trivial handle-pairs J .

An elementary expansion

(6) $$e : Y \longrightarrow Y'$$

is a projection-preserving map

(7) $$C\ell(Y'-J) \xrightarrow{c(e)} Y$$

satisfying the following conditions

(a) $c(e)$ is a diffeomorphism (after the corners are smoothed) on

$$C\ell(Y'-J) - \partial J .$$

(b) $c(e)$ maps handles to handles and commutes with the attaching data.

(c) $c(e)$ is onto.

Remark. $C\ell(Y'-J)$ is in general not the same as $Y' - \mathrm{Int}\, J$, e.g., Y' might be equal to J and hence $C\ell(Y'-J) = \phi$. Therefore, if we make the convention that ϕ is a rigid handlebody of any dim, then there is a canonical expansion $\phi \longrightarrow J$. Here is another example:

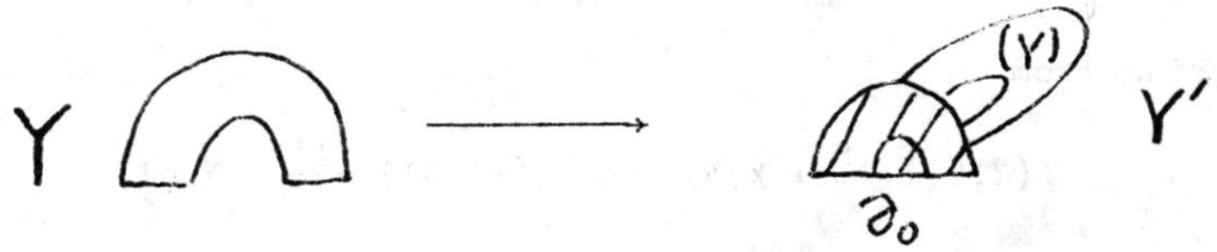

We define the morphisms of E^n_k to be identities and compositions of isomorphisms and elementary expansions.

The obvious definition of face and degeneracy operators now make $E^n_\cdot$ a simplicial category. Finally, we stabilize by multiplying all diagrams by D^1. Then, a handle $D^j \times D^{n-j}$ is converted to a handle $D^j \times D^{n+1-j}$, and ∂_0 is replaced by $\partial_0 \times D^1$. On expansions, we let $c(e \times D^1) = c(e) \times id_{D^1}$. This defines a simplicial functor, the "suspension."

(8) $$\textstyle\sum : E^n_\cdot \longrightarrow E^{n+1}_\cdot .$$

(Cf. (1) of the Introduction.)

Remark. If $Y \xrightarrow{e} Y'$ is a morphism, there is a diagram

(9)
$$\begin{array}{c} C\ell(Y'-\cup\{H\}) \subset Y' \\ \Big\downarrow c(e) \\ Y \end{array}$$

where $\cup\{H\}$ is the set of handles involved in the elementary expansions, and $c(e)$ has (essentially) the properties (a)-(c) of $c(e)$ for an elementary expansion. We shall denote $C\ell(Y'-\cup\{H\})$ by $Y(e)$.

2.2. The Bisimplicial Category S.E.$(X;\xi)$.

Let X be a space and let ξ be a stable vector bundle over X (i.e., $\xi \in KO(X)$) .

The categories $E_k(X;\xi)^n$, $k,n = 0,1,2,\ldots$ are defined as follows: The objects are diagrams

$$f:(Y,\partial_0) \longrightarrow X \tag{10}$$

where (Y,∂_0) is an object in E_k^n and $f:Y \longrightarrow X$ is a continuous map, together with a stable bundle isomorphism

$$\psi:tY \longrightarrow f^*\xi \tag{11}$$

where tY is the tangent bundle.

Remark. ψ determines a bundle map over f and vice versa. We shall denote both by ψ .

A morphism from

$$((Y,\partial_0) \xrightarrow{f} X,\psi) \text{ to } ((Y',\partial_0') \xrightarrow{f'} X,\psi')$$

is a morphism $e:(Y,\partial_0) \longrightarrow (Y',\partial_0')$ in E_k^n such that

$$\begin{array}{ccc} Y(e) & \hookrightarrow & Y' \\ {\scriptstyle c(e)}\downarrow & & \downarrow{\scriptstyle f'} \\ Y & \xrightarrow{f} & X \end{array} \tag{12}$$

commutes and is covered by the commutative bundle diagram

$$\begin{array}{ccc} tY(e) & \hookrightarrow & tY' \\ {\scriptstyle t(c(e))}\downarrow & & \downarrow{\scriptstyle \psi'} \\ tY & \xrightarrow{\psi} & \xi \end{array} \tag{13}$$

It is easy to see that the composition of two such morphisms satisfies the same property, and we get a category $E_k(X;\xi)^n$. Note that $E_k(X;\xi)^n$ has a composition law "+" -- disjoint union of the (Y,∂_0)'s -- and hence the classifying space has an infinite delooping in the sense of Γ-spaces [8]. We shall define an explicit delooping on the category level, using Waldhausen's idea [9] [10].

First we define cofibrations. Let $((Y,\partial_0) \xrightarrow{f} X,\psi)$ and $((Y',\partial_0') \xrightarrow{f'} X,\psi')$ be objects of $E_k(X;\xi)^n$ such that

(a) $Y \subset Y'$ and $\partial_0 = \partial_0' \cap Y$,

(b) $Y \cup \partial_0'$ is a lower filtrations of Y' ,

(c) f and ψ are restrictions of Y' .

Then we write

$$(14) \qquad \begin{array}{ccc} (Y,\partial_0) & \rightarrowtail & (Y',\partial_0') \\ & {\scriptstyle f}\searrow \quad \swarrow{\scriptstyle f'} & \\ & X & \end{array}$$

(or sometimes just $(Y \rightarrowtail Y')$) and call this arrow a cofibration. (In particular, we allow $Y = \phi$.) Composition of cofibrations is defined the obvious way. If $((Y,\partial_0) \xrightarrow{f} X,\psi) \rightarrowtail ((Y',\partial_0') \xrightarrow{f'} X,\psi')$ is a cofibration, we define the <u>cofiber</u> $f/f':(Y',\partial_0')/(Y,\partial_0) \longrightarrow X$ to be the object

$$(15) \qquad (C\ell(Y'-Y),\partial(C\ell(Y'-Y)-(\partial Y'-\partial_0')) \xrightarrow{f|C\ell(Y'-Y)} X$$

with the obvious restriction as the bundle identification. If $Y_1 \rightarrowtail Y_2 \rightarrowtail Y_3$ are cofibrations, there is a new cofibration $Y_2/Y_1 \rightarrowtail Y_3/Y_1$, and a natural identification of $(Y_3/Y_1)/(Y_2/Y_1)$ with Y_3/Y_2 .

Let $S_\ell E_k(X;\xi)^n$ be the category with filtrations

$$(16) \qquad Y_1 \rightarrowtail Y_2 \rightarrowtail \cdots \rightarrowtail Y_\ell$$

as objects, and where morphisms are diagrams

$$(17) \qquad \begin{array}{ccccccc} Y_1 & \rightarrowtail & Y_2 & \rightarrowtail & \cdots & \rightarrowtail & Y_\ell \\ \downarrow & & \downarrow & & & & \downarrow \\ Y_1' & \rightarrowtail & Y_2' & \rightarrowtail & \cdots & \rightarrowtail & Y_\ell' \end{array}$$

satisfying the following conditions:

(a) The vertical arrows are expansions.

(b) Let J_i be the union of handles involved in the expansion $e_i:Y_i \longrightarrow Y_i'$. Then $J_{i+1} \cap Y_i' = J_i$, and the induced diagram

$$(18)\qquad \begin{array}{ccc} Y'_i & \hookrightarrow & Y'_{i+1} \\ \uparrow & & \uparrow \\ Y(e_i) & \hookrightarrow & Y(e_{i+1}) \\ {\scriptstyle c(e_i)}\downarrow & & \downarrow{\scriptstyle c(e_{i+1})} \\ Y_i & \hookrightarrow & Y_{i+1} \end{array}$$

commutes.

With these definitions, a diagram

$$(19)\qquad \begin{array}{ccc} Y_1 & \rightarrowtail & Y_2 \\ \downarrow & & \downarrow \\ Y'_1 & \rightarrowtail & Y'_2 \end{array}$$

in $S_2E_k(X;\xi)^n$ induces a morphism $Y_2/Y_1 \longrightarrow Y'_2/Y'_1$.

$S_\ell E_k(X;\xi)^n$ is a simplicial category for each ℓ, and we wish to assemble these into a bisimplicial category $S.E.(X;\xi)^n$. Hence we must define face and degeneracy operations (functors) in the ℓ-direction.

We define face and degeneracy functors as follows:

Let $Y_1 \rightarrowtail Y_2 \rightarrowtail \dots \rightarrowtail Y_\ell$ be an object in $S_\ell E_k(X;\xi)^n$. (Note that some of the Y_i's may be empty.) We define

$$(20)\qquad s_j(Y_1 \rightarrowtail \dots \rightarrowtail Y_\ell) = \begin{cases} Y_1 \rightarrowtail \dots Y_j = Y_j \rightarrowtail \dots \rightarrowtail Y_\ell & \text{if } j > 0 \\ \phi \rightarrowtail Y_1 \rightarrowtail \dots \rightarrowtail Y_\ell & \text{if } j = 0 . \end{cases}$$

(Note that $Y = Y$ is a cofibration.)

$$(21)\qquad d_j(Y_1 \rightarrowtail \dots \rightarrowtail Y_\ell) = \begin{cases} Y_1 \rightarrowtail \dots \rightarrowtail Y_{j-1} \longrightarrow Y_{j+1} \rightarrowtail \dots \rightarrowtail Y_\ell & \text{if } j>0 \\ Y_2/Y_1 \rightarrowtail \dots \rightarrowtail Y_\ell/Y_1 & \text{if } j = 0 . \end{cases}$$

Thus, the s_j's and d_j's are just repetition and deletion of the filtrations, except for d_0, which we should think of as "quotient by Y_1".

The extension of s_j and d_j to the morphisms is now immediate and we can easily check the simplicial identities. Hence $S.E.(X;\xi)^n$ is a bisimplicial category.

Moreover, if $(Y,\partial_0) \xrightarrow{f} X$ is an object in $E_k(X;\xi)^n$, we can multiply by D^1 and get an object $(Y\times I,\partial_0\times I) \xrightarrow{\text{proj}} (Y,\partial_0) \xrightarrow{f} X$ in $E_k(X;\xi)^{n+1}$.

This induces a functor

$$\sum : S.E.(X;\xi)^n \longrightarrow S.E.(X;\xi)^{n+1} \tag{22}$$

which we shall call the "suspension functor", and we define

$$BE(X;\xi) = \varinjlim_n |S.E.(X;\xi)^n| . \tag{23}$$

Let $E^h_k(X;\xi)^n \subset E_k(X;\xi)^n$ be the full subcategory of objects such that $\partial_0 \subset Y$ is a homotopy equivalence. Since all the handles have index > 2 and $< n-2$, this is equivalent to $(Y;\partial_0 Y, C\ell(\partial Y - \partial_0 Y))$ being an h-cobordism and to $H_*(Y,\partial_0;\underline{\rho}) = 0$ for any local coefficient system $\underline{\rho}$ over Y.

Lemma 2.1. *Suppose* $(Y,\partial_0) \rightarrowtail (y_0', \partial_0')$ *is a cofibration with both* (Y,∂_0), $(Y',\partial_0') \in E^h_k(X;\xi)^n$. *Then,* $(Y'/Y, \partial_0(Y'/Y)) \in E^h_k(X;\xi)^n$.

(Again, $\phi \in E^h_k(X;\xi)^n$ *by convention.)*

Proof. By excision,

$$H_*(Y'/Y, \partial_0(Y'/Y);\underline{\rho}) \cong H_*(Y', Y \cup \partial_0';\underline{\rho}) . \tag{24}$$

We then use the long exact homology sequence to the triple $(Y', Y \cup \partial_0', \partial_0')$ and the isomorphism $H_*(Y',\partial_0';\underline{\rho}) = 0$ to get

$$H_*(Y', Y \cup \partial_0';\underline{\rho}) \cong H_{*-1}(Y \cup \partial_0', \partial_0';\underline{\rho}) . \tag{25}$$

Excision again implies

$$H_{*-1}(Y \cup \partial_0', \partial_0';\underline{\rho}) \cong H_{*-1}(Y,\partial_0;f) = 0 . \tag{26}$$

Hence $H_*(Y'/Y, \partial_0(Y'/Y);\underline{\rho}) = 0$, and the assertion follows. □

Therefore, we can perform the S. construction on $E^h_k(X;\xi)^n$ as well, and define $S.E^h_.(X;\xi)^n$. But since $H_*(Y\times I, \partial_0 \times I;\underline{\rho}) = H_*(Y,\partial_0;\underline{\rho})$, suspension restricts to a functor $\sum : S.E^h_.(X;\xi)^n \longrightarrow S.E^h_.(X;\xi)^{n+1}$. So we can define

$$BE^h(X;\xi) = \varinjlim_n S.\, E^h_.(X;\xi)^n . \tag{27}$$

This space we call "the rigid handlebody space."

We shall end this section by noting what kind of functionality BE and BE^h satisfy. Let C be the category where the objects are pairs $(X;\xi)$ with X a space and ξ a stable bundle over X, and where a morphism from (X,ξ) to (Y,η) is a continuous map $f:X \longrightarrow Y$ and a stable bundle isomorphism $\psi:\xi \simeq f^*\eta$.

Proposition 2.2. *BE and BE^h are functors from $\mathcal{C}$ to infinite loop-spaces.*

Proof. If $(f,\psi):(X,\xi) \longrightarrow (Y,\eta)$ is a morphism in $\mathcal{C}$, we get morphism $E_k^h(X;\xi) \longrightarrow E_k^h(Y;\eta)$ by composition

(28) $$((Z,\partial_0) \longrightarrow X) \longrightarrow ((Z,\partial_0) \longrightarrow X \longrightarrow Y)$$

etc. This commutes with all the constructions above, and we get map $BE^h(X;\xi) \longrightarrow BE^h(Y;\eta)$. $|S.E.^h(X;\xi)^n|$ is connected, and is the underlying space of a Γ-space by composition law "disjoint union". Hence, by [8], it is an infinite loop space. $\sum$ becomes an infinite loop space map. Similarly for $E_k(X;\xi)^n$ $E_k(Y;\eta)^n$ and the proposition follows. □

Remark. We shall see later that up to homotopy, this "induced" map does not depend on the particular choice of bundle isomorphism ψ . In fact, there is a kind of functoriality even without such ψ . (Cf. Corollary 2.4.).

2.3. Comparison with $Wh^{Comb}(X)$.

Let $\overline{E}^h(X)$ be Waldhausen's expansion category(2) [10,pp. 48-50]. So $|S.\overline{E}.^h(X)| = Wh^{Comb}(X)$. We shall define and study a functor

(29) $$b_n:E_k^h(X;\xi)^n \longrightarrow \overline{E}_k^h(X) .$$

First we fix refractions

(30) $$\rho_{k,i}:D^k \times D^i \longrightarrow \partial D^k \times D^i \cup D^k \times \{0\}$$

for each k and i such that

(31) $$\rho_{k,i} = \rho_{k,i+1}|D^k\times(D^i\times\{0\})$$

(where 0 denotes the center of D^1). Let $(Y,\partial_0) \xrightarrow{f} X$ be an object in $E_k(X;\xi)^n$ with filtration

(32) $$Y^{(0)} = \partial_0 \subset Y^{(1)} \subset\ldots\subset Y^{(\ell)} = Y .$$

---

(2) We shall assume that the attached cells are of dim > 2 . It is easy to see that this extra assumption does not affect the results of [10].

Assume that

$$Y^{(i)} = Y^{(i-1)} \cup_{g_i} D^{j_i} \times D^{n-j_i} \times \Delta^k \tag{33}$$

where

$$g_i : \partial D^{j_i} \times D^{n-j_i} \times \Delta^k \longrightarrow \partial Y^{(i-1)} \tag{34}$$

as in §2.1. We now define $\overline{Y}^{(i)}$ $i = 0,\ldots,\ell$ and maps

$$r_i : Y^{(i)} \longrightarrow \overline{Y}^{(i)} \tag{35}$$

inductively as follows.

(a) $\overline{Y}^{(0)} = X \times \Delta^k$ and $r_0 = (f|\partial_0) \times \pi_0$.

(b) Assume $\overline{Y}^{(i-1)}$, r_{i-1} already defined. Let

$$\overline{Y}^{(i)} = \overline{Y}^{(i-1)} \cup_{\overline{g}_i} D^{j_i} \times \Delta^k \tag{36}$$

where $\overline{g}_i$ is the composition:

$$D^{j_i} \times \Delta^k = D^{j_i} \times \{0\} \times \Delta^k \xrightarrow{g_i} Y^{(i-1)} \xrightarrow{r_{i-1}} \overline{Y}^{(i-1)} , \tag{37}$$

and define r_i by

$$r_i = \left\{ \begin{array}{l} r_{i-1} \text{ on } \overline{Y}^{(i-1)} , \\ \rho_{j_i,n-j_i} \times \Delta^k \text{ on } \\ (\rho_{j_i,n-j_i})^{-1}(\overset{o}{D}{}^{j_i}\times\{0\}) \times \Delta^k , \text{ and} \\ r_{i-1} \circ (\rho_{j_i,n-j_i} \times \Delta^k) \text{ on } \\ (\rho_{j_i,n-j_i})^{-1}(\partial D^{j_i} \times D^{n-j_i} \times \Delta^k) . \end{array} \right\} \tag{38}$$

(What this really amounts to is the following, "collapsing the handles to the cores via ρ ", and "collapsing ∂_0 to $X \times \Delta^k$ via $f \times \pi_0$ ".) This defines a filtration

$$Y^{(0)} = X \times \Delta^k \subset \overline{Y}^{(1)} \subset \ldots\ldots \subset \overline{Y}^{(\ell)} \tag{39}$$

of parametrized cell complexes with a retraction induced by f , and it is clear that $\overline{Y}^{(i)} \simeq Y^{(i)} \cup_{f|\partial_0} X$. In particular, $\overline{Y}^{(\ell)} \supset X \times \Delta^k$ is a homotopy equivalence and we have an object in Waldhausen's expansion category. The construction is compatible with morphisms, hence we get a functor b_n of (29). b_n preserves cofibrations and therefore induces

$$b_n : S.E^h_\cdot(X;\xi)^n \longrightarrow S.\overrightarrow{E}^h_\cdot(X) . \tag{40}$$

Moreover, we easily see that b_n commutes with suspension, i.e. $b_n = b_{n+1}\Sigma$. Therefore, we obtain a map

$$b : BE^h(X;\xi) \longrightarrow Wh^{Comb}(X) . \tag{41}$$

The main result and the reason for introducing the rigid handle space is the following:

<u>Theorem 2.3</u>. <u>Let</u> X <u>be a finite complex. Then</u>

$$b : BE^h(X;\xi) \longrightarrow Wh^{Comb}(X)$$

<u>is a weak homotopy equivalence.</u>

<u>Proof</u>. Let $S.E.^h(X;\xi) = \varinjlim_n S.E.^h(X;\xi)^n$ (as categories) and let $N.$ be the nerve functor. It then suffices to prove that

$$N_j(S_\ell E^h_\cdot(X;\xi)) \longrightarrow N_j(S_\ell \overrightarrow{E}^h_\cdot(X)) \tag{42}$$

is a weak homotopy equivalence for all ℓ and j, and we shall do this by constructing a homotopy inverse of a finite subcomplex of $N_j(S_\ell \overline{E}^h_\cdot(X))$. (It is very unlikely that this can be done at the category level, e.g., by adjoint functors.)

For simplicity and illustrating the idea, let us first consider the case $\ell = 1$ and $j = 0$. Then, $N_j S_\ell E^h_\cdot(X;\xi)^n = E^h_\cdot(X;\xi)^n$. Let $\coprod \overrightarrow{E}^h_k(X) \times \Delta^k/\sim$, $\coprod E^h_k(X,\xi)^n \times \Delta^k/\sim$ be the usual realizations and let $Z \subset \coprod \overrightarrow{E}^h_k(X) \times \Delta^k/\sim$ be a finite subcomplex. We wish to construct a map

$$c : Z \longrightarrow \coprod E^h_k(X,\xi)^n/\sim \tag{43}$$

(for n large) which approximates the homotopy inverse.

Fix a manifold neighborhood M of X such that stably $tM \simeq r^*\xi$, where $r : M \longrightarrow X$ is a retraction. This can be done by choosing a regular neighborhood of X in a higher dimensional Euclidean space and then take the disc bundle of the induced bundle from ξ over this regular neighborhood. Replace X by M and let

(44) $$M \times \Delta^k \subset \overline{Y}^{(1)} \subset \ldots \subset \overline{Y}^{(s)} = Y \xrightarrow{\overline{\pi}} M \times \Delta^k$$

be the cell filtration of an object in $\overline{E}^h_k(M)$ whose realization is in Z. We have

(45) $$\overline{Y}^{(j)} = \overline{Y}^{(j-1)} \cup_{f_j} e^{m_j} \times \Delta^k ,$$

where $f_j : S^{m_j - 1} \times \Delta^k \longrightarrow \overline{Y}^{(j-1)}$. In particular, $f_1 : S^{m_1 - 1} \times \Delta^k \longrightarrow M \times \Delta^k$. We define g_1' as

(46) $$g_1' : S^{m_1 - 1} \times D^{n-m_1} \times \Delta^k \xrightarrow{\text{proj}} S^{m_1-1} \times \Delta^k \xrightarrow{f_1} M \times \Delta^k$$

where $n = \dim M + 1 >> \max(m_1, \ldots, m_s)$.

$\overline{\pi}|e^{m_1} \times \Delta^k$ determines a null homotopy of g_1', and hence a trivialization of $(g_1')^* tM$. Using the canonical trivialization of $t(S^{m_1-1} \times D^{n-m_1})$, we get a k-parameter family of stable bundle representations $t(S^{m_1-1} \times D^{n-m_1}) \rightarrow tM$ over g_1'. By Hirsch's immersion theorem, this determines a k-parameter family of immersions, and then by general position (n large !) a k-parameter family of embeddings

(47) $$g_1 : S^{m_1-1} \times D^{n-m_1} \times \Delta^k \longrightarrow M \times \Delta^k .$$

We use this to attach m_1-dim handles to $M \times I \times \Delta^k$ along $M \times \{1\} \times \Delta^k$. Thus we obtain a manifold $W^{(1)}$ equipped with the following:

(a) A retraction

(48) $$\pi^{(1)} : W^{(1)} \longrightarrow M \times \Delta^k \approx M \times \{0\} \times \Delta^k$$

coming from $\overline{\pi}|\overline{Y}^{(1)}$, and such that the composition

(49) $$W^{(1)} \longrightarrow M \times \Delta^k \longrightarrow \Delta^k$$

is a differentiable fibration with fibers $\approx M \times I \cup D^{m_i} \times D^{n-m_i}$.

(b) A stable bundle map

(50) $$b^{(1)} : tW^{(1)} \longrightarrow t(M \times \Delta^k)$$

over $\pi^{(1)}$.

(c) A fiberwise homotopy equivalence $W^{(1)} \sim Y^{(1)}$ rel $M \times \{0\} \times \Delta^k$.

Now, we compose f_2 with projection and the homotopy equivalence in (c) to get

(51) $$g_2':S^{m_2-1}\times D^{n-m_2}\times\Delta^k\longrightarrow W^{(1)}\ .$$

If n is large enough, g_2' can be deformed into $\partial W^{(1)}-M\times\{0\}\times\Delta^k$, preserving the projection to Δ^k. We continue as above to obtain $W^{(2)}$ with retraction $\pi^{(2)}:W^{(2)}\longrightarrow M\times\Delta^k$, bundle map $b^{(2)}:tW_2\longrightarrow t(M\times\Delta^k)$ and homotopy equivalence $W^{(2)}\simeq\overline{Y}^{(2)}$ with properties like (a)-(c) for $W^{(1)}$. By induction, we then construct a filtration

(52) $$M\times I\times\Delta^k\subset W^{(1)}\subset\ldots\subset W^{(s)}=W$$

and a projection

(53) $$\pi:W\longrightarrow M\times\Delta^k$$

such that

(54) $$W^{(j)}=W^{(j-1)}\cup\text{ k-parameter of } m\text{-dim handles},$$

and $W\simeq\overline{Y}$ rel $M\times\Delta^k$ and preserving the projection to Δ^k, and with a stable isomorphism

(55) $$tW\simeq\pi^*t(M\times\Delta^k)\ .$$

Define $Y^{(i)}=\overline{W^{(i)}-M\times I\times\Delta^k}$ and $\partial_0=Y^{(s)}\cap(M\times I\times\Delta^k)=$ $=\partial Y^{(s)}\cap(M\times\{1\}\times\Delta^k$ and let $Y=Y^{(s)}$. Then (Y,∂_0) is an object in E_k^n (n large), and if f is the composition

(56) $$Y\subset W\xrightarrow{\ \pi\ }M\times\Delta^k\xrightarrow{r\times\Delta^k}X\times\Delta^k\xrightarrow{\ \text{proj}\ }X\ ,$$

then $(Y,\partial_0)\xrightarrow{f}X$ with the stable isomorphism $tY\simeq f^*\xi$ coming from (55) and $tM\simeq r^*\xi$ is an object in $E_k^h(X;\xi)^n$. One checks that this construction has the following properties:

(a) Different choices give homotopic results, in the sense that if $Y^{(0)'}\subset\ldots\subset Y^{(s)'}\longrightarrow X$ is obtained by other choices, then there is a family parametrized by $\Delta^k\times I$ which restricts to Y and Y' over the two ends.

(b) There is a family of cell complexes parametrized by $\Delta^k\times I$ from b_nY to $\overline{Y}$ where $b_n:E_k^h(X;\xi)^n\longrightarrow\overline{E}_k^h(X)$.

(c) The construction is compatible with the suspension $\sum$ in the sense of (a).

(d) The relative version holds -- i.e., if we already have Y over $\partial\Delta^k$, the construction of Y over Δ^k can be made to extend the construction over $\partial\Delta^k$.

Using (c) and (d), we can define c on Z if n is large enough. From (a) and (b), we see that $b_n c(Z)$ and Z are homotopic in $\coprod_k \overline{E}^h_k(X) \times \Delta^k/\sim$ and that if Z' is a finite subcomplex of $\coprod_k E^h_k(X,\xi)^n$, then $cb_n(Z')$ and Z' are homotopic in $\coprod_k E^h_k(X;\xi)^{n'}$ for $n' >> n$. This procedure now extends immediately in the $S.$ direction. To extend in the nerve direction, one only has to observe that cancelling pairs of cells can be simultaneously thickened to cancelling pairs of handles. From these constructions, we see that b is a weak homotopy equivalence. $\square$

Corollary 2.4. *The weak homotopy type of* $BE^h(X;\xi)$ *does not depend on* ξ.

Proof. In fact, if $f:X \longrightarrow Y$ is a continuous map between finite complexes, $\xi \in KO(X)$ and $\eta \in KO(Y)$, then for any finite subcomplexes $Z \subset BE^h(X;\xi)$ and $Z' \subset BE^h(Y;\eta)$, we have maps $c:Z \longrightarrow BE^h(X;\xi)$ and $c':Z' \longrightarrow BE^h(Y;\xi)$ such that the domains of c' and c contain the ranges of c and c' respectively and $c'c(Z)$ is homotopic to Z in $BE^h(X;\xi)$ and $cc'(Z')$ is homotopic to Z' in $BE^h(Y;\xi)$. This follows from Theorem 2.3 and the functorial property of $Wh^{Comb}(X)$.

Remark. If $\xi = f^*\eta$, Corollary 2.4 follows from Proposition 2.2.

III. S. AND THE Q-CONSTRUCTION.

3.1. Definition of $QE.(X;\xi)^n$.

If $Y_1 \rightarrowtail Y_2$ in $E_k(X;\xi)^n$, we shall write $Y_2 \longrightarrow Y_2/Y_1$. Observe that if $Y_3 \rightarrowtail Y_2/Y_1$, there is a unique Y_4 and an arrow $Y_4 \rightarrowtail Y_2$ such that $Y_1 \rightarrowtail Y_2$ factors through $Y_1 \rightarrowtail Y_4 \rightarrowtail Y_2$ and $Y_4/Y_1 = Y_3$. We can think of this as a pull back (restriction)

(57)

$$
\begin{array}{ccc}
Y_1 & =\!=\!=\!= & Y_1 \\
\downarrow & & \downarrow \\
Y_4 & \rightarrowtail & Y_2 \\
\downarrow & & \downarrow \\
Y_3 & \rightarrowtail & Y_2/Y_1
\end{array}
$$

and we note that it is functorial under both vertical and horizontal "composition". Explicitly, define $Y_4 = Y_1 \cup Y_3$ — union as subsets of Y_2 , with $\partial_0 Y_4 = (\partial_0 Y_2) \cap Y_4$. Then we can define $QE.(X;\xi)^n$ as the category with objects rigid handlebodies, and where a morphism from Y_1 to Y_2 is a diagram

(58) $Y_1 \twoheadleftarrow Y_0 \rightarrowtail Y_2$.

Composition is defined in the usual way, using the pullback (57) above. Adding the expansion in an obvious way, we turn $QE.(X;\xi)^n$ into a bicategory and we can try to prove that $Q \simeq S$. just as in the additive case. However, we shall need the result for some other categories as well, so we introduce a more general setting that will include all the cases we need.

3.2. The General Setting.

We first formalize the most important properties of the quotient. This is done as follows. If $\underline{C}$ is a category, we define another category $\underline{M}(\underline{C})$ by

(59) $Ob\ \underline{M}(\underline{C}) = Mor\ \underline{C}$

(60) $Mor(\rho,\theta) = \begin{cases} \phi & \text{if source } (\rho) \neq \text{source } (\theta) , \\ \{\psi \mid \theta = \psi\rho\} & \text{otherwise,} \\ \text{i.e., diagrams} \end{cases}$

$$
\begin{array}{ccc}
 & & C_2 \\
 & \overset{\rho}{\nearrow} & \ \searrow^{\psi} \\
C_1 & \underset{\theta}{\longrightarrow} & C_3
\end{array}
$$

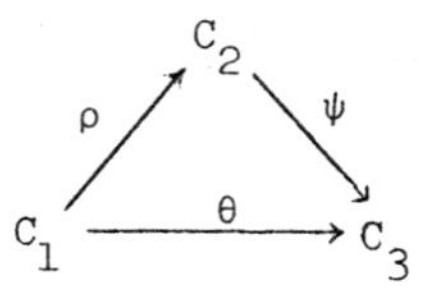

Composition is composition of ψ's. $\underline{\underline{M}}$ is functorial -- i.e. a functor $f:\underline{\underline{C}} \longrightarrow \underline{\underline{D}}$ induces a functor $M(f):\underline{\underline{M}}(\underline{\underline{C}}) \longrightarrow \underline{\underline{M}}(\underline{\underline{D}})$. Let

(61) $$\mu:\underline{\underline{M}}(\underline{\underline{M}}(\underline{\underline{C}})) \longrightarrow \underline{\underline{M}}(\underline{\underline{C}})$$

be the functor defined by

(62) $$\mu\left(C_1 \nearrow C_2 \searrow \atop \longrightarrow C_3\right) = (C_2 \longrightarrow C_3) \ .$$

Let us now define the structure necessary to define $S.\underline{\underline{C}}$ and $Q\underline{\underline{C}}$ and prove that they are homotopically equivalent.

A cofibration category $\underline{\underline{C}}$ is a category with initial object 0 and a functor $q:\underline{\underline{M}}(\underline{\underline{C}}) \longrightarrow \underline{\underline{C}}$ (quotient) satisfying the following:

(a)

(63) $$\begin{array}{ccc} \underline{\underline{M}}(\underline{\underline{M}}(\underline{\underline{C}})) & \xrightarrow{\mu} & \underline{\underline{M}}(\underline{\underline{C}}) \\ {\scriptstyle M(q)}\downarrow & & \downarrow{\scriptstyle q} \\ \underline{\underline{M}}(\underline{\underline{C}}) & \xrightarrow{q} & \underline{\underline{C}} \end{array}$$

commutes.

(b)

(64) $$q(id) = 0 \ ,$$ $$q(B \rightarrow C) = C \iff B = 0 \ .$$

In addition, there is a set of commutative squares.

(65) $$\begin{array}{ccc} A & \xrightarrow{\alpha} & B \\ {\scriptstyle \gamma}\downarrow & & \downarrow{\scriptstyle \beta} \\ C & \xrightarrow{\delta} & D \end{array}$$

making $\underline{\underline{C}}$ into a bicategory and such that each square induces morphisms $q(\alpha) \longrightarrow q(\delta)$ and $q(\gamma) \longrightarrow q(\beta)$, and such that

(c)

(66) $$q(q(\alpha) \longrightarrow q(\delta)) = q(q(\gamma) \longrightarrow q(\beta)) \ .$$

We call these squares admissible.

(d) If

(67) $$\begin{array}{ccc} A & \longrightarrow & B \\ \downarrow & & \downarrow \\ C & \longrightarrow & D \end{array}$$

is admissible, then

(68)
$$\begin{array}{ccc} A & \longrightarrow & C \\ \downarrow & & \downarrow \\ B & \longrightarrow & D \end{array}$$

is also admissible.

(e) $\underline{\underline{C}}$ satisfies the following pullback property:

If $A \xrightarrow{\alpha} B$ and $C \longrightarrow q(\alpha)$ are morphisms, there is a unique admissible square

(69)
$$\begin{array}{ccc} A & \xrightarrow{=} & A \\ \beta\downarrow & & \downarrow\alpha \\ D & \longrightarrow & B \end{array}$$

which induces $C = q(\beta) \longrightarrow q(\alpha)$.

Remarks. (i) (a) is a fancy way of saying that quotients exist and satisfy the usual functorial properties, e.g. $(A/B)/(C/B) = A/C$.

(ii) Think of admissible squares as diagrams of inclusions such that $B \cap C = A$ (as subsets of D).

(iii) If $A \xrightarrow{\alpha} B$ is a morphism, it is convenient to introduce the notation $B \twoheadrightarrow q(\alpha)$. The pullback is then a completion of the diagram

(70)
$$\begin{array}{ccc} A & \xrightarrow{=} & A \\ \downarrow & & \downarrow\alpha \\ D & \dashrightarrow & B \\ \downarrow & & \downarrow \\ C & \longrightarrow & q(\alpha) \end{array} .$$

It follows that all diagrams of the form

(71)
$$\begin{array}{ccc} A & \xrightarrow{=} & A \\ \downarrow & & \downarrow \\ D & \longrightarrow & B \end{array}$$

are admissible, and hence also

(72)
$$\begin{array}{ccc} A & \longrightarrow & D \\ \downarrow= & & \downarrow \\ A & \longrightarrow & B \end{array}$$ by (d) .

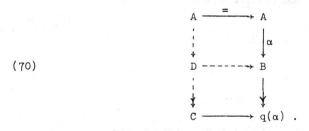

(iv) The pullback property actually implies that $\underline{\underline{C}}$ also can be made a category with the $\twoheadrightarrow$ arrow as morphisms. In fact, if we have $A \xrightarrow{\alpha} B$ and $C \xrightarrow{\beta} q(\alpha)$, we have $B \twoheadrightarrow q(\alpha)$ and $q(\alpha) \twoheadrightarrow q(\beta)$. But if we take pullback of

$$\begin{array}{ccc} & & B \\ & & \downarrow \\ C & \longrightarrow & q(\alpha) \end{array} \tag{73}$$

we get $D \xrightarrow{\gamma} B$ and $q(\beta) = q(\gamma)$ by (a) . Hence we can define $B \twoheadrightarrow q(\gamma)$ as the composition.

(v) It follows from the properties of the pullback that it is functorial both "horizontally" (i.e. with respect to $\longrightarrow$) and "vertically" (i.e. with respect to $\downarrow$) . In other words, in the pullback diagram

(74)

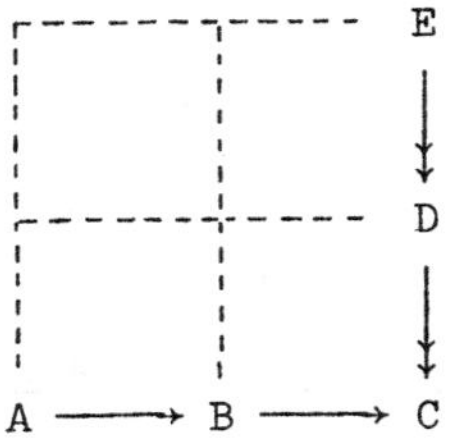

we can fill in the square in a number of ways, and they all give the same result.

(vi) When no confusion should arise, we write $q(A \to B) = B/A$.

Now we are ready to define $S.\underline{\underline{C}}$ and $Q\underline{\underline{C}}$ for a cofibration category $\underline{\underline{C}}$.

$S.\underline{\underline{C}}$: This is a simplicial set with

$$S_k\underline{\underline{C}} = \text{diagrams}$$

$$C_1 \longrightarrow C_2 \quad \cdots \longrightarrow C_k \quad \text{in } \underline{\underline{C}} , \tag{75}$$

and faces and degeneracy maps are defined by

$$d_i(C_1 \longrightarrow \cdots \longrightarrow C_k) = \begin{cases} C_1 \to \cdots \to C_{i-1} \to C_{i+1} \to \cdots \to C_k & \text{for } i > 0 , \\ C_2/C_1 \to C_3/C_1 \to \cdots \to C_k/C_1 & \text{for } i = 1 . \end{cases} \tag{76}$$

$$s_i(C_1 \longrightarrow \cdots \longrightarrow C_k) = \begin{cases} 0 \longrightarrow C_1 \longrightarrow \cdots \longrightarrow C_k , & \text{for } i = 0 , \\ C_1 \to \cdots \to C_i = C_i \longrightarrow \cdots \longrightarrow C_k & \text{for } i > 0 . \end{cases} \tag{77}$$

Remark. To define $S.\underline{\underline{C}}$ we only need (a) and (b) .

$Q\underline{\underline{C}}$: This is the category with the same objects as $\underline{\underline{C}}$, but where a morphism from C to D is a diagram

$$C \twoheadleftarrow D_0 \longrightarrow D \,. \tag{78}$$

Composition is defined by pullback

$$\begin{array}{ccccc} E_1 & \longrightarrow & E_0 & \longrightarrow & E \\ \downarrow & & \downarrow & & \\ D_0 & \longrightarrow & D & & \\ \downarrow & & & & \\ C & & & & \end{array} \tag{79}$$

which is well-defined by fuctoriality of $\longrightarrow$, $\twoheadrightarrow$ (Cf. (iv) and (v) of the remark above.)

3.3. Proof of $S.\underline{\underline{C}} \simeq Q\underline{\underline{C}}$.

Theorem 3.1. $S.\underline{\underline{C}}$ and the nerve of $Q\underline{\underline{C}}$ have the same homotopy type.

Proof. Actually, we shall do the opposite; we shall prove that the category of simplices of $S.\underline{\underline{C}}$ is homotopically equivalent to $Q\underline{\underline{C}}$. Let $\overline{S.\underline{\underline{C}}}$ be the category where the objects are simplices of $S.\underline{\underline{C}}$ and where a morphism from $\sigma \in S_k\underline{\underline{C}}$ to $\iota \in S_\ell\underline{\underline{C}}$ is a non-decreasing function

$$[k] = [0,\dots,k] \xrightarrow{\ \mu\ } [\ell] = [0,\dots,\ell] \tag{80}$$

such that $\sigma = \mu^*\iota$: If $\sigma = (C_1 \longrightarrow \dots \longrightarrow C_k)$ and $\iota = (D_1,\dots,D_\ell)$, this means that $C_j = D_{\mu(j)}/D_{\mu(0)}$, where we set $D_0 = 0$. Clearly, $\overline{S.\underline{\underline{C}}}$ is homotopically equivalent to $S.\underline{\underline{C}}$. ([8, Appendix A]). Now, we define a functor

$$\alpha : \overline{S.\underline{\underline{C}}} \longrightarrow Q\underline{\underline{C}} \tag{81}$$

by $\alpha(C_1 \longrightarrow \dots \longrightarrow C_k) = C_k$, and if $\mu : (C_1 \longrightarrow \dots \longrightarrow C_k) \longrightarrow (D_1,\dots,D_\ell)$, then $\alpha(\mu) = (C_k \twoheadleftarrow D_{\mu(k)} \longrightarrow D_\ell)$. By Quillen's Theorem A of [6], it now suffices to prove that α/X is contractible for all X . This will be done in three steps.

(A) $\alpha^{-1}(X) \longrightarrow \alpha/X$ *is a homotopy equivalence*. In fact, we shall construct a left adjoint to ι. An object in α/X is a diagram

$$\begin{array}{ccc} & & X_0 \longrightarrow X \\ & & \downarrow \\ C_1 \longrightarrow \cdots \longrightarrow & & C_n \end{array} \tag{82}$$

By pullback, we get a new diagram

$$\begin{array}{ccccccccc} C_0' & \longrightarrow & C_1' & \longrightarrow \cdots \longrightarrow & C_{n-1}' & \longrightarrow & X_0 & \longrightarrow & X \\ & & \downarrow & & \downarrow & & \downarrow & & \\ & & C_1 & \longrightarrow \cdots \longrightarrow & C_{n-1} & \longrightarrow & C_n & & \end{array} \tag{83}$$

such that $C_j = C_j'/C_0'$. Set

$$D_i = \begin{cases} C_{i-1}' & i = 1,\dots,n \\ X_0 & i = n+1 \\ X & i = n+2\,. \end{cases} \tag{84}$$

Then $(D_1 \longrightarrow \cdots \longrightarrow D_{n+2})$ is an object in $\alpha^{-1}(X)$ and there is a universal morphism

$$\mu \left(\begin{array}{ccc} & & X_0 \longrightarrow X \\ & & \downarrow \\ C_1 \to \cdots \to & & C_n \end{array} \right) \longrightarrow (D_1 \longrightarrow \cdots \longrightarrow D_{n+2} = X) \tag{85}$$

defined by $\mu(j) = j+1$.

(B) *Let* $\alpha^{-1}(X)_{n.d} \subset \alpha^{-1}(X)$ *be the full subcategory of non-degenerate objects, i.e. objects* $C_1 \longrightarrow \cdots \longrightarrow C_n$ *such that* $C_i \neq C_j$ *if* $i < j$. *Then* $\alpha^{-1}(X)_{n\cdot d} \simeq \alpha^{-1}(X)$. In fact, let ι be the inclusion and consider $\sigma\backslash\iota$ where $(C_1 \longrightarrow \cdots \longrightarrow C_m \longrightarrow X) = \sigma$ is an object in $\alpha^{-1}(X)$. Then $\sigma = v^*\sigma'$ where σ' is non-degenerate, v is surjective, and σ', v are unique. It follows that $\sigma\backslash\iota$ has exactly one object and one morphism.

(C) $\alpha^{-1}(X)_{n.d.}$ *has the initial object* $C_1 = X$: If

$$\mu : (C_1 = X) \longrightarrow (D_1 \longrightarrow \cdots \longrightarrow D_n = X) \tag{86}$$

is a morphism, we have $D_{\mu(1)}/D_{\mu(0)} = X$; hence $D_{\mu(0)} = 0$ and $D_{\mu(1)} = X$. By non-degeneracy, it follows that $\mu(0) = 0$ and $\mu(1) = n$.

3.4. The Additivity Theorem [6] [9].

Let $\underline{\underline{C}}$ be a cofibration category as above. We shall define a new category $\varepsilon(\underline{\underline{C}})$ as follows:

Objects of $\underline{\varepsilon}(\underline{\underline{C}})$ are morphisms in $\underline{\underline{C}}$. Morphisms from $(A\to B)$ to $(C\to D)$ are admissible squares

$$\phi = \begin{array}{ccc} A & \longrightarrow & B \\ \downarrow & & \downarrow \\ C & \longrightarrow & D \end{array}$$

If we define $q(\phi) = (C/A \longrightarrow D/B)$ and admissible "squares" to be "cubes" of admissible squares in $\underline{\underline{C}}$, we can make $\underline{\varepsilon}(\underline{\underline{C}})$ into a new cofibration category (with $(0 \longrightarrow 0)$ as the initial object). Now there are functors:

$$s,q:\underline{\varepsilon}(\underline{\underline{C}}) \longrightarrow \underline{\underline{C}} \tag{87}$$

taking $(A \longrightarrow B)$ to A, B/A respectively. Assume now that $\underline{\underline{A}},\underline{\underline{B}}$ are <u>cofibration subcategories</u> of $\underline{\underline{C}}$, i.e. they are closed under quotients, and if ϕ is a square in the subcategory which is admissible in $\underline{\underline{C}}$, then it is admissible in the subcategory. If we form the pullback

$$\begin{array}{ccc} \underline{\varepsilon}(\underline{\underline{C}};\underline{\underline{A}},\underline{\underline{B}}) & \longrightarrow & \underline{\varepsilon}(\underline{\underline{C}}) \\ {\scriptstyle s\times q}\downarrow & & \downarrow{\scriptstyle s\times q} \\ \underline{\underline{A}} \times \underline{\underline{B}} & \longrightarrow & \underline{\underline{C}} \times \underline{\underline{C}} \end{array} \tag{88}$$

then $\varepsilon(\underline{\underline{C}};\underline{\underline{A}},\underline{\underline{B}})$ is again a cofibration category (subcategory of $\underline{\varepsilon}(\underline{\underline{C}})$), and $s\times q:\underline{\varepsilon}(\underline{\underline{C}};\underline{\underline{A}},\underline{\underline{B}}) \longrightarrow \underline{\underline{A}} \times \underline{\underline{B}}$ preserves all the structure -- hence induces functors

$$\begin{array}{c} S.\varepsilon(\underline{\underline{C}};\underline{\underline{A}},\underline{\underline{B}}) \longrightarrow S.A \times S.B \\ Q\varepsilon(\underline{\underline{C}};\underline{\underline{A}},\underline{\underline{B}}) \longrightarrow Q\underline{\underline{A}} \times Q\underline{\underline{B}} \end{array} \tag{89}$$

which we also denote by $s\times q$.

Before we proceed, we shall add more structure to our cofibration categories. $\underline{\underline{C}}$ is a cofibration category with sum, if there is a composition law $+$, preserving all the structures such that the diagrams

$$\begin{array}{ccc} A & \longrightarrow & A+Y \\ \uparrow & & \uparrow \\ X & \longrightarrow & X+Y \end{array} \qquad\qquad \begin{array}{ccc} C & \twoheadleftarrow & X+C \\ \downarrow & & \downarrow \\ Y & \longrightarrow & X+Y \end{array} \tag{90}$$

are pushout diagrams in $Q\underline{\underline{C}}$. (So the second diagram also reads from Y to $X+C$.) We shall call a cofibration category with sums <u>quasi-exact</u>. Using the sum, we can define a splitting

$$(91) \qquad \chi : Q\underline{A} \times Q\underline{B} \longrightarrow Q\underline{\varepsilon}(\underline{C};\underline{A},\underline{B})$$

of $s \times q$ by $\chi(X,Y) = (X \rightarrowtail X+Y)$.

<u>Theorem 3.2. (The Additivity Theorem).</u>

If $\underline{C}$ is a quasi-exact category, and $\underline{A}$, $\underline{B}$ are subcategories, then χ <u>and</u> $s \times q$ <u>are homotopy inverse of each other</u>.

Before giving the proof, we need a little preparation. Call the morphism of the type $A \rightarrowtail B$ in $Q\underline{C}$ <u>injective</u>, and a morphism of the type $A \twoheadleftarrow B$ <u>surjective</u>. Thus every morphism in $Q\underline{C}$ has a unique decomposition $\alpha = \tau\sigma$, where τ is injective and σ is surjective. But we can also factor α as $\alpha = \sigma'\tau'$ where τ' is injective and σ is surjective: Let α be $A \twoheadleftarrow B_0 \rightarrowtail B$ such that $A = q(B_1 \rightarrowtail B_0)$. Then, there is an admissible square

$$(92) \qquad \begin{array}{ccc} B_1 & \xrightarrow{=} & B_1 \\ \downarrow & & \downarrow \\ B_0 & \rightarrowtail & B \end{array}$$

inducing

$$(93) \qquad \begin{array}{ccc} B_0 & \xrightarrow{\tau} & B \\ \sigma \downarrow & & \downarrow \sigma' \\ A & \xrightarrow[\tau']{} & B/B_1 \end{array}$$

That σ' and τ' are unique follows from the fact that

$$(94) \qquad \begin{array}{ccc} B_0 & \rightarrowtail & B \\ \downarrow & & \downarrow \\ A & \rightarrowtail & C \end{array}$$

commutes in $Q\underline{C}$ if and only if it is a "pullback" diagram in the sense of the definition of cofibration categories.

<u>Proof of the Additivity Theorem</u>. By Quillen's Theorem A of [6], it suffices to prove that the category $\underline{D} = \chi/\phi$ is contractible for every object ϕ in $Q\underline{\varepsilon}(\underline{C};\underline{A},\underline{B})$. The objects of $\underline{D}$ are triples (X,Y,u) where μ is a morphism in $Q\underline{\varepsilon} = Q\underline{\varepsilon}(\underline{C};\underline{A},\underline{B})$ from $(X,X+Y)$ to ϕ . We can think of u as a diagram

$$
(95)\qquad
\begin{array}{ccccc}
s\phi & \longrightarrow & t\phi & \twoheadrightarrow & q\phi \\
\uparrow & & \uparrow & & \uparrow \\
L & \longrightarrow & M & \twoheadrightarrow & N \\
\downarrow & & \downarrow & & \downarrow \\
X & \longrightarrow & X+Y & \twoheadrightarrow & Y
\end{array}
$$

where the upper lefthand squre is admissible and the rest are induced by such squares. Hence $N = M/L$. Here we write $\phi = (s\phi \to t\phi)$, and we shall use the notation u_s, u_t, u_q for the vertical morphisms in $Q\underline{\underline{A}}$, $Q\underline{\underline{C}}$, $Q\underline{\underline{B}}$ respectively.

Let now $\underline{\underline{D}}' \subset \underline{\underline{D}}$ be the full subcategory of objects with u_q injective, and $\underline{\underline{D}}'' \subset \underline{\underline{D}}'$ the full subcategory of objects where u_s is surjective. The theorem follows from the following two lemmas.

<u>Lemma 3.3</u>. <u>$\underline{\underline{D}}' \subset \underline{\underline{D}}$ has a left adjoint</u>.

<u>Proof.</u> By the pushout property of

$$
(96)\qquad
\begin{array}{ccc}
X+N & \twoheadrightarrow & N \\
\downarrow & & \downarrow \\
X & \twoheadrightarrow & Y
\end{array}\ ,
$$

u factors uniquely as

$$
(97)\qquad
\begin{array}{ccccc}
s\phi & \longrightarrow & t\phi & \twoheadrightarrow & q\phi \\
\uparrow & & \uparrow & & \uparrow \\
L & \longrightarrow & M & \twoheadrightarrow & N \\
\downarrow & & \downarrow & & \| \\
X & \longrightarrow & X+N & \twoheadrightarrow & N \\
\| & & \downarrow & & \downarrow \\
X & \longrightarrow & X+Y & \twoheadrightarrow & Y
\end{array}
\qquad \Big\uparrow\ \bar{u}
$$

Let $\overline{(X,Y,u)} = (X,N,\bar{u})$ and let $(X,Y,u) \longrightarrow \overline{(X,Y,u)}$ be the obvious morphism (i.e. $(X=X,\ Y \leftarrowtail N)$). We shall prove that this is universal. So, suppose we have a morphism $(X,Y,u) \longrightarrow (X',Y',u')$ with (X',Y',u') in $\underline{\underline{D}}'$. This means that we have a diagram

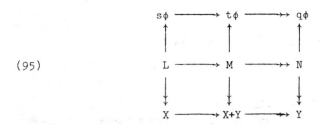

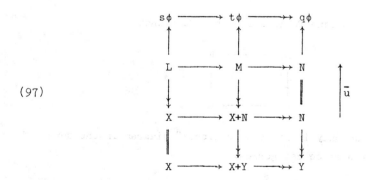

(98)
$$\begin{array}{ccccc}
s\phi & \longrightarrow & t\phi & \twoheadrightarrow & q\phi \\
\uparrow & & \uparrow & & \uparrow \\
N' & \longrightarrow & M' & \longrightarrow & N' \\
\downarrow & & \downarrow & & \| \\
X' & \longrightarrow & X'+Y' & \twoheadrightarrow & Y' \\
\uparrow & & \uparrow & & \uparrow \\
X_0 & \longrightarrow & X_0+Y_0 & \twoheadrightarrow & Y_0 \\
\downarrow & & \downarrow & & \downarrow \\
X & \longrightarrow & X+Y & \twoheadrightarrow & Y
\end{array}$$

such that the upper half is u' , the lower half is a sum, and the composition is u . But the lower half factors uniquely as

(99)
$$\begin{array}{ccccc}
X' & \longrightarrow & X'+Y' & \twoheadrightarrow & Y' \\
\uparrow & & \uparrow & & \uparrow \\
X_0 & \longrightarrow & X_0+Y_0 & \twoheadrightarrow & Y_0 \\
\downarrow & & \downarrow & & \downarrow \\
X & \longrightarrow & X+N & \twoheadrightarrow & N \\
\| & & \downarrow & & \downarrow \\
X & \longrightarrow & X+Y & \twoheadrightarrow & Y
\end{array}$$

and the lemma follows.

Lemma 3.4. $\underline{\underline{D}}'' \subset \underline{\underline{D}}'$ has a left adjoint.

Proof. Let (X,Y,u) be an object in $\underline{\underline{D}}'$ -- i.e. such that $N = Y$ with the notation as in the proof of the previous lemma. Now represent u as $\sigma\tau$ where σ is surjective and τ injective (see the discussion preceeding the proof). We have a representation of u as

(100)
$$\begin{array}{ccccc}
s\phi & \longrightarrow & t\phi & \twoheadrightarrow & q\phi \\
\downarrow & & \downarrow & & \downarrow \\
P & \longrightarrow & Q & \twoheadrightarrow & R \\
\uparrow & & \uparrow & & \uparrow \\
X & \longrightarrow & X+Y & \twoheadrightarrow & Y .
\end{array}$$

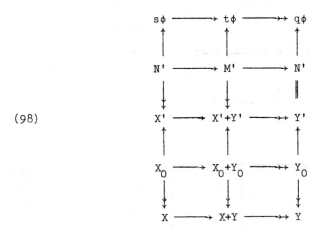

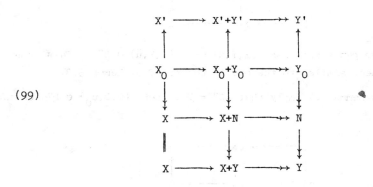

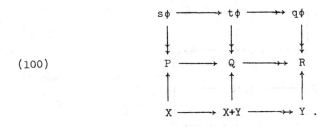

By pushout property of

$$
(101)\qquad
\begin{array}{ccc}
P & \longrightarrow & P+Y \\
\uparrow & & \uparrow \\
X & \longrightarrow & X+Y ,
\end{array}
$$

this factors uniquely as

$$
(102)\qquad
\begin{array}{ccccc}
s\phi & \longrightarrow & t\phi & \twoheadrightarrow & q\phi \\
\downarrow & & \downarrow & & \downarrow \\
P & \longrightarrow & Q & \twoheadrightarrow & R \\
\| & & \uparrow & & \uparrow \\
P & \longrightarrow & P+Y & \twoheadrightarrow & Y \\
\uparrow & & \uparrow & & \| \\
X & \longrightarrow & X+Y & \twoheadrightarrow & Y
\end{array}
\quad \Big\uparrow u
$$

and consequently we get a morphism $(X,Y,u) \longrightarrow (P,Y,\bar{u}) \in \underline{\underline{D}}''$. This is universal by an argument similar to the one in the proof of Lemma 3.3.

Lemma 3.3 and Lemma 3.4 imply that $\underline{\underline{D}}'' \simeq \underline{\underline{D}}$. Let $(0,0,u_0) \in \underline{\underline{D}}''$ be the object with

$$
(103)\qquad
\begin{array}{ccccc}
s\phi & \longrightarrow & t\phi & \twoheadrightarrow & q\phi \\
\| & & \uparrow & & \uparrow \\
s\phi & = & s\phi & \twoheadrightarrow & 0 \\
\downarrow & & \downarrow & & \downarrow \\
0 & = & 0 & \twoheadrightarrow & 0
\end{array}
\quad \Big\uparrow u_0 \qquad .
$$

A morphism from this to another object (X,Y,u) is a pair of morphisms $0 \twoheadleftarrow X_0 \longrightarrow X$, $0 \twoheadleftarrow Y_0 \longrightarrow Y$ such that the composite

$$
(104)\qquad
\begin{array}{ccccc}
s\phi & \longrightarrow & t\phi & \twoheadrightarrow & q\phi \\
\| & & \uparrow & & \uparrow \\
s\phi & \longrightarrow & L & \twoheadrightarrow & Y \\
\downarrow & & \downarrow & & \| \\
X & \longrightarrow & X+Y & \twoheadrightarrow & Y \\
\uparrow & & \uparrow & & \uparrow \\
X_0 & \longrightarrow & X_0+Y_0 & \twoheadrightarrow & Y_0 \\
\downarrow & & \downarrow & & \downarrow \\
0 & \longrightarrow & 0 & \twoheadrightarrow & 0
\end{array}
$$

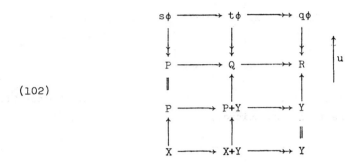

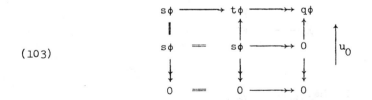

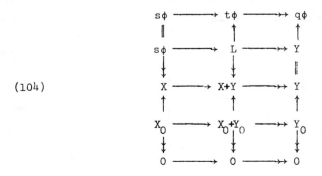

is u_0 . It follows that $X_0 = X$, $Y_0 = 0$. On ther other hand, these choices determine a unique morphism $(0,0,u_0) \longrightarrow (X,Y,u)$. So $(0,0,u_0)$ is an initial object in $\underline{\underline{D}}''$. Therefore, D" is contractible and the theorem follows. □

Remark. This proof is very similar to Quillen's in the additive case [6]. The main difference is that he works with $s \times q : Q\underline{\underline{\varepsilon}} \longrightarrow Q\underline{\underline{A}} \times Q\underline{\underline{B}}$ (actually only for the case $\underline{\underline{A}} = \underline{\underline{B}} = \underline{\underline{C}}$, but that makes no difference). He needs pushout for arbitrary injective and surjective morphism, which we don't have in the category to which we shall apply this. It is somewhat surprising that working with χ instead, we only need the much weaker pushout property.

IV. APPLICATIONS OF THE ADDITIVITY THEOREM

4.1. Some Consequences of the Additivity Theorem.

In this chapter, all categories are quasi-exact. An *exact functor* $f:\underline{\underline{C}} \longrightarrow \underline{\underline{D}}$ is a functor commuting with quotient and preserving admissible squares and sums.

Theorem 4.1. Let F_1, F_2 *be two exact functors* $\underline{\underline{C}} \longrightarrow \underline{\underline{D}}$ *and let* $F_1 \longrightarrow F_2$ *be a natural transformation such that*

$$\begin{array}{ccc} F_1(X) & \longrightarrow & F_1(Y) \\ \downarrow & & \downarrow \\ F_2(X) & \longrightarrow & F_2(Y) \end{array}$$

is admissible for every $X \longrightarrow Y$. *Then* $F_2/F_1 = q(F_1 \longrightarrow F_2)$ *is also exact,* F_1, F_2 *and* F_2/F_1 *induce functors* $Q\underline{\underline{C}} \longrightarrow Q\underline{\underline{D}}$, *and*

$$QF_2 \simeq QF_1 + Q(F_2/F_1)$$

where $+$ *comes from the sum in* $\underline{\underline{D}}$.

Proof. One has to check that F_2/F_1 exists and preserves all the structure. This is slightly tedious, but not difficult. We shall prove that it commutes with quotient. The rest is left out. Let $A \longrightarrow B$ be a morphism in $\underline{\underline{C}}$. Since

$$(105) \qquad \begin{array}{ccc} F_1(A) & \longrightarrow & F_1(B) \\ \downarrow & & \downarrow \\ F_2(A) & \longrightarrow & F_2(B) \end{array}$$

is admissible, the induced morphisms $F_2/F_1(A) \longrightarrow F_2/F_1(B)$ and $F_1(B)/F_1(A) \longrightarrow F_2(B)/F_2(A)$ have the same quotients. But $F_i(B)/F_i(A) = F_i(B/A)$ $(i = 1,2)$, since F_i is exact. For the last statement, we refer to the proof in Quillen [6, §3, Corollary 1]. □

Remark. What we have proven for $Q\underline{\underline{C}}$ is valid for $S.\underline{\underline{C}}$ in view of the homotopy equivalence $Q\underline{\underline{C}} \simeq S.\underline{\underline{C}}$.

Next we observe that some of Waldhausen's constructions in [4] [9] [10] extend to our situation.

Let $F.\underline{\underline{C}}$ be the simplicial set nerve (C) . So $F_n\underline{\underline{C}} = S_{n+1}\underline{\underline{C}}$. Then there is a simplicial map

(106) $$F.\underline{\underline{C}} \longrightarrow S.\underline{\underline{C}}$$

defined by $(C_0 \longrightarrow \ldots \longrightarrow C_n) \longrightarrow (C_1/C_0 \longrightarrow \ldots \longrightarrow C_n/C_0)$. In fact, $F.\underline{\underline{C}}$ and $S.\underline{\underline{C}}$ are also (simplicial) quasi-exact categories. A morphism is a diagram

(107) $$\begin{array}{ccc} C_1 & \longrightarrow \ldots \longrightarrow & C_n \\ \downarrow & & \downarrow \\ C_1' & \longrightarrow \ldots \longrightarrow & C_n' \end{array}$$

of admissible squares, and sum is defined in the obvious way. Then $F.\underline{\underline{C}} \longrightarrow S.\underline{\underline{C}}$ is a simplicial exact functor, and an exact functor $f:\underline{\underline{C}} \longrightarrow \underline{\underline{D}}$ induces exact functors $F.\underline{\underline{C}} \longrightarrow F.\underline{\underline{D}}$ and $S.\underline{\underline{C}} \longrightarrow S.\underline{\underline{D}}$. Now define a new simplicial quasi-exact category $F.(f)$ as the pullback

(108) $$\begin{array}{ccc} F.(f) & \longrightarrow & S.\underline{\underline{C}} \\ \downarrow & & \downarrow \\ F.\underline{\underline{D}} & \longrightarrow & S.\underline{\underline{D}} \end{array} .$$

If we consider $\underline{\underline{D}}$ as a simplicial category in the trivial way, there is a simplicial functor $\underline{\underline{D}} \longrightarrow F.(f)$ defined by $D \longmapsto (D = D = \ldots = D, 0)$.

Theorem 4.2. The induced sequence of simplicial categories

$$Q\underline{\underline{D}} \longrightarrow QF.(f) \longrightarrow QS.\underline{\underline{C}}$$

is a fibration up to homotopy.

Proof. The composite is constant and $QS_n\alpha$ $(\alpha = \underline{\underline{C}}, \underline{\underline{D}}$ or $F.(f))$ is connected, so it suffices to prove that

(109) $$Q\underline{\underline{D}} \longrightarrow QF_n(f) \longrightarrow QS_n\underline{\underline{C}}$$

is a fibration up to homotopy for each n . The trick is to identify $F_n(f)$ with $\underline{\underline{\varepsilon}}(F_n(f);\underline{\underline{D}},S_n\underline{\underline{C}})$ by

(110) $$\begin{array}{l} (A_0 \longrightarrow A_1 \longrightarrow \ldots \longrightarrow A_m, B_1 \longrightarrow \ldots \longrightarrow B_m) \\ \longmapsto ((A_0 = \ldots = A_0, 0 = \ldots = 0) \longrightarrow (A_0 \longrightarrow \ldots \longrightarrow A_m, B_1 \longrightarrow \ldots \longrightarrow B_m)) . \end{array}$$

(Te second cofibration has quotient $(f(B_1 \longrightarrow \ldots \longrightarrow B_m), B_1 \longrightarrow \ldots \longrightarrow B_m))$. By The Additivity Theorem, we then get that

(111) $$QF_n(f) \longrightarrow Q\underline{\underline{D}} \times QS_n\underline{\underline{C}}$$

$(A_0 \longrightarrow \ldots \longrightarrow A_m, B_1 \longrightarrow \ldots \longrightarrow B_m) \longmapsto (A_0, B_1 \longrightarrow \ldots \longrightarrow B_n)$ is a homotopy equivalence. The result now follows from the diagram

(112) $$\begin{array}{ccccc} Q\underline{\underline{D}} & \longrightarrow & QF_n(f) & \longrightarrow & QS_n\underline{\underline{C}} \\ \| & & \downarrow & & \| \\ Q\underline{\underline{D}} & \longrightarrow & Q\underline{\underline{D}}\times QS_n\underline{\underline{C}} & \longrightarrow & QS_n\underline{\underline{C}} \end{array} . \square$$

Corollary 4.3. (A) *The square*

$$\begin{array}{ccc} Q\underline{\underline{C}} & \longrightarrow & QF.(Id_{\underline{\underline{C}}}) \\ \downarrow & & \downarrow \\ Q\underline{\underline{D}} & \longrightarrow & QF.(f) \end{array}$$

is homotopy cartesian, and $QF.(Id_{\underline{\underline{C}}}) \simeq 0$.

(B) S. *is a delooping in the sense that*

$$\Omega|S.S.\underline{\underline{C}}| \simeq |S.\underline{\underline{C}}| .$$

Proof. (A) follows from the map of homotopy fibrations

(113) $$\begin{array}{ccccc} Q\underline{\underline{C}} & \longrightarrow & QF.(id_{\underline{\underline{C}}}) & \longrightarrow & QS.\underline{\underline{C}} \\ \downarrow & & \downarrow & & \| \\ Q\underline{\underline{D}} & \longrightarrow & QF.(f) & \longrightarrow & QS.\underline{\underline{C}} \end{array}$$

that $QF.(id_{\underline{\underline{C}}}) = 0$ follows from the fact that $Q_mF.(id_{\underline{\underline{C}}})$ can be considered the nerve of the category $Q_m(\underline{\underline{C}})$ where $ob(Q_m(\underline{\underline{C}}))$ is the nerve of $Q\underline{\underline{C}}$ in degree m and the morphisms are induced by admissible squares. But this category has an initial object $0 = 0 \ldots = 0$.

(B) follows from the fibration $QC \longrightarrow QF.(id_{\underline{\underline{C}}}) \longrightarrow QS.\underline{\underline{C}}$, (A) and $Q \simeq S$. $\square$

Remark. (A) implies that the fibration in Theorem 4.2 can be continued to the left by $Qf: Q\underline{\underline{C}} \longrightarrow Q\underline{\underline{D}}$. Clearly we have similar results with Q replaced by S .

Now we come to an analogue of Waldhausen's localization theorem [5], [9], [10]. Let $\underline{\underline{C}}$ be a quasi-exact category and let $w\underline{\underline{C}}$ be a subcategory such that $ob(w\underline{\underline{C}}) = ob(\underline{\underline{C}})$. Then, we can form the simplicial category $wS.\underline{\underline{C}}$ where morphisms are diagrams

$$\begin{array}{ccccccc} C_1 & \longrightarrow & C_2 & \longrightarrow \cdots \longrightarrow & C_n \\ \downarrow & & \downarrow & & \downarrow \\ C_1' & \longrightarrow & C_2' & \longrightarrow \cdots \longrightarrow & C_n' \end{array} \tag{114}$$

where vertical maps are in w and squares are admissible. Let C^w denote the full category of $\underline{\underline{C}}$ of objects X such that $(0 \to X) \in w\underline{\underline{C}}$.

We say that $w\underline{\underline{C}}$ is detected by quotients if it satisfies the following property:

(a) $A \longrightarrow B \in w\underline{\underline{C}} \Longleftrightarrow B/A \in \underline{\underline{C}}^w$.

(b) $0 \longrightarrow 0 \in w\underline{\underline{C}}$.

(c) $\alpha\beta \in w\underline{\underline{C}}$ and $\beta \in w\underline{\underline{C}}$ (or $\alpha \in w\underline{\underline{C}}$) $\Longrightarrow \alpha \in w\underline{\underline{C}}$ (or $\beta \in w\underline{\underline{C}}$).

Note that there is a map

$$S.\underline{\underline{C}} \longrightarrow wS.\underline{\underline{C}} \tag{115}$$

where we consider $S.\underline{\underline{C}}$ a category with only identities as morphisms. Then we can form the square

$$\begin{array}{ccc} S.\underline{\underline{C}}^w & \longrightarrow & wS.\underline{\underline{C}}^w \\ \downarrow & & \downarrow \\ S.\underline{\underline{C}} & \longrightarrow & wS.\underline{\underline{C}} \end{array} \tag{116}$$

<u>Theorem 3.4</u>. (<u>The Localization Theorem</u> [5], [9], [10]).

<u>If $w\underline{\underline{C}}$ is detected by cones, this square is homotopy cartesian, and</u> $wS.\underline{\underline{C}}^w \simeq 0$.

<u>Proof</u>. To prove the first statement we use Corollary 3.3 with $f: \underline{\underline{C}}^w \longrightarrow \underline{\underline{C}}$ (and $S.$ instead of Q). In fact, since $S_n\underline{\underline{C}}^w \subset S_n\underline{\underline{C}}$ we can identify objects in $F_n(f)$ with diagrams $C_0 \longrightarrow \cdots \longrightarrow C_n$ such that $C_i/C_0 \in C^w$, $i = 1,\ldots,n$. Since $w\underline{\underline{C}}$ is detected by cones this is equivalent to $(C_0 \to C_i) \in w\underline{\underline{C}}$ for all i, and hence $(C_i \to C_{i+1}) \in w\underline{\underline{C}}$ for all i. Therefore we can identify $F_n(f)$ with $N_n(w\underline{\underline{C}})$, where $N.(-)$ is the nerve functor. This gives a homotopy equivalence

$$S.F_n(f) \simeq N_n(wS.\underline{\underline{C}}) \tag{117}$$

for all n and hence

$$S.F.(f) \simeq wS.\underline{\underline{C}}. \tag{118}$$

Similarly, $S.F.(id_{\underline{\underline{C}}^w}) \simeq wS.\underline{\underline{C}}^w$. □

4.2. Applications to $E_k(X;\xi)^n$.

Let $\mathcal{H}$ be the set of cofibrations $Y \overset{g}{\rightarrowtail} Y'$ in $E_k(X;\xi)^n$ such that the induced map $g_*: Y \cup_{\partial_0} X \longrightarrow Y' \cup_{\partial_0'} X$ is a homotopy equivalence rel X . We form the bicategory $\mathcal{H}E_k(X;\xi)^n$ with $E_k(X;\xi)^n$ as vertical category, $\mathcal{H}$ as a horizontal morphism, and where bimorphisms are diagrams

$$\begin{array}{ccc} Y_1 & \rightarrowtail & Y_2 \\ \downarrow & & \downarrow \\ Y_1' & \rightarrowtail & Y_2' \end{array} \tag{119}$$

which define morphisms in $S_2E_k(X;\xi)^n$. More generally, we define $\mathcal{H}S_\ell E_k(X;\xi)^n$ by requiring morphisms and bimorphisms to preserving filtration and quotients. These assemble to a bisimplicial bicategory $\mathcal{H}S.E.(X;\xi)^n$. Now we do the same construction for $E^h_.(X;\xi)^n$, and we have a diagram

$$\begin{array}{ccc} S.E^h_.(X;\xi)^n & \longrightarrow & \mathcal{H}S.E^h_.(X;\xi)^n \\ \downarrow & & \downarrow \\ S.E.(X;\xi)^n & \longrightarrow & \mathcal{H}S.E.(X;\xi)^n \end{array} \tag{120}$$

where the horizontal functors are induced by the inclusion of the identities in $\mathcal{H}$. This looks very much like the diagram in the Localization Theorem (Theorem 4.4), and indeed we have

Theorem 4.4. The sequence

$$S.E^h_.(X;\xi)^n \longrightarrow S.E.(X;\xi)^n \longrightarrow \mathcal{H}S.E.(X;\xi)^n$$

is a homotopy fibration after geometric realization.

Proof. Let $N^v_.$ denote vertical nerve. Then for every p,q we have simplicial categories $\mathcal{H}N^v_pS.E_k(X;\xi)^n$ and by [9], it suffices to prove that

$$N^v_pS.E^h_k(X;\xi)^n \longrightarrow N^v_pS.E_k(X;\xi)^n \longrightarrow \mathcal{H}N^v_pS.E_k(X;\xi)^n \tag{121}$$

is a homotopy fibration after geometric realization. Let $\underline{\underline{C}} = N^v_pE_k(X;\xi)^n$. Cofibrations in $E_k(X;\xi)^n$ induce on this a structure of a quasi-exact category such that admissible squares are

$$\begin{array}{c}\phi_0\\ \downarrow\\ \vdots\\ \downarrow\\ \phi_p\end{array} \tag{122}$$

where ϕ_i has form

$$\begin{array}{ccc} Y_1 & \rightarrowtail & Y_2 \\ \downarrow & & \downarrow \\ Y_3 & \rightarrowtail & Y_4 \end{array} \tag{123}$$

where $Y_3 \cap Y_2 = Y$. (Hence $\partial_0 Y_3 \cap \partial_0 Y_2 = \partial_0 Y_1$.) Quotients are defined by the obvious extension of the definition of Y_2/Y_1 for $Y_1 \rightarrowtail Y_2$ in $E_k(X;\xi)^n$ and the sum is a disjoint union of Y's . Properties (a) - (d) of the definition of cofibration category (§3.2) are more or less immediate, but we need to check the pullback property (e) . It clearly suffices to do the case $p = 0$. So let $Y_1 \rightarrowtail Y_2$ and $Y_3 \rightarrowtail Y_2/Y_1$ be cofibrations. Then we can form $Y_4 = Y_1 \cup Y_3$ -- considered as the union of subsets of Y_2 , and we let $\partial_0(Y_4) = Y_4 \cap \partial_0(Y_2)$. Then clearly $Y_1 \rightarrowtail Y_4 \rightarrowtail Y_2$ and

$$(Y_4/Y_1 \rightarrowtail Y_2/Y_1) = (Y_3 \rightarrowtail Y_2/Y_1) \; . \tag{124}$$

Now we let $w\underline{\underline{C}}$ be the subcategory where all cofibrations are in H . Then,

$$\underline{\underline{C}}^w = N_p^v E_k^h(X;\xi)^n \; , \tag{125}$$

and (121) becomes

$$S.\underline{\underline{C}}^w \longrightarrow S.\underline{\underline{C}} \longrightarrow wS.\underline{\underline{C}} \; . \tag{126}$$

But this is a fibration by The Localization Theorem. □

Let us define

$$\begin{aligned} A(X;\xi)^n &= \Omega|HS.E.(X;\xi)^n| \\ A(X;\xi) &= \varinjlim_n A(X;\xi)^n \end{aligned} \tag{127}$$

where $A(X;\xi)^n \longrightarrow A(X;\xi)^{n+1}$ by suspension.

Corollary 4.5. *There is a fibration up to homotopy*

$$\Omega|BE(X;\xi)| \longrightarrow A(X;\xi) \longrightarrow BE^h(X;\xi)$$

which is weakly homotopically equivalent to the homotopy fibration (3)

$$\Omega^{fr}(X) \longrightarrow A(X) \longrightarrow WH^{Comb}(X)$$

(if X *is a finite complex) of the Introduction.*

Proof. It follows from Theorem 4.4 that we have

$$\Omega|BE(X;\xi)^n \longmapsto A(X;\xi) \longrightarrow BE^h(X;\xi) \ . \tag{128}$$

Taking limits of the fibrations (128) with respect to n, we have

$$\Omega|BE(X;\xi)| \longrightarrow A(X;\xi) \longrightarrow BE^h(X;\xi) \ . \tag{129}$$

Replacing handles by cells, we have a fibration

$$\Omega|S.\overline{E}.(X)| \longrightarrow \Omega|hS.\overline{E}.(X)| \longrightarrow Wh^{Comb}(X) \ , \tag{130}$$

where $\overline{E}(X)$ is the expansion category of Waldhausen [10, pp. 48-50] and h denotes the weak equivalence of $\overline{E}(X)$ given by homotopy equivalence, and $Wh^{Comb}(X) = |S.\overline{E}.^h(X)|$. As we did in §2.3, (129) is mapped into (130) by functors and we may argue as we did in §2.3 that (120) and (130) are weakly homotopically equivalent. This is tedious but not difficult and we leave it to the reader. On the other hand, it follows from [10] [1] that $\Omega|hS.\overline{E}.(X)|$ is $A(X)$. □

V. INVOLUTIONS.

We shall construct involutions up to homotopy on $BE(X;\xi)$, $BE^h(X;\xi)$ and $A(X;\xi)$, commuting up to homotopy with the maps in the fibration of Corollary 4.5.

5.1. Involutions on $BE(X;\xi)$

The basic observation is that if Y is a k-parameter family of rigid handlebodies on ∂_0 , we can also consider Y a family of handlebodies on $C\ell(\partial Y-\partial_0)$ by interchanging cores and cocores and attaching all handles in the reverse order, denoted by $\overline{Y}$. This clearly defines a functor

(131) $$\tau_n':E_k(X;\xi)^n \longrightarrow E_k(X;\xi)^n$$

such that $\tau_n'\tau_n' = \text{id}$, and we shall show how to extend this to $QE_k(X;\xi)^n$, preserving $QE_k^h(X;\xi)^n$. Note that a cofibration $Y_1 \rightarrowtail Y_2 \twoheadrightarrow Y_2/Y_1$ induces $\overline{Y_2/Y_1} \rightarrowtail \overline{Y}_2 \twoheadrightarrow \overline{Y}_1$ -- i.e., it "turns the arrow around and interchanges the rôles of $\rightarrowtail$ and $\twoheadrightarrow$ ". Hence, if $Y \twoheadleftarrow Z_0 \rightarrowtail Z$ is a morphism from Y to Z in $QE_k(X;\xi)^n$, we obtain a diagram $\overline{Y} \rightarrowtail \overline{Z}_0 \twoheadleftarrow \overline{Z}$ which is a composition of an injective morphism $\overline{Y} \longrightarrow \overline{Z}_0$ and a surjective morphism $\overline{Z}_0 \longrightarrow \overline{Z}$. Clearly, this induces a simplicial functor

(132) $$\tau_n:QE_k(X;\xi)^n \longrightarrow QE_k(X;\xi)^n$$

which preserves $QE_k^h(X;\xi)^n$; hence induces a functor τ_n on this as well. Moreover, $\tau_n^2 = \text{id}$.

Next we stabilize and we have to examine how τ_n behaves under suspension. First we define the negative suspension

(133) $$\textstyle\sum_-:E_k(X;\xi)^n \longrightarrow E_k(X;\xi)^{n+1}$$

by

(134) $$\textstyle\sum_-((Y,\partial_0) \xrightarrow{f} X) = ((Y\times I,\partial_0\times I\cup Y\times\partial I) \xrightarrow{\pi} (Y,\partial_0) \xrightarrow{f} X) .$$

If $H \simeq D^k \times D^{n-k}$ is a handle in Y attached along $S^{k-1} \times D^{n-k}$, $\sum_- H$ is a handle $\simeq D^k \times D^{n-k} \times I$ attached along $S^{k-1} \times D^{n-k} \times I \cup D^k \times D^{n-k} \times \partial I \simeq$ $\simeq S^k \times D^{n-k}$. Hence $\sum_-$ increases the index of every handle by one. It is easy to check that $\sum_-$ is an exact functor taking expansion to expansions -- hence induces a functor

(135) $$\textstyle\sum_-:QE.(X;\xi)^n \longrightarrow QE.(X;\xi)^{n+1} .$$

Lemma 5.1. $\tau'_{n+1}\Sigma = \Sigma_-\tau'_n$ and

$$\tau'_{n+1}\Sigma_- = \Sigma\,\tau'_n\ .$$

Proof. Simply check the definition.

Lemma 5.2. $\Sigma + \Sigma_- \simeq 0\colon QE.(X,\xi)^n \longrightarrow QE.(X;\xi)^{n+1}$.

Proof. Let $(Y,\partial_0) \xrightarrow{f} X$ be an object in $E_k(X;\xi)^n$. Form the object

$$(136)\qquad (Y\times[0,2],\partial_0\times[0,2]\cup Y\times\{0\}) \xrightarrow{\pi} (Y,\partial_0) \xrightarrow{f} X\ ,$$

with the handlebody structure which is $\Sigma(Y,\partial_0)$ on $(Y\times[1,2],\partial_0\times[0,2])$ and $\Sigma_-(Y,\partial_0)$ on $(Y\times[0,1],\partial_0\times[0,1]\cup Y\times\{0,1\})$ (the handles in $Y\times[1,2]$ attached first). This construction induces a functor

$$(137)\qquad F\colon QE.(X,\xi)^n \longrightarrow QE.(X;\xi)^{n+1}\ .$$

We claim that $F \simeq 0$. If H is a handle in (Y,∂_0) , $(\Sigma H\cup\Sigma_- H)$ is a cancelling pair of handles in $F(Y,\partial_0)$. Therefore, we can construct a natural transformation from F to the constant functor 0 in each degree. Hence $F \simeq 0$. Observe now that if we forget expansions, there is a natural transformation $\Sigma \rightarrowtail F$ as in Theorem 4.1 with $F/\Sigma = \Sigma_-$. The theorem does not apply directly, since we have another morphism direction and another simplicial direction (and maps that are homotopic on all skeleta are not necessarily homotopic). However, the proof does, and we conclude that $\Sigma + \Sigma_- \simeq F \simeq 0$. □

Corollary 5.3. τ_n commutes up to homotopy with Σ^2 , i.e.,

$$\Sigma^2\tau_n \simeq \tau_{n+2}\Sigma^2\ .$$

Standard mapping cylinder constructions now give a map τ on $BE(X;\xi)$. This map is not necessarily unique up to homotopy and it may not be an involution, but it induces a unique involution on homotopy and homology. We shall say that τ is algebraically unique and call it an algebraic involution. Similarly we get an algebraically unique involution on $BE^h(X;\xi)$.

5.2. Extension of τ_n to $|Q\mathcal{H}E.(X;\xi)^n|$.

This case is a lot more complicated than $BE(X;\xi)$. In fact, τ is no longer induced by a functor, and the construction does not work unstably (i.e., on $\mathcal{H}QE.(X;\xi)^n$, n finite). The idea is this:

Applying the involution above to a morphism $h\colon Y_0 \rightarrowtail Y_1$ in $\mathcal{H}$, we obtain $\overline{Y}_1 \twoheadrightarrow \overline{Y}_0$. This is no longer a morphism in $\mathcal{H}$, so we need some kind of mapping cylinder construction to turn it into a cofibration. But such construction actually exists if we allow ourselves to increase dimension by one. Let

$$(137)\qquad \begin{aligned} T'(h) &= Y_1 \times [0,2] \cup Y_0 \times [2,3] \subset Y_0 \times [0,3] \\ \partial_0 T'(h) &= \partial_0\overline{Y}_1 \times [0,1] \cup \partial_0\overline{Y}_0 \times [2,3] \cup Y_2/Y_1 \times \{1\}\ . \end{aligned}$$

The handle structure is given as follows: first build up $Y_1\times[0,1]\cup Y_0\times[2,3]$ as $\sum\overline{Y}_1 \coprod \sum\overline{Y}_0$ and then add $Y_1 \times [1,2]$ as $\sum_- \overline{Y}_1$.

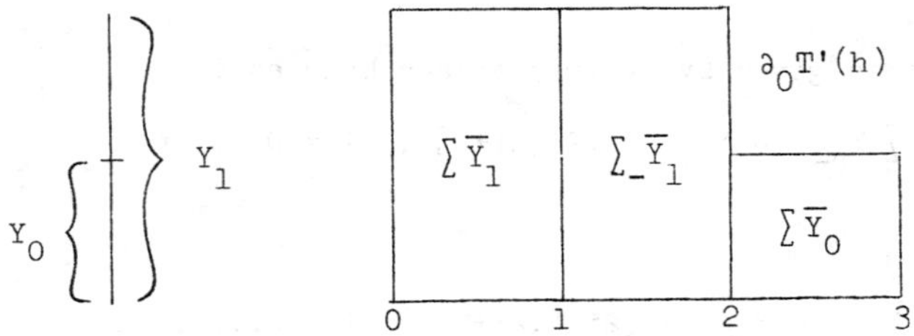

This is an obvious H-morphism $\sum\overline{Y}_1 \rightarrowtail T'(h)$. Moreover, the handles in $Y_0 \times [1,3] \cup C\ell(Y_1-Y_0) \times [0,2]$ come in cancelling pairs, so there is an expansion $\sum \overline{Y}_0 \longrightarrow T'(h)$

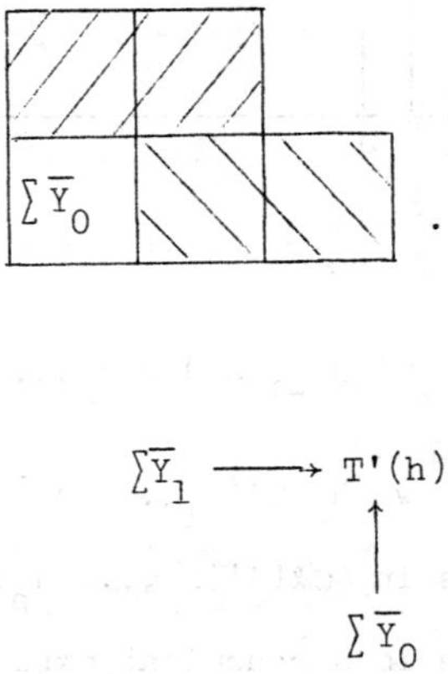

We then have the diagram

$$(138)\qquad \begin{array}{ccc} \sum\overline{Y}_1 & \longrightarrow & T'(h) \\ & & \uparrow \\ & & \sum\overline{Y}_0 \end{array}$$

which, loosely speaking, is an H-map from $\sum\overline{Y}_1$ to $\sum\overline{Y}_0$ "up to a canonical expansion". This construction is functorial with respect to Q-morphisms and expansions, but not H-morphisms. Therefore we wish to get rid of the latter, so we take nerves. Let $QH.E.(X;\xi)^n$ be the resulting bisimplicial bicategory -- $QH.E^n_.$ for short. The objects are sequences

$$(139)\qquad Y_0 \overset{h_1}{\rightarrowtail} Y_1 \overset{h_2}{\rightarrowtail} \cdots \overset{h_p}{\rightarrowtail} Y_p$$

of morphisms in $\mathcal{H}$. We shall extend the diagram (138) to this case. Let

$$T'(h_1,\ldots,h_p) = \bigcup_{i=0}^{p-1} Y_{p-i} \times [2i,2i+2] \cup Y_0 \times [2p,2p+1]$$

as a subspace of $Y_0 \times [0,2p+1]$, and

$$\partial T'(h_1,\ldots,h_p) =$$

$$(140) \quad = \bigcup_{i=0}^{p-1} \partial_0 \overline{Y}_{p-i} \times [2i,2i+2] \cup \bigcup_{i=0}^{p-1} C\ell(Y_{p-i}-Y_{p-i-1}) \times \{2i+2\}$$

$$\cup\ \partial_0\overline{Y}_0 \times [2p,2p+1] .$$

The handle structure is given by: first attach handles in

$$(141) \qquad \sum \overline{Y}_{p-i} = Y_{p-i} \times [2i,2i+1] , \quad i = 0,\ldots,p$$

and then add

$$(142) \qquad \sum_- \overline{Y}_{p-i} = Y_{p-i} \times [2i+1,2i+2] , \quad i = 0,\ldots,p-1 .$$

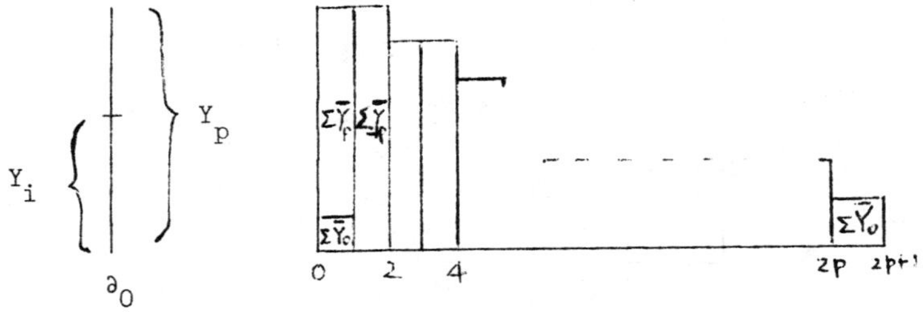

Then clearly we have the $\mathcal{H}$-maps

$$(143) \quad \sum\overline{Y}_p \rightarrowtail T'(h_p) \rightarrowtail T'(h_{p-1},h_p) \rightarrowtail \ldots \longrightarrow T'(h_1,\ldots,h_p)$$

where we identify $T'(h_j,\ldots,h_p)$ with $T'(h_1,\ldots,h_p) \cap Y_p \times [0,2(p-j)+3]$. Moreover, we see that the handles in $C\ell(T'(h_j,\ldots,h_p)-Y_j\times[0,1])$ come in cancelling pairs, such that there is a canonical expansion $\sum\overline{Y}_j \to (h_j,\ldots,h_p)$. Hence we can think of (143) as a string of $\mathcal{H}$-morphism.

$$(144) \qquad \sum \overline{Y}_p \longrightarrow \sum \overline{Y}_{p-1} \longrightarrow \ldots \longrightarrow \sum \overline{Y}_0$$

"up to canonical expansions". The trouble is that, as remarked above, (143) defines functors $Q\mathcal{H}_p E^n_\cdot \longrightarrow Q\mathcal{H}_p E^{n+1}_\cdot$ that do not assemble to a functor $Q\mathcal{H}.E^n_\cdot \longrightarrow Q\mathcal{H}.E^{n+1}_\cdot$. Our plan now is to use the construction to define a functor on a bigger space $Q\hat{H}.E.^{n+1}$ which contains $QH.E.^{n+1}$ as a deformation retract. First, we need some notations and definitions. Let [p] be the category $\{0\leq 1\leq 2\leq \ldots \leq p\}$, considered as a bicategory the trivial way -- with only identities as vertical morphisms. Then we define <p> to be the bicategory with

<u>objects</u>: all relations $k \in S$ where $S \subset \{0,1,\ldots,p\}$.

<u>horizontal morphisms</u>: arrows

$$(k \in S) \rightarrowtail (k' \in S) \qquad (\text{same } S!) \tag{146}$$

with $k \leq k'$.

<u>vertical morphisms</u>: arrows

$$\begin{array}{c}(k \in S)\\ \downarrow \\ (k \in S')\end{array} \tag{147}$$

(same k!) with $S \subset S'$.

<u>bimorphisms</u>: diagrams

$$\begin{array}{ccc}(k \in S) & \rightarrowtail & (k' \in S)\\ \downarrow & & \downarrow\\ (k \in S') & \rightarrowtail & (k' \in S')\end{array} \tag{148}$$

<u>Examples</u>: <1> :

$$\begin{array}{ccccccc} 0\in\{0\} & & 0\in\{0,1\} & & 1\in\{0,1\} & & 1\in\{1\}\\ \bullet & \longrightarrow & \bullet & \rightarrowtail & \bullet & \longleftarrow & \bullet \end{array}$$

<2> :

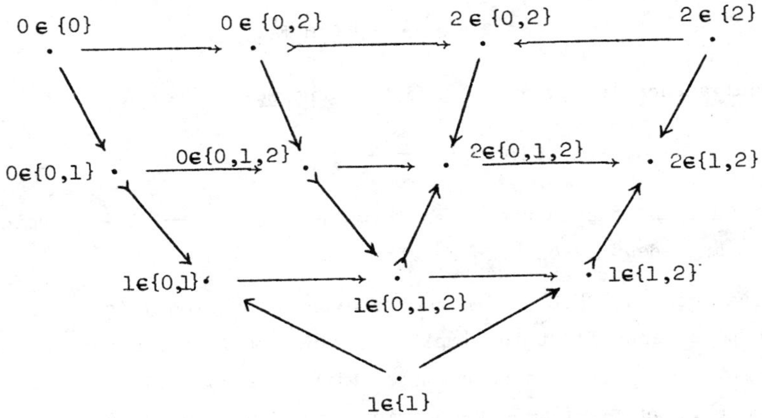

Let $\underline{p}$ denote the set $\{0,1,\ldots,p\}$. There are functors $\alpha:\langle p\rangle \longrightarrow [p]$ and $\beta:[p] \longrightarrow \langle p\rangle$ defined (on objects) by $\alpha(k\in S) = k$, $\beta(k) = k \in \underline{p}$. Then, clearly $\alpha \circ \beta = \mathrm{id}$, and there is a <u>vertical</u> natural transformation

$$\varepsilon:\mathrm{id} \longrightarrow \beta\alpha\ , \quad \varepsilon(k\in S):(k\in S) \longrightarrow (k\in \underline{p}) \tag{149}$$

taking horizontal morphisms to bidiagrams. Now let $v:[p] \longrightarrow [q]$ be a functor (i.e. a non-decreasing map $\underline{p} \longrightarrow \underline{q}$). Then v induces a functor $v:\langle p\rangle \longrightarrow \langle q\rangle$ by $v(k\in S) = v(k) \in v(S)$. Note that v commutes with α, but <u>not</u> with β. (In fact, $v(\alpha(k\in S)) = v(k) = \alpha(v(k) \in v(S))$ but $\beta(v(k)) = v(k) \in \underline{q}$ which in general is <u>not</u> equal to $v(\beta(k)) = (v(k)\in v(\underline{p}))$.) Now, we are ready to define $Q\hat{\mathcal{H}}.E^n_.$. Note that $Q\mathcal{H}_pE^n_.$ may be thought of as the bicategory of functors $[p] \longrightarrow \mathcal{H}E^n_.$, where $\mathcal{H}E^n_.$ is a bicategory as in Chapter IV. Morphisms are natural transformations with respect to Q and expansions in $Q\mathcal{H}.E.$ such that the squares induced by morphisms in $[p]$ satisfy the usual conditions (of being bimorphisms, admissible or induced by such).

We define $Q\hat{\mathcal{H}}_pE^n_.$ by replacing $[p]$ by $\langle p\rangle$ in the above description of $Q\mathcal{H}_pE^n_.$. Then functors $\langle p\rangle \xrightarrow{v} \langle q\rangle$ induces functors $v^*:Q\hat{\mathcal{H}}_qE^n_. \longrightarrow Q\hat{\mathcal{H}}_pE^n_.$, making $Q\hat{\mathcal{H}}.E^n_.$ into a simplicial object (actually a simplicial bicategory -- just like $Q\hat{\mathcal{H}}.E^n_.$. Moreover, $\alpha:\langle p\rangle \longrightarrow [p]$ induces a functor $\alpha^*:Q\mathcal{H}.E^n_. \longrightarrow Q\hat{\mathcal{H}}.E^n_.$, and we have

<u>Lemma 5.4</u>. α^* <u>is a weak homotopy equivalence</u>.

<u>Proof</u>. It suffices to prove that

$$\alpha^*:Q\mathcal{H}_pE^n_. \longrightarrow Q\hat{\mathcal{H}}_pE^n_. \tag{150}$$

is an equivalence for all p, and taking nerves in the Q-direction, we only have to prove that

$$\alpha^*:Q_\ell\mathcal{H}_pE^n_. \longrightarrow Q_\ell\hat{\mathcal{H}}_pE^n_. \tag{151}$$

is a weak equivalence for every p and ℓ. But now we have

$$\beta^*:Q_\ell\hat{\mathcal{H}}_pE^n_. \longrightarrow Q_\ell\mathcal{H}_pE^n_. \ , \tag{152}$$

with $\beta^*\alpha^* = \mathrm{id}$, and the natural transformation $\varepsilon:\mathrm{id} \longrightarrow \beta\alpha$ inducing a homotopy from the identity on $Q_\ell\hat{\mathcal{H}}_pE^n_.$ to $\alpha^*\beta^*$. $\square$

The extra room in $Q\hat{\mathcal{H}}.E^n_.$ (due to the extra expansions) is still not quite enough to extend T' to a simplicial map, but it turns out that we can define a Δ-map, i.e. a map which commutes with face maps [7]. But this gives a map on $|Q\hat{\mathcal{H}}.E^n|$ well-defined up to homotopy and we use Lemma 5.4 to get a map on $|QH.E^n_.|$. Before we do this, we need some more notations. Let $H^{op}E^n_.$ be $HE^n_.$ with the horizontal category replaced by its opposite, similarly for $\hat{H}^{op}E^n_.$. Then $QH^{op}_.E^n_.$, $Q\hat{H}^{op}_.E^n_.$ are defined in the obvious way. Then there are natural cellular homeomorphisms $|QH^{op}_.E^n_.| \simeq |QH.E^n_.|$ and $\|QH^{op}_.E^n_.\| \simeq \|QH.E^n_.\|$ where $\|\ \|$ is the realization of Δ-set, and similarly for $Q\hat{H}^{op}_.E^n_.$. We now extend T' to a functor

$$T = T_n : Q\hat{H}_p E^n_. \longrightarrow Q\hat{H}^{op}_p E^{n+1}_. \tag{153}$$

as follows. Let $u: \langle p\rangle \longrightarrow HE^n_.$ be an object in $Q\hat{H}_p E^n_.$ and let $(k \in S)$ be an object in $\langle p\rangle$. We label the elements of S that are $\geq k$ as $k = k_j < k_{j-1} < \ldots < k_0$. If $k = k_0$, we let $T(u)(k \in S) = \sum u(k \in S)$. Otherwise we have horizontal morphisms

$$u(k_j \in S) \xrightarrow{h_j} u(k_{j-1} \in S) \xrightarrow{h_{j-1}} \ldots \xrightarrow{h_1} u(k_0 \in S) \tag{154}$$

and we define $T(u)(k \in S) = T'(h_j, \ldots, h_1)$. Suppose $k, k' \in S$, and $k < k'$. Then $k' = k_{j'}$ for some $j' < j$, and we know already that there is an H-map

$$\begin{aligned} T(u)(k' \in S) &= T'(h_{j'}, \ldots, h_1) \longrightarrow \\ T(u)(k \in S) &= T'(h_j, \ldots, h_1) \end{aligned} \tag{155}$$

and this is clearly functorial on the horizontal category of $\langle p\rangle$. Suppose now $S \subset S' \subset \underline{p}$, and label the elements of $S' \geq k$, $k = k'_{j'} < k'_{j'-1} < \ldots < k'_0$. Then $k_i \leq k'_i$, $i = 0, \ldots, j$. We have

$$\begin{aligned} T(u)(k \in S') = &\bigcup_{0 \leq i \leq j'} u(k'_i \in S') \times [2i, 2i+2] \\ &\cup u(k \in S') \times [2j', 2j'+1] . \end{aligned} \tag{156}$$

Let

$$\begin{aligned} T(u)(k \in S')|S = &\bigcup_{0 \leq i < j} u(k_i \in S') \times [2i, 2i+2] \\ &\cup u(k \in S') \times [2j, 2j+1] \end{aligned} \tag{157}$$

with handlebody structure as $T'(g_j, \ldots, g_1)$ where

$$u(k \in S') \xrightarrow{g_j} u(k_{j-1} \in S') \xrightarrow{g_{j-1}} \ldots \xrightarrow{g_1} u(k_0 \in S') . \tag{158}$$

This makes sense, since $S' \supset S$ and since $u(k_i \in S') \subset u(k'_i \in S')$. Then

$$
\begin{aligned}
(159)\qquad & C\ell(T(u)(k\in S') - T(u)(k\in S')|S) \\
&= \bigcup_{0\le i\le j} C\ell(u(k_i'\in S') - u(k_i\in S')) \times [2i,2i+2] \\
&\quad \bigcup_{j<i<j'} C\ell(u(k_i'\in S') - u(k\in S')) \times [2i,2i+2] \\
&\quad \cup\ u(k\in S') \times [2j+1,2j'+1] \ .
\end{aligned}
$$

But each $C\ell(u(k_i'\in S') - u(k_i\in S')) \times [2i,2i+2]$, etc. is a union of pairs of cancelling handles, hence there is a canonical expansion $T(u)(k\in S')|S \longrightarrow T(u)(k\in S')$.

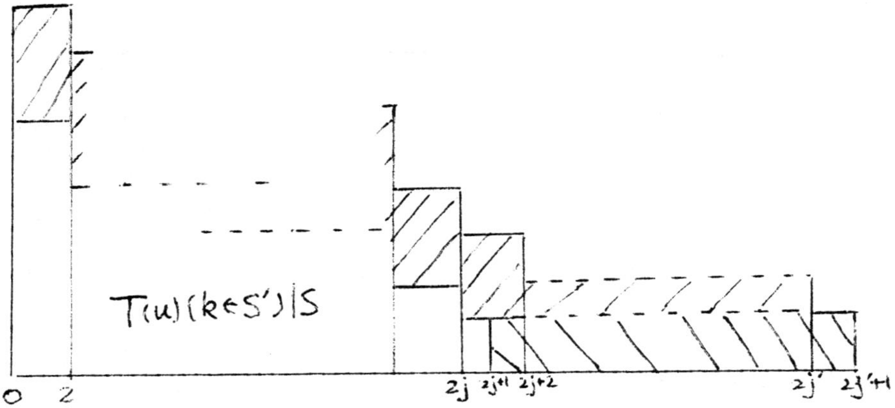

Now we compare $T(u)(k\in S')|S$ with

$$
(160)\qquad T(u)(k\in S) = \bigcup_{0\le i\le j} u(k_i\in S) \times [2i,2i+2] \cup u(k\in S) \times [2j,2j+1] \ ,
$$

and observe that the difference is

$$
(161)\qquad \bigcup_{0\le i\le j} J_i \times [2i,2i+2] \cup J_j \times [2j,2j+1]
$$

where J_i is the union of the handles involved in the expansion $u(k_i\in S) \longrightarrow u(k_i\in S')$. Thus these expansions induce a canonical expansion $T(u)(k\in S) \longrightarrow T(u)(k\in S')|S$. Composing the two, we get the desired expansion $T(u)(k\in S) \longrightarrow T(u)(k\in S')$. $T(u)$, so defined, takes bimorphisms to bimorphisms. Therefore, it defines a functor $\mathrm{ob}Q\hat{H}_p E^n_{\cdot} \longrightarrow \mathrm{ob}Q\hat{H}_p^{op} E^{n+1}_{\cdot}$

which is simplicial with respect to parameters. To see that it actually defines a functor with respect to Q and expansions, it suffices to observe that each $T(u)(k \in S)$ is a union of $\sum\overline{Y}$'s and $\sum_{-}\overline{Y}$'s which both obey this kind of functoriality.

Hence, what remains to be proved is that T commutes with face maps on $\{Q\hat{H}_p E^n_{\cdot}\}_p$. Let $v:\langle p\rangle \longrightarrow \langle q\rangle$ be an injective functor, i.e. a strictly increasing functor $\underline{p} \longrightarrow \underline{q}$. We shall show that the diagram

(162)
$$\begin{array}{ccc} Q\hat{H}_q E^n_{\cdot} & \xrightarrow{T} & Q\hat{H}^{op}_q E^{n+1}_{\cdot} \\ \downarrow & & \downarrow \\ Q\hat{H}_p E^n_{\cdot} & \xrightarrow{T} & Q\hat{H}^{op}_p E^{n+1}_{\cdot} \end{array}$$

commutes.

Let $u:\langle q\rangle \longrightarrow HE^n_{\cdot}$ be an object in $Q\hat{H}_q E^n_{\cdot}$, and let $(k \in S) \in ob\langle p\rangle$. Then,

(163)
$$\begin{aligned} T(v^*u)(k \in S) &= T(u \circ v)(k \in S) \\ &= \bigcup_{i=0}^{j-1} u(vk_i \in vS) \times [2i,2i+2] \cup \\ &\quad u(vk \in vS) \times [2j,2j+1] . \end{aligned}$$

with the same notation as above. On the other hand

(164)
$$\begin{aligned} v^*T(u)(k \in S) &= T(u)(vk \in vS) \\ &= \bigcup_{i=0}^{\ell-1} u(k'_i \in vS) \times [2i,2i+2] \\ &\quad \cup\ u(vk \in vS) \times [2\ell,2\ell+1] , \end{aligned}$$

where $vk = k'_\ell < k'_{\ell-1} < \ldots < k'_0$ are the elements $\geq vk$ in vS. But since v is injective, we must have $\ell = j$ and $k'_i = vk_i$. Hence $T(v^*u)(k \in S) = v^*T(u)(k \in S)$. A similar comparison shows that the diagram commutes on morphisms

Remark. If we do not assume that v is injective, we only get a natural transformation (expansion) $v^*T \longrightarrow Tu^*$.

Let $\|Q\hat{H}.E^n_{\cdot}\|$ be the geometric realization where we only identify according to face maps in the $\hat{H}_{\cdot}$ direction. Then there is a homotopy equivalence

(165)
$$\|Q\hat{H}.E^n_{\cdot}\| \longrightarrow |Q\hat{H}.E^n_{\cdot}|$$

(further collapse of degenerate simplices).

Define

$$\tau'_n : |QH.E^n_.| \xrightarrow{\simeq} \|Q\hat{H}.E^n_.\| \xleftarrow{\simeq} \|QH.E^n_.\|$$

(166)
$$\xrightarrow{T_n} \|Q\hat{H}^{op}_. E^{n+1}_.\| \xrightarrow{\simeq} |Q\hat{H}^{op}_. E^{n+1}_.|$$

$$\xleftarrow{\simeq} |QH^{op}_. E^{n+1}_.| \simeq |QH.E^{n+1}_.| \ .$$

This is not quite what we want, however. The reason is that this is essentially an extension of $Y \longmapsto \sum\overline{Y}$, which we know does not commute with stabilization (suspension). But $Y \longrightarrow \sum^2\overline{Y}$ does, so we define

(167)
$$\tau_n = \sum\tau^m : |QH.E^n_.| \longrightarrow |QH.E^{n+2}_.| \ .$$

This map is now well-defined up to homotopy.

5.3. Stabilization to a Map on $A(X;\xi)$.

The suspension $\sum$ and $\sum_-$ extend immediately to all categories $QH.E^n_.$, $Q\hat{H}.E^n_.$ and their opposite, and they commute with the natural maps between them. However, when we compare with T , we have

Lemma 5.5. $T_{n+1}\sum = \sum_- T_n : Q\hat{H}_p E^n_. \longrightarrow Q\hat{H}^{op}_p E^{n+2}_.$.

Proof. We shall show that $T\sum$ and $\sum_- T$ coincide on objects -- the rest follows easily. Let $u \in ob(\hat{H}_p E^{n+1}_.)$ and $(k \in S) \in ob\langle p\rangle$. If k is the largest element of S , then

(168)
$$(T\textstyle\sum(u))(k \in S) = \sum(\overline{\sum u(k \in S)})$$

$$= \textstyle\sum(\sum_-(\overline{u(k \in S)})) = \sum_-(T(u)(k \in S))$$

$$= \textstyle\sum_-(\overline{\sum u(k \in S)}) = \sum(\sum_- \overline{u(k \in S)}) \ .$$

The remaining cases are reduced to the following: Let $Y_0 \overset{h_1}{\rightarrowtail} \dots \overset{h_\ell}{\rightarrowtail} Y_\ell$ be a sequence of morphisms in H . Then $T'(\sum h_1,\dots,\sum h_\ell) = \sum_- T'(h_1,\dots,h_\ell)$. The proof of this fact consists of a careful writing down of both sides and comparing the results. It is easy but tedious, and we leave out the details. The following picture illustrates for $p = 1$:

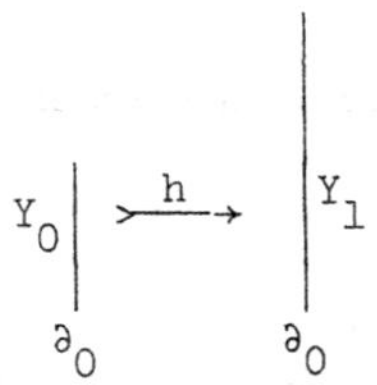

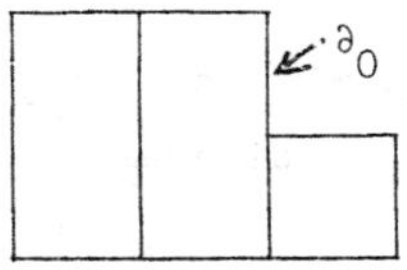

T'(h)

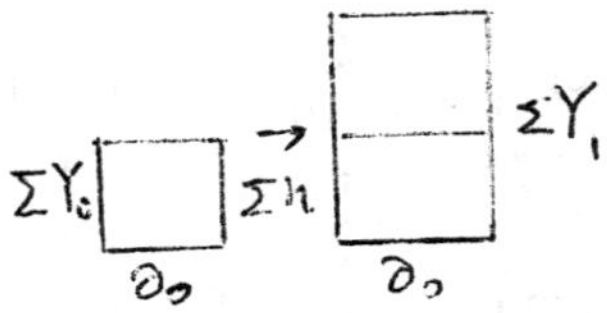

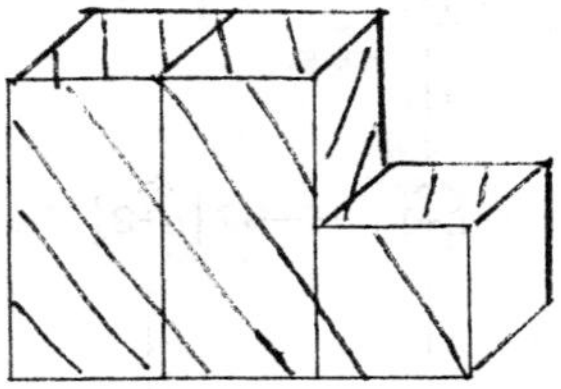

$T'(\sum h) = \sum_- T'(h)$

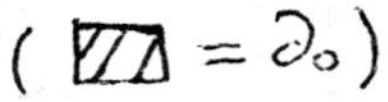

□

Corollary 5.6. τ_n commutes up to homotopy with double suspension, i.e.

$$\textstyle\sum^2 \tau_n \simeq \tau_{n+2} \sum^2 : |QH.E^n_.| \longrightarrow |QH.E^{n+4}_.| \ .$$

Proof. Consider the diagram

$$(168)\qquad \begin{array}{ccccccccc}
|QH.E^n_.| \ldots \|Q\hat{H}.E^n_.\| & \xrightarrow{T_n} & \|Q\hat{H}^{op}_. E^{n+1}_.\| \ldots |QH.E^{n+1}_.| & \xrightarrow{\Sigma} & |QH.E^{n+2}_.| \\
\downarrow \Sigma^2 \qquad \downarrow \Sigma^2 & & \downarrow \Sigma^2_- \qquad \downarrow \Sigma^2_- & & \downarrow \Sigma^2_- \\
|QH.E^{n+2}_.| \ldots \|Q\hat{H}.E^n_.\| & \xrightarrow{T_{n+2}} & \|Q\hat{H}^{op}_. E^{n+3}_.\| \ldots |QH.E^{n+3}_.| & \xrightarrow{\Sigma} & |QH.E^{n+4}_.|
\end{array}$$

where ... denotes the succession of homotopy equivalences involved in the definition of τ_n , τ_{n+2} . This diagram actually commutes, the middle square by Lemma 5.5, and the rest by the remarks at the beginning of this section. If we replace ... by a homotopy equivalence, we get $\tau_{n+2}\sum^2 \simeq \sum^2_- \tau$, and the lemma follows from a generalization of Lemma 5.2. □

Taking loops and stabilizing as for $BE(X;\xi)$, we now get a map $\tau : A(X;\xi) \longrightarrow A(X;\xi)$, which is algebraically unique in the sense of §5.1. Similarly the maps τ commute algebraically with the maps between the spaces of Corollary 4.5. It is obvious in the case of $BE(X;\xi)$ and $BE^h(X;\xi)$ and for $A(X;\xi)$ it follows from the fact that $QE.(X;\xi)^n \simeq QH_0E.(X;\xi)^n \simeq$ $\simeq Q\hat{H}_0E.(X;\xi)^n$ and on $Q\hat{H}_0E(X;\xi)^n$, $T(Y) = \sum\overline{Y}$.

5.4. Proof that τ is algebraically an involution.

We shall prove that for every n, $\tau_{n+2}\tau_n \simeq \sum^4$ from which the assertion follows. Consider the following diagram:

(170)

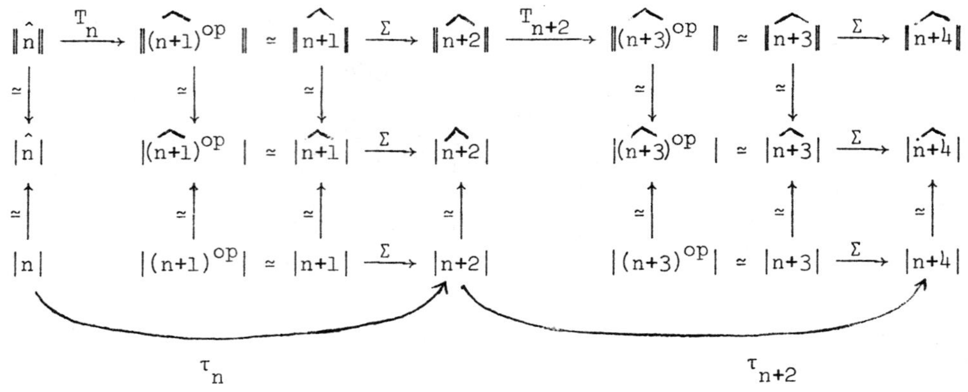

$$\begin{array}{ccccccccccccc}
\|\hat{n}\| & \xrightarrow{T_n} & \|\widehat{(n+1)}^{op}\| & \simeq & |\widehat{n+1}| & \xrightarrow{\Sigma} & \|\widehat{n+2}\| & \xrightarrow{T_{n+2}} & \|\widehat{(n+3)}^{op}\| & \simeq & |\widehat{n+3}| & \xrightarrow{\Sigma} & |\widehat{n+4}| \\
\downarrow\simeq & & \downarrow\simeq & & \downarrow\simeq & & & & \downarrow\simeq & & \downarrow\simeq & & \\
|\hat{n}| & & |\widehat{(n+1)}^{op}| & \simeq & |\widehat{n+1}| & \xrightarrow{\Sigma} & |\widehat{n+2}| & & |\widehat{(n+3)}^{op}| & \simeq & |\widehat{n+3}| & \xrightarrow{\Sigma} & |\widehat{n+4}| \\
\uparrow\simeq & & \uparrow\simeq & & \uparrow\simeq & & \uparrow\simeq & & \uparrow\simeq & & \uparrow\simeq & & \uparrow\simeq \\
|n| & & |(n+1)^{op}| & \simeq & |n+1| & \xrightarrow{\Sigma} & |n+2| & & |(n+3)^{op}| & \simeq & |n+3| & \xrightarrow{\Sigma} & |n+4|
\end{array}$$

with curved arrows $\tau_n : |n| \to |n+2|$ and $\tau_{n+2} : |n+2| \to |n+4|$ along the bottom row.

where we write $|n|$ for $|QH.E^n_{.}|$, $|\hat{n}|$ for $|Q\hat{H}.E^n_{.}|$ etc. It suffices to prove that the top line is $\simeq \sum^4$. But the composite of the four middle maps in this line from $\|(n+1)^{op}\|$ to $\|(n+3)^{op}\|$ has a simpler description, if we observe that T_{n+2} can also be considered as a map T^{op}_{n+2} from $Q\hat{H}^{op}_{.}E^{n+2}_{.} \longrightarrow Q\hat{H}E^{n+3}_{.}$. Then we get a commutative diagram

$$(171)\qquad \begin{array}{ccccc}
\|(n+1)^{op}\| & \xrightarrow{\Sigma} & \|(n+2)^{op}\| & \xrightarrow{T^{op}_{n+2}} & \|n+3\| \\
\uparrow\simeq & & \uparrow\simeq & & \uparrow\simeq \\
\|n+1\| & \xrightarrow{\Sigma} & \|n+2\| & \xrightarrow{T_{n+2}} & \|n+3^{op}\| .
\end{array}$$

Hence we are left to prove that

$$(172)\qquad \sum T^{op}_{n+2} \sum T_n \simeq \sum^4 .$$

Now we use Lemma 5.5 which is also valid for T^{op}. Thus (171) is equivalent to

$$(173)\qquad \sum\sum_- T^{op}_{n+1} T_n \simeq \sum^4 .$$

Since $\sum_-\sum = \sum_-\sum$ and $\sum^2_- \simeq \sum^2$, this will follow from

Lemma 5.6. $T^{op}_{n+1}T_n \simeq \sum\sum_- : \|Q\hat{H}.E^n\| \longrightarrow \|Q\hat{H}.E^{n+2}_{.}\|$.

Proof. We shall construct a natural transformation of expansions $\sum\sum_- \longrightarrow T^{op}_{n+1}T_n$. Let $u:\langle p\rangle \longrightarrow H.E^n_{.}$ be an object of $Q\hat{H}_pE^n_{.}$, and let $(k \in S)$ be an object of $\langle p\rangle$. Then recall the canonical expansion

$$(174) \qquad T(u)(k\in S)\,|\,\{k\} \longrightarrow T(u)(k\in S)\ .$$

But $T(u)(k\in S)\,|\,\{k\}$ is just $\overline{\Sigma u(k\in S)}$, so we have an expansion

$$(175) \qquad \Sigma\overline{u(k\in S)} \longrightarrow T(u)(k\in S)\ .$$

<u>Example.</u> $p = 2$, $S = \{0,1,2\}$, $k = 0$

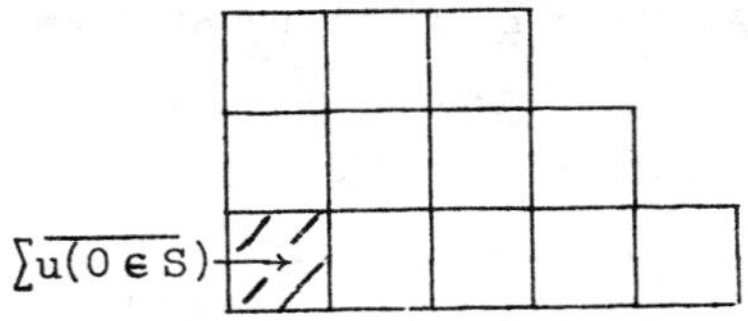

We can also turn this expansion "upside down" as in §5.1 and get an expansion

$$(176) \qquad \Sigma_- u(k\in S) = \overline{\Sigma\overline{u(k\in S)}} \longrightarrow \overline{T(u)(k\in S)}\ .$$

Now take (175) with $T(u)$ instead of u and compare it with the suspension of (176). The result is the desired expansion

$$(177) \qquad e(u)\colon \Sigma\Sigma_- u(k\in S) \longrightarrow \Sigma(T(u)(k\in S) \longrightarrow T(T(u))(k\in S)\ .$$

First we show that $e(u)$ is a natural transformation of functors $\langle p\rangle \longrightarrow \mathcal{H}E^{p+2}_{\cdot}$ (and hence defines a map $\mathrm{ob}\,Q\hat{\mathcal{H}}.E^{n}_{\cdot} \longrightarrow$ expansions in $Q\hat{\mathcal{H}}.E^{n+2}_{\cdot}$) . Observe that

$$(178) \qquad T(T(u))(k\in S) = T(u)(k_0\in S)\times[0,1]\cup\ldots$$

$$= u(k_0\in S)\times[0,1]\times[0,1]\cup\ldots$$

(where k_0 is the largest element of S), and the handle structure on the part $u(k_0\in S)\times[0,1]\times[0,1]$ is as in $\Sigma\Sigma_- u(k_0\in S)$. The expansion $e(u)$ thus comes from cancelling pairs outside $\Sigma\Sigma_- u(k\in S)\subset\Sigma\Sigma_- u(k_0\in S)$. If $k\leq k'\in S$, we have

$$(179) \qquad T(T(u))(k\in S) \rightarrowtail T(T(u))(k'\in S)$$

and the diagram

$$(180) \qquad \begin{array}{ccc} T(T(u))(k\in S) & \subseteq & T(T(u))(k'\in S) \\ \cup\!\mid & & \cup\!\mid \\ \Sigma\Sigma_- u(k\in S) & \subseteq & \Sigma\Sigma_- u(k'\in S)\ . \end{array}$$

Hence we have a bimorphism $\begin{smallmatrix}\bullet & \rightarrowtail & \bullet\\ \uparrow & & \uparrow\\ \bullet & \rightarrowtail & \bullet\end{smallmatrix}$, so $e(u)$ is a natural transformation with respect to horizontal morphisms. But (176) and (177) are natural transformations with respect to vertical morphisms, so $e(u)$ is a natural transformation of functors $\langle p\rangle \longrightarrow \mathcal{H}E_{\cdot}^{n+2}$.

Hence e takes objects of $Q\hat{\mathcal{H}}_pE_{\cdot}^n$ to expansions in $Q\hat{\mathcal{H}}_pE_{\cdot}^{n+2}$ and it remains to show that this is a natural transformation with respect to Q and expansion, and that it commutes with face map in the $\hat{\mathcal{H}}_{\cdot}$ direction. The first part follows from a similar argument as we used above to prove that $e(u)$ is a natural transformation. The second part means that if $v:[p] \longrightarrow [q]$ is an injective functor, the diagram

$$\begin{array}{ccc} ob(Q\hat{\mathcal{H}}_qE_{\cdot}^n) & \xrightarrow{\ e\ } & \text{expansions }(Q\hat{\mathcal{H}}_qE_{\cdot}^{n+2}) \\ \big\downarrow{\scriptstyle v^*} & & \big\downarrow{\scriptstyle v^*} \\ ob(Q\hat{\mathcal{H}}_pE_{\cdot}^n) & \xrightarrow{\ e\ } & \text{expansions }(Q\hat{\mathcal{H}}_pE_{\cdot}^{n+2}) \end{array} \tag{181}$$

commutes. But $v^*e(u)$ and $ev^*(u)$ are both expansions $\sum\sum\_u(vk \in vS) \longrightarrow T(T(u))(vk \in vS)$ (for each $(k \in S)$) , and the definition does not depend on whether we think of the left-hand side as $v^*u(k \in S)$ or $u(vk \in vS)$ (and similarly for the right-hand side) . Taking nerves in the Q direction, we have natural transformations of functors $\sum\sum\_ \longrightarrow T^2 : Q_\ell\hat{\mathcal{H}}_pE_k^n \longrightarrow Q_\ell\hat{\mathcal{H}}_pE_k^{n+2}$ for each ℓ, p, k -- compatible with face and degeneracy maps in ℓ, k , and face maps in p . Then, the natural transformation defines a homotopy between $\sum\sum\_$ and T^2 from $\|Q\hat{\mathcal{H}}.E_{\cdot}^n\|$ to $\|Q\hat{\mathcal{H}}.E_{\cdot}^{n+2}\|$. $\square$

BIBLIOGRAPHY

1. Z. Fiedorowicz, Classifying spaces of topological monoids, preprint.

2. A. E. Hatcher, Higher simple homotopy theory, Ann. of Math. (2), 102 (1975), 101-137.

3. A. E. Hatcher, Concordance spaces, higher simple-homotopy theory, and applications, Proc. Symp. Pure Math. 32, Part I (1978), 3-21.

4. K. Igusa, Higher singularity of smooth functions are unnecessary, reprint.

5. J.-L. Loday, Homotopie des espaces de concordances [d'après F. Waldhausen], Séminaire Bourbaki, Springer Lecture Notes in Math. 710 (1979), 187-205.

6. D. Quillen, Higher algebraic K-theory I, Higher K-theories, Springer Lecture Notes in Math. 341 (1973), 85-147.

7. C. P. Rourke and B. J. Sanderson, Δ-sets I: homotopy theory, Quart. J. Math. Oxford (2), 22 (1971), 321-338.

8. G. Segal, Categories and cohomology theories, Topology 13 (1974), 293-312.

9. F. Waldhausen, Algebraic K-theory of generalized free products, Part I, Ann. of Math. 108 (1978), 135-204.

10. F. Waldhausen, Algebraic K-theory of topological spaces I, Proc. Symp. Pure Math. 32, Part I (1978), 35-60.

DEPARTMENT OF MATHEMATICS
PRINCETON UNIVERSITY
PRINCETON, NEW JERSEY 08544-0037

MATHEMATICAL INSTITUTE
UNIVERSITY OF OSLO
OSLO, NORWAY

Department of Mathematics
Box 37 -- Fine Hall
Princeton University
Princeton, NJ 08544-0037

Canadian Mathematical Society
Conference Proceedings
Volume 2, Part 2 (1982)

IMMERSIONS OF HOMOTOPY-EQUIVALENT MANIFOLDS

Norman Levitt[1]

ABSTRACT. Consider a fixed immersion $j:M^n \to V^{n+k}$, $k \geq 3$, and a simple homotopy equivalence $h:N^n \to M^n$ where M, N, V are PL manifolds. Then $j \circ h$ is homotopic to an immersion. We ask when the structure of this immersion, i.e. its induced stratification by multiple point sets, may be made the same, up to simple homotopy, as that for the original j. We find that the obstructions lie in a certain direct sum of Wall groups, and are determined by the normal invariant of the homotopy equivalence as well as the stratification determined by j. This leads to an exact sequence for the set of homotopy structures on an immersion.

§0. The notion of stratified surgery was introduced by W. Browder and F. Quinn [B-Q] as a tool for the study of group actions on manifolds. In fact, it is usually thought of in that context. However, the theory of stratified surgery, of itself, stands apart from transformation groups. It is the purpose of this paper to study a rather natural geometric problem, not arising from the theory of group actions, in which stratified surgery theory plays a natural, important role.

At this point, the author wishes to thank Bill Browder and Andrew Ranicki for useful and insightful discussions which clarified much which might otherwise still be obscure to him.

We shall study, specifically, the "stratified homotopy type" of stratifications in a closed manifold M^n which arise from an immersion $j:M^n \to V^{n+k}$. Here, it is assumed that M and V are PL manifolds and that j is in general position. This means, quite clearly, that M is stratified by submanifolds of points of varying multiplicity under the immersion j. That is, if we define $M^{(i)}$ to be $\{x \in M | \# j^{-1}(j(x)) = i\}$ we find, in general,

1980 Mathematics Subject Classification -- Primary 57Q25, 57Q35
[1]Supported by NSF Grant No. MCS80-26053

© 1982 American Mathematical Society
0731-1036/82/0000-0477/$09.75

that $M^{(i)}$ is a submanifold of M of dimension $n-(i+1)k$, and that $\mathrm{cl}\, M^{(i)} = \bigcup_{s \geq i} M^{(s)}$. Of course, there is much more to be said concerning the structure of this stratification of M by strata $M^{(i)}$ than is contained in these preliminary observations, but they will serve to establish the context of the problem.

Given a specific immersion $j:M^n \to V^{n+k}$, we define the __multiplicity__ $m(j)$ to be the largest i such that $M^{(i)} \neq 0$.

We assume throughout that the codimension k is ≥ 3.

We now consider a (simple) homotopy equivalence $h:N^n \to M^n$, where N is likewise a PL manifold. Our first observation is rather trivial, in view of the theorem of Haefliger [H_2].

0.1 PROPOSITION. $j \circ k:M^n \to V^{n+k}$ is homotopic to an immersion.

PROOF. Let W^{n+k} denote the total space of the normal (block) bundle of the immersion j. We regard M^n as a submanifold of W^{n+k} in the natural way. Virtually by definition, the immersion $j:M^n \to V^{n+k}$ extends to a codimension-0 immersion $J:W^{n+k} \to V^{n+k}$.

On the other hand, if we consider the composition $N^n \xrightarrow{h} M^n \subset W^{n+k}$, then the theorem of Haefliger implies that it may be deformed to a PL embedding $H:N^n \subset W^{n+k}$. Thus, we obtain $j_0 = J \circ H:M_0^n \to V^{n+k}$ which is obviously an immersion homotopic to $j \circ h$.

0.2 COROLLARY. For a suitable choice of J, H, we will have $m(j_0) = m(j)$. In fact, up to homotopy, h has the property $h^{-1}(M^{(i)}) = N^{(i)}$

In fact, we extend this corollary a bit. Think of W^{n+k} as containing both M and N (the latter under the embedding H). Let $M^{(0)} = W-M$, $N^{(0)} = W-N$. Then the identity map on W is homotopic to a map $\hat{\iota}_H$ with $\hat{\iota}_H^{-1} M^{(0)} = N^{(0)}$ and $\iota_H^{-1} M^{(i)} = N^{(i)}$, $\iota|N = h$ (h having the stratum preserving properties indicated in 0.2 above).

We shall not prove the facts stated above at this time; such a proof will be subsumed in a more categorical discussion in §§1-2 below. The emphasis now is on setting the context.

With the immersions j, J, $J \circ H$ of the respective manifolds M, W, N into V in mind we adopt some notation: the use of a subscript asterisk, e.g. "M_*", "$M_*^{(i)}$", "W_*" etc., denotes the image of the corresponding space (e.g. M, $M^{(i)}$, W) under the immersion. We also observe that, with judicious choice of the immersion $J:W \to V$, we shall have $m(J) = m(j)$. It will then be seen that the stratified map $\hat{\iota}_H : W \to W$ is the "pullback" (in a sense later to be made clear) of a map $\iota_H : W_* \to W_*$ with $\iota_H^{-1} : M^{(i)} = N_*^{(i)}$.

The crux of our analysis is that ι_H is, in fact, a degree-one normal map of stratified spaces, taking W_*, $N_*^{(1)}$, $N_*^{(2)}, \ldots N^{(m(j))}$ to W_*, $M_*^{(1)}$, $M_*^{(2)}, \ldots M^{(m(j))}$. We claim that to say that $J \circ H$ induces on N a stratification simply homotopy equivalent to that induced by j on M is properly understood to mean that ι_H is a simple homotopy equivalence of stratified spaces. In particular, the problem of controlling the stratification of $J \circ H$, relative to that of j, is a matter of changing H within its concordance class so that ι_H becomes bordant to a simple homotopy equivalence. One notes, in this context, that the normal bordism class of ι_H depends only on the class of $h:N \to M$ in the structure set $\mathcal{S}(M)$ of cobordism classes of homotopy structures on M.

The reason that it is natural to undertake to study the relation between the respective stratifications on M and N by a map on the level of the image space W_* is that for j, $J \circ H$ to have precisely the same stratification up to homotopy means not merely that $M^{(i)}$, $N^{(i)}$ be simply homotopy equivalent but that they be equivalent as i-fold covering spaces of their respective images. We also point out at this stage that in the more formal discussion of §2 and §3 below, ι_H will be specifically defined as a map $W_* \to \overline{W}$, where $\overline{W}$ is a stratified space constructed so that

(1) $\overline{W}$ with its stratification is simplified homotopy equivalent to W_* with $M_*^{(1)}$, $M_*^{(2)}$, etc.

(2) The map ι_H is transverse (in the PL sense) to the stratification of $\overline{W}$, and the stratification induced by ι_H on W_* is $N_*^{(1)}$, $N_*^{(2)} \ldots$.

Thus, though on a homotopy level, ι_H may be understood as a map $W_* \to W_*$, we do not claim that the stratification of $J \circ H$ is the transverse inverse image of that j induced by this particular map. Rather, $\overline{W}$ is a sort of bastard object since, stratum-wise, it is simply homotopy equivalent to W_*, $M_*^{(1)}, \ldots M_*^{(m(j))}$, but this equivalence is merely "Poincare" rather than strictly transverse to the structure $M_*^{(i)}$. On the other hand, the "normal data" of the strata of $\overline{W}$ are precisely chosen so that $\iota_H : W_* \to \overline{W}$ will be PL transverse and induce the stratification $N_*^{(1)}, \ldots N_*^{(m(j))}$ on W_*.

We shall assume throughout that $n-(m(j)-1)k \geq 5$, i. e., that the lowest non-void stratum of the stratification of M is a closed manifold of dimension at least 5. Our key result is then the following.

0.3 THEOREM A. Suppose that the stratified surgery problem ι_H is stratified-normal bordant to a stratified (simple) homotopy equivalence. Then the embedding H is concordant to $H_1 : N \subset W$ so that ι_{H_1} is a stratified simple homotopy equivalence.

This result says, in effect, that if the surgery problem of making ι_H a stratified simple homotopy equivalence can be solved abstractly, then it can be solved by altering H within its concordance class.

Theorem A of itself leaves open the question of determining the stratified surgery obstruction $\sigma(\iota_H)$. It seems likely, on intuitive grounds, that $\sigma(\iota_H)$, which, since it depends only on h, will be relabeled $\sigma(h)$, should itself depend only on the normal invariant of h, i.e. the element of $[M, G/PL]$ which is the image of $[h] \varepsilon$ (M) in the usual exact sequence of structure theory. Indeed, this turns out to be the case.

We consider a connected component $A^{(i)}$ of $M_*^{(i)}$, and, with $\pi_1 A^{(i)}$ the usual fundamental group, we let $w(A^{(i)}) : \pi_1 A^{(i)} \to \{\pm 1\}$ denote the orientation character. Let $q(i) = \dim M^{(i)} = \dim M_*^{(i)} = \dim A^{(i)} = n-(i-1)k$ $L^s_{q(i)}(\pi_1 A^{(i)}, wA^{(i)})$ is the usual Wall surgery obstruction group. We adopt the notation

$$L^{(i)} = \bigoplus_{A^{(i)}} L^s_{q(i)}\left(\pi_1 A^{(i)}, w(A^{(i)})\right).$$

(Note $L^{(i)}$ depends on the original immersion j.)

Then, by the general theory of Browder and Quinn, it follows in this case that $\sigma(h) \in \bigoplus_{i=2}^{m(j)} L^{(i)}$. Note that the summation starts at 2 because of the fact that $\iota_H : W_* \to W_*$ is a homotopy equivalence, while on the highest stratum $N^{(1)}_* \to M^{(1)}_*$, the contribution to the stratified surgery obstruction vanishes essentially because $h:N \to M$ is a simple homotopy equivalence.

Thus we may, in some sense, claim to have characterized $\sigma(h)$ if we can determine its coordinates in the various Wall group summands $L^s_{q(i)}\left(\pi_1 A^{(i)}, wA^{(i)}\right)$. To this end we remind the reader of some definitions.

Let $p:\tilde{X} \to X$ be a finite m-fold cover of the C-W complex $\tilde{X}$ (X not necessarily connected) and let ξ^r be an r-block bundle over $\tilde{X}$. We wish to define the <u>transfer</u> <u>bundle</u> $p_! \xi^r$ which will be an mr-dimensional block bundle over X.

Without loss of generality, assume X is a simplicial complex and let t be a typical simplex. Let $t_p, \ldots, t_m$ be the components of $p^{-1}t$, and let $u_i:t \overset{\cong}{\to} t_i$ be the inverse of $p|t_i \to t$. We specify that the block $B(t)$ of $p_! \xi^r$ over t is to be $\bigoplus_{i=1}^{m} p_i{}^*(\xi^r|t_i)$. In other words, if $U \subset X$ has $p|p^{-1}U$ the trivial m-fold cover, then $p_! \xi^r|U$ looks like the Whitney sum of summands $p_i^{-1}{}^* \xi^n|U_i$ where U_i is a component of $p^{-1}U$ and $p_i = p|U_i$. However, if the cover $p:\tilde{X} \to X$ is non-trivial, $p!\xi^r$ is not a global Whitney sum because of the "twisting" of summands by $\pi_1 X$.

Parenthetically, we remind the reader of the analogous transfer construction for r-vector bundles, [A], since we will have some occasion subsequently to discuss the smooth version of certain of our results. The picture for vector bundles may also help the reader clarify his intuitive grasp of the block-bundle case.

Thus, if ξ^r is a vector bundle, $p_!\xi^r$ is a bundle over X whose total space

$$E(p_!\xi^r) = \{(a_1,\ldots,a_m) \in E(\xi^r)\times\cdots\times E(\xi^r) \mid p\pi a_1 = p\pi a_2 \cdots = p\pi a_m, \ \pi a_1,\ldots \pi a_m \text{ all distinct}\}.$$

Here π is the projection $\pi:E(\xi) \to \tilde{X}$. The projection map π' of $E(\xi^r)$ is $\pi'(a_1,\ldots a_m) = p\pi a_1 = p\pi a_2 \cdots = p\pi a_m \in X$. It is easily seen that $p_!\xi^r$ is naturally an mr-vector bundle and that the fiber of $p_!\xi^r$ over $x \in X$ is naturally to be identified, as a vector space, with the sum $\bigoplus_{px_i=x} \xi^r_{x_i}$, where $\xi^r_{x_i}$ is the fiber of ξ^r over $x_i \in \tilde{X}$.

Having defined transfer of block-bundles (and vector bundles) we are in a position to define a certain transfer map $p_!^{\oplus}:[\tilde{X},{}^G/PL] \to [X,{}^G/PL]$ [respectively $p_!^{\oplus}:[\tilde{X},{}^G/O] \to [X,{}^G/O]$]. We do this as follows: It is well known that an element of $[\tilde{X},{}^G/Top]$ may be represented in the following way: two block bundles α^r, β^r over $\tilde{X}$, together with a fiber-homotopy equivalence $f:\alpha^r \to \beta^r$. It is clear that we obtain a transfer operation $p_!^{\oplus}$ by setting

$$p_!^{\oplus}[\alpha^r,\beta^r,f] = [p_!\alpha^r,p_!\beta^r,f']$$

where f' is the obvious induced fiber homotopy equivalence. Note that the transfer on $[\tilde{X},{}^G/PL]$ is that corresponding to the Whitney-sum H-space structure on ${}^G/PL$. It is further to be noted that this construction may be extended to the functor $[\tilde{X},{}^G/Top]$ as well, and is, in this context, the transfer map associated to the Whitney sum H-space structure on ${}^G/Top$. It is known [R] that ${}^G/Top$ supports a distinct H-space structure whose corresponding transfer operation is "well behaved" with respect to surgery obstructions, which is certainly not the case for $p_!^{\oplus}$ (see [W]).

In practice, however, we shall omit the "$\oplus$" notation henceforth, since no confusion will arise.

The transfer construction just outlined comes into play in our present context as follows:

We have already spoken of the spaces $M^{(i)} \subset M$, of points of multiplicity i under the immersion j. But in some sense, the space $M^{(i)}$ of itself gives an incomplete picture of the i-fold point behavior of the immersion j. Therefore, we resort to the following definition.

0.4 DEFINITION. The i-fold <u>resolution</u> of j is a space $\tilde{M}^{(i)}$ whose underlying set is as follows: A point of $\tilde{M}^{(i)}$ consists of an unordered i-tuple, $\{a_1 \ldots a_i\} \subset M$ such that $a_r \neq a_s$ for $r \neq s$ and $pa_1 = pa_2 \cdots = pa_i$, together with a distinguished point $a_r \in \{a_1 \ldots a_i\}$.

As for the topology of $\tilde{M}^{(i)}$ two points $(\{a_1 \cdots a_i\}, a_r)$ and $(\{b_1, b_2 \ldots b_i\}, b_s)$ are considered ε-close if the set $\{a_1, \ldots, a_i\}$ is ε-close to $\{b_1, \ldots, b_i\}$ (in some metric for M) and if, in particular, a_r is ε-close to b_s.

Note that $\tilde{M}^{(i)}$ is a manifold and clearly contains $M^{(i)}$ since the map $y \to (j^{-1}(y), y)$ is clearly an inclusion. Moreover $\tilde{M}^{(i)}$ is an i-fold cover of the space $\underline{M}^{(i)} = \{\{a_1 \cdots a_i\} | a_r \neq a_s$ for $r \neq s$ and $ja_1 = ja_2 \cdots = ja_i\}$. The projection map is: $p^{(i)}: (\{a_1, \cdots a_i\}, a_r) \to \{a_1 \cdots a_i\}$. Furthermore, the restriction $p^{(i)}|M^{(i)}$ may be identified with the map $j|M^{(i)} \to M_*^{(i)}$. In fact, there is a natural map $r^{(i)}: \tilde{M}^{(i)} \to M$ given by $r^{(i)}: (\{a_1 \cdots a_i\}, a_r) \to a_r$ and there is a unique map $r_*^{(i)} \underline{M} \to M_*$ such that

$$\begin{array}{ccc} \tilde{M}^{(i)} & \xrightarrow{r^{(i)}} & M \\ {\scriptstyle p^{(i)}} \downarrow & & \downarrow \\ \underline{M}^{(i)} & \xrightarrow{r_*^{(i)}} & M_* \end{array}$$

is strictly commutative.

Now let $S \in [M, {}^G/PL]$ be the normal invariant of h. $S^{(i)}$ shall mean $(r^{(i)})^* S$, i.e., the composition $\widetilde{M}^{(i)} \xrightarrow{r(i)} M \xrightarrow{S} {}^G/PL$. Let $\underline{A}^{(i)}$ denote a component of $\underline{M}^{(i)}$. Each such $\underline{A}^{(i)}$ contains a unique component $A^{(i)}$ of $M_*^{(i)}$ and all components of $M_*^{(i)}$ are so contained. It is easily seen that under this correspondence $\pi_1(A^{(i)}) \to \pi_1(\underline{A}^{(i)})$ is an isomorphism and the character $w(\underline{A}^{(i)})$ of $\pi_1(\underline{A}^{(i)})$ as the fundamental group of a manifold is merely the orientation character $w(A^{(i)})$ mentioned above. Thus, if $\widetilde{A}^{(i)}$ covers $\underline{A}^{(i)}$ under $p^{(i)}$, we may look at $(p^{(i)}|\widetilde{A}^{(i)})_! (S^{(i)}|\widetilde{A}^{(i)})$ which is an element in $\{\underline{A}^{(i)}, {}^G/PL\}$. Since $\underline{A}^{(i)}$ is a closed manifold we obtain from the above a well-defined surgery obstruction

$$\sigma\big((p^{(i)}|\widetilde{A}^{(i)})_! (S^{(i)}|\widetilde{A}\ )\big) = \sigma_h(A^{(i)}) \in L^s_{q(i)}(\pi_1(A^{(i)}, wA^{(i)})$$

(since $\pi_1 \underline{A}^{(i)}$ is identified with $\pi_1 A^{(i)}$ and $w\underline{A}^{(i)}$ with $wA^{(i)}$). Let $\sigma_h(M^{(i)}) = \sum_{A^{(i)}} \sigma_h(A^{(i)})$. Then $\sum_{i=2}^{m(j)} \sigma_h(M^{(i)})$ is an element of $\sum_{i=2}^{m(j)} L^{(i)}$.

0.5 THEOREM B. $\sum_{i=2}^{m(j)} \sigma_h(M^{(i)}) = \sigma(h)$.

0.6 COROLLARY. $\sigma(h)$ depends only on the normal invariant of h, viz $S \in [M, {}^G/PL]$.

In §§1-2 below, we supply the details of Theorems A and B, as well as justifying some of the preliminary claims which may not have been self-evident in this outline. More specifically, in §1 immediately below, we consider separately the case when $m(j) = 2$, i.e. the highest multiplicity for the immersion is 2. We make this restriction for two reasons: First of all, the flavor of the constructions and proofs in the case of arbitrarily high multiplicity may be realized through a study of the simpler double point case without being overwhelmed by technicalities. Secondly, if we replace the assumption that M, N, V, j are PL, $k \geq 3$ by the assumption that

M, N, V, j are smooth and $2k > n$ then all our arguments carry over virtually intact, and thus our theory becomes useful for the study of smooth immersions in the metastable range.

In §2, we then indicate how the essential ideas of §1 carry over to the case of arbitrary multiplicity.

It is possible to reformulate and slightly extend theorems A and B.

Let M, j, ξ^k, W^{n+k}, J, etc. be as above, with, as usual $n-m(j)k \geq 5$, $k \geq 3$.

0.7 DEFINITION. A <u>homtopy structure</u> on (M,j) consists of a manifold N^n and an embedding $H:N \subset W$ which is a simple homotopy equivalence, such that ι_H is a stratified simple homotopy equivalence.

Two such structures N_1,H_1, N_2,H_2 will be considered equivalent if they are concordant, that is, if H_1 is concordant to H_2 in such a way that ι_{H_1} is concordant to ι_{H_2}. We leave to the reader the task of making this notion quite precise.

We let $\mathcal{J}(M,j)$ denote the set of equivalence classes of homotopy structures on (M,j).

Now let $L_0(M,j)$ denote $\bigoplus_{i=1}^{m(j)} L^{(i)}$ where $L^{(i)}$ has the meaning assigned above for $i \geq 2$ and $L^{(1)}$ merely means $L^s_n(\pi_1 M^n, w(M^n))$.

Let $L_1(M,j)$ be defined similarly to L_0 in that $L_1(M,j)$ shall be $\bigoplus_{i=1}^{m(j)} L_1^{(i)}$ where the summands of $L_1^{(i)}$ correspond to those of $L^{(i)}$, with the summand $L^s_{q(i)}\left(\pi_1 A^{(i)}, wA^{(i)}\right)$ replaced by $L^s_{q(i)+1}\left(\pi_1 A^{(i)}, wA^{(i)}\right)$ [i.e. the summands of $L_1(M,j)$ are the same as for $L_0(M,j)$ save that dimensions have been increased by 1 throughout].

Now let $S \in [M, {}^G/PL]$. We define $\sigma(S) = \sum_{i=1}^{m(j)} \sigma^{(i)}(S)$, where $\sigma^{(i)}S = \sum_{A^{(i)}} \left(p^{(i)}|\tilde{A}^{(i)}_{!} S^{(i)}|\tilde{A}_i\right)$. In particular, $\sigma^{(1)}$ is just the ordinary surgery obstruction for the normal invariant S, $\sigma^{(1)} \in L^s_n\left(\pi_1 M, w(M)\right)$.

0.8 THEOREM C. There is a structure exact sequence

$$[\Sigma M, G/PL] \xrightarrow{\tau} L_1(M,j) \xrightarrow{\alpha} \mathcal{S}(M,j) \xrightarrow{S} [M, G/PL] \xrightarrow{\sigma} L_0(M,j).$$

Here S is the usual normal invariant map, i.e. $S[N,H] = S\{N,H\}$ where $\{N,H\}$ denotes the class of N,H in the familiar structure set $\mathcal{S}(M)$ of homotopy structures on N (H being, to all intents and purposes, a simple homotopy equivalence to M^n). The map σ is, of course the homomorphism described in the paragraph above. We are content to understand the map $L_1(M,j) \xrightarrow{\alpha} (M,j)$ to be an action on the trivial structure. We shall not specify α at this point in any detail.

§1. THE DOUBLE POINT CASE

Before proving Theorems A and B in the special case where the immersion $j:M^n \to V^{n+k}$ has double points, but no points of greater multiplicity, we shall enlarge the context somewhat. In §0 above, we merely supposed that M^n,V^{n+k} were PL, as was the immersion j, and that $k \geq 3$. It is also natural to consider the case where M^n,V^{n+k},j are smooth, and $2k > n$, i.e. the metastable range. Of course, with $2k > n$, it is clear that $m(j) \leq 2$ a fortiori. In the smooth case, we shall naturally assume that N^n is smooth as well. If we consider $W^{n+k} = D(\xi^k)$, where ξ^k is now the normal k-vector bundle of the immersion j, then the composite $N^n \xrightarrow{h} M^n \subset W^{n+k}$ may be deformed into a smooth embedding $H:N^n \subset W^{n+k}$. Here, we rely on Haefliger's metastable smooth embedding theorem $[H_1]$, rather than on the codimension 3 Haefliger $[H_2]$ result. Therefore, by the argument of 0.1 above, $j \circ h:N^n \to V^{n+k}$ is homotopic to a smooth immersion, essentially $J \circ H$ where H is the embedding $N^n \subset W^{n+k}$ and $J:W^{n+k} \to V^{n+k}$ a suitable codimension 0 immersion extending j. The various other observations of §0 apply as well, mutatis mutandus.

Thus, for the record, the constructions and results below apply in either of the cases:

(a) M^n, V^{n+k}, N^n, j PL; $k \geq 3$, $m(j) = 2$.

(b) M^n, V^{n+k}, N^n, j smooth, $2k > n$.

We may now be more specific about certain of the constructions alluded to in §0 without a particular description. First of all we construct a specific immersion $J: W^{n+k} \to V^{n+h}$ with double points but no points of higher multiplicity. First, note that $M^{(2)}$ is a closed submanifold of M of dimension $n-k$. The restriction $p = j|M^{(2)} \to M_*^{(2)}$ is a double cover; we let τ denote the corresponding free involution $\tau: M^{(2)} \leftarrow$ so that $M_*^{(2)} = M^{(2)}/\tau$. Let $\nu_M(M^{(2)})$ denote the normal (block or vector) bundle of $M^{(2)}$ in M. Clearly, we have $\nu_M(M^{(2)}) = \tau^*(\xi^k|M^{(2)})$. It is then clear that, regarding $M_*^{(2)}$ as a submanifold of V^{n+k}, $\nu_V(M_*^{(2)}) = p_!(\xi^k|M^{(2)})$.

Note now the following general fact. If $p:\tilde{X} \to X$ is a 2-fold cover $\tau:\tilde{X} \leftarrow$ the deck transformation and α^r a bundle (block or vector) over $\tilde{X}$, and if D denotes, respectively, the total block-bundle space or the total space of the associated disc bundle, we have a natural extension B^r of $\tau^*\alpha$ to D since $\tilde{X} \to D$ is a homotopy equivalence. Now, let $\tilde{E}$ denote the total space of B^r. We claim that the double cover $p:\tilde{X} \to D$ naturally extends to a double cover $p':\tilde{E} \to E$ where E is respectively the block-bundle space or the disc bundle of the transfer $p_!\alpha^r$ over X. This assertion is merely an extension of the categorical fact that $p^*(p_!\alpha) = \alpha \oplus \tau^*\alpha$.

Thus, in the case at hand let $P \subset M$ denote a tubular neighborhood of $M^{(2)}$ which realizes the block-bundle space (or disc bundle) of $\nu_M(M^{(2)}) = \tau^*(\xi^k|M^{(2)})$. Let U denote the block (disc) bundle space of $\xi^k|P$, as a submanifold of W. Then, thinking of the block (disc) bundle space of $p_!\xi^k$ as realized by a tubular neighborhood U_* of $M_*^{(2)}$, we have an extension J_0 of p to a double cover $J_0: U \to U_*$. It is easily seen that J_0 coincides with j on P.

Now we modify our picture of W slightly; rather than visualizing it merely as the block (disc) bundle of ξ^k over M we think of it as $U \quad W_-$ where U is as above, namely the block (disc) bundle of $\xi^k|P$ and where W_- is a smaller tubular neighborhood of $M_- = cl(M-P)$ in the block (disc) bundle of space of $\xi^k|M_-$. (There is a negligible problem in making ∂W smooth in the smooth case, which we may safely ignore.) Thus, under J, ∂P embeds in ∂U_*, with tubular neighborhood the embedded image of $W_- \cap U$. $J|\partial P = j|\partial P$ extends the embedding to $j|M_-$ (easily seen to avoid U_*) and this in turn extends to J_-, an embedding of $W_- \to V$, agreeing with J_0 on $W_- \cap U$. We let $J = J_0 \cup J_-: U \cup W_- = W \to V$. The diagram below illustrates the construction of J schematically

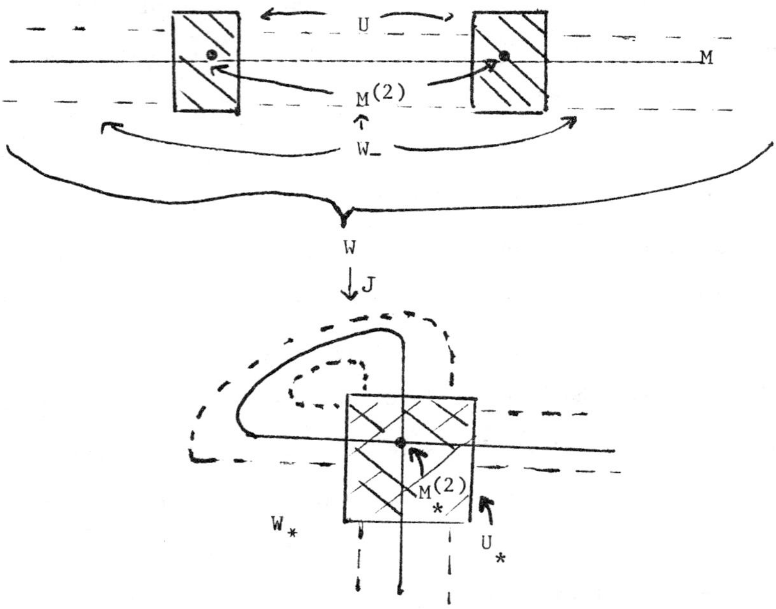

Fig. 1.1

Now, we show how to justify certain assumptions concerning the embedding H, which will give us a reasonable amount of control over the structure of the double point set of the immersion $J \circ H: N^n \to V^{n+k}$.

First of all, note the splitting of M^n into codimension 0 submanifolds P, M_- with common boundary ∂P. Let $Q = h^{-1}P$, $N_- = h^{-1}M_-$ with common boundary $\partial Q = h^{-1}\partial P$. By Wall's extension [W] of the Browder splitting theorem [Br] we see that it may be assumed that $h|Q \to P$, $h|N \to M$, $h|\partial Q \to \partial P$ are all (simple) homotopy equivalences. Moreover, it is easily seen, whether in the smooth metastable case via the appropriate relative version of the smooth Haefliger theorem [H_1] or the PL codimension 3 case, via the relativazation of the PL Haefliger theorem [H_2], that we may assume that the embedding $H:N \to W$ respects this splitting. That is we may assume $H^{-1}U = Q$, $H^{-1}W_- = N_-$, $H^{-1}(U \cap W_-) = \partial Q$. (This is the "splitting" property necessary for the precise definition of ι_H below.)

Several consequences follow from seeing that the conditions above may be imposed on J and H respectively. First of all, it is clear that the double point set $N^{(2)}$ of the immersion $J \circ H:N \to V$ lies in Q, and, moreover there are no points of higher multiplicity than 2. This follows since $J \circ H|N_-$ is clearly an embedding, while $J \circ H|Q$ is an embedding $Q \_ U$ followed by a double cover $U \to U_*$.

Moreover, in either the smooth or PL case, an elementary general position argument suffices to show that $N^{(2)} \subset Q \subset N$ is a manifold. More specifically, if T denotes the obvious extension of the free involution τ on $M^{(2)}$ to a free involution $T:U \circlearrowleft$, then $N^{(2)} = Q \cap TQ \subset \text{int}\, U$, with Q meeting TQ transversally. Thus,

$$\nu_U(N^{(2)}) = \nu_U(Q)|N^{(2)} \oplus \nu_Q N^{(2)} = \nu_U(Q)|N^{(2)} \oplus (T|N^{(2)})^*(\nu_U(Q)|N^{(2)}).$$

If we denote $\nu_U(Q)$ by η^k and the restriction $T|N^{(2)}$ by τ', we may write the above as $\nu_U(N^{(2)}) = (1 \oplus \tau'^*)(\eta^k|N^{(2)})$. Thus, passing to quotient under the action of T, we obtain

$$\nu_V(N_*^{(2)}) = p_!(\eta^k|N^{(2)}).$$

In §0 above, we indicated that to analyze the homotopy type of $N^{(2)}$ relative to $M^{(2)}$, and in particular, whether the two are equivariantly homotopy equivalent, we must set up a certain stratified surgery problem on W_*. We do so now.

For our purposes, it is convenient to take note of the fact that, up to simple homotopy type $M^{(2)}$, P, Q, and U are identifiable in a specific way. For the sake of concreteness we shall take U to be the "model" of this simple homotopy type. Thus, in particular, we may find a bundle over U

corresponding to the bundle $\eta^k|Q$. We refer to this bundle as $\overline{\eta}^k$. Consider the transferred bundle $p_!\overline{\eta}^k$ over U_*. Let E denote the block (resp. disc) bundle of $p_!\overline{\eta}^k$.

We shall construct a certain map $\iota_2: U_* \to E$, transverse to $U_* \subset E$. It is perhaps easier to do this in the smooth case, with its familiar vector bundle language, than in the PL case. Therefore, in the interests of expository clarity, we handle this case first.

Assuming that we are in the smooth case, U may be thought of as the disc bundle $D(\eta^k)$ of the vector bundle $\eta^k_Q = \eta^k|Q$. Thus there is a projection map $\pi: U \to Q$ whose fibers are Euclidean discs of dimension k. Thus $\overline{\eta}^k = \pi^*\eta^k$. Then the associated disc bundle $D(\overline{\eta}^k)$ over U may be written $\{(x,y) \in U\times U | \pi x = \pi y\}$. Here, projection onto U is $(x,y) \to x$ and the fiber over x is $\pi^{-1}(\pi x)$.

Now let us look at the transferred bundle $p_!\overline{\eta}^k$ over U_*. A typical point in $D(p_!\overline{\eta}^k) = E$ may be written $(x, \{y_1,y_2\})$ where $x \in U_*$ and the unordered pair $\{y_1,y_2\}$ has the property $\pi\{y_1,y_2\} = \pi(p^{-1}x)$. I.e., if x_1, x_2 are the two points of $p^{-1}x$ in some order, then $\pi y_1 = \pi x_1$, $\pi y_2 = \pi x_2$. The map $\iota_2: U_*: \to E$ is given by $x \to (x, \{x_1,x_2\})$ where $\{x_1,x_2\} = p^{-1}x$.

Note that ι_2 is transverse to $U_* \subset E$ and, in fact,

$$\iota_2^{-1}U_* = \{x \in U_* | x_1, x_2 \in Q\} = \{x \in U_* | x_1, x_2 \in N^{(2)}\} = N_*^{(2)}.$$

Moreover, if we let $\phi^k = \overline{\eta}^k|M^{(2)}$ we have a natural map $E = D(p_!\overline{\eta}^k) \to D(p_!\phi^k)$. Letting ι_2' be the composition of this with ι_2, then ι_2' is, in essence a fiber homotopy equivalence from $D(p_!\xi^k|M_*^{(2)})$ to $D(p_!\phi^k)$; it is, in fact, the transfer of the obvious homotopy equivalence

$$\begin{array}{ccc} \xi^k|M^{(2)} & \xrightarrow{f} & \phi^k \\ & \searrow \quad \swarrow & \\ & M^{(2)} & \end{array}$$

which represents $S|M^{(2)} \in [M^{(2)}, G/O]$ where S, recall, is the normal invariant of $h: N \to M$.

Thus, the map $N_*^{(2)} \to U_* \sim M_*^{(2)}$, given by restricting ι_2', is in a natural way well known to standard surgery theory [Br, W, R] to be a degree-one normal map, i.e., by definition, the surgery problem representing $p_!(S|M^{(2)}) \in [M_*^{(2)}, G/O]$.

Retracing our steps, we shall proceed to make the analogous construction in the PL case. The technicalities of dealing with block bundle theory make the analogy somewhat cumbersome. We now visualize U as the total space of a block-bundle η^k over U. First of all, this means that Q is decomposed into cells c and U into corresponding blocks b(c). We now further assume that U has been triangulated, with Q a subcomplex, so that the cells c and the blocks b(c) are themselves subcomplexes, and so that the involution $T:U \circlearrowleft$ is a simplicial map.

For each simplex σ of U in this triangulation, we let $\beta(\sigma)$ denote the smallest block b(c) such that $\sigma \in b(c)$. We now note that, since $Q \subset U$ is a homotopy equivalence, there is a unique block bundle $\overline{\eta}^k$ over U with $\overline{\eta}^k|Q \cong \eta^k_Q$. Thus a copy U_1 of U lies in $D(\overline{\eta}^k)$ with $U \cap U_1 = Q$. We use $b_1(c)$ to denote the block of $\eta^k|Q$ corresponding to b(c). We use $b^2(c)$ to denote the block of $\overline{\eta}^k$ lying over the "cell" $b(c) \subset U$. Thus $b_1(c) \subset b^2(c)$.

We now consider a simplex τ of U_* in the obvious triangulation of U_* whose simplices are the images under p of simplices in U. Thus $p^{-1}\tau$ is the disjoint union $\tau_1 \cup \tau_2$ of two simplices of U. We now let

$$K(\tau) = \tau_1 \times \tau_2 \times \beta(\tau_1) \times \beta(\tau_2)$$

and form the union $K = \bigcup_\tau K(\tau)$ with the obvious identifications. There is a natural inclusion $U_* \xrightarrow{\alpha} K$ given by $x \to (x_1, x_2, x_1, x_2) \in K(\tau)$ where $x \in \tau$, $p^{-1}x = \{x_1, x_2\}$, $x_i \in \tau_i$. We claim that this inclusion is a simple homotopy equivalence.

Moreover, there is a natural k-block bundle ρ^{2k} over K. Let $\beta^2(\sigma) = b^2(c)$ for σ a simplex of U, $\beta(\sigma) = b(c)$. Let

$$E(\tau) = \tau_1 \times \tau_2 \times \beta^2(\tau_1) \times \beta^2(\tau_2).$$

Then $E(\tau) \supset K(\tau)$, and, forming the union $E = D(\rho^{2k}) = \underset{\tau}{} E(\tau)$ we obtain we obtain the total space whose blocks are $E(\tau)$ over cells $K(\tau)$. We claim that under the inclusion $\alpha: \tilde{U}_* \to K$, $\rho^{2k}|U_* = p_!\overline{\eta}^k$.

Now we construct $\iota_2: U_* \to E$ by $\iota_2: x \dashrightarrow (x_1, x_2, x_{1,1}, x_{2,1})$ where, as before, $\{x_1, x_2\} = p^{-1}x$ and $x_{1,1}, x_{2,1}$ are the points of $b_1(\sigma) \subset U_1$ corresponding to x_1, x_2 respectively. It is clear that $\iota_2^{-1}K = N_*^{(2)} \subset U_*$, for, to have $(x_1, x_2, x_{1,1}, x_{2,1})$ in K we must have $x_1, x_2 \in Q$ whence $x_{1,1} = x_1, x_{2,1} = x_2$, hence $x_1, x_2 \in Q \cap TQ = N^{(2)}$ whence $x = p(x_1) = p(x_2) \in N_*^{(2)}$. We also claim that $\iota_2: U_* \to E$ is transverse to the base space K.

Analogously to the smooth case, if we let $\phi^k = \overline{\eta}^k|M^{(2)}$ then skirting technicalities, there is a "block-bundle map" $\rho^{2k} \to p_!\phi^k$ and, regarding U as $D(\xi^k|M^{(2)} \oplus \tau^*(\xi^k|M^{(2)}))$, we may think of the composition $p_!\xi^k|M^{(2)}) = U_* \xrightarrow{\iota_2} E \to D(p\,\phi^k)$ as the fiber homotopy equivalence representing $p_!(S|M^{(2)}) \in [M_*^{(2)}, G/PL]$. Thus, just as in the smooth case, $N_*^{(2)} \to K \sim M_*^{(2)}$ is a degree-one normal map, the surgery problem representing the element $p_!(S|M^{(2)})$.

Having constructed ι_2, we wish to extend both domain and range. Once more, we first advert to the smooth case. Note that, with $\iota_2: U_* \to E$ as constructed above, we have, in E, the space

$$X = \{(x, \{y_1, y_2\} \in E | y_1 \in Q \text{ or } y_2 \in Q)\}.$$

Note further that $\iota_2^{-1} X = Q_* \subset U_*$. Moreover, if we let $\dot{E}$ denote the boundary sphere bundle of $p_!\overline{\eta}^k$ contained in E, we see that $X \cap \dot{E}$ lies in $\dot{E}$ as a submanifold and $\iota_2|\partial U \to \dot{E}$ is transverse to $X \cap \dot{E}$ with $\iota_2^{-1}(X \cap \dot{E}) = (\partial Q)_* \tilde{=} \partial Q$. In fact, $X \cap \dot{E}$ is easily seen to be homotopically equivalent to ∂Q via $\iota_2|(\partial Q)_*$.

Now $X \cap \dot{E}$ has, as may be further observed, a bundle neighborhood in $\dot{E}$ realizing η^k where η^k corresponds to $\eta|\partial Q = \nu_{\partial U}(\partial Q_*)$. Let $Y \supset X \cap \dot{E}$ where Y is constructed so that $\partial Q \sim X \cap \dot{E}$ extends to N_-, $\partial Q \sim Y, X \cap \dot{E}$. The bundle $\eta_- = \eta^k|N_-$ corresponds to η_- over Y, and there is an obvious bundle map $\eta_- \to \eta_-$. Thus, if we let $Z = D(\eta_-)$ and form $E \cup Z$, the union being along $D(\eta)$ = the tubular neighborhood of $X \cap \dot{E}$ in E, we obtain a space $\overline{W}$. Note that, since $N_- \to Y$ extends to $\iota_1 : W_- = D(\eta_-) \to D(\eta_-) = Z$, we may form $\iota_H = \iota_1 \cup \iota_2 : W_* = U_* \cup W_- \to E \cup Z = \overline{W}$. Now note that $\overline{W}$ may be thought of as a stratified space with strata U_* (as a subset of E), and $Y \quad X - U$.

1.2 LEMMA: The map ι_H has the following properties:

(1) ι_H is a simple homotopy equivalence;

(2) ι_H is transverse to the stratification of $\overline{W}$ with $\iota_H^{-1} U_* = N_*^{(2)}$,

$\iota_H^{-1}(Y \cup X - U) = N_*^{(1)} \cong N^{(1)}$, $\iota_H^{-1}[W - (Y \cup X)] = N_*^{(0)} \cong N^{(0)}$.

N.B. The notation ι_H was used bcause the construction implicitly depends on the initial embedding $H : N \subset W$. The goal is to see what obstructions arise to making ι_H a stratified simple homotopy equivalence via a change in H preserving concordance class (thereby giving $J \circ H$ the same stratification, up to homotopy, as j).

Briefly, in the PL case the same sort of construction may be put through. Here, be it remembered, E is a block bundle over $K \sim U_*$. We define X as $X(\tau)$ where $X(\tau) \subset E(\tau) = \tau_1 \times \tau_2 \times \beta^2(\tau_1) \times \beta^2(\tau_2)$ is given by

$$X(\tau) = (\tau_1 \times \tau_2 \times \beta(\tau_1) \times \beta^2(\tau_2)) \quad (\tau_1 \times \tau_2 \times \beta^2(\tau_1) \times \beta(\tau_2)).$$

$\dot{E}$ is now defined as the bounding sphere-block bundle of $E = E(\rho^{2k})$ and the and the rest of the construction proceeds analogously, with a stratified map $\hat{W}_*$, $M_*^{(1)}$, $M_*^{(2)} \to \hat{W}$, $X \cup Y - K$, K the result, satisfying Lemma 1.2.

Let us recapitulate: We have a map $\iota_H : W_* \longrightarrow \overline{W}$, in fact, of itself, a homotopy equivalence, and transverse to a certain stratification of $\overline{W}$, so that the induced stratification on W_* is the "natural one" for the immersion $N \subsetneq W \to W_*$. On the other hand, the stratification on $\overline{W}$ is itself stratum-wise homotopy-equivalent to that on W_* induced by the immersion $M \subset W \to W_*$. We leave it to the reader to confirm this observation.

1.3 REMARK. When we say that $\overline{W}$, with its stratification, is stratum-wise homotopy equivalent to the stratification of W_* coming from the immersion j of M, we do not mean to imply that the homotopy equivalence is strictly transverse, in the smooth or PL sense, but merely so in the sense of spherical fibrations. That is, the normal bundle of a stratum of W_* (with respect to j) is merely fiber-homotopy equivalent to the normal bundle of its corresponding stratum in $\overline{W}$. The map $\iota_H W_* \to \overline{W}$ inducing the stratification associated to the immersion of N is, however, strictly transverse.

Moreover, simply viewed as a map of a manifold (with boundary) to a Poincare´ pair, ι_H may be thought of as a normal map, where the stable bundle over $\overline{W}$ into which the stable normal bundle of W maps is merely the pullback to $\overline{W}$ of that bundle under the homotopy inverse of ι_H. Thus, ι_H is a stratified normal map, and the context is thus established for applying Browder-Quinn Theory.

Note that we may "resolve" $\overline{W}$ to form a new space $\hat{W}$ as follows: First, let $\tilde{E}$ be the double cover of E corresponding to the double cover $p:U \to U_*$. Thus $\iota_2 :U_* \to E$ lifts to $\hat{\iota}_2 :U \to \tilde{E}$. Now we may attach Z to $\tilde{E}$.

That is, if we consider $\hat{\iota}_2$, it is a stratified map where the two lower strata are: a copy of U (or a double cover of K in the PL case); and a double cover of $X - U_*$ (resp. X-K). Let D denote the double cover of $X \cap \dot{E}$. This is a trivial covering and D is made up of disjoint pieces D_0, D_1 so that $\hat{\iota}_2^{-1} D_0 = \partial Q$, $\hat{\iota}_2^{-1} D_1 = \partial(TQ)$. We attach $Z = D(\eta_-)$ to $\tilde{E}$ along the tubular neighborhood of D_0 in the double cover of $\dot{E}$, and we extend $\hat{\iota}_2$ to $\hat{\iota}_H :W \to \hat{W}$ by the obvious map. Now $\hat{W}$ has a distinguished codimension -k stratum $\hat{N}$. In the smooth case, this is easily described as $Y \cup R$ where R is the total space of $D(\overline{\eta}^k)$ as it sits in $D(\overline{\eta}^k \oplus T^*(\overline{\eta}^k)) = \tilde{E}$. In the PL case we leave a particular description to the reader. In any event, $\hat{\iota}_H$ is transverse to N with

$$\hat{\iota}_H^{-1}\,\hat{N} = H(N) \cong N,\ \hat{\iota}_H^{-1}\,\hat{R} = Q, \iota^{-1}\,Y = N_-.$$

Thus ι_H^{-1} is a normal map of stratified spaces as well. Of course, $\iota_H : N \to N \sim M$ is a simple homotopy equivalence.

Recall that $Z \quad \dot{E} \subset \dot{E} \subset \overline{W}$ is the disc-(block) bundle neighborhood of $X \quad \dot{E} = Y \quad \dot{E}$. Let C_1 be the closure of the complement of $Z \cap \dot{E}$ in $\dot{E}$ Let C_2 be the bounding sphere bundle of $Z = D(\eta_-)$. Set $C_0 = C_1 \cap C_2$. Then by construction $\iota_2^{-1}(C_1 \cup C_2) = \partial W_*$, and $\iota | \partial W_* \to C_1 \cup C_2$ is a simple homotopy equivalence. Moreover $\iota^{-1}C_1$ is the closure of the complement of a tubular neighborhood of ∂Q_* in ∂U_*, $\iota^{-1}C_2$ is closure $(\partial W_- - U_*)$ and ι restricted to these subspaces is a simple homotopy equivalence onto the respective images C_1, C_2. Similarly $\iota^{-1}(C_0)$ is homotopically equivalent to C_0 via the restriction of ι.

Consider now a normal bordism of the stratified normal map ι, that is a normal map $I : A \to \overline{W}$ where A is an n+k-manifold with $W_* \quad \partial A$, and where there is another (n+k)-manifold $B \quad \partial A$ disjoint from W_*. We assume that $I^{-1}(C_1 \cup C_2) = cl(\partial A - W_* - B)$ and that I is transverse to the stratification of $\overline{W}$. Let $\iota_1 = I | B$. In this case we say that W_*, ι is normally bordant to B, ι_1 via A, I.

Note that A, and the map I may be "resolved" to produce a manifold $\hat{A}$ and a normal map $\hat{I} : \hat{A} \to \hat{W}$ extending ι. There is a natural map $\hat{A} \to A$ which is a double cover on $\hat{I}^{-1}\tilde{E} \to I^{-1}E$, and a diffeomorphism (resp. PL homeomorphism) on the complement. B, ι_1 is simultaneously resolved as $\hat{B}, \hat{\iota}_1$, and $\hat{A}, \hat{I}$ is a normal bordism of stratified maps from W, $\hat{\iota}_H$ to $\hat{B}, \hat{\iota}_1$. We leave details of this resolution to the reader, noting only that what is involved merely mimics the resolution of $\overline{W}$ to $\hat{W}$.

1.4 THEOREM A_2. If W, ι_H is normally bordant to W_1, ι_H where ι_1 is a stratified simple homotopy equivalence then $H: N \subset W$ is concordant to an embedding G such that ι_G is a stratified homotopy equivalence.

We remark that theorem 1.4 is merely theorem A in the case $m(j) = 2$.

PROOF OF 1.4. Assume that W_*, ι_H is normally bordant to the stratified simple homotopy equivalence W_1, ι_1, via the stratified normal bordism A, I. It follows that the resolution $\hat{W}_1, \hat{\iota}_1$ is a stratified simple homotopy equivalence of $\hat{W}_1, \hat{N}_1$ to $\hat{W}, \hat{N}$ where $N_1 = \iota_1^{-1} N$.

STEP 1. First, we consider the resolution $\hat{I}: \hat{A} \to \hat{W}$. This may be taken to be a normal stratified map $\hat{A}, V \to \hat{W}, \hat{N}$ where $V = \hat{I}^{-1}(\hat{N})$. Noting that $\hat{\iota}_H, \hat{\iota}_1$ are stratified simple homotopy equivalences $W, N \to \hat{W}, \hat{N}$, $\hat{W}_1, \hat{N}_1 \to \hat{W}, \hat{N}$ respectively $(\hat{N}_1 = \iota_1^{-1} \hat{N})$, we ask whether stratified surgery may be done on $\hat{I}$, rel$(W \cup \hat{W}_1)$ so as to convert it to a stratified trivial cobordism. The obstruction is a single element $a \in L^s_{n+1}(\pi_1 \hat{N}, w(\hat{N}))$. Assuming this is non-zero, we may modify I as follows: Note that, if set $W_1^- = \iota_1^{-1}(Z)$, $N_1^- = \iota_1^{-1}(Y)$ then $\iota_1 | W_1$ is a stratified simple homotopy equivalence $W_1^-, N_1' \to Z, Y$ as is its restriction to $W_1^- \cap (\iota_1^{-1} \dot{E})$, $N_1 \cap (\iota_1^{-1} \dot{E})$ $\longrightarrow Z \cap \dot{E}, Y \cap \dot{E}$. We may then do stratified surgery on W_1^-, $\iota_1 | W_1^-$ (rel $W_1^- \cap \iota_1^{-1}(\dot{E})$) to convert it to yet another stratified simple homotopy equivalence, $\iota' : W' \to Z$. Let the trace of the surgery be A_1, I_1. This may be chosen so that the obstruction to doing surgery on A_1, I_1 (i.e. on the interior) to make it the trivial bordism turns out to be the element $-a \in L^s_{n+1}(\pi_1 Y, w(Y)) \cong L^s_{n+1}(\pi_1, \hat{N}, w(\hat{N}))$. Now take I_2 to be $I \cup I_1$ on $A_2 = A_{\overline{W_1}} A_1$, and ι_2 to be $\iota' \cup (\iota_1 | \iota_1^{-1}(E))$. Then we may replace I by $\hat{I}_2$, W_1 by $W' \cup \iota_1^{-1}(E) = W_2$. We then find that the obstruction to making I_2 to a (stratified) trivial bordism by surgery will be $a + (-a) = 0$. So we may therefore relabel, and assume $a = 0$ to begin with.

STEP 2. We may assume that $\iota_1|\partial W \to C_1 \cup C_2$ is a simple homotopy equivalence, and the restrictions to $\iota_1^{-1} C_1$, $\iota_1^{-1} C_2$, $\iota_1^{-1} C_0$ simple homotopy equivalences to their respective images. For it is easily seen that $\iota_1^{-1}(X \cup Y)$ has a regular neighborhood, W_1' such that the restriction $\iota_1|W_1'$ is homotopic to a map ι_1' with the asserted property. Clearly W_1, ι_1 is normally bordant to W_1', ι_1' so we may extend A, I by this normal bordism, relabel the result as A, I while relabeling W_1', ι_1' as W_1, ι_1 without loss of geneerality (and without affecting the supposition of Step 1). It should also then be noted that we may assume that $\iota_1^{-1} X$, $\iota_1^{-1} Y$, $\iota_1^{-1}(X \cap Y)$ are simple-homotopy equivalent to their respective images under ι_1.

STEP 3. Consider $I^{-1} E$ with its subspaces $I^{-1}\dot{E}$, $I^{-1} C_1$, $I^{-1} C_3$ (where C_3 is the closure of $\dot{E} - C_1$, i.e. the tubular neighborhood of $X \quad \dot{E}$). Clearly, one may assume these are submanifolds of A with $I^{-1}\dot{E} = \partial I^{-1} E$. $I^{-1} C_1$, $I^{-1} C_3$ are submanifolds of $I^{-1}\dot{E}$ with common boundary $I^{-1} C_0$. We now assert that we may do normal surgery on A, I, rel $W_* \cup W_1$, so that the restrictions of I to the various spaces above are simple homotopy equivalences to their respective images. The argument merely involves routine surgery theory and thus will be omitted.

STEP 4. Consider the "resolution" of I, $\hat{I}:\hat{A} \to \hat{W}$, mentioned previously We see that if $\hat{E}$ is the double cover of E, $\hat{E} \quad \hat{W}$, then $\hat{I}^{-1}\hat{E} \to \hat{E}$ is a simple homotopy equivalence, as is $\mathrm{cl}\,(\partial \hat{I}^{-1}\hat{E} - W - \hat{W}_1) \to \dot{\hat{E}}$. Here W is, of course, the "resolution" of $\hat{W}_*$, $\hat{W}_1$ of W_1, while $\dot{\hat{E}}$ is the double cover of $\dot{E}$. If We let $R \cup Y = N \subset W$ be the resolution of $X \cup Y$, then we see that $\iota_H^{-1} R = Q \to X$, $\iota_1^{-1} R \to \hat{X}$ are simple homotopy equivalences (mod boundary, where $X \cap Y \cong R \cap Y$ is the "boundary" of R). We may then do surgery $\mathrm{rel}(W \quad \hat{W}_1)$ on $\hat{A}, \hat{I}$ so that $\hat{I}^{-1} R$, $\hat{I}^{-1}(R \cap Y) \to R$, $R \cap Y$ is a simple homotopy equivalence. Moreover, this may be done so that the bordism is a a trivial one on $\hat{A}$ itself (though not on $\hat{I}^{-1} R$). So we may relabel, by slight abuse of notation, calling the result of this alteration $\hat{A}, \hat{I}$. Finally, we may assume that $\hat{I}^{-1}\hat{E}$, $\hat{I}^{-1}\dot{\hat{E}} \to \hat{E}, \dot{\hat{E}}$. In effect, this comes about because in showing that the trace of the surgery from the "old" $\hat{A},\hat{I}$ to the "new" is a trivial bordism, we have, a sub-bordism, also trivial from the "old" $I^{-1}\hat{E}$ to the new.

STEP 5: We now do surgery on $\hat{I}^{-1}Z$, $\hat{I}^{-1}Y$ (rel $W, \hat{W}_1$ and intersections with $\hat{I}^{-1}E$), so that the result, $J_-:B_- \to Z$ has B_-, $J_-^{-1}Y$, $J_-^{-1}C_2$ going to their respective images by simple homotopy equivalences. Thus, letting $B = B_- \cup \hat{I}^{-1}\hat{E}$, and $J = J_- \cup \hat{I}|\hat{I}^{-1}\hat{E}$, we see that $J^{-1}\hat{N} = V$ is a trivial cobordism from $N \quad W$ to $N_1 = \hat{\iota}_1^{-1}\hat{N}$. Now $J^{-1}\hat{E} = \hat{I}^{-1}\hat{E}$ still double covers $I^{-1}E$. Consequently, if B_* is B mod the involution on $J^{-1}E$, we may parameterize $B \to B_*$ as $W \times I \to W_* \times I$. Likewise V is a concordance from the original embedding $H:N \to W$ to $G:N \to W$. By construction, the stratified normal map $\iota_G:W_* \to \overline{W}$ is a stratified simple homotopy equivalence, in fact, it is clearly ι_1.

This completes the proof of 1.4.

To avoid confusion, we shall let $\overline{N}^{(1)}$, $\overline{N}^{(2)}$, $\overline{N}_*^{(2)}$ denote the various multiplicity sets of $J \circ G$ and their respective images under this immersion.

1.5 COROLLARY. Under the hypothesis of 1.4, there is an embedding $G:N \varepsilon W$ such that $W_*, \overline{N}_*^{(0)}, \overline{N}_*^{(1)}, \overline{N}_*^{(2)}$ is simple-homotopy equivalent to W_*, $M_*^{(0)}$, $M_*^{(1)}$, $M_*^{(2)}$.

This follows directly from 1.4 and its proof.

1.6 THEOREM B_2. The sole obstruction to making W_*, ι normally bordant to a stratified simple homotopy equivalence is the element

$$\sigma\left(p_!(S|M^{(2)})\right) \varepsilon L^s_{n-k}\left(\pi_1 M_*^{(2)}, wM_*^{(2)}\right).$$

[If $M^{(2)}$ is not connected, this is to mean a collection of elements, one for each component, in the Wall groups of the respective fundamental groups, with their respective orientation characters.]

We remark that this is merely the special case of Theorem B when $m(j) = 2$. Theorem B_2 follows, in fact, from the general considerations of

[B-Q]. We shall, however, remind the reader of the general outline of the proof in the particular case at hand.

First of all, if W_*, ι_H is normally bordant to a stratified simple homotopy equivalence then, a fortiori, the surgery problem $N_*^{(2)} \to U_* \sim M_*^{(2)}$ (resp, $K_* \sim M_*^{(2)}$) is solvable and hence the surgery obstruction, which we have seen above to be that of $\sigma(p_!(S|M^{(2)})) \in [M_*^{(2)}, G/O]$ (resp. $[M_*^{(2)}, G/PL]$) must vanish.

The converse follows from the following steps. Suppose $\sigma(p_!(S|M^{(2)})) \in L^s_{n-k}(\pi_1 M_*^{(2)}, wM_*^{(2)})$ is zero. (We remind the reader that if $M^{(2)}$ is not connected, then $L^s_{n-k}(\pi_1 M_*^{(2)}, wM^{(2)})$ must be interpreted as $\bigoplus_{A^{(2)}} L^s_{n-k}(\pi_1 A^{(2)}, wA^{(2)})$ where the $A^{(2)}$ range over the components of $M_*^{(2)}$ and $wA^{(2)}$ is the respective orientation character.)

1.7 LEMMA. W_*, ι_H is bordant to W_1, ι_1 where $\iota_1|\iota_1^{-1}E$ is a stratified simple homotopy equivalence to E, X, U_* (resp. E, X, K).

PROOF. Let Λ, λ be a normal bordism of $N_*^{(2)}$, $\iota|N_*^{(2)} \to U_* \to M_*^{(2)}$ to a simple homotopy equivalence L, ℓ. Clearly the map of the normal bundle Λ to the appropriate bundle over M_* splits off a summand $\alpha^{2k} \to p_!\overline{\eta}^k$. Now note that we therefore may extend Λ, λ to a normal map $d: D(\alpha^{2k}) \to E$, transverse to the stratification of E. Moreover, if $\beta^{2k} = \alpha^{2k}|L$, then $d|D(\beta^{2k})$ is a stratified simple homotopy equivalence. Let E_1 be a slightly smaller copy of E, lying with E, so that $E = E_1 \cup$ (collar on $\dot{E}_1$) (as a stratified space). By abuse of notation, we regard d as a map to $\dot{E}_1$, and we may identify $D(\alpha^{2k}|N_*^{(2)})$ with a tubular neighborhood Θ of $N_*^{(2)}$ in U_*, while identifying $d|D(\alpha^{2k}|N_*^{(2)})$ with ι_H restricted to this tubular neighborhood Θ. Thus we may form the space

$$A_0 = W_* \times [0,1] \cup D(\alpha^{2k})$$

identifying $D(\alpha^{2k}|N_*^{(2)})$ with $\Theta\times\{1\}$. Then $I_0 = \iota_H \times \mathrm{id} \cup d: A_0 \to W^*$ is a normal map.

Let

$$B_0 = (W-\Theta)\times\{1\} \cup S(\alpha^{2k}) \cup D(\alpha^{2k}|L) \subset A_0.$$

We claim the following, leaving many details to the reader. There is a submanifold $B_1 \subset B_0$ such that $d|B_1$ deforms to a stratified map $e:B_1 \to \overline{W}$ such that $e|e^{-1}E$ is a stratified homotopy equivalence. Briefly, the trick to seeing that is to find a smaller copy $\overline{W}_1$ of $\overline{W}$ lying within $\overline{W}$, so so that $d^{-1}\overline{W}_1 = B$. $\overline{W}_1$ will contain a space E_1', a smaller copy of E with $U_* \subset E_1' \subset E_1 \subset E$, so that $d|d^{-1}E_1'$ is a stratified simple homotopy equivalence. Moreover, ∂B will be $d^{-1}C$ where C is the copy in $\overline{W}_1$ of $C_1 \cup C_2 \subset \overline{W}$. d then obviously deforms to $e:B \to \overline{W}$ with $\partial B = e^{-1}(C_1 \cup C_2)$ and e a stratified homotopy equivalence of $e^{-1}(E)$ to E.

Let $f:B_1\times I \to \overline{W}$ be the homotopy of $d|B$ to e. Then form $A_1 = A_0 \cup B\times I$, and extend I_0 to $I_1 = I_0 \cup f$. Now we may deform I_1, rel $W_*\times\{0\} \cup B\times\{1\}$ to a normal bordism from W, ι_H to B, e $(B = B\times\{1\})$. This completes the outline of the proof of 1.7.

The remaining difficulty is that, if $B_- = e^{-1}Z$, then $e|B_-$ is not a stratified homotopy equivalence. We may assume that $e|e^{-1}C_3, e^{-1}(N\cap\dot{E})$ is a simple homotopy equivalence; so the problem is to do stratified surgery on B_-, staying away from $e^{-1}(C_3)$, to make it so. The obstruction to so doing is an element $a \in L_n^s(\pi_1 Y, wY)$. We shall show that $a = 0$. This comes about as follows: First, we resolve B, e to $\hat{B}$, $\hat{e}:\hat{B} \to W$. We note that $\hat{e}|\hat{e}^{-1}\tilde{E}$ is a stratified simple homotopy equivalence on $\hat{e}^{-1}\tilde{E}, \hat{e}^{-1}\hat{N} \cap \tilde{E}$. Thus a is the surgery obstruction of the normal map $\hat{e}|\hat{e}^{-1}\hat{N} \to \hat{N}$ since $\hat{e}^{-1}(\hat{N}\cap\hat{E})$ is, homotopically, of codimension $k \geq 3$ in $\hat{e}^{-1}\hat{N}$. But $\hat{e}|\hat{e}^{-1}\hat{N} \to \hat{N}$ is normally bordant to the simple homotopy equivalence $\hat{\iota}_H|\hat{N} \to \hat{N}$. Thus $a = 0$.

This completes the proof of 1.6.

1.8 COROLLARY. $\sigma p_!(S|M^{(2)}) = 0$ if and only if H is concordant to G such that $\overline{N}_*, \overline{N}^{(1)}_*, \overline{N}^{(2)}_*$ is simple homotopy equivalent to $M_*, M^{(1)}_*, M^{(2)}_*$.

We leave the special case of theorem C when $m(j) = 2$ as an exercise.

§2. THE GENERAL CASE

We now show, or at least sketch in reasonable detail, how the basic idea of §1 above may be extended to cover the case where $m(j)$ is arbitrarily large, $k \geq 3$, and the set $M^{(m(j))}$ of points of highest multiplicity has dimension $n - (m(j)-1)k \geq 5$.

For expository reasons, we make certain factitious assumptions, with the aim of simplifying the picture in the reader's mind without compromising, in any essential way, the conceptual core of the proof. The assumptions are these:

(1) $m(j) = 3$.

(2) $M^{(1)}_*, M^{(2)}_*, M^{(3)}_*$ are all connected.

(3) The normal bundle ξ^k of $j:M^n \to V^{n+k}$ and the normal bundle η^k of $H:N^n \subset W^{n+k}$ are vector bundles rather than PL block-bundles.

The respective justifications for these simplifying assumptions are as follows.

First of all, the assumption that $m(j) = 3$, i.e. that $j:M \to V$ has, at worst, triple points allows the proof to go forward without a bewildering mass of notational detail. On the other hand, the reader should have no trouble in seeing that for $m(j)$ arbitrarily large, the present proof easily adapts into an inductive argument.

Secondly, the assumption that $M^{(1)}_*$ $M^{(2)}_*$ $M^{(3)}_*$ all be connected is merely to avoid the pedantry of continually speaking of the various components, the respective fundamental groups of these components, and the respective Wall groups of these fundamental groups. It will easily be seen that the proof can be modified, through appropriate additional language, to cover the case where the connectivity assumption is dropped.

Finally, the fiction that ξ^k, η^k are vector bundles is resorted to merely to remain within the language of vector bundles which, the author assumes, is familiar and comfortable for most readers. The adaptation of the relevant constructions and arguments to the more general case where ξ^k, η^k are merely PL k-block bundles is much in the spirit of §1. That is to say, the technique for adopting a vector bundle construction to a block bundle construction that was evident in the course of §1 will be found to apply in this section as well; further details are omitted.

Thus, with simplifications as above, let us set the context for analyzing the obstructions to having $H:N^n \to W^{n+k}$ be such, up to concordance, that the stratification of N_* via the multiple-point sets of the immersion $J \circ H$, coincides, up to simple homotopy equivalence, with the stratification of M_* via the multiple points sets of j.

The first step is to observe how to extend j to a codimension 0 immersion J of W^{n+k} into V. As in §1, W is a disc bundle $D(\xi^k)$. Let $U^{(3)} = D(\xi^k | P^{(3)})$ where P^3 is a tubular neighborhood of $M^{(3)}$ in M. Let α_x denote the fiber of α at x when α is a bundle with x in the base space. Then, in a specific way, we may identify $D(\nu_{P^{(3)}}(M^{(3)})_{x_1}$ as $D(\xi)_{x_2} \oplus D(\xi)_{x_3}$, where $\{x_1, x_2, x_3\} = j^{-1}(jx_1)$. Thus an arbitrary point in U may be specified as $(a_1, a_2, a_3) \in (D\xi)_{x_1} \oplus (D\xi)_{x_2} \oplus (D\xi)_{x_3}$ where (a_1, a_2, a_3) is understood to be in the fiber of $[D(\nu_{U(3)}(M^{(3)})_{x_1}]$. For the correct choice of J, $U_*^{(3)} = J(U^{(3)})$ may be described as $\{(x, \{a_1, a_2, a_3\} | x \in M_*^{(3)},\ j^{-1}x = \{x_1, x_2, x_3\}, a_i \in D(\xi)_{x_i},\}$ with J the obvious map $(a_1, a_2, a_3) \to (x, \{a_1, a_2, a_3\})$ where $a_1, a_2, a_3 \in D\xi_{x_1} + D\xi_{x_2} + D\xi_{x_3}\}$ $\{x_1, x_2, x_3\} = j^{-1}x$.

Clearly, $\mathrm{cl}\, M^{(2)} \cap U^{(3)}$ is described as $\{(a_1, a_2, a_3) | a_1 = 0$ and $a_2 = 0$ or $a_3 = 0\}$. We consider a tubular neighborhood of $M^{(2)} \cap \partial U^{(3)}$ in $\partial U^{(3)}$. Now $\partial U^{(3)} = \{a_1, a_2, a_3) \big| |a_1| = 1$ or $|a_2| = 1$ or $|a_3| = 1\}$. The tubular neighborhood of $M^{(2)} \cap \partial U^{(3)}$ may therefore be written

$\{a_1, a_2, a_3) \mid |a_1| \le \frac{1}{2}; (|a_2| = 1, |a_3| \le \frac{1}{2})$ or $(|a_2| \le \frac{1}{2}, |a_3| = 1)\}$.

Now let $\underline{M}^{(2)} = d(M^2 - P^{(3)})$ and $P^{(2)}$ a regular neighborhood of $\underline{M}^{(2)}$ in $cl(M-P^3)$ so that $P^{(2)} \cap \partial U^{(3)}$ is $M \cap R$ where R is the tubular neighborhood of $M^{(2)} \cap \partial U^{(3)}$ in $\partial U^{(3)}$ mentioned above. Let $U^{(2)}$ be $\overset{1}{D^2}(\xi | P^{(2)})$, so $U^{(2)} \cap U^{(3)} = R$. Here $\overset{1}{D^2}$ means the union of discs of radius $\frac{1}{2}$ in D.

Note that $P^{(2)}$ may be parameterized as a disc bundle $D(\nu_{P^{(2)}}(\underline{M}^{(2)})$ where $D(\nu_{P^{(2)}}(M^{(2)})_{x_1}$ is thought of as $(\overset{1}{D^2}\xi)_{x_2}$. Thus a typical point of $U^{(2)}$ may be written $(a_1, a_2) \in \overset{1}{D^2}\xi_{x_1} \oplus \overset{1}{D^2}\xi_{x_2}$ with $P^{(2)} = \{(0, a_2)\}$. We claim that for the right choice of J, $U^{(2)}_*$ may be written $\{(x, \{a_1, a_2\} | x \in \underline{M}^{(2)}, a_1 \in \overset{1}{D^2}\xi_{x_1}, a \in \overset{1}{D^2}\xi_{x_2}, j^{-1}x = \{x_1, x_2\})$ with

$J|U^{(2)}_*: (a_1, a_2) \to (x, \{a_1, a_2\})$ where $a_1 \in \overset{1}{D^2}\xi_{x_1}$, $a_2 \in \overset{1}{D^2}\xi_{x_2}$, $j\text{-}1x = \{x_1, x_2\}$.

$U^{(2)}_*$ $U^{(3)}_* = J(R)$, i.e., we may fit $U^{(2)}_*$ together with $U^{(3)}_*$ by letting y in $U^{(2)}_*$ be identified with z in $U^{(3)}_*$ when and only when there is some point x in R where $J(x) = z$ (in the first description of J as a map $U^{(3)} \to U^{(3)}_*)$ and $J(x) = y$ (in the second description of J as a map $U^{(2)} \to U^{(2)}_*)$. Now let $M_- = cl(M-P^{(2)}-P^{(3)})$ and let $W_- = \overset{1}{D^3}(\xi | M_-)$. For the right choice of J, $J|W_-$ is merely a homeomorphism.

We let $W = W_- \cup U^{(2)}_* \cup U^{(3)}_*$ and $W_* = W_- \cup U^{(2)}_* \cup U_*^{3)}$ so that $J: W \to W_*$ is a map with $J|U^{(3)} \to U^{(3)}_*$ a triple cover, $J|U^{(2)} \to U^{(2)}_*$ a double cover, $J|W_- \to W_-$ a homeomorphism.

Now consider the simple homotopy equivalence $h: N^n \to M^n$. By repeated application of the Browder-Wall splitting theorem we may assume that

$$h|h^{-1}P^{(3)}, h^{-1}\partial P^{(3)} \to P^{(3)}, \partial P(3)$$

$$h|h^{-1}P^{(2)}, h^{-1}\partial P^{(2)} \to P^{(2)}, \partial P^{(2)}$$

$$h|h^{-1}M_-, h^{-1}\partial M_- \to M_-, \partial M_-$$

are all simple homotopy equivalences.

Let $Q^{(3)} = h^{-1}P^{(3)}$, $Q^{(2)} = h^{-1}P^{(2)}$ $N_- = h^{-1}M_-$. Then we may assume that under the embedding $H: N \subset W$, we shall have

$$H(Q^{(3)}) \subset U^{(3)}$$

$$H(Q^{(2)}) \subset U^{(2)}$$

$$H(N_1) \subset W_- \quad .$$

This is the "splitting" condition needed for the precise definition of ι_H below.

Thus it is clear that, after a small isotopy of H to put $J \circ H$ in general position, $m(J \circ H) = 3$, with $N^{(3)} \subset U^{(3)}$, $N^{(2)} \subset U^{(3)} \cup U^{(2)}$.

As in §1 above, let η^k denote $\nu_W(N)$. Then $U^{(3)}$ may be viewed as the disc bundle $D(\eta|Q^{(3)})$ while $U^{(2)}$ is $D(\eta|Q^{(2)})$. Let $\bar{\eta}$ over W correspond to η under the homotopy equivalence H.

Let us now consider the following spaces: First, let $p^{(3)} = J|U^{(3)}$, $p^{(2)} = J|U^{(2)}$ so that $p^{(3)}$ is a 3-fold cover of its image and $p^{(2)}$ a 2-fold cover. Let $E^{(3)} = D(p_!^{(3)}(\eta|U^{(3)}))$ while $E^{(2)} = D(p_!^{(2)}(\bar{\eta}|U^{(2)}))$ and $Z = D(\bar{\eta}|W_-)$.

Specifically, $E^{(3)}$ may be written

$$E^{(3)} = \{x, \{a_1, a_2, a_3\} | x \in U_*^{(3)}; a_1, a_2, a_3 \in U^{(3)}\} \quad x, a_i \in \eta q_i$$

where $(p^{(3)})^{-1} x = \{x_1, x_2, x_3\}$ and $q_i \in Q^{(3)}$.

Thus, there is a natural map $\iota_3 : U_*^{(3)} \to E^{(3)}$ given by $x \to (x, \{x_1, x_2, x_3\})$ where $\{x_1, x_2, x_3\} = (p^{(3)})^{-1} x$.

Similarly, let

$$E^{(2)} = D(p_!^{(2)}(\bar{\eta}|U^{(2)})) = \{(x, \{a_1, a_2\} | x \in U_*^{(2)},\ a_1, a_2 \in U^{(2)}\} \quad x, a_i \in \eta_{q_i}$$

$p^{(2)^{-1}} x = \{x_1, x_2\}$, $q_i \in Q^{(2)}$. ι_2 is defined by $x \to (x, \{x_1, x_2\})$.

Finally $Z = D|(\bar{\eta}|W_-) = \{(x,a) \mid x, a \in W_-,\ x, a \in \eta_n,\ n \in N_-\}$ and $\iota_1 : W_- \to Z$ is given by $x \to (x,x)$.

We wish to form a space $\overline{W} = E^{(3)} \cup E^{(2)} \cup Z$ where identifications are to be made as follows: Let $(x, \{a_1, a_2\})$ be a point in $E^{(2)}$ where

$x \in U_*^{(2)} \cap U_*^{(3)}$. I.e. $(p^{(3)})^{-1} x = \{x_1, x_2, x_3\}$; by the choice of a_1, a_2 we have implicitly picked out x_1, x_2 with $x_i, a_i \in \eta_{q_i}$ for some $q_i \in Q^{(2)} \cap Q^{(3)}$. We identify this with $(x, \{a_1, a_2, q_3\}$ where $q_3 \in Q^{(3)}$ is defined by $x_3 \in \eta_{q_3}$.

Similarly, let (x,a) be a point in Z where $x \in W_- \cap U_*^{(2)}$. Then $(p^{(2)})^{-1} x = \{x, x_2\}$, and we identify (x,a) with $(x, \{a, q_2\})$ where $q_2 \in Q^{(2)}$ is defined by $x_2 \in \eta_{q_2}$.

Finally, if $(x,a) \in Z$, $x \in W_- \cap U_*^{(3)}$, then $(p^{(3)})^{-1} x = \{x, x_2, x_3\}$, and we identify (x,a) with $(x, \{a, q_2, q_3\})$ where q_2, q_3 are specified by $x_i \in \eta_{q_i}$, $i = 2,3$.

We wish to distinguish certain subspaces of $\overline{W}$. First, $K^{(3)}$ is merely $U_*^{(3)}$ regarded as the base space of $E^{(3)}$. Let $X \subset E^{(3)}$ be the space $X = \{(x, a_1, a_2, a_3\} \mid$ at least two of a_1, a_2, a_3 are in $Q^{(3)}\}$.

Let $K^{(2)} = U_*^{(2)} \cup (X - K^{(3)})$ where the copy of $U_*^{(2)}$ in question is the base of $E^{(2)}$.

Let $Y = \{(x, a_1, a_2, a_3) \in E^{(3)} \mid$ at least one of a_1, a_2, a_3 is in $Q^{(3)}\}$.
Let $Y' = \{(x, a_1, a_2) \in E^{(2)} \mid$ at least one of a_1, a_2 is in $Q^{(2)}\}$.
Let $K^{(1)} = W_- \cup Y \cup Y' - K^{(2)} - K^{(3)}$. We now claim:

2.1 LEMMA $\overline{W}, K^{(1)}, K^{(2)}, K^{(3)}$ is stratum wise simple homotopy equivalent to $W_*, M_*^{(1)}, M_*^{(2)}, M_*^{(3)}$.

The proof is left to the reader.

We have already noted the maps $\iota_1 : W_- \to Z$, $\iota_2 : U_*^{(2)} \to E^{(2)}$, $\iota_3 : U_*^{(3)} \to E^{(3)}$.

2.2 LEMMA (a) $\iota_1 : W_- \to Z$ is transverse to $W_- \subset Z$ and $(\iota_1)^{-1} W_- = N_-$

(b) $\iota_2 : U_*^2 \to E(2)$ is transverse to the stratification $Y', U_*^{(2)}$ of $E^{(2)}$ with $(\iota_2)^{-1} Y' = Q_*^{(2)}$, $(\iota_2)^{-1} U_*^{(2)} = N_*^{(2)} \cap Q_*^{(2)}$.

(c) $\iota_3 : U^{(3)}_* \to E^{(3)}$ is transverse to the stratification $Y, X, U^{(3)}_* = K^{(3)}$ of $E^{(3)}$ with $(\iota_3)^{-1} Y = Q^{(3)}$, $(\iota_3)^{-1} X = \mathrm{cl} N^{(2)}_* \cap Q^{(3)}_*$, $(\iota_3)^{-1} U^{(3)}_* = N^{(3)}$.

It would be nice to define $\iota_H : W_* \to \overline{W}$ as $\iota_1 \cup \iota_2 \cup \iota_3$; unfortunately, these are inconsistent on the various intersections of W_-, $U^{(2)}_*$, $U^{(3)}_*$. However, we claim the following.

2.3 LEMMA. There is a map $\iota_H : W_* \to \overline{W}$, transverse to the stratification $K^{(1)}, K^{(2)}, K^{(3)}$ of $\overline{W}$ with $\iota_H^{-1} K^{(i)} = N^{(i)}_* \subset W_*$.

Moreover:

(a) $\iota_H | W_- \to Z \subset \overline{W}$ is ι_1.

(b) $\iota_H | U^{(2)}_* \to E^{(2)} \subset \overline{W}$ is homotopic to ι_2 via

$$I_2 : U^{(2)}_* \times [0,1] \to E^{(2)} \text{ such that } (I_2)^{-1} Y' = Q^{(2)} \times [0,1]$$

$$(I_2)^{-1} U^{(2)}_* = [N^{(2)}_* \cap Q^{(2)}_*] \times [0,1].$$

(c) $\iota_H | U^{(3)}_* \to E^{(3)} \subset W$ is homotopic to $\iota^{(3)}$ via

$$I_3 : U^{(3)}_* \times [0,1] \to E^{(3)} \text{ such that } (I_3)^{-1} Y = Q^{(3)}_* \times [0,1]$$

$$(I_3)^{-1} X = [\mathrm{cl}(N^{(2)}_* \cap Q^{(3)}_*)] \times [0,1],\ (I_3)^{-1} U^{(3)}_* = N^{(3)}_* \times [0,1].$$

The proof is straightforward; details are omitted.

It follows that $\iota_H : W_* \to \overline{W}$ is transverse to the stratification $K^{(1)}, K^{(2)}, K^{(3)}$ of W with $\iota_H^{-1} K^{(i)} = N^{(i)}_*$. Moreover, ι_H may be viewed as a normal map. We also note that under ι_H, ∂W_* goes to the "boundary" of $\overline{W}$ via a simple homotopy equivalence. Again, details are left to the reader.

Let us first consider the induced normal map $N^{(3)}_* \to K^{(3)} = U^{(3)}_* \sim M^{(3)}_*$.

2.4 LEMMA. $N^{(3)} \to L^{(3)}$ as a surgery problem represents $p_!(S|M^{(3)})$ where S is the normal invariant of $h: N^n \to M^n$ in $[M, G/PL]$.

This observation is quite straightforward.

2.5 THEOREM A. ι_H is normally bordant to a stratified simple homotopy equivalence if and only if H is concordant to an embedding $G:N \subset W$ so that ι_G is a stratified simple homotopy equivalence.

The proof follows closely that of the special case Theorem A_2 and will not be given here.

2.6 THEOREM B. The obstructions to making ι_H normally bordant to a stratified simple homotopy equivalence consist of the two elements

$$\sigma\left(p_!^{(3)}(S|M^{(3)})\right) \in L^s_{n-2k}\left(\pi_1 M_*^{(3)}, w(M_*^{(3)})\right)$$

and

$$\sigma\left(p_!^{(2)} S^{(2)}\right) \in L^s_{n-k}\left(\pi_1 M_*^{(2)}, w(M_*^{(2)})\right)$$

where, as in §0, $S^{(2)}$ is the pullback of S to $\tilde{M}^{(2)}$ under the natural map $r^{(2)}:\tilde{M}^{(2)} \to M$, $p^{(2)}$ is the natural cover $\tilde{M}^{(2)} \to \underline{M}^{(2)}$, thus making $p_!^{(2)} S^{(2)}$ an element of $[\underline{M}^{(2)}, G/PL]$ whose surgery obstruction $\sigma\left(p_!^{(2)} S^{(2)}\right)$ lies in $L^s_{n-k}\left(\pi_1 M^{(2)}, wM^{(2)}\right) = L^s_{n-2k}\left(\pi_1 M_*^{(2)}, wM_*^{(2)}\right)$.

PROOF. We note the trivial fact that the bordism of ι_H to a stratified simple homotopy equivalence implies the vanishing of the stated obstructions.

Conversely, suppose that $p_!\left(S|M^{(3)}\right)$ has vanishing surgery obstruction in $L^s_{n-2k}\left(\pi_1 M_*^{(3)}, w(M_*^{(3)})\right)$. Then we may do surgery to get a stratified map $\iota_0:W_1 \to W_*$ so that $\iota_0^{-1}K^{(3)} \to K^{(3)}$ is a simple homotopy equivalence; if $I_0:A_0 \to W_*$ is the trace of this surgery, we may assume that $I_0^{-1}(E^{(3)}) \sim E^{(3)}$ and, as well $\iota_0^{-1}\left(E^{(3)}\right) \sim E^{(3)}$. Now let us study the surgery problem $\iota_0^{-1}[\mathrm{cl}(K^{(2)} - E^{(3)})] \to \mathrm{cl}\left(K^{(2)} - E^{(3)}\right)$ which may be assumed to be a homotopy equivalence on the boundary. The idea is to compare this surgery problem with a suitable closed surgery problem. Without going into details, the first thing to notice is that $\overline{W}$ may be "resolved" to $\hat{W} = Z \cup \tilde{E}^{(2)} \cup \tilde{E}^{(3)}$ where the map $\hat{W} \to \overline{W}$ consists of a homeomorphism $Z \to Z$, a 2-fold cover $\tilde{E}^{(2)} \to E^{(2)}$ and a 3-fold cover $\tilde{E}^{(3)} \to E^{(3)}$. Of course, the map $\iota_{\hat{H}}:W_* \to \overline{W}$ is simultaneously resolved to a map $\iota_{\hat{H}}:W \to \hat{W}$. If we let $\hat{K}^{(1)}, \hat{K}^{(2)}, \hat{K}^{(3)}$ denote the stratification of W induced from this resolution, we shall have $\hat{W}, \hat{K}(1), \hat{K}(2), \hat{K}(3)$ stratum-wise simple homotopy equivalent to $W, M^{(1)}, M^{(2)}, M^{(3)}$ and $\iota_H^{-1}\hat{K}^{(i)} = N^{(i)}$.

But it also may be observed that $I_0 : A_0 \to \overline{W}$ and its restriction $\iota_0 : W_0 \to \overline{W}$ may be resolved to $\hat{I}_0 : \hat{A}_0 \to \hat{W}$, $\hat{\iota}_0 : \hat{W}_0 \to \hat{W}$. I.e., these will be a diagram

$$\begin{array}{ccc} \hat{A}_0 & \xrightarrow{\hat{I}_0} & \hat{W} \\ \downarrow & & \downarrow \\ A_0 & \xrightarrow{I_0} & \overline{W} \end{array}$$

where $\hat{A}_0 \to A_0$ is an immersion.

Now $\iota_H^{-1}(\hat{K}^{(2)} \cup \hat{K}^{(3)}) = N^{(2)} \cup N^{(3)}$ is the set of points of multiplicity ≥ 2 for the immersion $N \subset W \to W_*$. Likewise, if we set

$$\Theta = \hat{I}_0^{-1}(\hat{K}^{(1)} \cup \hat{K}^{(2)} \cup \hat{K}^{(3)}),$$

Θ is a submanifold of $\hat{A}_0$ and a bordism from N to $N_0 = \iota_0^{-1}(\hat{K}^{(1)} \cup \hat{K}^{(2)} \cup \hat{K}^{(3)})$. Likewise, the immersions $\Theta \subset \hat{A}_0 \to A_0$, $N_0 \to \hat{W}_0 \to W_0$ will have double and triple points. Consider the resolved double point set $\tilde{\Theta}^{(2)}$ of the immersion $\Theta \subset \hat{A}_0 \to A_0$. This is immersed in $\hat{A}_0$ and is a bordism from $\tilde{N}^{(2)} \to N$ to $\tilde{N}_0^{(2)} \to N_0$. Now note that $\tilde{N}^{(2)}$ maps naturally into the resolved double point set $\tilde{M}^{(2)}$ of j, and that it is, in fact, a degree-one normal map, representing, in fact, $S^{(2)}$, the pullback to $\tilde{M}^{(2)}$ of the normal invariant $S \in [M, G/PL]$ of h. Thus, $\tilde{\Theta}^{(2)} \to \tilde{M}^{(2)}$ is a normal bordism of this surgery problem to $\tilde{N}_0^{(2)} \to \tilde{M}^{(2)}$. But let L be the inverse image of a regular neighborhood of $M^{(3)}$ in $\tilde{M}^{(2)}$. Likewise, let L' be the inverse image of a regular neighborhood of $N_0^{(3)}$ in $\tilde{N}^{(2)}$. It is easy to show that the normal map $\tilde{N}_0^{(2)} \to \tilde{M}^{(2)}$, which we shall denote v, may be assumed to have $v^{-1}L = L'$, with $v|L', \partial L' \to L, \partial L$ a homotopy equivalence.

Recall, now, that $\tilde{\Theta}^{(2)}$ covers $\underline{\Theta}^{(2)}$ which immerses in A_0 as a bordism from $\underline{N}^{(2)} \to W_*$ to $\underline{N}_0^{(2)} \to W_0$. The normal map $\tilde{N}_0^{(2)} \to \tilde{M}^{(2)}$ transfers down to $\underline{N}^{(2)} \to \underline{M}^{(2)}$ and thus $\underline{\Theta}^{(2)} \to \underline{M}^{(2)}$ is a normal bordism to $p_! v : \underline{N}_0^{(2)} \to \underline{M}^{(2)}$. More explicitly, $p_! v : \underline{N}_0^{(2)} \to \underline{M}^{(2)}$ is seen to be a representative of $p_!^{(2)}(S^{(2)})$. But note now that $p_! v|L'_*(\partial L')_* \to L_*, \partial L_*$ is a simple homotopy equivalence. Thus, the surgery obstruction for $p_!(S^{(2)})$ is equal to that for the mod boundary problem $p_! v|\underline{N}^2 - L'_* \to \underline{M}^{(2)} - L_*$, since up to homotopy, L_* is of codimension $k \geq 3$ in $\underline{M}^{(2)}$. But now it is readily seen that this mod boundary problem is precisely the problem $\iota_0^{-1}(cl(K^{(2)}-E^{(3)})) \to cl(K^{(2)}-E^{(3)})$. Thus if the obstruction $\sigma(p_!^{(2)}S^{(2)}) \in L^s_{n-k}(\pi_1 M_*^{(2)}, w(M_*^{(2)}))$ vanishes, then further surgery can be done to make $(N^{(2)})_* \to M_*^{(2)} \sim K^{(2)}$ a simple homotopy equivalence. By the Browder-Quinn theory, there will then be no further obstructions to making ι_H bordant to a stratified simple homotopy equivalence. Thus, at least in outline, the proof of 2.5 is complete.

2.7 REMARK. The removal of the special assumptions $m(j) = 3$, $M^{(i)}$ connected, η^k, ξ^k vector bundles is, as we have observed, merely a matter of technique involving no further essential insights. Thus we will content ourselves with the outlines of Theorems A (via A_2) and B in these special cases as sufficient indication of how to proceed in the general case.

2.8 REMARK. We remove all special assumptions. We note that we began our discussion with the consideration of a general position immersion $j: M^n \to V^{n+k}$, $k \geq 3$ and a simple homotopy equivalence $h: N^n \to M^n$. Suppose, instead, we can consider an arbitrary element $t \in [M^n, {}^G/PL]$. Let ξ be the normal block bundle of j. By the stability theorem for ${}^G/PL$, t may be thought of as an element in $[M^n, G(k)/\widetilde{PL}(k)]$ and represented by a block bundle η^k over M^n and a fiber homotopy equivalence $\xi^k \to \eta^k$. Thinking of this as a map a from W to the total space of η^k, we make it transverse to the base and find $N^n \subset W^{n+k}$, and thus a surgery problem $N^n \to M^n$ (not necessarily with vanishing obstruction). Now notice that for $i \geq 2$, $\sigma(M^{(i)})$ still has the same meaning as in §0, since it depended only on the normal invariant of h, not on h itself, and is therefore defined for any

$t \in [M, {}^{G}/PL]$. We now may define $\sigma(M^{(1)})$ as the surgery obstruction of $N^n \to M^n$ lying in $L_n^s(\pi_1 M, wM)$.

2.9 COROLLARY. $\sum_{i=1}^{m(j)} \sigma(M^{(j)})$ vanishes if and only if $a: W \to E(\eta^k_H)$ is homotopic, rel boundary, to a_0 so that $N_0 = a_0^{-1}(\text{base } \eta^k) \subset W$ is a simple homotopy equivalence such that, with respect to the immersion $J \circ H$, ι_H is a stratified simple homotopy equivalence

$$W_*, (N_0^{(1)}), (N_0^{(2)})_* \cdots (N_0^{m(j)})_* \to W_*, M_*^{(1)}), M_*^{(2)} \cdots M_*^{(m(j))}.$$

REMARK: With 2.9 in view, the extension of Theorems A and B, in the form of Theorem C is an exercise in Browder-Quinn stratified surgery theory and thus will be left to the reader.

References

[A] M.F. Atiyah. Characters and cohomology of finite groups, Publ. Math. Inst. des Hautes Etudes Sci. (Paris) 9(1961), 23-64.

[Br] W. Browder. Surgery on Simply Connected Manifolds, Springer, 1972.

[B-Q] W. Browder and F. Quinn. A surgery theory for G-manifolds and stratified spaces, in Manifolds-Tokyo, Univ. of Tokyo Press, 1973.

[H_1] A. Haefliger, Plongements differentiable de varietes dans varietes, Comment. Math. Helv. 36 (1961) 47-82.

[H_2] A. Haefliger, Knotted spheres and related geometric problems, Proc. Internat. Congress Math., Moscow 1966(1968), 437-445.

[R] A. Ranicki, Exact Sequences in the Algebraic Theory of Surgery, Princeton University Press (also, University of Tokyo Press), 1981.

[R-S] C.P. Rourke and B.J. Sanderson, Block bundles-I, Ann. of Math. 87 (1968) 1-28.

[W] C.T.C. Wall. Surgery on Compact Manifolds, Academic Press, 1970.

RUTGERS UNIVERSITY
DEPARTMENT OF MATHEMATICS
NEW BRUNSWICK, NJ 08903

Canadian Mathematical Society
Conference Proceedings
Volume 2, Part 2 (1982)

Homotopy Equivalent Manifolds by Pasting

Shmuel Weinberger[1]

It is well known that in dimension at least five, every (smooth) homotopy sphere can be obtained by cutting the standard sphere along the equator and pasting the hemispheres together by a diffeomorphism homotopic to the identity. We study here to what extent similar statements hold for other homotopy equivalences between manifolds. Throughout, we work in the topological category.

Karras, Kreck, Neumann, and Ossa have studied the equivalence relation generated by the following procedure. A manifold M is identified with any manifold that can be obtained by cutting M along a separating codimension one submanifold, and glueing the pieces together by an orientation preserving diffeomorphism. In the smooth category, their result is that two manifolds are equivalent iff they have the same Euler characteristic and signature. In particular, the homotopy type is modified wildly by this procedure. On the other hand, if the glueing map is homotopic to the identity, clearly the homotopy type is preserved.

More precisely, a homotopy equivalence $f:M'\to M$ is called cut-pastable (abbreviated CP) if there is a codimension one submanifold N of M separating M into two components, M_+ and M_- say, and a homeomorphism $h:N\to N$ (with h the identity on the boundary), a homotopy $H:M\times I\to M\times I$ from h to the identity, 1_N, (relative to the boundary) and a homeomorphism G such that:

1980 Mathematics Subject Classification 57N99, Secondary 57R67
1. The author is supported by an NSF graduate fellowship.

© 1982 American Mathematical Society
0731-1036/82/0000-0478/$03.75

$$\begin{array}{ccc} M_+ \cup_{1_N} N\times I \cup_{h^{-1}} M_- & \xrightarrow{\;1_{M_+} \cup H \cup 1_{M_-}\;} & M \\ & {\scriptstyle G}\searrow \quad \nearrow{\scriptstyle f} & \\ & M' & \end{array}$$

commutes up to homotopy (relative to the boundary). We say that $f:M' \to M$ is specially cut-pastable (abbreviated SCP) if we can take N above to be such that $\pi_1 N \to \pi_1 M$ is an isomorphism.

In this paper, we study and classify those homotopy equivalences which are CP, SCP, or the result of a sequence of such operations. Some related qualitative problems are to determine the "most efficient" fundamental group possible for the codimension one submanifold for an arbitrary CP (non-SCP) homotopy equivalence, the minimum number of CP's necessary to produce a given homotopy equivalence, and whether there is any difference in the theory in different dimensions. For instance we will see that the theory in dimension four is radically different from the general high dimensional theory.

The results found here form part of the author's Ph.D. thesis and were, for the most part, announced at the 1981 Topology conference in London, Ontario. The author would like to thank his advisor, Sylvain Cappell, for many useful conversations and much encouragement when this work was done. He would also like to thank the University of Western Ontario and the Kochman family for their hospitality.

We start with a classification of SCP homotopy equivalences. First recall that to any homotopy equivalence $f:M' \to M$, there is an associated map $\nu(f):M \to G/Top$ called the normal invariant of f. Also, the topology of G/Top is well understood and in particular,

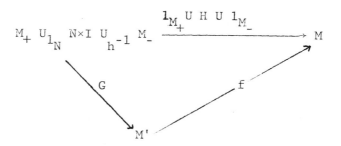

$H^2(G/Top;Z_2)=Z_2$. We will denote the nonzero element by k_2. This is the so-called Kervaire class. There are similar classes, k_{4n+2} in $H^{4n+2}(G/Top;Z_2)$, but the statements of our theorems only require the two dimensional class. (For information on G/Top see the books of Kirby-Siebenmann and Madsen-Milgram.)

Theorem A: *Let* $h:M'\to M^n$, n *at least five*, *be a homotopy equivalence between closed manifolds*, *then* h *is SCP iff*:

1) h *is a simple homotopy equivalence*,

2) $\nu(h)$ *lifts to* $\Sigma\Omega(G/Top)$,

$$\begin{array}{ccc} & & \Sigma\Omega(G/Top) \\ & \nearrow & \Big\downarrow \hat{1}_{\Omega(G/Top)} \\ M & \xrightarrow{\nu(h)} & G/Top \end{array}$$

and 3) $\nu(h)^*(k_2) = 0$ *in* $H^2(M;Z_2)$.

Indeed, the first two conditions are necessary for h to be CP. Thus, if h is CP, there is only a two dimensional charactersitic class which measures the obstruction to h being SCP. This suggests a basic interplay between $\pi_1(N)$ and $\nu(h)^*(k_2)$, where N is the codimension one submanifold CP along to obtain h.

Definition: A homotopy equivalence is *twisted* if $\nu(h)^*(k_2)\neq 0$, and lies in the image of $H^2(\pi_1 M;Z_2)\to H^2(M;Z_2)$. Otherwise h is *untwisted*.

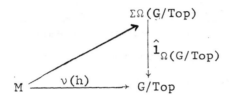

A geometric interpretation of twistedness can be given as follows: If $h:M'\to M$ is CP, then N can be taken to divide M into components M_+ and M_- such that $\mathrm{Ker}\,\pi_1(M_{\pm})\to\pi_1(M)$ have order at most two. (Actually there are five normal forms for the possible "reduced" fundamental group situations, but this need not concern us here.) For closed manifolds, both kernels can be taken to have the same order. h is twisted iff the homomorphisms above are nontrivial Z_2 extensions. Extensions $1 \to Z_2 \to E \to \pi_1(M) \to 1$ are classified by elements of $H^2(\pi_1 M; Z_2)$. For twisted homotopy equivalences, the preimage of $\nu(h)^*(k_2)$ in $H^2(\pi_1 M; Z_2)$ exactly classifies the extensions (they're both the same) $1 \to Z_2 \to \pi_1(M_{\pm}) \to \pi_1(M) \to 1$. For untwisted homotopy equivalences there is a quite satisfactory theorem:

Theorem B: *Let $h: M' \to M$ be a homotopy equivalence between manifolds which restricts to a homeomorphism between ∂M and $\partial M'$. Suppose that h is untwisted and that $\pi_1 M$ is:*

a) *finite with 2-sylow subgroup a product of elementary abelian and dihedral groups*, $\dim M \geq 5$ (*or just abelian $\times$ dihedral and dimension six*),

or b) *abelian*, $\dim M > \max(5, \operatorname{rank} \pi_1(M)+3)$

or c) *such that* $H_*(\pi_1 M; Z_{(2)})=0$ *for* $* > \dim M - 2 \geq 3$,

then h is CP iff:

1) *h is a simple homotopy equivalence*,

and 2) $\nu(h)$ *lifts to* $\Sigma\Omega(G/Top)$.

(*If the fundamental group is finite, then 2) is equivalent to* $M' \times S^1 \times S^1 \to M \times S^1 \times S^1$ *being CP*.)

This theorem sometimes extends to the twisted case:

Theorem C: *If $\pi_1 M$ is cyclic, or $Sq^2: H^2(M;Z_2) \to H^4(M;Z_2)$ is injective, then the characterization of theorem B still holds. If only the 2-sylow subgroup is cyclic, h is normally cobordant to a CP homotopy equivalence iff it is normally cobordant to a simple homotopy equivalence and condition 2 holds.*

The last statement suggests that there are obstructions in the twisted case to the normal cobordism invariance of the property of being CP. This is in fact the case, and the obstruction lies in a quotient of cok: $L^s_{n+1}(E) \to L^s_{n+1}(\pi_1(M))$, where E is the Z_2 extension determined by the Kervaire class. A key step in this development is proving that CP homotopy equivalences can be put in normal form. This requires ambient surgery on homeomorphisms homotopic to the identity, which extends the familiar handle trading of other codimension one contexts.

This obstruction leads to an example of a non-CP homotopy equivalence satisfying conditions 1) and 2) of theorem B. The ambient manifold in this example is the boundary of a regular neighborhood of the 2-skeleton of the seven torus T^7. The homotopy equivalence is twisted; no analogous untwisted examples are known.

For working out specific examples, the above theorems often suffice. For example, for highly connected manifolds we have:

Theorem D: Let M^{2n} be a closed n-1 connected manifold, n greater than eight. Then, $h:M'\to M$ is CP iff it is SCP. This is always the case unless n is a multiple of four. On the other hand, if n=4k, and M^{8k} is not the sphere, there always exists a non-CP homotopy equivalence to M. Moreover, if the quadratic form given by cup product on middle dimension cohomology is:

a) _definite, h is CP iff h is homotopic to a homeomorphism._

b) _indefinite, then every homotopy equivalence is the result of a sequence of CP's._

In lower dimensions there is a similar theorem but the calculations are slightly complicated by maps of odd Hopf invariant. For other low dimensional results see below. Theorem D easily yields examples of non-CP homotopy equivalences that are the result of a sequence of CP's. A class of spaces for which this does not happen include certain quotients of spheres by free group actions.

Theorem E: *Let M^n be a homotopy real, complex or quaternionic projective space or a homotopy lens space with the order of π_1 squarefree, n at least five. Let $h:M'\to M$ be a homotopy equivalence, then the following are equivalent:*

1) *h is CP,*
2) *h is SCP,*
3) *h is the result of a sequence of CP's (SCP's)*
4) *h is a simple homotopy equivalence, and the first half of the splitting invariants (cf. [Sullivan] for CP^n, [Lopez de Medrano] for RP^n, and [Homer] for lens spaces. The case of quaternionic projective space is analogous and easier.) of M and M' coincide,*

and 5) *h is a simple homotopy equivalence and $\nu(h)|_{M^{[n/2]}}$ is nullhomotopic,*

where M^k is the k-skeleton of M^n.

Surprisingly, if the fundamental group of the lens space has a square factor, the conclusion of theorem E fails:

Theorem E': *For $L^{8k+1}_{p^2}$ with 8k+1 less than 2p+1, condition 1) does not imply condition 5) in the above theorem; however, for an arbitrary manifold (of dimension at least five) 5) implies 1) and 2).*

The above results suggest that stability for sequences of CP's may be an interesting question. Is there a number k such that whatever homotopy equivalence is the result of a sequence of CP's is the result of a sequence of not more than k CP's? This is the case, but it is not trivial since there are manifolds, such as $RP^{4k+1}\#RP^{4k+1}$, for which the group of homotopy equivalences obtainable by a sequence of CP's is infinitely generated. (For a group structure on this set, see [Kirby-Siebenmann Essay V]. That the group is infinitely generated uses the UNil result of

[Cappell 2].)

Theorem F: For every manifold of dimension at least five, M, there is a (finite) number k(M) such that any homotopy equivalence to M that is the result of a sequence of CP's is the result of a sequence of no more than k(M) CP's.

It is also of interest to calculate the minimum number that will work for M. We reserve k(M) for this number.

Example: $k(S^4 \times S^4 \times \cdots S^4) = m$ if there are m factors. In particular, k(M) can be arbitrarily large.

Up to this point, the results have treated manifolds of dimension at least five. In dimension four almost all of the above results fail. It is easy to see that Theorem A fails, since most four manifolds do not have codimension one submanifolds with the same fundamental group. A much harder result is the following:

Theorem G: Let L_p and L'_p be homotopy equivalent three dimensional lens spaces, then $S^1 \times L_p$ cannot be obtained from $S^1 \times L'_p$ by any number of CP's.

This result is anomalous for several reasons. For instance, the homotopy equivalence $S^1 \times (L_p \to L'_p)$ is normally cobordant to the identity, and if p is odd Sq^2 is injective, so the analog of theorem C for dimension four would predict that the homotopy equivalence is CP. Furthermore, in general normally cobordant homotopy equivalences are simoulaneously the result or not the result of a sequence of CP's, so theorem G is anomalous even for p even.

The proof of theorem G actually yields an invariant for many fundamental groups π lying in cok $H_1(\pi;Q) \to L_1(\pi) \otimes Q$ to the normal cobordism invariance of a four dimensional homotopy equivalence's being CP. If a strong version of the Novikov Higher Signature Conjecture were known to be true for all three-manifold groups, this would be an obstruction for all π. Despite the unsettled status of this conjecture, Cappell has verified enough cases of it to allow the proof to go through in case $\pi = Z \times Z_p$, ie. the case of theorem G.

Under the additional hypothesis that the codimension one submanifold lie in the Poincare category an essentially low dimensional proof can be given.

We close with a sketch proof of theorems A-C.

A calculation using the topological invariance of Whitehead torsion shows that all CP homotopy equivalences are simple. Also, the top line of p. 2 shows that N divides M into two pieces on each of which the homotopy equivalence is homotopic to a homeomorphism; in particular, M is the union of two pieces on each of which the normal invariant is nullhomotopic. This implies that $\nu(h)$ lifts to $\Sigma\Omega(G/Top)$. Furthermore, a homological calculation shows that for h SCP, $\nu(h)_*: H_2(M;Z_2) \to H_2(G/Top;Z_2) = Z_2$ is trivial, which is equivalent to condition 3) of theorem A.

To complete the proof of theorem A one uses the following lemma:

<u>Lemma</u>: Suppose that $f: M \to G/Top$ lifts to $\Sigma\Omega(G/Top)$ and that $f^*(k_2)$ vanishes, then f lifts to $\Sigma\Omega((G/Top)^2)$, where $(G/Top)^2$ is the homotopy theoretic fiber of the Kervaire map $G/Top \to K(Z_2, 2)$

Note that no claim is made that the lift to $\Sigma\Omega((G/Top)^2)$ lifts the given lift to $\Sigma\Omega(G/Top)$.

Using this lemma, one takes the transverse inverse image of $\Omega((G/Top)^2)$. (Transversality only requires a "normal structure", so it does not matter that $\Omega((G/Top)^2)$ is not a manifold. However some care is necessary in dimension five because of difficulties with four dimensional topological transversality.) Ambient surgery allows one to assume that the inverse image is connected and that the inclusion induces an isomorphism of fundamental groups. The π-π theorem of surgery theory and the s-cobordism theorem combine to show that h can be obtained by cutting and pasting along the submanifold just constructed.

Remark: A version of theorem A holds in the smooth category, and is proven by similar techniques. To state this theorem, one introduces a space, BSCP, which factors the map $\Sigma\Omega((G/O)^2)\to\Sigma\Omega(G/O)$ which plays the role of "the classifying space for SCP homotopy equivalences" (although it is not an H-space). A homotopy equivalence is SCP iff it is simple and $\nu(h)$ lifts to BSCP. From this point of view one can restate theorem A even in the piecewise linear category with a space BSCP(PL), but the lemma above shows that the lifting problem to BSCP(PL) is equivalent to conditions 2) and 3) of theorem A.

The proofs of theorems B and C are more complicated. The first step is to classify the possible fundamental group set-ups for the inverse image of $\Omega(G/Top)$. There are five normal forms, three of which correspond to $\nu(h)^*(k_2)=0$ and have been dealt with by theorem A. The remaining normal froms are exactly determined by twistedness. In the untwisted case, the fundamental group of the submanifold is $Z_2\times\pi_1(M)$. This yields a surgery obstruction in ker $L_n(Z_2\times\pi)\to L_n(\pi)$ which is actually the well-defined obstruction to obtaining h from this lift. To prove theorem B, we use the fact that this obstruction is coming from a problem on closed manifolds, and is therefore rather special. In fact, one can describe a homomorphism from group homology to L-theory, and the obstruction must lie in this image. We show how to geometrically modify the lift so that the obstruction will lie in the image of the high dimensional homology. Taylor-Williams have shown that for many finite groups this image vanishes. By applying their results and extending some of them to various infinite groups we complete the proof of theorem B in much the same way as theorem A. (There are groups for which all of the homology groups are mapped nontrivially to L-theory, see [Weinberger 3], answering a question of B. Williams, and for these a classification of CP homotopy equivalences seems remote.) For theorem C, we need very detailed understanding of the surgery obstruction in ker $L_n(Z_2 \times Z_{2m}) \to L_n(Z_m)$. By extending the method used for untwisted homotopy equivalences it follows that the obstruction can be made to lie in ker $L_n(Z_2 \times Z_{2m}) \to L_n(Z_{2m})$. While this group does not vanish, no nonzero elements arise from surgery problems on closed manifolds, aside from a codimension two arf invariant which can be made to vanish by other special arguments. At this point one argues as before to establish theorem C.

BIBLIOGRAPHY

1. W. Browder, Surgery on Simply Connected Manifolds, Spriger-Verlag, Berlin and New York, 1972.

2. S. Cappell, The Homotopy Invariance of the Higher Signatures, Inventiones Mathematicae 33(1976) 171-179

3. S. Cappell, On Connected Sums of Manifolds, Topology 13(1974) 395-400

4. W. Homer, Equivariant PL Embeddings of Spheres,Topology 19(1980) 51-63

5. U. Karras, M. Kreck, W. Neumann, and E. Ossa, Cutting and Pasting of Manifolds: SK Groups, Publish or Perish, Boston, Mass. 1973

6. R. Kirby, and L. Siebenmann, Foundational Essays on Topological Manifolds, Smoothings, and Triangulations, Ann. of Math Studies, Princeton University Press, Princeton 1977

7. S. Lopez de Medrano, Involutions on Manifolds, Springer-Verlag, Berlin and New York, 1971

8. I. Madsen and J. Milgram, The Classifying Spaces for Surgery and Cobordism of Manifolds, Ann. of Math Series, Princeton University Press, Princeton 1977

9. S.P. Novikov, Homotopically Equivalent Smooth Manifolds, AMS Trans. (2) 48(1965) 271-396.

10. D. Sullivan, Geometric Topology Seminar Notes, Princeton University 1967

11. L. Taylor and B. Williams, Surgery on Closed Manifolds, Notre Dame preprint, 1980

12. C.T.C. Wall, Surgery on Compact Manifolds, Academic Press New York, 1970

13. S. Weinberger, Homotopy Equivalent Manifolds by Pasting, Thesis, Courant Institute June 1982

14. S. Weinberger, The Novikov Conjecture and Low Dimensional Topology, Courant Institute preprint, February 1982

15. S. Weinberger, There Exist Finitely Presented Groups with Infinite Ooze, Courant Institute Preprint, October 1981

COURANT INSTITUTE OF THE MATHEMATICAL SCIENCES
251 MERCER STREET
NEW YORK, NEW YORK 10012
U.S.A.

Transfer

Canadian Mathematical Society
Conference Proceedings
Volume 2, Part 2 (1982)

UNIVERSAL BERNOULLI NUMBERS AND THE S^1 - TRANSFER

Haynes Miller

Several authors ([4], [6], [7], [8], [12]) have considered a stable "transfer" map

$$t\colon\ \mathbb{C}P_0^\infty \wedge S^1 \to S^0,$$

and it is of interest to develop techniques by which to compute its effect in stable homotopy. In this note we begin an attack on this problem via the Novikov spectral sequence ([2], [13]) $E_r(-)$.

We shall study t by means of its unique factorization through the Moore spectrum for $\mathbb{Q}/\mathbb{Z}$:

$$\begin{array}{ccc} \mathbb{C}P_0^\infty \wedge S^2 & \xrightarrow{\ \bar{u}\ } & S\mathbb{Q}/\mathbb{Z} \\ & \searrow^{t\wedge S^1} & \downarrow \partial \\ & & S^1 \end{array}$$

The behavior of ∂ in the Novikov spectral sequence is quite well understood [13], and our principal result here, Cor. 3.10, describes $\bar{u}_*$ on the standard generators in $MU_*\mathbb{C}P_0^\infty$. The analogous result for $K_*\mathbb{C}P_0^\infty$ appears in [8], and I understand that Knapp now has a proof for the present case also.

In [16], D. M. Segal gave generators for $E_2^0(\mathbb{C}P_0^\infty)$. In Section 4 we describe these elements together with a simplified proof of their properties. Their images under $\bar{u}_*$ turn out to be "universal Bernoulli numbers," in the sense that they are to the formal group for MU, which is universal, as the usual Bernoulli numbers are to the multiplicative formal group. A construction of these classes, and a computation of their denominators, is carried out in Section 1.

As a corollary, we recover the fact, due to Becker and Schultz, that for $k > 0$ neither μ_{8k+1} nor the generator of the image of the J-homomorphism in dimension $8k - 1$ lies in the image of t. It is to be hoped that this work will lead to a complete computation of the image of t_* in $E_2^2(S^0)$, providing a context for the germinal result of K. Knapp [7].

This research was supported in part by NSF Grant MCS 77-05414.
© 1982 American Mathematical Society
0731-1036/82/0000-0479/$04.25

I have been helped in this work by conversations and correspondence with A. Liulevicius, V. Giambalvo, N. Koblitz, and S. Mitchell, and I thank them all.

Section 1. BERNOULLI NUMBERS ATTACHED TO A FORMAL GROUP.

We assume the reader is familiar with the basic properties of formal groups as exposed for instance in [2] or [3]. As an illustration, for any element u in a ring A, one has a formal group

$$G_u(X,Y) = X + Y - uXY.$$

The additive formal group G_a is then the case $u = 0$, and the multiplicative formal group G_m is the case $u = 1$. If A is torsion-free, embed it in $A_{\mathbb{Q}} = A \otimes \mathbb{Q}$, let $\log_F: F \to G_a$ be the (unique) isomorphism of formal groups over $A_{\mathbb{Q}}$, and let $\exp_F(T)$ its inverse. For example, $\log_{G_m}(T) = -\ln(1 - T)$ and $\exp_{G_m}(T) = 1 - e^{-T}$.

DEFINITION 1.1. Let A be a torsion-free ring and F a formal group over A. The Bernoulli numbers associated to F are the coefficients $B_n(F) \in A_{\mathbb{Q}}$ in the powerseries expansion

$$\frac{T}{\exp_F(T)} = \sum_{n=0}^{\infty} \frac{B_n(F)}{n!} T^n.$$

The divided Bernoulli numbers are $B_n(F)/n$.

EXAMPLE 1.2. (a) $B_0(F) = 1$ for all F. (b) $B_n(G_a) = 0$ for all $n > 0$. (c) $B_n(G_m)$ is the usual Bernoulli number, occuring in

$$\frac{T}{1-e^{-T}} = \sum_{n=0}^{\infty} \frac{B_n(G_m)}{n!} T^n.$$

Recall [5] that the reduction of $B_n(G_m)/n$ in $\mathbb{Q}/\mathbb{Z}$ has order d_n, where

$$d_n = \begin{cases} \prod_{(p-1)|n} p^{\nu_p(n)+1} & \text{for even } n \\ 2 & \text{for } n = 1 \\ 1 & \text{for odd } n > 1. \end{cases}$$

These denominators are universal:

THEOREM 1.3. If F is a formal group over a torsion-free ring A, then

$$d_n \frac{B_n(F)}{n} \in A.$$

We begin our proof of this theorem with a lemma.

LEMMA 1.4. If F and F' are formal groups isomorphic over a torsion-free ring A, then $B_n(F)/n \equiv B_n(F')/n \mod A$.

PROOF. Let φ: $F \to F'$ be the isomorphism, so that $\exp_{F'}(T) = \varphi(U)$ where $U = \exp_F(T)$. Then

$$\frac{T}{\exp_{F'}(T)} = \frac{T}{\varphi(U)} = \frac{T}{U}\,\frac{U}{\varphi(U)} = \frac{T}{\exp_F(T)} + T\sum_{n=1}^{\infty} a_n U^{n-1}$$

where $U/\varphi(U) = \sum_{n=0}^{\infty} a_n U^n$, $a_0 = 1$, $a_n \in A$. The lemma is equivalent to the assertion that the value at 0 of the $i^{\underline{th}}$ derivative of U^n lies in A for all $i \geq 0$ and $n \geq 1$. This is well-known for $U = \exp_F(T)$, and follows by Leibnitz' formula for larger n. □

A ring-homomorphism f: $A \to B$ carries a formal group F over A to a formal group f_*F over B. There is an initial object in the category of pairs (A,F), namely [2] the Lazard group G over the Lazard ring L. We now recall a well-known integrality statement for L. Let $A = \mathbb{Z}[b_1, b_2, \ldots]$ and let $\varphi(T) = \sum_{i=0}^{\infty} b_i T^{i+1}$ with $b_0 = 1$. Let ${}^{\varphi}G_m(X,Y) = \varphi(G_m(\varphi^{-1}(X), \varphi^{-1}(Y)))$; this is a formal group over A. Let η: $L \to A$ classify ${}^{\varphi}G_m$. Then we have:

THEOREM 1.5. (Stong-Hattori) In this situation, the diagram

$$\begin{array}{ccc} L & \xrightarrow{\ \eta\ } & A \\ \downarrow & & \downarrow \\ L_{\emptyset} & \xrightarrow{\ \eta_{\emptyset}\ } & A_{\emptyset} \end{array}$$

is a Cartesian square. In particular, L is torsion-free. □

For a proof, see [10]. Theorem 1.3 follows as a corollary. For φ: $G_m \to {}^{\varphi}G_m$ is an isomorphism, so by Lemma 1.4 $B_n(G_m)/n$ and $B_n({}^{\varphi}G_m)/n$ have the same denominators; but $B_n({}^{\varphi}G_m)/n = \eta_* B_n(G)/n$, so the theorem holds for G by the Stong-Hattori theorem. The general case follows immediately. □

REMARK 1.6. Consequently, if F is a formal group over a ring A, possibly <u>with</u> torsion, then "Bernoulli numerators" $N_n(F) \in A$ are defined, viz.,

$$N_n(F) = \eta_*\left(d_n \frac{B_n(G)}{n}\right)$$

where η: $L \to A$ classifies F.

<u>Section 2</u>. THE TRANSFER.

We give a brief description of the transfer construction, focussing on examples useful to us here. The reader is referred to [12, 14] for more exhaustive accounts of the transfer. We end with a proof (due to S. Mitchell [14]) of the fact, stated in [7], that t: $\mathbb{C}P_0^{\infty} \wedge S^1 \to S^0$ is the cofiber of a natural collapse map.

Let $\pi: E \to B$ be a smooth map, and let $\xi \downarrow E$ and $\zeta \downarrow B$ be vector bundles. A <u>relative framing</u> is a lift $j: E \to \mathbb{R}^k_B$ of π to an embedding (with normal bundle $\nu(j)$) into a trivial vector bundle over B together with a bundle-isomorphism

$$\phi: \xi \oplus \mathbb{R}^k_E \to \pi^*\zeta \oplus \nu(j).$$

Given this data, application of the Pontrjagin-Thom collapse gives a stable map

$$t: B^\zeta \to E^\xi \tag{2.1}$$

of Thom spaces, called the <u>transfer</u>. An obvious modification allows us to suppose that ζ and ξ are merely virtual bundles.

EXAMPLE 2.2. Let L_n denote the complex dual of the tautologous complex line-bundle over $\mathbb{C}P^n$. For $-\infty < q \le r < \infty$, define

$$\mathbb{C}P^r_q = (\mathbb{C}P^{r-q})^{qL_{r-q}}.$$

For $r \le s$, the bundle-map $L_{r-q} \to L_{s-q}$ induces a map

$$i: \mathbb{C}P^r_q \to \mathbb{C}P^s_q$$

of Thom spaces. For $p \le q$, recall that the inclusion $j: \mathbb{C}P^{r-q} \to \mathbb{C}P^{r-p}$ has normal bundle $(q-p)L_{r-q}$. Taking $\xi = qL_{r-q}$ and $\zeta = pL_{r-p}$, we obtain a transfer map

$$c: \mathbb{C}P^r_p \to \mathbb{C}P^r_q.$$

It is not hard to see that if p, q, r, and s are all nonnegative then under the usual homeomorphism

$$\mathbb{C}P^r_q \cong (\mathbb{C}P^r/\mathbb{C}P^{q-1})$$

these maps coincide with the natural inclusion and collapse maps. Also, their obvious compatibility allows us to include the possibility of $r = \infty$ or $s = \infty$.

Given $i + j = q \in \mathbb{Z}$ and $k \ge 0$, the bundle map $\overline{\Delta}: qL_k \to iL_k \times jL_k$ covering the diagonal of $\mathbb{C}P^k$ induces

$$\Delta: \mathbb{C}P^{q+k}_k \to \mathbb{C}P^{i+k}_i \wedge \mathbb{C}P^{j+k}_j \tag{2.3}$$

These maps are clearly associative, unitary, commutative, and behave well with respect to i and c.

EXAMPLE 2.4. Let $\pi: S^{2n+1} \to \mathbb{C}P^n$ be the usual projection map. The bundle $\tau(\pi)$ of tangents along the fiber is complementary to the normal bundle of π, and is trivialized by the infinitesimal generator of the S^1 - action. Thus we obtain a stable transfer map

$$\bar{t}\colon\ \mathbb{C}P^n_0 \wedge S^1 \to S^{2n+1}_+ .$$

The projection to S^{2n+1} has degree 1 and is of no further interest, by Hopf's theorem, so we consider the other factor, $\mathbb{C}P^n_0 \wedge S^1 \to S^0$. These maps are compatible over n, and yield a stable map

$$t\colon\ \mathbb{C}P^\infty_0 \wedge S^1 \to S^0 .$$

This map was studied by J. C. Becker and R. E. Schultz, who proved:

THEOREM 2.5. [4] There is a stable map $j_{S^1}\colon\ U \to \mathbb{C}P^\infty_0 \wedge S^1$ such that the composite $t \circ j_{S^1}$ is adjoint to the composite

$$j_{\mathbb{C}}\colon\ U \xrightarrow{J_{\mathbb{C}}} Q_1S^0 \xrightarrow{*[1]} QS^0 . \qquad \square$$

Recall also

PROPOSITION 2.6. [12] If $\lambda\colon\ \mathbb{C}P^\infty_0 \wedge S^1 \to U$ carries (ℓ, z) to multiplication by z in the line ℓ, then the composite $j_{S^1} \circ \lambda$ is homotopic to the identity. $\square$

We end this section with:

LEMMA 2.7. [7] The diagram

$$\mathbb{C}P^\infty_{-1} \wedge S^1 \xrightarrow{c} \mathbb{C}P^\infty_0 \wedge S^1 \xrightarrow{t} S^0$$

is a cofibration sequence.

PROOF. [14] First note that $\bar{t}c\colon\ \mathbb{C}P^n_{-1} \wedge S^1 \to S^{2n+1}_+$ is the transfer associated to the composite $S^{2n+1} \to \mathbb{C}P^n \to \mathbb{C}P^{n+1}$. Since this composite is null-homotopic, $\bar{t}c$ factors up to homotopy through the Thom space S^{2n+1} of a bundle over a point. Since the null-homotopies are compatible as n increases, the composite $\bar{t}c$ factors through the contractible spectrum S^∞, and hence is null-homotopic. Therefore tc is null-homotopic.

There results a map

$$\alpha\colon\ \mathbb{C}P^\infty_{-1} \wedge S^2 \to C(t)$$

to the mapping-cone of t. It is clearly a homology-isomorphism in positive dimensions. In Remark 3.5(c) we shall see that P^1 is nontrivial on $H^0(\mathbb{C}P^\infty_{-1} \wedge S^2)$ (where $P^1 = Sq^2$ if $p = 2$). According to Theorem 2.5, the J-homomorphism $j_{\mathbb{C}}\colon\ U \to S^0$ factors through t, so the element α_1 of Hopf invariant 1 is carried on $C(t)$, and it follows that P^1 is nontrivial on $H^0(C(t))$ as well. Therefore $H_0(\alpha)$ is also an isomorphism. The map α is thus a homotopy-equivalence, and the result follows. □

REMARK 2.8. One may use ideas of Löffler and Smith [11] in place of the work of Becker and Schultz to see that P^1 detects t.

Section 3. TRANSFER, THOM CLASS, AND COACTION.

In this section we show how to extends Adams' treatment [2] of the complex bordism of $\mathbb{C}P^\infty$ to the case of $\mathbb{C}P_n^\infty$ for $n \in \mathbb{Z}$. We shall adopt the convention that the E-homology of X is $X_*E = \pi_*(X \wedge E)$, the stable homotopy of $X \wedge E$. (The homology (and cohomology) of a space is thus always reduced.) If E is a ring-spectrum (always associative and commutative) for which E_*E is flat over $E_* = \pi_*E$, then we have a right coaction map

$$\psi\colon X_*E \to X_*E \otimes_{E_*} E_*E.$$

This convention seems natural from several points of view; in particular, it fits well with [13].

Recall that an orientation of a ring-spectrum E is an element $x = x_E \in E^2(\mathbb{C}P^\infty)$ restricting to the canonical generator of $E^2(S^2)$. If E is oriented by x, it follows that $E^*(\mathbb{C}P_0^\infty) = E^*[[x]]$; and if $\mu\colon \mathbb{C}P_0^\infty \wedge \mathbb{C}P_0^\infty \to \mathbb{C}P_0^\infty$ is the multiplication map, then $\mu^* x = F(x \otimes 1, 1 \otimes x)$ defines a formal group F over E^*. The homeomorphism $\mathbb{C}P^\infty \cong MU(1)$ provides MU with a canonical orientation x_{MU}, and (MU, x_{MU}) is universal in the obvious sense. By a famous result of Quillen, the natural map $L \to MU^*$ is an isomorphism; see [2].

Now consider $\mathbb{C}P_n^\infty$ for $n \in \mathbb{Z}$. The diagonal (2.3) $\Delta\colon \mathbb{C}P_n^\infty \to \mathbb{C}P_0^\infty \wedge \mathbb{C}P_n^\infty$ makes $E^*(\mathbb{C}P_n^\infty)$ into a module over $E^*(\mathbb{C}P_0^\infty)$, which is free on one generator u by the Thom isomorphism theorem. If $n \geq 0$, we take u such that $c^*u = x^n \in E^{2n}(\mathbb{C}P_0^\infty)$, using the notation of (2.2). If $n < 0$, we take u such that $x^{-n}u = c^*1 \in E^0(\mathbb{C}P_n^\infty)$. In either case it makes good sense for $i \geq n$ to write x^i for $x^{i-n}u$, and henceforth we do so.

Let $\beta_i = \beta_i^E \in \pi_{2i}(\mathbb{C}P_n^\infty \wedge E)$ be dual to x^i, and write

$$\hat{\beta}(T) = \hat{\beta}_n(T) = \sum_{i \geq n} \beta_i T^i.$$

Then the module structure dualizes to give

$$\Delta_*\hat{\beta}(T) = \beta(T) \otimes \hat{\beta}(T) \tag{3.1}$$

where $\beta(T) = \hat{\beta}_0(T)$.

Our approach to the coaction for $\mathbb{C}P^\infty$ differs from that of Adams. We may write

$$\psi\hat{\beta}(T) = \sum_{j \geq n} \beta_j \otimes f_j(T)$$

for suitable power series $f_j(T)$; these power series are independent of n, in view of the maps i and c. Now the Cartan formula of [1], p.71, applied to $\hat{\beta}(T)$, asserts that

$$\sum \beta_i \otimes \beta_j \otimes f_{i+j}(T) = \sum \beta_i \otimes \beta_j \otimes f_i(T) f_j(T).$$

Together with the obvious fact that $f_0(T) = 1$, this yields $f_j(T) = f_1(T)^j$. Write

$$f_1(T) = b(T) = \sum_{i \geq 0} b_i T^{i+1}.$$

The unital property of ψ shows that $b_0 = 1$.

We must next recall some generalities concerning the Kronecker pairing. For a ring-spectrum E and an E-module-spectrum F with structure-map $\varphi\colon\ F \wedge E \to F$, we have a pairing

$$< , >:\ F^p(X) \otimes X_q E \to F_{q-p}:$$

for $f\colon\ X \to \Sigma^p F$, $e\colon\ S^q \to X \wedge E$,

$$<f,e>:\ S^q \xrightarrow{e} X \wedge E \xrightarrow{f\wedge 1} \Sigma^p F \wedge E \xrightarrow{\varphi} \Sigma^p F.$$

LEMMA 3.2. In addition to these notations, let $\mu\colon\ F_* \otimes_{E_*} E_*E \to F_*E$ be the evident multiplication. Assume that E_*E is flat over E_*. Then

$$f_* e = \mu<f,\psi e>.$$

PROOF. Since

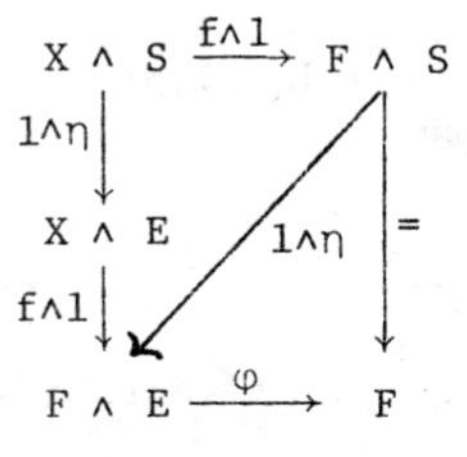

commutes, the top row of the following commutative diagram evaluates f_*.

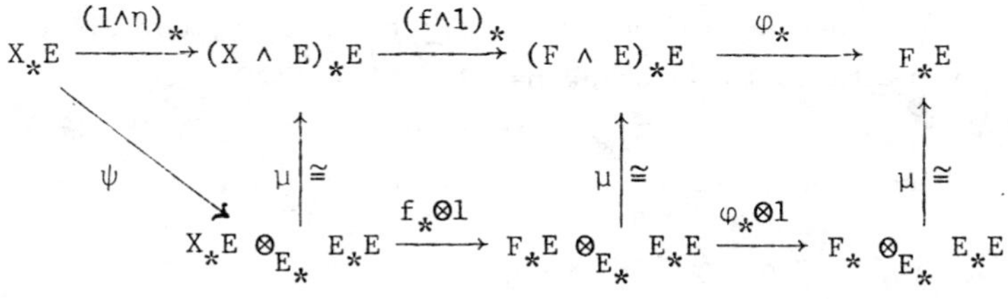

But the bottom row is $<f,->$. □

Now $b(T)$ is by definition $\langle x,\psi\beta(T)\rangle$; so we have proved:

PROPOSITION 3.3. In $\pi_*(\mathbb{C}P_n^\infty \wedge MU)$,

$$\psi\hat{\beta}(T) = \hat{\beta}(1 \otimes b(T)).$$

with $b(T) = x_*\beta(T)$. □

REMARK 3.4. This line of argument may be used to determine the form of the coaction in any space with polynomial integral cohomology. Consider $\mathbb{H}P^\infty$ for example. For the generator $y \in MU^4\mathbb{H}P_0^\infty$ we may take the second Connor-Floyd Chern class of the canonical bundle over $\mathbb{H}P^\infty$ thought of as a complex 2-plane bundle. Under the natural inclusion $i\colon \mathbb{C}P^\infty \to \mathbb{H}P^\infty$, this bundle pulls back to $L \oplus L^*$, so, with $[n] = [n]_G$ as in [15],

$$i^*y = cf_2(L \oplus L^*) = cf_1(L)cf_1(L^*) = x[-1](x).$$

Let γ_i be a dual to y^i, and form a power series

$$\gamma(T) = \sum_{i\geq 0} \gamma_i T^{2i}.$$

Since i_* is a coalgebra map, we find that

$$i_*\beta(T) = \gamma(c(T))$$

for some power series $c(T)$. We compute:

$$c(T) = \langle y,i_*\beta(T)\rangle = \langle i^*y,\beta(T)\rangle$$

$$= \langle x[-1](x),\beta(T)\rangle = T[-1](T).$$

For the coaction, we compute:

$$\psi\gamma(c(T)) = \psi i_*\beta(T) = i_*\psi\beta(T)$$

$$= i_*\beta(1 \otimes b(T)) = \gamma(1 \otimes c_L(b(T)))$$

where $c_L(T) = \eta_{L*}c(T)$. This implicitly determines the coaction in $\mathbb{H}P^\infty$; cf. [16].

REMARK 3.5. (a) Since we are using the <u>right</u> coaction, our elements $b_i \in BP_{2i}BP$ are <u>conjugate</u> to those of Adams [2 : I].

(b) Let $G_L = \eta_{L*}G$ and $G_R = \eta_{R*}G$; then by [2 : I (11.4)] the power series $b(T)$ is an isomorphism of formal groups from G_R to G_L; that is,

$$b(\exp_R(T)) = \exp_L(T) \tag{3.6}$$

where again $\exp_R(T) = \eta_{R*}\exp_G(T)$, etc. Also $\rho_*\colon MU_*MU \to H\mathbb{Q}_*MU$ carries

G_L to G_a, so

(3.7) $$\rho_* b(T) = \log_G(T).$$

(c) The same results hold in any oriented ring-spectrum E such that E_*E is flat over E_*. In particular we find in mod p homology that

$$\psi\hat{\beta}(T) = \hat{\beta}(1 \otimes \zeta(T)),$$

where $\zeta(T) = \sum \zeta_i T^{p^i}$ with $\zeta_i = \chi\xi_i$ for p odd and $\zeta_i = \chi\xi_i^2$ for $p = 2$.

Now let $t\colon \mathbb{C}P_0^\infty \wedge S^1 \to S^0$ be the transfer map constructed in Section 2. Since MU_* is evenly graded, $MU_*(t) = 0$; so the cofibration sequence

$$S^0 \xrightarrow{i} \mathbb{C}P_{-1}^\infty \wedge S^2 \xrightarrow{c} \mathbb{C}P_0^\infty \wedge S^2$$

induces a long exact sequence in $E_2(-)$, with boundary homomorphism

$$t_*\colon E_2^s(\mathbb{C}P_0^\infty \wedge S^2) \to E_2^{s+1}(S^0).$$

In [9] it is shown that if x is a permanent cycle in the first spectral sequence, then t_*x represents tx in the second.

Let $u\colon \mathbb{C}P_{-1}^\infty \wedge S^2 \to H\mathbb{Q} = S\mathbb{Q}$ represent the Thom class $x_{H\mathbb{Q}}^{-1}$. We then have a map of cofibration sequences:

(3.8) $$\begin{array}{ccccc} S^0 & \xrightarrow{i} & \mathbb{C}P_{-1} \wedge S^2 & \xrightarrow{c} & \mathbb{C}P_0 \wedge S^2 \\ \Big\downarrow{=} & & \Big\downarrow{u} & & \Big\downarrow{\bar{u}} \\ S^0 & \longrightarrow & S\mathbb{Q} & \longrightarrow & S\mathbb{Q}/\mathbb{Z} \end{array}.$$

This yields a factorization

$$\begin{array}{ccc} E_2^s(\mathbb{C}P_0 \wedge S^2) & \xrightarrow{t_*} & E_2^{s+1}(S^0) \\ & \searrow^{\bar{u}_*} \quad \nearrow_{\partial} & \\ & E_2^s(S\mathbb{Q}/\mathbb{Z}) & \end{array}$$

of t_*, in which ∂ is the boundary - homomorphism induced by the bottom sequence in (3.8).

According to the program of [13], it is via the map ∂ that elements in $E_2^s(S^0)$ are best described; so it is very natural to compute $u_*\colon \pi_*(\mathbb{C}P_{-1}^\infty \wedge S^2 \wedge MU) \to \mathbb{Q} \otimes MU_*$.

Since u factors as

$$\mathbb{C}P_{-1}^\infty \wedge S^2 \xrightarrow{x^{-1}} MU \xrightarrow{\rho} H\mathbb{Q},$$

the first step is to compute

$$x_*^{-1}\hat{\beta}(T) = \mu\langle x^{-1}, \psi\hat{\beta}(T)\rangle \qquad \text{by (3.2)}$$

$$= \mu\langle x^{-1}, \hat{\beta}(1 \otimes b(T))\rangle \qquad \text{by (3.3)}$$

$$= \mu(1 \otimes b(T)^{-1})$$

$$= b(T)^{-1}.$$

Now using (3.7), we find:

THEOREM 3.9. $u_* T\hat{\beta}(T) = \dfrac{T}{\log_G(T)} \in \mathbb{Q} \otimes MU_*[[T]]$.

Since $u_*\beta_{-1} = 1$, we have also:

COROLLARY 3.10. $\bar{u}_* T\beta(T) = \dfrac{T}{\log_G(T)} - 1 \in \mathbb{Q}/\mathbb{Z} \otimes MU_*[[T]]$. □

Section 4. THE IMAGE OF THE PRIMITIVES.

We begin by recalling the primitive generators in $\pi_*(\mathbb{C}P_0^\infty \wedge MU)$. Write $\exp(T)$ for $\exp_G(T)$.

PROPOSITION 4.1. (D. M. Segal [16]) If we define $p_n \in \pi_{2n}(\mathbb{C}P_0^\infty \wedge MU) \otimes \mathbb{Q}$ by means of the expansion

$$\beta(\exp(T)) = \sum_{n=0}^{\infty} \frac{p_n}{n!} T^n,$$

then p_n lies in $\pi_{2n}(\mathbb{C}P_0^\infty \wedge MU)$ and generates the subgroup of primitives.

PROOF. We first check that p_n is primitive, by means of the following calculation (due in different guise to Segal).

$$\psi\beta(\exp(T)) = \beta(1 \otimes b(\exp_R(T))) \qquad \text{by (3.3)}$$

$$= \beta(1 \otimes \exp_L(T)) \qquad \text{by (3.6)}$$

$$= \beta(\exp(T)) \otimes 1.$$

Next, note that $n!$ times the coefficient of T^n in $(\exp(T))^k$ is integral. It follows that p_n is integral. On the other hand, the coefficient $n!b_{n-1}$ of β_1 in p_n generates a summand in MU_*; this follows from the fact that its image in $\mathbb{Z}$ under the map classifying G_m is -1. Therefore p_n is a generator of $E_2^{0,2n}(\mathbb{C}P_0^\infty)$. But this group embeds into $E_2^{0,2n}(\mathbb{C}P_0^\infty \wedge S\mathbb{Q})$ since $\mathbb{C}P_0^\infty$ is torsion-free, and the latter group is just $\pi_{2n}(\mathbb{C}P_0^\infty \wedge S\mathbb{Q}) = \mathbb{Q}$ since $\mathbb{C}P_0^\infty \wedge S\mathbb{Q}$ is an MU-module-spectrum. □

REMARK 4.2. Since $\exp(x_{MU\mathbb{Q}}) \in MU\mathbb{Q}^2(\mathbb{C}P_0^\infty)$ reduces to $x_{H\mathbb{Q}} \in H\mathbb{Q}^2(\mathbb{C}P_0^\infty)$, we find that p_n reduces to $n!\beta_n^H \in \pi_{2n}(\mathbb{C}P_0^\infty \wedge H)$. In the H-structure of $\mathbb{C}P_0^\infty$, $n!\beta_n^H = (\beta_1^H)^n$; and it follows that

$$p_n = \beta_1^n.$$

That is,

$$\beta(\exp(T)) = e^{\beta_1 T}.$$

Therefore, incidentally, $\beta(\exp(S))\beta(\exp(T)) = \beta(\exp(S + T))$, and, replacing S and T by $\log(S)$ and $\log(T)$, we obtain the formula of Ravenel and Wilson [15]:

(4.3) $$\beta(S)\beta(T) = \beta(G(S,T)).$$

This line of argument may of course be reversed.

REMARK 4.4. Analogous primitive generators may be constructed for $\mathbb{H}P^\infty$. The power series $\frac{1}{2}\exp(T)\exp(-T)$ is even, so we may define, following [16], $q_n \in \pi_{4n}(\mathbb{H}P_0^\infty \wedge MU) \otimes \mathbb{Q}$ by

$$\frac{1}{2}\gamma_0 + \frac{1}{2}\gamma(\exp(T)\exp(-t)) = \sum_{n\geq 0} \frac{q_n}{(2n)!} T^{2n}.$$

An analogue of the above proof shows that q_n is integral and a primitive generator.

We now evaluate

$$\bar{u}_*: \ E_2^0(\mathbb{C}P_0^\infty \wedge S^2) \to E_2^2(S\mathbb{Q}/\mathbb{Z})$$

in terms of the Bernoulli numbers introduced in Section 1. Since MU_* is the Lazard ring, the universal Bernoulli number B_n lies in $MU_{2n} \otimes \mathbb{Q}$.

THEOREM 4.5.

$$\bar{u}_* p_n = -\frac{B_{n+1}}{n+1}.$$

PROOF. Replacing T by $\exp(T)$ in Corollary 3.10,

$$\bar{u}_*\beta(\exp(T)) = \frac{1}{T} - \frac{1}{\exp(T)}.$$

The result follows upon expanding both sides. □

This together with Theorem 1.3 implies that $\bar{u}_* p_n$ has order d_n. On the other hand, recall from [13] that (if $||$ means divides exactly)

$$E_2^{0,0}(S\mathbb{Q}/\mathbb{Z}) = \mathbb{Q}/\mathbb{Z}$$

$$\begin{aligned} E_2^{0,2(p-1)u}(S\mathbb{Q}/\mathbb{Z}) \otimes \mathbb{Z}_{(p)} &= \mathbb{Z}/p^{n+1} && \text{if } p^n \,||\, u > 0, \quad p > 2 \\ &= \mathbb{Z}/2 && \text{if } 2 \nmid u > 0, \quad p = 2 \\ &= \mathbb{Z}/4 && \text{if } u = 2, \quad p = 2 \end{aligned}$$

$$= \mathbb{Z}/2^{n+2} \quad \text{if} \quad 2^n \| u > 2,\ n > 0, \quad p = 2.$$

Furthermore, ∂: $E_2^0(S\mathbb{Q}/\mathbb{Z}) \to E_2^1(S^0)$ merely kills the $\mathbb{Q}/\mathbb{Z}$. Comparing these orders with the numbers d_n, we find

THEOREM 4.6. The image of t_*: $E_2^{0,2u}(\mathbb{C}P_0^\infty \wedge S^2) \to E_2^{1,2u}(S^0)$ is the subgroup of index 2 except when $u = 1$ or 2, when the map is surjective. □

From [13] we then easily recover the theorem of Becker and Schultz:

THEOREM 4.7. [4] For $k > 0$, neither μ_{8k+1} nor the generator j_{8k-1} of the image of the J-homomorphism in dimension $8k - 1$ lies in the image of t: $\pi_*(\mathbb{C}P_0^\infty \wedge S^1) \to \pi_*(S^0)$. □

REMARK 4.8. [4] Since $2j_{8k-1}$ is in the image of the usual complex J-homomorphism, it does lie in Im(t) by the result (Theorem 2.5 above) of Becker and Schultz.

January, 1980
Revised July, 1981

<u>Added in proof</u>: Many of these results occur explicitly in K. Knapp's Bonn Habilitationsschrift, "Some applications of K-theory to framed bordism: e-invariant and transfer," Bonner Mathematishe Schriften, Heft 118, 1979. For instance, Lemma 2.7 occurs there as Theorem 2.9, and Corollary 3.10 occurs there as (5.21).

BIBLIOGRAPHY

1. J. F. Adams, "Lectures on generalized homology", Lecture Notes in Math., vol. 99, Springer-Verlag, 1969, 1-138.

2. J. F. Adams, Stable Homotopy and Generalized Homology, Univ. of Chicago Press, 1974.

3. S. Araki, Typical Formal Groups in Complex Cobordism and K-Theory, Lectures in Mathematics 6, Kyoto Univ., Kinokuniya Book-Store Co., Ltd., n.d.

4. J. C. Becker and R. E. Schultz, "Equivariant function spaces and stable homotopy theory", Comm. Helv. Math., 49 (1974), 1-34.

5. Z. I. Borevich and I. R. Shafarevich, Number Theory, Academic Press, 1966.

6. I. Hansen, "Framed bordism of free Lie group actions and the transfer", preprint.

7. K. H. Knapp, "On the bi-stable J-homomorphism", Lecture Notes in Math., vol. 763, Springer-Verlag, 1979, 13-22.

8. K. H. Knapp, "On odd-primary components of Lie groups", Proc. Amer. Math. Soc., 79 (1980), 147-152.

9. D. C. Johnson, H. R. Miller, W. S. Wilson, and R. S. Zahler, "Boundary homomorphisms in the generalized Adams spectral sequence and the nontriviality of infinitely many γ_t in stable homotopy", Notas de Mat. y Symp., No. 1, Soc. Math. Mex., 1975.

10. P. S. Landweber, "Annihilator ideals and primitive elements in complex bordism", Ill. J. Math., 17 (1973), 273-284.

11. P. Löffler and L. Smith, "Line bundles over framed manifolds", Math. Zeit., 138 (1974), 35-52.

12. B. M. Mann, E. Y. Miller, and H. R. Miller, "S^1-equivariant function spaces and characteristic classes", Trans. Amer. Math. Soc., to appear.

13. H. R. Miller, D. C. Ravenel, and W. S. Wilson, "Periodic phenomena in the Adams-Novikov spectral sequence", Ann. of Math., 106 (1977), 469-516.

14. S. A. Mitchell, "Complex bordism and stable homotopy type of $B(\mathbb{Z}/p \times \mathbb{Z}/p)$", thesis, Univ. of Washington, 1981.

15. D. C. Ravenel and W. S. Wilson, "The Hopf ring for complex cobordism", J. of Pure and Appl. Alg., 9 (1977), 241-280.

16. D. M. Segal, "The cooperation on $MU_*(CP^\infty)$ and $MU_*(HP^\infty)$ and the primitive generators", J. of Pure and Appl. Alg., 14 (1979), 315-322.

DEPARTMENT OF MATHEMATICS
UNIVERSITY OF WASHINGTON
SEATTLE, WASHINGTON 98195

Canadian Mathematical Society
Conference Proceedings
Volume 2, Part 2 (1982)

TRANSFERS IN ALGEBRAIC K- AND L-THEORY INDUCED BY S^1-BUNDLES[1)]

Hans J. Munkholm[2)] and Erik K. Pedersen

1. Geometric transfers coming from S^1-bundles

Let $S^1 \longrightarrow E \xrightarrow{p} B$ be a fiber bundle with fundamental group sequence

(1.1) $\mathbb{Z} \xrightarrow{i} \pi \xrightarrow{\varphi} \rho \longrightarrow \{1\}$

and orientation map

(1.2) $\omega: \rho \longrightarrow \{\pm 1\} = \mathrm{Aut}(\mathbb{Z})$

Also suppose given orientation maps $w_B: \rho \longrightarrow \{\pm 1\}$, $w_E: \pi \longrightarrow \{\pm 1\}$ such that

(1.3) $w_E(g) = w_B(\varphi(g))\, \omega(\varphi(g))$, $g \in \pi$.

If B is a finitely dominated CW complex then so is E and there is a homomorphism

(1.4) $p^*_{K_o} : \widetilde{K}_o(\mathbb{Z}\rho) \longrightarrow \widetilde{K}_o(\mathbb{Z}\pi)$

mapping the finiteness obstruction $\widetilde{\sigma}(B)$ to $\widetilde{\sigma}(E)$, (see Ehrlich, [6]).

Let B and B_1 be finite complexes. If $f: B_1 \longrightarrow B$ is a homotopy equivalence then so is the pull back $\overline{f} : E_1 \longrightarrow E$ and there is a homomorphism (see Anderson, [2])

(1.5) $p^*_{Wh}: Wh(\rho) \longrightarrow Wh(\pi)$

mapping the Whitehead torsion $\tau(f)$ to $\tau(\overline{f})$.

Finally, if B is a (simple) Poincare duality space while $f: N \longrightarrow B$ is a surgery problem (we suppress the bundle data from the notation) then the pull back $\overline{f}: M \longrightarrow E$ is again a surgery problem and there is a homomorphism

(1.6) $p^*_L : L^{\varepsilon}_{\ell}(\rho; w_B) \longrightarrow L^{\varepsilon}_{\ell+1}(\pi; w_E)$

mapping the surgery obstruction for f to the one for $\overline{f}$. Here $\varepsilon = s$ or $\varepsilon = h$.

It follows from Pedersen [14] that the homomorphisms p^* of (1.4), (1.5), (1.6) depend only on the fundamental group data (1.1), and the orientation data ω, w_B, w_E .

---

1) Presented at the conference by the first named author.

2) Partially supported by Rektorkollegiet, Denmark.

© 1982 American Mathematical Society
0731-1036/82/0000-0480/$03.50

In section 2 of this note we announce a complete algebraic description of p^*_{Wh} and p^*_L (in terms of matrices). Details will appear in Munkholm and Pedersen [11], [12]. Various results on $p^*_{K_O}$ and p^*_{Wh} have been obtained earlier by Anderson [3,4], Pedersen and Taylor [15], Munkholm and Pedersen [9,10], and Ehrlich [7]. The formulas that are given in theorem 2.2 have been obtained independently by Ranicki using a theory of pseudo chain complexes which is so far in a preliminary form, [16].

The main part of the present paper is section 3 in which we prove some vanishing results for p^*_{Wh}.

2. Algebraic S^1-transfer maps

The proper generality for the definition of the various algebraic S^1-transfer maps seems to be the following:

We suppose given an associative ring R with unit as well as

- an anti involution $r \longrightarrow r^*$ of R
- an automorphism $r \longrightarrow r^t$ of R
- a unit $t \in R$

These data are subject to the following requirements:

(2.1) $\quad r^{*t} = r^{t*}$, $(t-1)r = r^t(t-1)$, $trt^{-1} = r^{t^2} (=(r^t)^t)$, $t^* = t^{-1}$, $t^t = t$

for all $r \in R$. Note that the purpose of $r \longrightarrow r^t$ is to pass a factor $(t-1)$ to the other side. We have the quotient map $\varphi: R \longrightarrow \bar{R} = R/(t-1)R$. Also $\bar{R}$ inherits an anti involution $\bar{*}$ and an automorphism $\bar{t}$. To fit with standard notation in our main example, we shall write $*$ for the composition $\bar{*}\,\bar{t}$ on R . Thus $\varphi(r)^* = \varphi(r^{t^*})$. We extend $*$ and t over the matrix rings $M_r(R)$ in the obvious way.

Main example Consider the exact sequence (1.1) with the orientation maps ω, w_B, w_E . Let $R = \mathbb{Z}\pi$ with $g^* = w_E(g)g^{-1}$, $t = i(1)$ and

$$g^t = \begin{cases} g & \omega(\varphi(g)) = 1 \\ -gt^{-1} & \omega(\varphi(g)) = -1 \end{cases}$$

$(g \in \pi)$. Then $\varphi: R \longrightarrow \bar{R}$ becomes the projection $\varphi: R\pi \longrightarrow R\rho$.

Theorem 2.1 [11]. In the above situation there is a homomorphism $\varphi^!: K_1(\bar{R}) \longrightarrow K_1(R)$ defined by: Let $A \in G\ell_r(\bar{R})$. Then

$$\varphi^!([A]) = \begin{bmatrix} \tilde{A} & -\tilde{C} \\ t-1 & \tilde{B}^t \end{bmatrix}$$

where $\tilde{A}$, $\tilde{B}$, $\tilde{C}$ are arbitrary matrices over R with

$$\varphi(\widetilde{A}) = A, \quad \widetilde{A}\widetilde{B} = 1 - \widetilde{C}(t-1)$$

In case $R = \mathbb{Z}\pi$ then $\varphi^!$ induces $\varphi^! : Wh(\rho) \longrightarrow Wh(\pi)$ and if $p : E \longrightarrow B$ is a fiber bundle as in section 1 then

$$p^*_{Wh} = \varphi^! : Wh(\rho) \longrightarrow Wh(\pi)$$

(here $\varphi = p_* : \pi \longrightarrow \rho$)

<u>Theorem 2.2 [12]</u> In the above situation there is a homomorphism

$$\varphi^! : L^{\varepsilon}_{\ell}(\overline{R}) \longrightarrow L^{\varepsilon}_{\ell+1}(R)$$

defined as follows:

Let $\ell = 2k$ be even and let $\alpha \in M_r(\overline{R})$ represent an element (a quadratic form) $[\alpha] \in L^{\varepsilon}_{2k}(\overline{R})$. Also choose an arbitrary $\widetilde{\alpha} \in M_r(R)$ with $\varphi(\widetilde{\alpha}) = \alpha$. Then there is a matrix $\widetilde{A} \in U^{\varepsilon}_r(R)$ of the form

$$\widetilde{A} = \begin{pmatrix} ? & (-1)^k(t^{-1}-1) \\ ? & \widetilde{\alpha}^t + (-1)^k\widetilde{\alpha}^* t \end{pmatrix} \tag{2.3}$$

and $\varphi^!([\alpha]) = [\widetilde{A}]$.

Let $\ell = 2k+1$ and let $A \in U^{\varepsilon}_r(\overline{R})$ represent an element of $L^{\varepsilon}_{2k+1}(R)$. Choose an $\widetilde{A} \in M_r(R)$ with $\varphi(\widetilde{A}) = A$ and write $\widetilde{J}$ for the $2r \times 2r$ matrix $\begin{pmatrix} 0 & 1 \\ 0 & 0 \end{pmatrix}$. Then there exist $2r \times 2r$ matrices $\widetilde{\Theta}$ and $\widetilde{X}$ over R so that

$$\widetilde{A}^*\widetilde{J}\widetilde{A}^t = t^{-1}\widetilde{\Theta}^t - (-1)^k\widetilde{\Theta}^* + \widetilde{X}(1-t^{-1}) + \widetilde{J} \tag{2.4}$$

For any such choice the matrix

$$W = \begin{pmatrix} (1-t)\widetilde{J} & \widetilde{A} \\ 0 & \widetilde{X} \end{pmatrix} \tag{2.5}$$

represents an element of $L^{\varepsilon}_{2k+2}(R)$ and

$$\varphi^!([A]) = [W] .$$

Moreover, if $p : E \longrightarrow B$ is a fiber bundle as in section 1 and $R = \mathbb{Z}\pi$ then

$$\varphi^! = p^*_L : L^{\varepsilon}_{\ell}(\rho; w_B) \longrightarrow L^{\varepsilon}_{\ell+1}(\pi; w_E)$$

(Again $\varphi = p_* : \pi \longrightarrow \rho$)

Concerning proofs let us just remark the following: The algebraic part is proved by writing down explicit matrix identities. As for the topology, we follow Wall's book [18] very closely, e.g., to prove that the matrix W of (2.5) represents the relevant surgery obstruction we simply construct a basis for the relevant kernel group consisting of immersed spheres and compute their self- and mutual intersections.

Note that in (2.3) the choice of the ?'s is immaterial (Wall, [18] chapter 6).

In [11] we also prove that the maps $p^*_{K_o}$ and $(p \times S^1)^*_{Wh}$ makes the diagram

$$\begin{array}{ccc} \widetilde{K}_o(\mathbb{Z}\pi) & \xleftarrow{\ \bar{h}\ } & Wh(\pi \times \mathbb{Z}) \\ \uparrow{\scriptstyle p^*_{K_o}} & & \uparrow{\scriptstyle (p \times S^1)^*_{Wh}} \\ \widetilde{K}_o(\mathbb{Z}\rho) & \xrightarrow[\ h\ ]{} & Wh(\rho \times \mathbb{Z}) \end{array}$$

commute. Here h and $\bar{h}$ are the Bass-Heller-Swan-maps. In view of theorem 2.1 this amounts to a complete algebraic description of $p^*_{K_o}$. As a corollary we obtain

Corollary 2.3 [11] If $S^1 \longrightarrow E \longrightarrow B$ is an orientable fibration with B finitely dominated and $\pi_1(B)$ finite then E has the homotopy type of a finite complex.

3. Some vanishing results for $\varphi^!:K_1(\mathbb{Z}\rho) \longrightarrow K_1(\mathbb{Z}\pi)$

Throughout this section we consider an extension $\mathbb{Z} \xrightarrow{i} \pi \xrightarrow{\varphi} \rho \longrightarrow \{1\}$ with $\omega \equiv 1$. We do not know of any case where the corresponding $\varphi^!:K_1(\mathbb{Z}\rho) \longrightarrow K_1(\mathbb{Z}\pi)$ is non trivial. In [11] we proved two kinds of vanishing results:

Proposition 3.1 If ρ is finite and π is infinite then $\varphi^! = 0 : K_1(\mathbb{Z}\rho) \longrightarrow K_1(\mathbb{Z}\pi)$.

Proposition 3.2 If ρ is finite and $h:\pi \longrightarrow \pi_1$ is any homomorphism into an abelian group then $h_*\varphi^! = 0:K_1(\mathbb{Z}\rho) \longrightarrow K_1(\mathbb{Z}\pi)$.

It is proved in [11] that $\varphi^!(K_1(\mathbb{Z}\rho)) \subseteq Cl_1(\mathbb{Z}\pi)$ ($= \mathrm{Ker}(SK_1(\mathbb{Z}\pi) \longrightarrow \bigoplus_p SK_1(\hat{\mathbb{Z}}_{(p)}\pi)$, see [13]). For π abelian $Cl_1(\mathbb{Z}\pi)$ is well understood by work of Alperin, Dennis and Stein, [1], [17]. Thus one might hope to detect non triviality of $\varphi^!$ by mapping into an abelian group. However, proposition 3.2 rules out such a hope.

The only explicit computation of $Cl_1(\mathbb{Z}\pi)$ for non abelian π that we are aware of are those by Oliver, [13]. For $\pi = \mathbb{Z}_\ell \times G$, where $\mathbb{Z}_\ell$ is cyclic of odd order and G is a 2-primary quaternionic, dihedral or semi-dihedral group of order at least 8, he shows that $Cl_1(\mathbb{Z}(\mathbb{Z}_\ell \times G)) = \mathbb{Z}_2^{\tau(\ell)-1}$ where $\tau(\ell)$ = number of divisors of ℓ. Concerning these groups we prove

Proposition 3.3 Let $\pi = \mathbb{Z}_\ell \times G$ as above, and let t be a central element of G. Then the projection $\varphi : \pi \to \rho = \pi/\langle t\rangle$ has $\varphi^! = 0 : K_1(\mathbb{Z}\rho) \to K_1(\mathbb{Z}\pi)$.

The proof of proposition 3.1 given in [11] relies on the geometric interpretation of $\varphi^!$. Hence it really only proves that the composition $K_1(\mathbb{Z}\rho) \longrightarrow K_1(\mathbb{Z}\pi) \longrightarrow Wh(\pi)$ vanishes. Partly for that reason, but mainly because of its intrinsic interest, we present below a completely algebraic proof of proposition 3.1.

Proposition 3.2 is essentially an easy corollary of 3.1 (see [11]).

Proposition 3.3 leaves open for somebody else's consideration the case when t involves a component in $\mathbb{Z}_\ell$.

Proof of proposition 3.3

If $|G|>8$ then G admits an epimorphism onto $D8$, the dihedral group of order 8. By [13] the induced map $C\ell_1(\mathbb{Z}(\mathbb{Z}_\ell \times G)) \longrightarrow C\ell_1(\mathbb{Z}(\mathbb{Z}_\ell \times D8))$ is an isomorphism. By naturality of $\varphi^!$ we can therefore assume that $|G| = 8$, i.e., from now on G is either D8 or Q8, the quaternionic group.

Next, let ζ_d be a primitive d^{th} root of 1. In [13] it is also shown that the obvious map

$$C\ell_1(\mathbb{Z}\pi) \longrightarrow \bigoplus_{d/\ell,\, d>1} C\ell_1(\mathbb{Z}[\zeta_d][G])$$

is an isomorphism. By naturality it suffices, therefore, to prove that the projection $\varphi : \mathbb{Z}[\zeta_d][G] \longrightarrow \mathbb{Z}[\zeta_d][G/\langle t\rangle]$ has $\varphi^! = 0$.

If $t = 1$ this is obvious. The only other possibility is $t = z$, the unique central involution in G. Then $G/\langle t\rangle = \mathbb{Z}_2^2$ and we have the cartesian square $(R = \mathbb{Z}[\zeta_d]$, d odd, $d > 1)$

$$(3.4)\qquad \begin{array}{ccc} RG & \longrightarrow & RG/\langle z-1\rangle = R[\mathbb{Z}_2^2] \\ \downarrow & & \downarrow \\ RG/\langle z+1\rangle & \longrightarrow & RG/\langle z-1, z+1\rangle = R/2[\mathbb{Z}_2^2] \end{array}$$

Oliver, [13], shows that $C\ell_1(RG)$ is the cokernel of the corresponding map

$$K_2(R[\mathbb{Z}_2^2]) + K_2(RG/\langle z+1\rangle) \longrightarrow K_2(R/2[\mathbb{Z}_2^2])$$

and that this cokernel is isomorphic to $\mathbb{Z}_2$ under a map induced by the composition

$$\Phi : K_2(R/2[\mathbb{Z}_2^2]) \xrightarrow{\partial} K_1(RG/\langle 1+z\rangle, \langle 1-z\rangle) \longrightarrow$$

$$K_1(RG/\langle (1-z)^2, 1+z\rangle, \langle 1-z\rangle) \overset{T}{\cong} R/2[\mathbb{Z}_2^2] \xrightarrow{\varepsilon} R/2 \xrightarrow{tr} \mathbb{Z}_2 .$$

The definitions of ε and tr can, of course, be found in [13]. We do not need them here. The isomorphism τ is explained below.

Now let $x \in K_1(R[\mathbb{Z}_2^2])$ so that $\varphi^!(x) \in K_1(RG)$. To check that $\varphi^!(x) = 0$ we must lift it back to $y \in K_2(R/2[\mathbb{Z}_2^2])$ through the boundary map ∂_2 of the cartesian square (3.4) and then find $\Phi(y)$. The key observation is the existence of the following commutative diagram

(3.5)

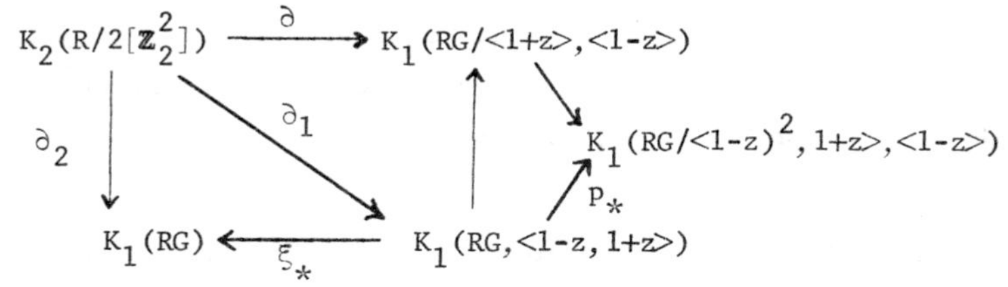

Here the unlabelled maps and p_* are induced by projections, so the right hand triangle commutes. ∂ and ∂_1 are boundary maps in exact sequences for an ideal so the middle triangle commutes. Finally ξ_* is induced by the map ξ making

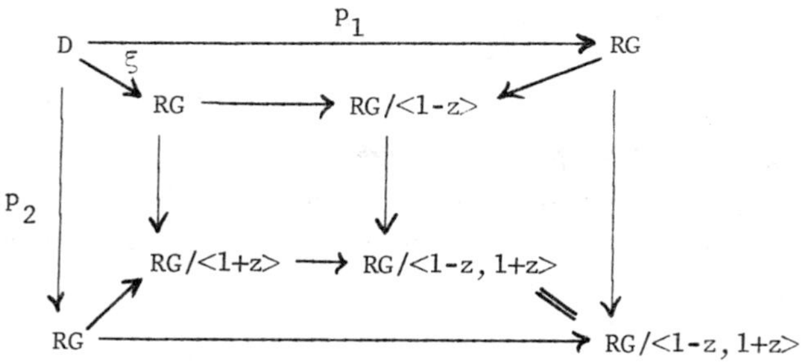

commute. Here both squares are cartesian, and, by definition (see [8]),

$$K_1(RG, \langle 1-z, 1+z\rangle) = \operatorname{Ker}(p_{1*}: K_1(D) \longrightarrow K_1(RG))$$

If $(u,v) \in D$ then there are $a, b \in RG$ such that $v = u + a(1-z) + b(1+z)$ and then $\xi(u,v) = v - b(1+z) = u + a(1-z)$. It easily follows that ξ_* is given by:

(3.6) $\xi_*(I + X(1-z) + Y(1+z)) = I + X(1-z), \; X, Y \in M_r(RG)$

Since $\varphi^!(x) \in \operatorname{Im}(\xi_*\partial_1)$ we see that there exist matrices X_0, Y_0 so that $I + X_0(1-z) + Y_0(1+z)$ is simple and $I + X_0(1-z)$ represents $\varphi^!(x)$. For any other representative of $\varphi^!(x)$ of the form $I + X(1-z)$ there is a simple matrix S so that $I + X(1-z) = S(I + X_0(1-z))$. With $Y = SY_0$

it follows that $I + X(1-z) + Y(1+z)$ is simple. It therefore represents $\partial_1 y$ for some choice of y with $\partial_2 y = \varphi^!(x)$. Since $p_*[I+X(1-z) + Y(1+z)] = [p(I+X(1-z)]$ in $K_1(RG/\langle(1-z)^2, 1+z\rangle, \langle 1-z\rangle)$ we can finish the proof by showing

(3.7) $\varphi^!(x)$ is represented by some $I + X(1-z)$ for which $I + p(X)(1-z) \in RG/\langle(1-z)^2, 1+z\rangle$ represents 0 in $K_1(RG/\langle(1-z)^2, 1+z\rangle, \langle 1-z\rangle)$

By lemma 15 of [13] the latter group is isomorphic to the ideal $\langle 1-z\rangle$ in $RG/\langle(1-z)^2, 1+z\rangle$ under a map taking $I + M(1-z)$ to $\det(I + M(1-z)) - 1$. But since we compute here modulo $(1-z)^2$ this becomes the trace $tr(M)(1-z)$. Thus we just have to represent $\varphi^!(x)$ by some $I + X(1-z)$ for which $tr(X)$ is divisible by $1-z$.

Now let $x = [A]$, $A \in M_r(R(\mathbf{Z}_2^2))$ and let $\widetilde{A},\widetilde{B},\widetilde{C}$ be as in 2.1. Adding to $\widetilde{B}$ a suitable multiple of $(z-1)$ we can make sure that $\widetilde{C}$ is divisible by $(z-1)$ so that $\widetilde{A}\widetilde{B} = I - \widetilde{C}_1(z-1)^2$. It follows that there exists a matrix $\widetilde{D}_1$ so that $\widetilde{B}\widetilde{A} = I - \widetilde{D}_1(z-1)^2$.

Now

$$\begin{pmatrix} \widetilde{B} & -\widetilde{D}_1(z-1)^2 \\ I & \widetilde{A} \end{pmatrix} = \begin{pmatrix} I & \widetilde{B} \\ 0 & I \end{pmatrix}\begin{pmatrix} I & 0 \\ -\widetilde{A} & I \end{pmatrix}\begin{pmatrix} 0 & -I \\ I & 0 \end{pmatrix}$$

is simple. Hence $\varphi^!(x)$ is represented by

$$\begin{pmatrix} I & 0 \\ -z\widetilde{A} & I \end{pmatrix}\begin{pmatrix} \widetilde{B} & -D_1(z-1)^2 \\ I & \widetilde{A} \end{pmatrix}\begin{pmatrix} \widetilde{A} & -\widetilde{C} \\ z-1 & \widetilde{B} \end{pmatrix}$$

which is easily computed and seen to have the desired property.

Proof of proposition 3.1

The basic step is the following

Lemma 3.8 Let $\mathbf{Z} \longrightarrow \pi \longrightarrow \rho \longrightarrow \{1\}$ be an orientable extension and let $j:\pi_2 \longrightarrow \pi$ be the inclusion of a subgroup of finite index. If $\varphi j:\pi_2 \longrightarrow \rho$ is onto then

$$(\varphi j)^! = j^*\varphi : K_1(\mathbf{Z}\rho) \longrightarrow K_1(\mathbf{Z}\pi_2)$$

where j^* is the usual transfer map.

Proof The kernel of φj is generated by the element $t_2 = t^m$ where m is the index of π_2 in π. Let $[A] \in K_1(\mathbf{Z}\rho)$ and let $\widetilde{A},\widetilde{B},\widetilde{C} \in Mr(\mathbf{Z}\pi_2)$ have $\varphi j(\widetilde{A}) = A$, $\widetilde{A}\widetilde{B} = I - \widetilde{C}(t_2-1)$. Then also $\widetilde{A}\widetilde{B} = I - \widetilde{C}(1+t+\ldots+t^{m-1})(t-1)$ so $\varphi^!([A])$ is represented by the matrix

$$M = \begin{pmatrix} \widetilde{A} & -\widetilde{C}(1+t+\ldots+t^{m-1}) \\ t-1 & \widetilde{B} \end{pmatrix}$$

Let $e_1,\ldots,e_{2r}$ be the standard basis for $(\mathbf{Z}\pi)^{2r}$ and let $\mu\colon(\mathbf{Z}\pi)^{2r} \longrightarrow (\mathbf{Z}\pi)^{2r}$ be the linear map having matrix M, with respect to $\{e_i\}$ (we follow Cohen's convention, [5], i.e., we have left modules and the ith row of the matrix specifices $\mu(e_i)$). Then $j^*\varphi^!([A])$ is represented by the matrix M_2 of μ with respect to some $\mathbf{Z}\pi_2$-basis of $(\mathbf{Z}\pi)^{2r}$. As a basis we choose

$$e_1,e_2,\ldots,e_{2r},te_1,te_2,\ldots,te_{2r},\ldots,t^{m-1}e_1,t^{m-1}e_2,\ldots,t^{m-1}e_{2r}$$

We now take $m = 3$. The reader can easily supply the dots needed to cover the general case. We have

$$M_2 = \begin{pmatrix} \widetilde{A} & 0 & 0 & -\widetilde{C} & -\widetilde{C} & -\widetilde{C} \\ 0 & \widetilde{A} & 0 & -t_2\widetilde{C} & -\widetilde{C} & -\widetilde{C} \\ 0 & 0 & \widetilde{A} & -t_2\widetilde{C} & -t_2\widetilde{C} & -\widetilde{C} \\ -1 & 1 & 0 & \widetilde{B} & 0 & 0 \\ 0 & -1 & 1 & 0 & \widetilde{B} & 0 \\ t_2 & 0 & -1 & 0 & 0 & \widetilde{B} \end{pmatrix}$$

Subtract row nr. $(i-1)$ from row nr. i (for $i = m, m-1,\ldots,2$) and add rows nr. $m+1, m+2,\ldots,2m-1$ to row nr. $2m$ to get

$$\begin{pmatrix} \widetilde{A} & 0 & 0 & -\widetilde{C} & -\widetilde{C} & -\widetilde{C} \\ -\widetilde{A} & \widetilde{A} & 0 & (1-t_2)\widetilde{C} & 0 & 0 \\ 0 & -\widetilde{A} & \widetilde{A} & 0 & (1-t_2)\widetilde{C} & 0 \\ -1 & 1 & 0 & \widetilde{B} & 0 & 0 \\ 0 & -1 & 1 & 0 & \widetilde{B} & 0 \\ t_2-1 & 0 & 0 & \widetilde{B} & \widetilde{B} & \widetilde{B} \end{pmatrix}$$

Subtract column nr. $2m$ from columns nr. $m+1, m+2,\ldots,2m-1$ and add column nr. i to column nr. $(i-1)$ (for $i = m,m-1,\ldots,2$) to get

$$\begin{pmatrix} \widetilde{A} & 0 & 0 & 0 & 0 & -\widetilde{C} \\ 0 & \widetilde{A} & 0 & (1-t_2)\widetilde{C} & 0 & 0 \\ 0 & 0 & \widetilde{A} & 0 & (1-t_2)\widetilde{C} & 0 \\ 0 & 1 & 0 & \widetilde{B} & 0 & 0 \\ 0 & 0 & 1 & 0 & \widetilde{B} & 0 \\ t_2-1 & 0 & 0 & 0 & 0 & \widetilde{B} \end{pmatrix}$$

Subtract (column nr. i) $\times \tilde{B}$ from column nr. m+i-1 (for i = 2,3,...,m), and add $A \times$ (column nr. m+i-1) to column nr. i (for i = 2,3,...,m) to get to

$$\begin{pmatrix} \tilde{A} & 0 & 0 & 0 & 0 & -\tilde{C} \\ 0 & 0 & 0 & -1 & 0 & 0 \\ 0 & 0 & 0 & 0 & -1 & 0 \\ 0 & 1 & 0 & 0 & 0 & 0 \\ 0 & 0 & 1 & 0 & 0 & 0 \\ t_2-1 & 0 & 0 & 0 & 0 & \tilde{B} \end{pmatrix}$$

Finally note that in $K_1(\mathbf{Z}\pi_2)$ this represents the same element as does

$$\begin{pmatrix} \tilde{A} & -\tilde{C} \\ t_2-1 & \tilde{B} \end{pmatrix} ,$$

i.e., it represents $(\varphi j)^{!}([A])$.

With the lemma established one proves proposition 3.1 as follows: Let the order of ρ be m . Then $mH^2(\rho;\mathbf{Z}) = 0$ so there is a commutative diagram of extensions

$$\begin{array}{ccccccccc} 0 & \longrightarrow & \mathbf{Z} & \longrightarrow & \pi & \overset{\varphi}{\longrightarrow} & \rho & \longrightarrow & \{1\} \\ & & \downarrow m & & \downarrow j & & \| & & \\ 0 & \longrightarrow & \mathbf{Z} & \longrightarrow & \rho \times \mathbf{Z} & \underset{\varphi_1}{\longrightarrow} & \rho & \longrightarrow & \{1\} \end{array}$$

Since φ_1 is split epic $\varphi_1^{!} = 0$ (lift $A \in Gl_r(\mathbf{Z}\rho)$ to an isomorphism $\tilde{A}$, take $\tilde{B} = \tilde{A}^{-1}$, $\tilde{C} = 0$). By the lemma it follows that $\varphi^{!} = 0$.

REFERENCES

1. R. Alperin, R. K. Dennis and M. Stein, The non-triviality of $SK_1(\mathbf{Z}\pi)$, Lecture Notes in Math., Vol. 353, Springer Verlag (1973), 1-7

2. D. R. Anderson, The Whitehead torsion of a fiber homotopy equivalence, Michigan Math. J. 21 (1974), 171-180

3. D. R. Anderson, The Whitehead torsion of the total space of a fiber bundle, Topology 11 (1972), 179-194

4. D. R. Anderson, The obstruction to finiteness of the total space of a flat bundle, Amer. J. Math. 95 (1973), 281-293

5. M. M. Cohen, A course in simple-homotopy theory, Springer Verlag, New York (1973)

6. K. Ehrlich, Fibrations and a transfer map in algebraic K-theory, J. Pure Appl. Alg. 14 (1979), 131-136

7. K. Ehrlich, Ph.D. thesis, Cornell U. 1977.

8. J. W. Milnor, Introduction to Algebraic K-theory, Ann. of Math. Studies, Vol. 72, Princeton, 1971

9. H. J. Munkholm, Transfer on algebraic K-theory and Whitehead torsion for PL-fibrations, J.Pure Appl. Alg. 20 (1981) 195-225

10. H. J. Munkholm and E. K. Pedersen, On the Wall finiteness obstruction for the total space of certain fibrations, Trans. Amer. Math. Soc. 261 (1980), 529-545

11. H. J. Munkholm and E. K. Pedersen, Whitehead transfers for S^1-bundles. An algebraic description, to appear Comm. Math. Helv.

12. H. J. Munkholm and E. K. Pedersen, Wall group transfers for S^1-bundles. An algebraic description, preprint from Odense University, Denmark, 1981

13. R. Oliver, SK_1 for finite group rings, I., Inv. Math. 57 (1980), 183-204

14. E. K. Pedersen, Universal geometric examples for transfer maps in algebraic K- and L-theory, to appear J. Pure Appl. Alg. 1981

15. E. K. Pedersen and L. Taylor, The Wall finiteness obstruction for a fibration, Amer. J. Math. 100 (1978), 887-896

16. A. Ranicki, Exact sequences in the algebraic theory of surgery, Princeton University Press, 1981

17. M. Stein, Whitehead groups of finite groups, Bull. Amer. Math. Soc. 84 (1978), 201-212

18. C. T. C. Wall, Surgery on compact manifolds, Academic Press, London 1970

DEPARTMENT OF MATHEMATICS
ODENSE UNIVERSITY
DK 5230 ODENSE
DENMARK

and

INSTITUTE FOR ADVANCED STUDY
PRINCETON, N.J. 08540
U.S.A.

Current Address:
Department of Mathematics
Odense University
DK 5230 Odense
Denmark

Note (added in proof) Concerning the map $p^*_{K_o}$ see also the first named author's joint note with A. Ranicki in these proceedings.

Canadian Mathematical Society
Conference Proceedings
Volume 2, Part 2 (1982)

THE PROJECTIVE CLASS GROUP TRANSFER INDUCED BY AN S^1-BUNDLE

Hans J. Munkholm and Andrew A. Ranicki[1]

Introduction

This note gives an explicit algebraic description of the geometric transfer map induced in the (reduced) projective class groups by an S^1-bundle $S^1 \longrightarrow E \xrightarrow{p} B$

$$p^*_{K_0} : \widetilde{K}_0(\mathbb{Z}[\rho]) \longrightarrow \widetilde{K}_0(\mathbb{Z}[\pi])$$

with $\pi = \pi_1(E)$, $\rho = \pi_1(B)$. This is the transfer map (1.4) of the preceding paper, Munkholm and Pedersen [4], to which we refer for terminology and background material. In particular,
$t \in \pi$ is the canonical generator of the cyclic group $\ker(p_*:\pi \twoheadrightarrow \rho)$ represented by the inclusion $S^1 \longrightarrow E$ of a fibre,
$\phi : \mathbb{Z}[\pi] \twoheadrightarrow \mathbb{Z}[\pi]/(t-1) = \mathbb{Z}[\rho]$; $r \longmapsto \bar{r}$ is the projection of fundamental group rings induced by $p_*:\pi \twoheadrightarrow \rho$, and
$\mathbb{Z}[\pi] \xrightarrow{\sim} \mathbb{Z}[\pi]$; $r \longmapsto r^t$ is a ring automorphism determined by the orientation class $w_1(p) \in H^1(B;\mathbb{Z}_2)$ such that $(t-1)r = r^t(t-1)$. In the orientable case $w_1(p) = 0$, $t \in \pi$ is central and $r^t = r$.

Our main results are:

Proposition 2.1 The projection of rings $\phi:\mathbb{Z}[\pi] \twoheadrightarrow \mathbb{Z}[\rho]$ gives rise to an algebraic transfer map in the projective class groups

$$\phi^!_0 : K_0(\mathbb{Z}[\rho]) \longrightarrow K_0(\mathbb{Z}[\pi]) \; ; \; [\mathrm{im}(\bar{X})] \longmapsto [\mathrm{im}(X^!)] - [\mathbb{Z}[\pi]^n] \, .$$

Here $\bar{X} \in M_n(\mathbb{Z}[\rho])$ is a projection (i.e. an $n \times n$ matrix $\bar{X}$ with entries in $\mathbb{Z}[\rho]$ such that $\bar{X}^2 = \bar{X}$) and $X^! \in M_{2n}(\mathbb{Z}[\pi])$ is the projection defined by

$$X^! = \begin{pmatrix} X & Y \\ t-1 & 1-X^t \end{pmatrix} \in M_{2n}(\mathbb{Z}[\pi])$$

for any $X, Y \in M_n(\mathbb{Z}[\pi])$ such that $\phi(X) = \bar{X}$, $X(1-X) = Y(t-1)$, $XY = YX^t$.
[]

1980 Mathematics Subject Classification. 57Q12, 18F25.
[1]Partially supported by NSF grants.

© 1982 American Mathematical Society
0731-1036/82/0000-0481/$07.00

Proposition 4.1 The algebraic and geometric transfer maps in the reduced projective class groups coincide, that is if B,E are finitely dominated CW complexes

$$\widetilde{\Phi}_0^! = p^*_{K_0} : \widetilde{K}_0(\mathbb{Z}[\rho]) \longrightarrow \widetilde{K}_0(\mathbb{Z}[\pi]) ;$$
$$[B] \longmapsto \widetilde{\Phi}_0^!([B]) = p^*_{K_0}([B]) = [E]$$

with [B],[E] the Wall finiteness obstructions.

[]

We should like to thank the Nassau Inn, Princeton for the hospitality of its back steps.

*

Contents

§1. Rings with pseudostructure

Let R be an associative ring with 1. We shall be using the following conventions regarding matrices and morphisms over R.

Given (left) R-modules M,N let $\mathrm{Hom}_R(M,N)$ denote the additive group of R-module morphisms

$$f : M \longrightarrow N \; ; \; x \longmapsto f(x) \; .$$

For $m,n \geqslant 1$ let $M_{m,n}(R)$ be the additive group of $m\times n$ matrices $X = (x_{ij})$ $(1 \leqslant i \leqslant m \, , \, 1 \leqslant j \leqslant n)$ with entries $x_{ij} \in R$, and use the isomorphism of abelian groups

$$M_{m,n}(R) \xrightarrow{\sim} \mathrm{Hom}_R(R^m, R^n) \; ;$$
$$X = (x_{ij}) \longmapsto (f:(r_1, r_2, \ldots, r_m) \longmapsto (\sum_{i=1}^{m} r_i x_{i1}, \sum_{i=1}^{m} r_i x_{i2}, \ldots, \sum_{i=1}^{m} r_i x_{in}))$$

to identify

$$M_{m,n}(R) = \mathrm{Hom}_R(R^m, R^n) \; .$$

If the R-module morphisms $f \in \mathrm{Hom}_R(R^m,R^n)$, $g \in \mathrm{Hom}_R(R^n,R^p)$ have matrices $X = (x_{ij}) \in M_{m,n}(R)$, $Y = (y_{jk}) \in M_{n,p}(R)$ the composite R-module morphism

$$gf : R^m \xrightarrow{f} R^n \xrightarrow{g} R^p \; ; \; r \longmapsto g(f(r))$$

has the product matrix

$$XY = (\sum_{j=1}^{n} x_{ij} y_{jk}) \in M_{m,p}(R) \; .$$

The $n \times n$ matrix ring $M_n(R) = M_{n,n}(R)$ is thus identified with the endomorphism ring $\mathrm{Hom}_R(R^n,R^n)$ of the f.g. free R-module R^n of rank n, as usual.

A <u>projection over R</u> is a matrix $X \in M_n(R)$ such that

$$X(1-X) = 0 \in M_n(R) \; ,$$

so that $\mathrm{im}(X) \subseteq R^n$ is a f.g. projective R-module with

$$\mathrm{im}(X) \oplus \mathrm{im}(1-X) = R^n$$

and $\mathrm{im}(1-X)$ is a f.g. projective inverse of $\mathrm{im}(X)$. Let

$$P_n(R) = \{X \in M_n(R) \,|\, X(1-X) = 0\} \subseteq M_n(R)$$

denote the subset of $M_n(R)$ consisting of projections. Every f.g. projective R-module P is isomorphic to $\mathrm{im}(X)$ for some $X \in P_n(R)$.

A <u>pseudostructure</u> $\phi = (\alpha, t)$ on the ring R consists of an automorphism

$$\alpha : R \xrightarrow{\sim} R \; ; \; r \longmapsto r^t$$

and an element $t \in R$ such that

$$t^t = t \; , \; (t-1)r = r^t(t-1) \; .$$

Let ϕ also denote the projection onto the quotient of R by the two-sided principal ideal $(t-1) \triangleleft R$

$$\phi : R \twoheadrightarrow \bar{R} = R/(t-1) \; ; \; r \longmapsto \bar{r} \; .$$

An S^1-bundle $S^1 \longrightarrow E \xrightarrow{p} B$ with $p_* = \phi : \pi_1(E) = \pi \twoheadrightarrow \pi_1(B) = \rho$ determines a pseudostructure $\phi = (\alpha, t)$ on $R = \mathbb{Z}[\pi]$ with $\bar{R} = \mathbb{Z}[\rho]$ (cf. Munkholm and Pedersen [3],[4]).

Let then (R, ϕ) be a ring R with pseudostructure $\phi = (\alpha, t)$.

A <u>pseudoprojection over (R,ϕ)</u> is a pair of matrices over R

$$(X, Y) \in M_n(R) \times M_n(R)$$

such that

$$X(1-X) = Y(t-1) \; , \; XY = YX^t \in M_n(R) \; ,$$

where $X^t = \alpha(X) = (x^t_{ij}) \in M_n(R)$. The pseudoprojection (X,Y) gives rise to a projection over $\bar{R}$

$$\bar{X} \in P_n(\bar{R})$$

with $\bar{X} = \phi(X) = (\bar{x}_{ij}) \in M_n(\bar{R})$, and also to a projection over R

$$X^! = \begin{pmatrix} X & Y \\ t-1 & 1-X^t \end{pmatrix} \in P_{2n}(R) \; .$$

Let

$$P_n(R,\phi) = \{(X,Y) \in M_n(R) \times M_n(R) \mid X(1-X) = Y(t-1) \, , \, XY = YX^t\}$$

denote the subset of $M_n(R) \times M_n(R)$ consisting of the pseudoprojections over (R, ϕ).

<u>Proposition 1.1</u> Every projection $\bar{X} \in P_n(\bar{R})$ over $\bar{R}$ lifts to a pseudoprojection $(X,Y) \in P_n(R,\phi)$ (non-uniquely), with $\phi(X) = \bar{X}$.

<u>Proof</u>: Every matrix $\bar{X} \in M_n(\bar{R})$ lifts to some $X \in M_n(R)$, with any two such lifts X_1, X_2 differing by

$$X_1 - X_2 = W(t-1) \in M_n(R)$$

for some $W \in M_n(R)$. Thus if $X \in M_n(R)$ is a lift of a projection $\bar{X} \in P_n(\bar{R})$ there exists $W \in M_n(R)$ such that

$$X(1-X) = W(t-1) \in M_n(R) \; .$$

Define the matrix

$$Z = \begin{pmatrix} X & W \\ t-1 & 1-X^t \end{pmatrix} \in M_{2n}(R) .$$

Now

$$Z(1-Z) = \begin{pmatrix} 0 & WX^t - XW \\ 0 & 0 \end{pmatrix} \in M_{2n}(R) ,$$

so that $(Z(1-Z))^2 = 0$ and

$$Z^2 + (1-Z)^2 = 1 - 2Z(1-Z) \in M_{2n}(R)$$

is invertible, with inverse

$$(Z^2 + (1-Z)^2)^{-1} = 1 + 2Z(1-Z) \in GL_{2n}(R) ,$$

so that there is defined a projection

$$X^! = (Z^2 + (1-Z)^2)^{-1} Z^2 \in P_{2n}(R) .$$

(The principal ideal $(Z(1-Z))$ of the matrix ring $M_{2n}(R)$ is nilpotent, and $X^! \in P_{2n}(R) \subset M_{2n}(R)$ is an idempotent (= projection) lifting the idempotent $[Z] \in M_{2n}(R)/(Z(1-Z))$ - cf. Bass [0,III.2.10], Swan [9,5.17]). Substituting the relation $Z^4 = 2Z^3 - Z^2$ we have

$$\begin{aligned} X^! &= (1 + 2Z(1-Z))Z^2 \\ &= (1 + 2Z)Z^2 - 2(2Z^3 - Z^2) \\ &= 3Z^2 - 2Z^3 \in P_{2n}(R) , \end{aligned}$$

with

$$\begin{aligned} X^! - Z &= (2Z-1)Z(1-Z) \\ &= \begin{pmatrix} 2X-1 & 2W \\ 2t-2 & 1-2X^t \end{pmatrix} \begin{pmatrix} 0 & WX^t - XW \\ 0 & 0 \end{pmatrix} \\ &= \begin{pmatrix} 0 & (2X-1)(WX^t - XW) \\ 0 & 0 \end{pmatrix} \in M_{2n}(R) . \end{aligned}$$

Defining

$$Y = W + (2X - 1)(WX^t - XW) \in M_n(R) ,$$

we have

$$X^! = \begin{pmatrix} X & Y \\ t-1 & 1-X^t \end{pmatrix} \in P_{2n}(R)$$

with $\phi(X) = \overline{X}$, $X(1-X) = Y(t-1)$, $XY = YX^t$. The projection $\overline{X} \in P_n(\overline{R})$ has been lifted to a pseudoprojection $(X,Y) \in P_n(R,\phi)$.

[]

Given an $\bar{R}$-module $\bar{M}$ let $\phi^!\bar{M}$ be the R-module with the same additive group as $\bar{M}$ and

$$R \times \phi^!\bar{M} \longrightarrow \phi^!\bar{M} \ ; \ (r,\bar{x}) \longmapsto \bar{r}\bar{x} \ .$$

An $\bar{R}$-module morphism $\bar{f} \in \mathrm{Hom}_{\bar{R}}(\bar{M},\bar{N})$ also defines an R-module morphism

$$\phi^! f : \phi^!\bar{M} \longrightarrow \phi^!\bar{N} \ ; \ \bar{x} \longmapsto \bar{f}(\bar{x}) \ .$$

Given a pseudoprojection $(X,Y) \in P_n(R,\phi)$ define the f.g. projective $\bar{R}$-module $\bar{P} = \mathrm{im}(\bar{X})$, and define the associated <u>pseudoresolution</u> of the restricted R-module $\phi^!\bar{P}$ to be the 1-dimensional f.g. projective R-module chain complex $C^!$ with

$$d_{C^!} = \begin{bmatrix} 1-X \\ 1-t \end{bmatrix} : C^!_1 = \mathrm{coker}(X^! = \begin{pmatrix} X & Y \\ t-1 & 1-X^t \end{pmatrix} : R^n \oplus R^n \longrightarrow R^n \oplus R^n)$$
$$\longrightarrow C^!_0 = R^n \ .$$

The homology R-modules of $C^!$ are given by

$$H_0(C^!) = \mathrm{coker}(\begin{bmatrix} 1-X \\ 1-t \end{bmatrix} : R^n \oplus R^n \longrightarrow R^n) = \phi^!\bar{P} \ ,$$

$$H_1(C^!) = \ker((t-1 \ \ 1-X^t) : R^n \longrightarrow R^n \oplus R^n) \ ,$$

and in many respects $C^!$ is like a f.g. projective R-module resolution of $\phi^!\bar{P}$. However, $C^!$ is a genuine resolution of $\phi^!\bar{P}$ (with $H_1(C^!) = 0$) if and only if $t-1 \in R$ is a non-zero-divisor. By Proposition 1.1 there exists a pseudoresolution $C^!$ of $\phi^!\bar{P}$ for any f.g. projective $\bar{R}$-module $\bar{P}$. As for uniqueness, we have:

<u>Proposition 1.2</u> Given pseudoprojections $(X,Y) \in P_n(R,\phi)$, $(X',Y') \in P_{n'}(R,\phi)$ and a morphism of f.g. projective $\bar{R}$-modules

$$\bar{f} : \bar{P} = \mathrm{im}(\bar{X}) \longrightarrow \bar{P}' = \mathrm{im}(\bar{X}')$$

there is defined an R-module chain map of the associated pseudoresolutions

$$f^! : C^! \longrightarrow C'^!$$

uniquely up to chain homotopy, such that

$$(f^!)_* = \phi^!\bar{f} : H_0(C^!) = \phi^!\bar{P} \longrightarrow H_0(C'^!) = \phi^!\bar{P}' \ .$$

The construction of $f^!$ is functorial up to chain homotopy, with

$$1^! = 1 \ , \ (f'f)^! = f'^! f^!$$

up to chain homotopy. In particular, if $\bar{f} \in \mathrm{Hom}_{\bar{R}}(\bar{P},\bar{P}')$ is an isomorphism then $f^! : C^! \longrightarrow C'^!$ is a chain equivalence.

Proof: Let $\overline{F} \in M_{n,n'}(\overline{R})$ be the matrix of the composite $\overline{R}$-module morphism

$$\overline{F} : \overline{R}^n \xrightarrow{\text{projection}} \text{im}(\overline{X}) = \overline{P} \xrightarrow{\overline{f}} \overline{P}' = \text{im}(\overline{X}') \xrightarrow{\text{inclusion}} \overline{R}^{n'} .$$

Choose a lift $F \in M_{n,n'}(R)$ of $\overline{F}$ and define

$$F^! = \begin{pmatrix} XFX' & XFY' - YF^tX'^t \\ 0 & X^tF^tX'^t \end{pmatrix} \in M_{2n,2n'}(R)$$

such that

$$X^!F^! = F^!X'^! \in M_{2n,2n'}(R) .$$

The R-module chain map $f^! : C^! \longrightarrow C'^!$ is defined by

$$\begin{array}{ccccc} C^! : & C_1^! = \text{coker}(X^!) & \xrightarrow{\begin{bmatrix} 1-X \\ 1-t \end{bmatrix}} & C_0^! = R^n \\ f^! \Big\downarrow & \Big\downarrow [F^!] & & \Big\downarrow XFX' \\ C'^! : & C_1'^! = \text{coker}(X'^!) & \xrightarrow{\begin{bmatrix} 1-X' \\ 1-t \end{bmatrix}} & C_0'^! = R^{n'} \end{array} .$$

If $F_1, F_2 \in M_{n,n'}(R)$ are two different lifts of $\overline{F}$ there exists $G \in M_{n,n'}(R)$ such that

$$F_1 - F_2 = G(t-1) \in M_{n,n'}(R) ,$$

and the R-module morphism

$$g^! = [0 \quad XGX'^t] : C_0^! = R^n \longrightarrow C_1'^! = \text{coker}(X'^!)$$

defines a chain homotopy

$$g^! : f_1^! \simeq f_2^! : C^! \longrightarrow C'^!$$

between the corresponding R-module chain maps $f_1^!, f_2^! : C^! \longrightarrow C'^!$.

If $(X,Y) = (X',Y') \in P_n(R,\phi)$ and $\overline{f} = 1 : \overline{P} = \text{im}(\overline{X}) \longrightarrow \overline{P} = \text{im}(\overline{X})$ then $F = X \in M_n(R)$ is a lift of the composite $\overline{R}$-module morphism

$$\overline{F} = \overline{X} : \overline{R}^n \xrightarrow{\text{projection}} \overline{P} \xrightarrow{\text{inclusion}} \overline{R}^n ,$$

so that

$$F^! = \begin{pmatrix} X^3 & 0 \\ 0 & (X^t)^3 \end{pmatrix} \in M_{2n}(R)$$

and the R-module morphism

$$h = [1+X+X^2 \quad 0] : C_0^! = R^n \longrightarrow C_1^! = \text{coker}(X^!)$$

defines a chain homotopy

$$h : f^! \simeq 1 : C^! \longrightarrow C^! .$$

Given pseudoprojections $(X,Y) \in P_n(R,\phi)$, $(X',Y') \in P_{n'}(R,\phi)$, $(X'',Y'') \in P_{n''}(R,\phi)$ and $\bar{R}$-module morphisms

$$\bar{f} : \bar{P} = \operatorname{im}(\bar{X}) \longrightarrow \bar{P}' = \operatorname{im}(\bar{X}') \ , \ \bar{f}' : \bar{P}' = \operatorname{im}(\bar{X}') \longrightarrow \bar{P}'' = \operatorname{im}(\bar{X}'')$$

let

$$\bar{f}'' = \bar{f}'\bar{f} : \bar{P} \xrightarrow{\bar{f}} \bar{P}' \xrightarrow{\bar{f}'} \bar{P}''$$

be the composite $\bar{R}$-module morphism. If $F \in M_{n,n'}(R)$ and $F' \in M_{n',n''}(R)$ are lifts of the composite $\bar{R}$-module morphisms

$$\bar{F} : \bar{R}^n \twoheadrightarrow \bar{P} \xrightarrow{\bar{f}} \bar{P}' \rightarrowtail \bar{R}^{n'}$$

$$\bar{F}' : \bar{R}^{n'} \twoheadrightarrow \bar{P}' \xrightarrow{\bar{f}'} \bar{P}'' \rightarrowtail \bar{R}^{n''}$$

then the product

$$F'' = FX'^2F' \in M_{n,n''}(R)$$

is a lift of the composite $\bar{R}$-module morphism

$$\bar{F}'' : \bar{R}^n \twoheadrightarrow \bar{P} \xrightarrow{\bar{f}''} \bar{P}'' \rightarrowtail \bar{R}^{n''}$$

such that

$$F''^! = F^!F'^! \in M_{2n,2n''}(R) ,$$

and so

$$f''^! = f'^!f^! : C^! \xrightarrow{f^!} C'^! \xrightarrow{f'^!} C''^! .$$

[]

§2. The projective class transfer

Proposition 2.1 Given a ring R with pseudostructure $\phi = (\alpha,t)$ there is defined an algebraic transfer map in the projective class groups

$$\phi_0^! : K_0(\bar{R}) \longrightarrow K_0(R) \ ; \ [\bar{P}] \longmapsto [\operatorname{im}(X^!)] - [R^n] ,$$

sending a f.g. projective $\bar{R}$-module $\bar{P} = \operatorname{im}(\bar{X})$ $(\bar{X} \in P_n(\bar{R}))$ to the projective class $[C^!] = [\operatorname{im}(X^!)] - [R^n] \in K_0(R)$ $((X,Y) \in P_n(R,\phi))$ of any pseudoresolution $C^!$ of $\phi^!\bar{P}$. If $\bar{P}$ is a (stably) f.g. free $\bar{R}$-module then $\phi_0^!([\bar{P}]) = 0 \in K_0(R)$, so that there is also defined an algebraic transfer map in the reduced projective class groups

$$\tilde{\phi}_0^! : \tilde{K}_0(\bar{R}) \longrightarrow \tilde{K}_0(R) \ ; \ [\bar{P}] \longmapsto [\operatorname{im}(X^!)] .$$

Proof: Given a f.g. projective $\overline{R}$-module $\overline{P}$ use Proposition 1.1 to lift a projection $\overline{X} \in P_n(\overline{R})$ such that $\overline{P} = \mathrm{im}(\overline{X})$ to a pseudoprojection $(X,Y) \in P_n(R,\phi)$, and let $C^! : \mathrm{im}(X^!) \longrightarrow R^n$ be the corresponding pseudoresolution of $\phi^!\overline{P}$. Up to R-module isomorphism

$$\mathrm{im}(X^!)\oplus\mathrm{coker}(X^!) = \mathrm{im}(X^!)\oplus\mathrm{im}(1-X^!) = R^{2n} ,$$

so that

$$\begin{aligned}[C^!] &= [R^n] - [\mathrm{coker}(X^!)] \\ &= [\mathrm{im}(X^!)] - [R^n] = \phi_0^!([\overline{P}]) \in K_0(R) .\end{aligned}$$

An element of $K_0(\overline{R})$ is the formal difference $[\overline{P}] - [\overline{P}']$, for some f.g. projective $\overline{R}$-modules $\overline{P} = \mathrm{im}(\overline{X})$, $\overline{P}' = \mathrm{im}(\overline{X}')$. Now $[\overline{P}] - [\overline{P}'] = 0 \in K_0(\overline{R})$ if and only if there exists an $\overline{R}$-module isomorphism $\overline{f} : \overline{P}\oplus\overline{Q} \xrightarrow{\sim} \overline{P}'\oplus\overline{Q}$ for some f.g. projective $\overline{R}$-module $\overline{Q}$, in which case Proposition 1.2 gives a chain equivalence $f^! : C^! \xrightarrow{\sim} C'^!$ of the corresponding pseudoresolutions of $\phi^!\overline{P},\phi^!\overline{P}'$. As the projective class of a chain complex is a chain homotopy invariant it follows that

$$\phi_0^!([\overline{P}] - [\overline{P}']) = [C^!] - [C'^!] = 0 \in K_0(R) ,$$

and so $\phi_0^! : K_0(\overline{R}) \longrightarrow K_0(R)$ is well-defined.

For $\overline{P} = \overline{R}^n$ take $\overline{X} = 1 \in P_n(\overline{R})$, $(X,Y) = (1,0) \in P_n(R,\phi)$, so that the projection

$$X^! = \begin{pmatrix} 1 & 0 \\ t-1 & 0 \end{pmatrix} : R^n\oplus R^n \longrightarrow R^n\oplus R^n$$

has $\mathrm{im}(X^!) \cong R^n$ and so

$$\phi_0^!([\overline{R}^n]) = [R^n] - [R^n] = 0 \in K_0(R) .$$

Thus $\widetilde{\phi}_0^! : \widetilde{K}_0(\overline{R}) \longrightarrow \widetilde{K}_0(R)$ is also well-defined.

[]

The original algebraic description in terms of matrices of the Whitehead group S^1-bundle transfer map

$$p^*_{Wh} = \widetilde{\phi}_1^! : Wh(\rho) \longrightarrow Wh(\pi)$$

due to Munkholm and Pedersen [3] was reformulated by Ranicki [6,§7.8] in terms of the theory of pseudo chain complexes. We shall now recall this theory, and show how it applies to the projective class group S^1-bundle transfer.

Given an R-module M let M^t denote the R-module with the same additive group and

$$R \times M^t \longrightarrow M^t \ ; \ (r,x) \longmapsto r^{-t}x \ ,$$

where $\alpha^{-1}: R \xrightarrow{\sim} R; r \longmapsto r^{-t}$ is the inverse of the ring automorphism $\alpha: R \xrightarrow{\sim} R; r \longmapsto r^t$ in the pseudostructure $\phi = (\alpha,t)$. An R-module morphism $f \in \mathrm{Hom}_R(M,N)$ also defines an R-module morphism

$$f^t : M^t \longrightarrow N^t \ ; \ x \longmapsto f(x) \ ,$$

such that

$$f(t-1) = (t-1)f^t : M^t \longrightarrow N$$

with $t-1 \in \mathrm{Hom}_R(M^t,M)$ defined by

$$t-1 : M^t \longrightarrow M \ ; \ x \longmapsto tx - x \ .$$

For $M = R^n$ use the R-module isomorphism

$$M^t \xrightarrow{\sim} R^n \ ; \ (r_1,r_2,\ldots,r_n) \longmapsto (r_1^t,r_2^t,\ldots,r_n^t)$$

to identify $M^t = R^n$, so that $t-1 \in \mathrm{Hom}_R(M^t,M)$ has matrix $t-1 \in M_n(R)$. If $f \in \mathrm{Hom}_R(R^m,R^n)$ has matrix $X = (x_{ij}) \in M_{m,n}(R)$ then $f^t \in \mathrm{Hom}_R((R^m)^t,(R^n)^t) = \mathrm{Hom}_R(R^m,R^n)$ has matrix $X^t = (x_{ij}^t) \in M_{m,n}(R)$.

A <u>pseudo chain complex over (R,ϕ)</u> $\mathcal{C} = (C,d,e)$ consists of a collection of R-modules $\{C_r \mid r \geqslant 0\}$ and two collections of R-module morphisms $\{d \in \mathrm{Hom}_R(C_r,C_{r-1}) \mid r \geqslant 1\}$, $\{e \in \mathrm{Hom}_R(C_r,C_{r-2}^t) \mid r \geqslant 2\}$ such that

$$d^2 = (t-1)e : C_r \longrightarrow C_{r-2} \ , \ d^t e = ed : C_r \longrightarrow C_{r-3}^t \ .$$

Note that $\mathcal{C}$ determines an $\bar{R}$-module chain complex $\bar{C}$ with

$$d_{\bar{C}} = 1 \otimes d : \bar{C}_r = \bar{R}\otimes_R C_r \longrightarrow \bar{C}_{r-1} = \bar{R}\otimes_R C_{r-1} \ ; \ a\otimes x \longmapsto a \otimes d(x) \ ,$$

and an R-module chain complex $C^!$ with

$$d_{C^!} = \begin{pmatrix} d & (-)^r e \\ (-)^r(t-1) & d^t \end{pmatrix}$$
$$: C_r^! = C_r \oplus C_{r-1}^t \longrightarrow C_{r-1}^! = C_{r-1}\oplus C_{r-2}^t \ ;$$
$$(x,y) \longmapsto (d(x) + (-)^r(t-1)(y) \ , \ (-)^r e(x) + d^t(y)) \ .$$

Proposition 7.8.8 of Ranicki [6] associates to an S^1-bundle of CW complexes $S^1 \longrightarrow E \xrightarrow{p} B$ with $p_* = \phi : \pi_1(E) = \pi \twoheadrightarrow \pi_1(B) = \rho$ a pseudo chain complex $\mathcal{C}(p) = (C,d,e)$ over $(\mathbb{Z}[\pi],\phi)$ with C_r $(r \geqslant 0)$ the f.g. free $\mathbb{Z}[\pi]$-module of rank the number of r-cells in B, such that the cellular chain complexes of the universal covers $\widetilde{B},\widetilde{E}$ of B,E are given by

$$C(\widetilde{B}) = \overline{C} \ , \ C(\widetilde{E}) = C^! \ .$$

If B is finitely dominated then so is E, and the Wall finiteness obstructions are given by the reduced projective classes

$$[B] = [C(\widetilde{B})] = [\overline{C}] \in \widetilde{K}_0(\mathbb{Z}[\rho]) \ ,$$
$$[E] = [C(\widetilde{E})] = [C^!] \in \widetilde{K}_0(\mathbb{Z}[\pi]) \ .$$

The <u>geometric transfer map</u> $p^*_{K_0} : \widetilde{K}_0(\mathbb{Z}[\rho]) \longrightarrow \widetilde{K}_0(\mathbb{Z}[\pi])$ is defined by

$$p^*_{K_0}([B]) = [E] \in \widetilde{K}_0(\mathbb{Z}[\pi]) \ ,$$

so that it will follow from the identification $p^*_{K_0} = \widetilde{\phi}^!_0$ in §4 below that

$$[C^!] = [C(\widetilde{E})] = [E]$$
$$= p^*_{K_0}([B]) = \widetilde{\phi}^!_0([\overline{C}]) \in \widetilde{K}_0(\mathbb{Z}[\pi]) \ .$$

In Ranicki [8] it will be shown algebraically that for any finitely dominated pseudo chain complex $\mathscr{C} = (C,d,e)$ over a ring with pseudostructure (R,ϕ) the algebraic transfer map $\phi^!_0 : K_0(\overline{R}) \longrightarrow K_0(R)$ sends the projective class $[\overline{C}] \in K_0(\overline{R})$ to

$$\phi^!_0([\overline{C}]) = [C^!] \in K_0(R)$$

(which will give an alternative proof of $p^*_{K_0} = \widetilde{\phi}^!_0$ on setting $R = \mathbb{Z}[\pi]$, $\mathscr{C} = \mathscr{C}(p)$). At any rate, for any pseudoprojection $(X,Y) \in P_n(R,\phi)$ there is defined a finitely dominated pseudo chain complex $\mathscr{C} = (C,d,e)$ over (R,ϕ) with

$$d = \begin{cases} 1-X : C_{2i+1} = R^n \longrightarrow C_{2i} = R^n \\ X : C_{2i+2} = R^n \longrightarrow C_{2i+1} = R^n \end{cases} \quad (i \geqslant 0)$$
$$e = Y : C_j = R^n \longrightarrow C^t_{j-2} = R^n \quad (j \geqslant 2)$$

for which

$$[\overline{C}] = [\mathrm{im}(\overline{X})] \in K_0(\overline{R}) \ ,$$
$$[C^!] = [\mathrm{im}(X^!)] - [R^n] = \phi^!_0([\overline{C}]) \in K_0(R) \ .$$

Note that $C^!$ is an infinite f.g. free R-module chain complex which is chain equivalent to the f.g. projective pseudoresolution $C^!$ of $\phi^!(\mathrm{im}(\overline{X}))$ associated to $(X,Y) \in P_n(R,\phi)$ in §1 above.

In the case when $t-1 \in R$ is a non-zero-divisor (which for a group ring $R = \mathbb{Z}[\pi]$ is equivalent to $t \in \pi$ being of infinite order) $\phi^!\overline{R}$ is an R-module of homological dimension 1, with a f.g. free R-module resolution

$$0 \longrightarrow R \xrightarrow{\ t-1\ } R \xrightarrow{\ \phi\ } \phi^!\overline{R} \longrightarrow 0 \ .$$

If $\overline{P}$ is a f.g. projective $\overline{R}$-module then $\phi^!\overline{P}$ is therefore an R-module of homological dimension 1, with a f.g. projective resolution

$$0 \longrightarrow P_1 \longrightarrow P_0 \longrightarrow \phi^!\overline{P} \longrightarrow 0 .$$

The <u>classical transfer map</u> in the projective class groups is defined by

$$\phi^! : K_0(\overline{R}) \longrightarrow K_0(R) \ ; \ [\overline{P}] \longmapsto [P_0] - [P_1] \ ,$$

and this definition extends by the Bass-Quillen resolution theorem to transfer maps in the higher K-groups

$$\phi^! : K_m(\overline{R}) \longrightarrow K_m(R) \quad (m \geqslant 1) \ .$$

(More generally, the classical methods give transfer maps $\phi^! : K_*(\overline{R}) \longrightarrow K_*(R)$ for any morphism of rings $\phi : R \longrightarrow \overline{R}$ such that $\phi^!\overline{R}$ is an R-module of finite homological dimension).

<u>Proposition 2.2</u> If (R,ϕ) is a ring with pseudostructure such that $t-1 \in R$ is a non-zero-divisor the projective class group transfer map $\phi_0^!$ defined above agrees with the classical transfer map

$$\phi_0^! = \phi^! : K_0(\overline{R}) \longrightarrow K_0(R) \ .$$

<u>Proof</u>: In this case the pseudoresolution $C^!$ of $\phi^!(\mathrm{im}(\overline{X}))$ associated to a pseudoprojection $(X,Y) \in P_n(R,\phi)$ in §1 above is a 1-dimensional f.g. projective R-module resolution of $\phi^!(\mathrm{im}(\overline{X}))$

$$0 \longrightarrow \mathrm{coker}(X^!) \longrightarrow R^n \xrightarrow{\overline{X}\phi} \phi^!(\mathrm{im}(\overline{X})) \longrightarrow 0 \quad ,$$

so that

$$\phi_0^!([\mathrm{im}(\overline{X})]) = [C^!] = \phi^!([\mathrm{im}(\overline{X})]) \in K_0(R) \ .$$

[]

For a group ring $R = \mathbb{Z}[\pi]$ the identification $\widetilde{\phi}_0^! = \widetilde{\phi}^! : \widetilde{K}_0(\mathbb{Z}[\rho]) \longrightarrow \widetilde{K}_0(\mathbb{Z}[\pi])$ given by Proposition 2.2 may also be obtained by combining the identifications $\widetilde{\phi}_0^! = p^*_{K_0}$ of §4 and $p^*_{K_0} = \widetilde{\phi}^!$ of Munkholm and Pedersen [2].

In Proposition 3.2 below the algebraic S^1-bundle transfer map $\phi_1^! : K_1(\overline{R}) \longrightarrow K_1(R)$ of Munkholm and Pedersen [3] in the case when $t-1 \in R$ is a non-zero-divisor will be similarly identified with the classical transfer map $\phi^! : K_1(\overline{R}) \longrightarrow K_1(R)$. It would be interesting to know if the definitions of $\phi_0^!$ and $\phi_1^!$ extend to algebraic transfer maps in the higher K-groups

$$\phi_m^! : K_m(\overline{R}) \longrightarrow K_m(R) \quad (m \geqslant 2)$$

in the case when $t-1 \in R$ is a zero divisor, so that $\phi^!\overline{R}$ is an R-module of infinite homological dimension and the classical methods fail.

§3. The Whitehead torsion transfer

The Whitehead torsion transfer map of Munkholm and Pedersen [3] was defined for any ring with pseudostructure (R,ϕ) to be

$$\phi_1^! : K_1(\overline{R}) \longrightarrow K_1(R) \ ; \ \tau(\overline{X}) \longmapsto \tau\begin{pmatrix} X & -Z \\ t-1 & Y^t \end{pmatrix}$$

with $X \in M_n(R)$ a lift of $\overline{X} \in GL_n(\overline{R})$ and $Y,Z \in M_n(R)$ such that

$$XY = 1 - Z(t-1) \in M_n(R) \ .$$

In Ranicki [6,§7.8] $\phi_1^!(\tau(\overline{X})) \in K_1(R)$ was interpreted as the torsion $\tau(C^!)$ of the based acyclic R-module chain complex

$$C^! : R^n \xrightarrow{(1-t \ \ X^t)} R^n \oplus R^n \xrightarrow{\begin{pmatrix} X \\ t-1 \end{pmatrix}} R^n$$

associated to the pseudo chain complex $\mathscr{C} = (C,d,e)$ with

$$d = X : C_1 = R^n \longrightarrow C_0 = R^n \ , \ C_r = 0 \ (r \geqslant 2) \ , \ e = 0 \ ,$$

for which

$$\tau(\overline{C}) = \tau(\overline{X} : \overline{R}^n \xrightarrow{\sim} \overline{R}^n) \in K_1(\overline{R}) \ .$$

(The identification $\phi_1^!(\tau(\overline{X})) = \tau(C^!) \in K_1(R)$ is immediate from the observation that $\begin{pmatrix} -Z \\ Y^t \end{pmatrix} : R^n \oplus R^n \twoheadrightarrow R^n$ is a splitting map for $(1-t \ \ X^t) : R^n \rightarrowtail R^n \oplus R^n$). It will be shown in Ranicki [8] that for any finite pseudo chain complex $\mathscr{C} = (C,d,e)$ over (R,ϕ) with each C_r $(r \geqslant 0)$ a based f.g. free R-module with $\overline{C}$ (and hence $C^!$) acyclic

$$\phi_1^!(\tau(\overline{C})) = \tau(C^!) \in K_1(R) \ .$$

We shall now interpret $\phi_1^!$ in terms of the pseudoresolution construction $(X,Y) \longmapsto C^!$ of §1.

<u>Proposition 3.1</u> The Whitehead torsion transfer map

$$\phi_1^! : K_1(\overline{R}) \longrightarrow K_1(R)$$

sends the torsion $\tau(\overline{f}) \in K_1(\overline{R})$ of an automorphism $\overline{f} \in \mathrm{Hom}_{\overline{R}}(\overline{P},\overline{P})$ of a f.g. projective $\overline{R}$-module $\overline{P}$ to the torsion

$$\phi_1^!(\tau(\overline{f})) = \tau(f^!) \in K_1(R)$$

of the induced self chain equivalence $f^! : C^! \xrightarrow{\sim} C^!$, with $C^!$ the pseudoresolution of $\phi^!\overline{P}$ associated to any pseudoprojection $(X,Y) \in P_n(R,\phi)$ with $\overline{P} = \mathrm{im}(\overline{X})$.

Proof: Stabilizing $\overline{f}$ by $1 \in \mathrm{Hom}_{\overline{R}}(\mathrm{im}(1-\overline{X}),\mathrm{im}(1-\overline{X}))$ it may be assumed that $\overline{P} = \overline{R}^n$ is a f.g. free $\overline{R}$-module, and $(X,Y) = (1,0) \in P_n(R,\phi)$, so that $C^! : R^n \xrightarrow{1-t} R^n$.

If $\overline{f} \in \mathrm{Aut}_{\overline{R}}(\overline{R}^n,\overline{R}^n)$ has matrix $\overline{X} \in GL_n(\overline{R})$ then

$$\begin{array}{ccccc} C^! & : & R^n & \xrightarrow{1-t} & R^n \\ f^! \downarrow \wr & & \downarrow X^t & & \downarrow X \\ C^! & : & R^n & \xrightarrow{1-t} & R^n \end{array}$$

for any lift $X \in M_n(R)$ of $\overline{X}$, so that

$$\tau(f^!) = \tau(C(f^!) : R^n \xrightarrow{(1-t \;\; X^t)} R^n \oplus R^n \xrightarrow{\begin{pmatrix} X \\ t-1 \end{pmatrix}} R^n)$$

$$= \phi_1^!(\tau(\overline{X})) = \phi_1^!(\tau(\overline{f})) \in K_1(R) \quad .$$

[]

By analogy with Proposition 2.2:

Proposition 3.2 If $t-1 \in R$ is a non-zero-divisor the Whitehead torsion transfer map $\phi_1^!$ agrees with the classical transfer map

$$\phi_1^! = \phi^! : K_1(\overline{R}) \longrightarrow K_1(R) \quad .$$

Proof: Given an automorphism $\overline{f} \in \mathrm{Aut}_{\overline{R}}(\overline{R}^n,\overline{R}^n)$ note that the self chain equivalence $f^! : C^! \xrightarrow{\simeq} C^!$ defined in the proof of Proposition 3.1 is a resolution of the automorphism $\phi^!\overline{f} \in \mathrm{Aut}_R(\phi^!\overline{R}^n,\phi^!\overline{R}^n)$, so that

$$\phi_1^!(\tau(\overline{f})) = \tau(f^!) = \phi^!(\tau(\overline{f})) \in K_1(R) \quad .$$

[]

For a group ring $R = \mathbb{Z}[\pi]$ the identification $\widetilde{\phi}_1^! = \widetilde{\phi}^! : Wh(\rho) \longrightarrow Wh(\pi)$ given by Proposition 3.2 may also be obtained by combining the identifications $\widetilde{\phi}_1^! = p^*_{Wh}$ of Munkholm and Pedersen [3] and $p^*_{Wh} = \widetilde{\phi}^!$ of Munkholm [1].

In §4 we shall make use of the following relation between the projective class group transfer $\phi_0^! : K_0(\overline{R}) \longrightarrow K_0(R)$ for a ring with pseudostructure (R,ϕ), the Whitehead torsion transfer $(\phi\times 1)_1^! : K_1(\overline{R}[z,z^{-1}]) \longrightarrow K_1(R[z,z^{-1}])$ for the polynomial extension ring with pseudostructure $(R[z,z^{-1}],\phi\times 1)$ and the canonical Bass-Heller-Swan injections

$$h_R : K_0(R) \rightarrowtail K_1(R[z,z^{-1}]) \; ; \; [P] \longmapsto \tau(z:P[z,z^{-1}] \xrightarrow{\sim} P[z,z^{-1}])$$

and $h_{\overline{R}} : K_0(\overline{R}) \rightarrowtail K_1(\overline{R}[z,z^{-1}])$ defined similarly.

Proposition 3.3 There is defined a commutative diagram

$$\begin{array}{ccc} K_0(\overline{R}) & \xrightarrow{\phi_0^!} & K_0(R) \\ \downarrow h_{\overline{R}} & & \downarrow h_R \\ K_1(\overline{R}[z,z^{-1}]) & \xrightarrow{(\phi\times 1)_1^!} & K_1(R[z,z^{-1}]) \end{array} .$$

Proof: Given a f.g. projective $\overline{R}$-module $\overline{P}$ let $(X,Y) \in P_n(R,\phi)$ be a pseudoprojection such that $\overline{P} = \mathrm{im}(\overline{X})$, and let $C^!$ be the corresponding pseudoresolution of $\phi^!\overline{P}$. Now

$$\begin{aligned} (\phi\times 1)_1^! h_{\overline{R}}([\overline{P}]) &= (\phi\times 1)_1^!(\tau(z:\overline{P}[z,z^{-1}] \xrightarrow{\sim} \overline{P}[z,z^{-1}])) \\ &= \tau(z:C^![z,z^{-1}] \xrightarrow{\sim} C^![z,z^{-1}]) \quad \text{(by Proposition 3.1)} \\ &= h_R([C^!]) = h_R\phi_0^!([\overline{P}]) \in K_1(R[z,z^{-1}]) , \end{aligned}$$

so that $(\phi\times 1)_1^! h_{\overline{R}} = h_R\phi_0^!$.

[]

§4. The algebraic and geometric transfer maps coincide

Let $S^1 \longrightarrow E \xrightarrow{p} B$ be an S^1-bundle with $p_* = \phi : \pi_1(E) = \pi \twoheadrightarrow \pi_1(B) = \rho$, and let $(R = \mathbb{Z}[\pi],\phi)$ be the corresponding ring with pseudostructure.

Proposition 4.1 The algebraic and geometric transfer maps in the reduced projective class groups coincide, that is

$$\tilde{\phi}_0^! = p^*_{K_0} : \tilde{K}_0(\mathbb{Z}[\rho]) \longrightarrow \tilde{K}_0(\mathbb{Z}[\pi]) .$$

Proof: We offer two proofs, in fact.

i) Given a pseudoprojection $(X,Y) \in P_n(\mathbb{Z}[\pi],\phi)$ and a number $m \geqslant 2$ the proof of Theorem F of Wall [10] gives an S^1-bundle of CW pairs

$$S^1 \longrightarrow (E,F) \xrightarrow{(p,q)} (B,K)$$

with K finite and B finitely dominated, such that $\pi_1(B) = \pi_1(K) = \rho$ and such that the relative pseudo chain complex $\mathcal{E}(p,q) = (C,d,e)$ is given by

$$C_r = \begin{cases} \mathbb{Z}[\pi]^n & \text{if } r \geqslant 2m \\ 0 & \text{if } r \leqslant 2m-1 \end{cases}$$

$$d = \begin{cases} 1-X : C_{2i+1} \longrightarrow C_{2i} \\ X : C_{2i+2} \longrightarrow C_{2i+1} \end{cases} \quad (i \geqslant m)$$

$$e = Y : C_r \longrightarrow C^t_{r-2} \quad (r \geqslant 2m+2) \ .$$

The finiteness obstruction of B (= the reduced projective class of $C(\widetilde{B}) = \bar{C}$) is given by

$$[B] = [\bar{C}] = [\mathrm{im}(\bar{X})] \in \widetilde{K}_0(\mathbb{Z}[\rho]) \ ,$$

and that of E by

$$[E] = [C^!] = [\mathrm{im}(X^!)] \in \widetilde{K}_0(\mathbb{Z}[\pi]) \ ,$$

so that

$$\begin{aligned} p^*_{K_0}([B]) &= [E] = [\mathrm{im}(X^!)] \\ &= \widetilde{\phi}^!_0([\mathrm{im}(\bar{X})]) = \widetilde{\phi}^!_0([B]) \in \widetilde{K}_0(\mathbb{Z}[\pi]) \ . \end{aligned}$$

ii) Consider the commutative diagram preceding Corollary 2.3 of Munkholm and Pedersen [4]

$$\begin{array}{ccc} \widetilde{K}_0(\mathbb{Z}[\pi]) & \xleftarrow{\ \bar{h}_\pi\ } & Wh(\pi\times\mathbb{Z}) \\ \uparrow p^*_{K_0} & & \uparrow (p\times 1)^*_{Wh} = (\widetilde{\phi\times 1})^!_1 \\ \widetilde{K}_0(\mathbb{Z}[\rho]) & \xrightarrow{\ h_\rho\ } & Wh(\rho\times\mathbb{Z}) \end{array}$$

in which $\bar{h}_\pi$ (resp. h_ρ) is the canonical Bass-Heller-Swan surjection (resp. injection). From Proposition 3.3 we have $(\widetilde{\phi\times 1})^!_1 h_\rho = h_\pi \widetilde{\phi}^!_0$, so that

$$\begin{aligned} p^*_{K_0} &= \bar{h}_\pi (\widetilde{\phi\times 1})^!_1 h_\rho \\ &= \bar{h}_\pi h_\pi \widetilde{\phi}^!_0 = \widetilde{\phi}^!_0 : \widetilde{K}_0(\mathbb{Z}[\rho]) \longrightarrow \widetilde{K}_0(\mathbb{Z}[\pi]) \ . \end{aligned}$$

[]

§5. The relative transfer exact sequence

A ring morphism $\phi: R \longrightarrow S$ induces morphisms in the algebraic K-groups

$$\phi_! : K_0(R) \longrightarrow K_0(S) \ ; \ [P] \longmapsto [\phi_! P] \ , \ \phi_! P = S \otimes_R P$$

$$\phi_! : K_1(R) \longrightarrow K_1(S) \ ; \ \tau(X) \longmapsto \tau(\phi(X)) \ , \ X \in GL_n(R)$$

which are related by a change of rings exact sequence

$$K_1(R) \xrightarrow{\ \phi_!\ } K_1(S) \xrightarrow{\ j\ } K_1(\phi_!) \xrightarrow{\ \partial\ } K_0(R) \xrightarrow{\ \phi_!\ } K_0(S)$$

with $K_1(\phi_!)$ the relative K-group of stable isomorphism classes of pairs (P,f) consisting of a f.g. projective R-module P and an S-module isomorphism $f : \phi_! P \xrightarrow{\sim} S^n$, with $(R^n,1) = 0 \in K_1(\phi_!)$ and

$$j : K_1(S) \longrightarrow K_1(\phi_!) \ ; \ \tau(Z) \longmapsto (R^n, Z) \ , \ Z \in GL_n(S)$$

$$\partial : K_1(\phi_!) \longrightarrow K_0(R) \ ; \ (P,f) \longmapsto [P] - [R^n] \ .$$

We shall now obtain an analogous exact sequence for the transfer maps

$$K_1(\bar{R}) \xrightarrow{\phi_1^!} K_1(R) \xrightarrow{j} K_1(\phi^!) \xrightarrow{\partial} K_0(\bar{R}) \xrightarrow{\phi_0^!} K_0(R) \ ,$$

relating the projective class group transfer $\phi_0^!$ of §2 to the Whitehead torsion transfer $\phi_1^!$ of §3.

A <u>base</u> (S,T) for a pseudoprojection $(X,Y) \in P_n(R,\phi)$ is a pair of matrices

$$S = \begin{pmatrix} S_1 \\ S_2 \end{pmatrix} \in M_{2n,m}(R) \quad , \quad T = (T_1 \ T_2) \in M_{m,2n}(R)$$

with $S_1, S_2 \in M_{n,m}(R)$, $T_1, T_2 \in M_{m,n}(R)$ such that

$$ST = X^! \in M_{2n}(R) \quad , \quad TS = 1 \in M_m(R) \ .$$

The factorization of R-module morphisms

$$X^! = \begin{pmatrix} X & Y \\ t-1 & 1-X^t \end{pmatrix} : R^n \oplus R^n \xrightarrow{S = \begin{pmatrix} S_1 \\ S_2 \end{pmatrix}} R^m \xrightarrow{T = (T_1 \ T_2)} R^n \oplus R^n$$

shows that a base (S,T) of (X,Y) determines a base (in the usual sense) of the f.g. projective R-module $\mathrm{im}(X^!) \subseteq R^n \oplus R^n$ consisting of m elements. Conversely, if $\mathrm{im}(X^!)$ is a f.g. free R-module of rank m then a choice of base for $\mathrm{im}(X^!)$ determines a factorization

$$X^! : R^n \oplus R^n \xrightarrow{S} R^m \xrightarrow{T} R^n \oplus R^n$$

with S onto and T one-one; it follows from the identity

$$\begin{aligned} S(TS-1)T &= ST(ST-1) \\ &= X^!(X^!-1) = 0 \in M_{2n}(R) \end{aligned}$$

that $TS = 1 \in M_m(R)$, and so (S,T) defines a base of (X,Y). There is thus a natural one-one correspondence between the bases (S,T) of the pseudoprojection (X,Y) and the bases of the f.g. projective R-module $\mathrm{im}(X^!)$, if any such exist. In dealing with bases of pseudoprojections we shall assume that (R,ϕ) satisfies the following two conditions:

i) f.g. free $\bar{R}$-modules have a well-defined rank,

ii) $\bar{\alpha}^2 : \bar{R} \xrightarrow{\sim} \bar{R} ; \bar{r} \longmapsto (\bar{r}^t)^t$ is an inner automorphism of $\bar{R}$,

in which case $m = n$ for any pseudoprojection base (S,T): by i) $[\bar{R}] \in K_0(\bar{R})$ generates an infinite cyclic subgroup of $K_0(\bar{R})$, and by ii) $\bar{\alpha}_! : K_0(\bar{R}) \xrightarrow{\sim} K_0(\bar{R}) ; [\bar{P}] \longmapsto [\bar{P}^t]$ is an involution of $K_0(\bar{R})$ fixing $[\bar{R}]$, so that if $(S,T) \in M_{2n,m}(R) \times M_{m,2n}(R)$ is a base for the pseudoprojection $(X,Y) \in P_n(R,\phi)$ the f.g. projective $\bar{R}$-module $\bar{P} = \mathrm{im}(1-\bar{X})$ is such that up to $\bar{R}$-module isomorphism

$$\bar{R}^m = \phi_!(\mathrm{im}(X^!)) = \mathrm{im}(\bar{X}) \oplus \bar{P}^t \quad , \quad \bar{R}^n = \mathrm{im}(\bar{X}) \oplus \bar{P} \quad ,$$

and it is clear from the action of $\bar{\alpha}_!$ on the identity

$$[\bar{P}] - [\bar{P}^t] = [\bar{R}^n] - [\bar{R}^m] \in K_0(\bar{R})$$

that $m = n$. In particular, the conditions i) and ii) are satisfied by the group rings with pseudostructure $(R = \mathbb{Z}[\pi], \phi)$ arising in topology.

A <u>based pseudoprojection</u> (X,Y,S,T) is a pseudoprojection $(X,Y) \in P_n(R,\phi)$ together with a base $(S,T) \in M_{2n,n}(R) \times M_{n,2n}(R)$. Given such an object define the associated <u>based pseudoresolution</u> of the R-module $\phi^!(\mathrm{im}(\bar{X}))$ to be the 1-dimensional based f.g. free R-module chain complex

$$D^! : R^n \xrightarrow{S_2} R^n$$

which is chain equivalent to the projective pseudoresolution $C^!$ of $\phi^!(\mathrm{im}(\bar{X}))$ associated to (X,Y) in §1. Explicitly, a chain equivalence $C^! \xrightarrow{\sim} D^!$ is defined by

$$\begin{array}{ccccc} C^! & : & \mathrm{coker}(X^!) & \xrightarrow{\begin{bmatrix} 1-X \\ 1-t \end{bmatrix}} & R^n \\ \downarrow{\scriptstyle \wr} & & \downarrow{\scriptstyle \begin{bmatrix} Y \\ -X^t \end{bmatrix}} & & \downarrow{\scriptstyle XS_1+YS_2} \\ D^! & : & R^n & \xrightarrow{\quad S_2 \quad} & R^n \end{array} \quad .$$

(This is the composite $C^! \xrightarrow{\sim} B^! \xrightarrow{\sim} D^!$ of the chain equivalence

$$\begin{array}{ccccc} C^! & : & \mathrm{coker}(X^!) & \xrightarrow{\begin{bmatrix} 1-X \\ 1-t \end{bmatrix}} & R^n \\ \downarrow{\scriptstyle \wr} & & \downarrow{\scriptstyle \begin{bmatrix} Y \\ -X^t \end{bmatrix}} & & \downarrow{\scriptstyle [X \;\; Y]} \\ B^! & : & R^n & \xrightarrow{[t-1 \;\; 1-X^t]} & \mathrm{im}(X^!) \end{array}$$

(defined for any pseudoprojection (X,Y)) and the chain isomorphism

$$\begin{array}{ccccc}
B^! & : & R^n & \xrightarrow{[t-1 \;\; 1-X^t]} & \mathrm{im}(X^!) \\
\Big\downarrow\simeq & & \Big\downarrow\simeq 1 & & \Big\downarrow\simeq \begin{bmatrix} S_1 \\ S_2 \end{bmatrix} \\
D^! & : & R^n & \xrightarrow{\quad S_2 \quad} & R^n
\end{array} \quad) .$$

A <u>morphism</u> of based pseudoprojections over (R,ϕ)

$$f : (X,Y,S,T) \longrightarrow (X',Y',S',T')$$

is just a morphism of the associated f.g. projective $\overline{R}$-modules

$$\overline{f} : \mathrm{im}(\overline{X}) \longrightarrow \mathrm{im}(\overline{X}') .$$

Replacing the projective pseudoresolutions $C^!, C'^!$ in the construction of Proposition 1.2 by the chain equivalent based pseudoresolutions $D^!, D'^!$ there is obtained an R-module chain map

$$f^! : D^! \longrightarrow D'^!$$

inducing the R-module morphism

$$(f^!)_* = \phi^!\overline{f} : H_0(D^!) = \phi^!(\mathrm{im}(\overline{X})) \longrightarrow H_0(D'^!) = \phi^!(\mathrm{im}(\overline{X}')) ,$$

uniquely up to chain homotopy. More precisely, $f^!$ is defined by

$$\begin{array}{ccccc}
D^! & : & R^n & \xrightarrow{\quad S_2 \quad} & R^n \\
\Big\downarrow f^! & & \Big\downarrow XFX' & & \Big\downarrow TF^!S \\
D'^! & : & R^{n'} & \xrightarrow{\quad S_2' \quad} & R^{n'}
\end{array}$$

with $F \in M_{n,n'}(R)$ the matrix of any R-module morphism $F \in \mathrm{Hom}_R(R^n, R^{n'})$ lifting the composite $\overline{R}$-module morphism

$$\overline{F} : \overline{R}^n \xrightarrow{\text{projection}} \mathrm{im}(\overline{X}) \xrightarrow{\quad \overline{f} \quad} \mathrm{im}(\overline{X}') \xrightarrow{\text{injection}} \overline{R}^{n'}$$

and

$$F^! = \begin{pmatrix} XFX' & XFY' - YF^tX'^t \\ 0 & X^tF^tX'^t \end{pmatrix} \in M_{2n,2n'}(R)$$

as before.

An <u>isomorphism</u> of based pseudoprojections is a morphism

$$f : (X,Y,S,T) \xrightarrow{\;\sim\;} (X',Y',S',T')$$

which is defined by an $\overline{R}$-module isomorphism $\overline{f} \in \mathrm{Hom}_{\overline{R}}(\mathrm{im}(\overline{X}), \mathrm{im}(\overline{X}'))$, in which case $f^! : D^! \xrightarrow{\;\simeq\;} D'^!$ is a chain equivalence of based

R-module chain complexes and the torsion of f is defined by

$$\tau(f) = \tau(f^!:D^! \xrightarrow{\sim} D'^!) \in K_1(R) .$$

In general, the torsion is an invariant of f but not of $\bar{f}$. However, if f is an automorphism (i.e. $(X,Y,S,T) = (X',Y',S',T')$) then the torsion $\tau(\bar{f}:\mathrm{im}(\bar{X}) \xrightarrow{\sim} \mathrm{im}(\bar{X})) \in K_1(\bar{R})$ is defined, and Proposition 3.1 shows that

$$\tau(f) = \tau(f^!) = \phi_1^!(\tau(\bar{f})) \in K_1(R) .$$

An isomorphism $f:(X,Y,S,T) \xrightarrow{\sim} (X',Y',S',T')$ is simple if

$$\tau(f) = 0 \in K_1(R) .$$

Define the relative transfer group $K_1(\phi^!)$ to be the abelian group with one generator for each simple isomorphism class of based pseudoprojections (X,Y,S,T) over (R,ϕ), with relations

$$(X,Y,S,T) + (X',Y',S',T') = (X\oplus X',Y\oplus Y',S\oplus S',T\oplus T') \in K_1(\phi^!) .$$

Proposition 5.1 The relative transfer group $K_1(\phi^!)$ fits into an exact sequence

$$K_1(\bar{R}) \xrightarrow{\phi_1^!} K_1(R) \xrightarrow{j} K_1(\phi^!) \xrightarrow{\partial} K_0(\bar{R}) \xrightarrow{\phi_0^!} K_0(R)$$

with

$$j : K_1(R) \longrightarrow K_1(\phi^!) ;$$

$$\tau(Z) \longmapsto \left(0,0,\begin{pmatrix} 0 \\ Z \end{pmatrix},(Z^{-1}(t-1)\ Z^{-1})\right) \quad (Z \in GL_n(R))$$

$$\partial : K_1(\phi^!) \longrightarrow K_0(\bar{R}) ; \ (X,Y,S,T) \longmapsto [\mathrm{im}(\bar{X})]$$

Proof: If $\bar{P},\bar{Q}$ are f.g. projective $\bar{R}$-modules such that

$$[\bar{P}] - [\bar{Q}] \in \ker(\phi_0^!:K_0(\bar{R}) \longrightarrow K_0(R))$$

let $-\bar{Q}$ be a f.g. projective inverse for $\bar{Q}$, so that $\bar{Q}\oplus -\bar{Q} = \bar{R}^m$ is a f.g. free $\bar{R}$-module, and let $(X,Y) \in P_n(R,\phi)$ be a pseudoprojection such that $\bar{P}\oplus -\bar{Q} = \mathrm{im}(\bar{X})$. Now

$$\begin{aligned}[\mathrm{im}(X^!)] - [R^n] &= \phi_0^!([\mathrm{im}(\bar{X})]) \\ &= \phi_0^!([\bar{P}] - [\bar{Q}] + [\bar{R}^m]) = 0 \in K_0(R) ,\end{aligned}$$

so that $\mathrm{im}(X^!)$ is a stably f.g. free R-module. Stabilizing $\bar{P},\bar{Q}$ if necessary it may be assumed that $\mathrm{im}(X^!)$ is an unstably f.g. free R-module. Choosing a base (S,T) for (X,Y) there is obtained an element $(X,Y,S,T) - \left(1,0,\begin{pmatrix} 1 \\ t-1 \end{pmatrix},(1\ 0)\right) \in K_1(\phi^!)$ $(1 \in GL_m(R))$ such that

$$\begin{aligned}[\overline{P}] - [\overline{Q}] &= [\overline{P}\oplus\overline{Q}] - [\overline{R}^m] \\ &= [\mathrm{im}(\overline{X})] - [\overline{R}^m] \\ &= \partial((X,Y,S,T) - (1,0,\begin{pmatrix}1\\ t-1\end{pmatrix},(1\ 0))) \\ &\in \mathrm{im}(\partial : K_1(\phi^!) \longrightarrow K_0(\overline{R})) ,\end{aligned}$$

verifying exactness at $K_0(\overline{R})$.

If $(X,Y,S,T),(X',Y',S',T')$ are based pseudoprojections such that

$$(X',Y',S',T') - (X,Y,S,T) \in \ker(\partial : K_1(\phi^!) \longrightarrow K_0(\overline{R}))$$

there exists a (stable) isomorphism

$$f : (X,Y,S,T) \xrightarrow{\sim} (X',Y',S',T') .$$

The torsion $\tau(f) \in K_1(R)$ is such that

$$\begin{aligned}(X',Y',S',T') - (X,Y,S,T) &= j(\tau(f)) \\ &\in \mathrm{im}(j : K_1(R) \longrightarrow K_1(\phi^!)) ,\end{aligned}$$

verifying exactness at $K_1(\phi^!)$.

If $Z \in GL_n(R)$ is such that $\tau(Z) \in \ker(j : K_1(R) \longrightarrow K_1(\phi^!))$ there exists a based pseudoprojection (X,Y,S,T) with a simple isomorphism

$$f : (X,Y,S,T)\oplus j\tau(Z) \xrightarrow{\sim} (X,Y,S,T) .$$

The automorphism of based pseudoprojections

$$g : (X,Y,S,T) \xrightarrow{\sim} (X,Y,S,T)$$

defined by the automorphism $\overline{f} \in \mathrm{Hom}_{\overline{R}}(\mathrm{im}(\overline{X}),\mathrm{im}(\overline{X}))$ is such that

$$\begin{aligned}\tau(Z) = \tau(g^!) &= \phi_1^!(\tau(\overline{f})) \\ &\in \mathrm{im}(\phi_1^! : K_1(\overline{R}) \longrightarrow K_1(R)) ,\end{aligned}$$

verifying exactness at $K_1(R)$.

[]

For the group ring with pseudostructure $(R = \mathbb{Z}[\pi],\phi)$ associated to an S^1-bundle $S^1 \longrightarrow E \xrightarrow{p} B$ with $p_* = \phi : \pi_1(E) = \pi \longrightarrow \pi_1(B) = \rho$, $\overline{R} = \mathbb{Z}[\rho]$ there is also defined a reduced version of the exact sequence of Proposition 5.1

$$Wh(\rho) \xrightarrow{\tilde{\phi}_1^!} Wh(\pi) \xrightarrow{\tilde{j}} Wh(\phi^!) \xrightarrow{\tilde{\partial}} \tilde{K}_0(\mathbb{Z}[\rho]) \xrightarrow{\tilde{\phi}_0^!} \tilde{K}_0(\mathbb{Z}[\pi])$$

in the Whitehead and reduced projective class groups, with $Wh(\phi^!)$ defined by

$$\mathrm{Wh}(\phi^!) = K_1(\phi^!)/j(\pm\pi)+(1,0,\begin{pmatrix}1\\t-1\end{pmatrix},(1\ 0)) .$$

See Ranicki [7,§7] for the geometric interpretation of this sequence.

Appendix: Connection with L-theory

We note the following connection between the algebraic K-theory S^1-bundle transfer maps

$$\widetilde{\phi}_0^! : \widetilde{K}_0(\mathbb{Z}[\rho]) \longrightarrow \widetilde{K}_0(\mathbb{Z}[\pi]) \quad , \quad \widetilde{\phi}_1^! : \mathrm{Wh}(\rho) \longrightarrow \mathrm{Wh}(\pi)$$

and the algebraic L-theory S^1-bundle transfer maps of Munkholm and Pedersen [3],[4] and Ranicki [6],[8]

$$\phi_L^! : L_n^X(\rho) \longrightarrow L_{n+1}^{\widetilde{\phi}_m^!(X)}(\pi) \quad (m = 0 \text{ or } 1)$$

which are defined for duality-invariant subgroups $X \subseteq \widetilde{K}_0(\mathbb{Z}[\rho])$ $(m=0)$ and $X \subseteq \mathrm{Wh}(\rho)$ $(m=1)$. (The geometric interpretation of $\phi_L^!$ for $m=1$ in terms of finite surgery obstruction theory extends to $m=0$ using the projective surgery obstruction theory of Pedersen and Ranicki [5]). The duality involutions on the algebraic K-groups are defined by

$$* : \widetilde{K}_0(\mathbb{Z}[\pi]) \longrightarrow \widetilde{K}_0(\mathbb{Z}[\pi]) ; \ [\mathrm{im}(X)] \longmapsto [\mathrm{im}(X^*)]$$

$$* : \mathrm{Wh}(\pi) \longrightarrow \mathrm{Wh}(\pi) ; \ \tau(X) \longmapsto \tau(X^*)$$

$$* : \mathrm{Wh}(\phi^!) \longrightarrow \mathrm{Wh}(\phi^!) ;$$

$$(X,Y,\begin{pmatrix}S_1\\S_2\end{pmatrix},(T_1\ T_2)) \longmapsto -(1-X^*,-t^{-1}Y^*,\begin{pmatrix}-t^{-1}T_2^*\\T_1^*\end{pmatrix},(-tS_2^*\ S_1^*)) ,$$

using the group ring involution

$$* : \mathbb{Z}[\pi] \longrightarrow \mathbb{Z}[\pi] ; \ \sum_{g\in\pi} n_g g \longmapsto \sum_{g\in\pi} w(g)n_g g^{-1} \quad (w = \text{orientation})$$

and the corresponding matrix ring involutions

$$* : M_n(\mathbb{Z}[\pi]) \longrightarrow M_n(\mathbb{Z}[\pi]) ; \ X = (x_{ij}) \longmapsto X^* = (x_{ji}^*) .$$

The maps in the exact sequence of §5

$$\mathrm{Wh}(\rho) \xrightarrow{\widetilde{\phi}_1^!} \mathrm{Wh}(\pi) \xrightarrow{\widetilde{j}} \mathrm{Wh}(\phi^!) \xrightarrow{\widetilde{\partial}} \widetilde{K}_0(\mathbb{Z}[\rho]) \xrightarrow{\widetilde{\phi}_0^!} \widetilde{K}_0(\mathbb{Z}[\pi])$$

are such that

$$\widetilde{\phi}_m^!* = -*\widetilde{\phi}_m^! \ (m = 0,1) \ , \ \widetilde{j}* = *\widetilde{j} \ , \ \widetilde{\partial}* = *\widetilde{\partial} .$$

The short exact sequence of $\mathbb{Z}[\mathbb{Z}_2]$-modules

$$0 \longrightarrow \operatorname{coker}(\tilde{\phi}_1^!) \xrightarrow{\tilde{j}} \mathrm{Wh}(\phi^!) \xrightarrow{\tilde{\partial}} \ker(\tilde{\phi}_0^!) \longrightarrow 0$$

gives rise to connecting maps in the Tate $\mathbb{Z}_2$-cohomology groups

$$\phi_H^! = \delta : \hat{H}^n(\mathbb{Z}_2;\ker(\tilde{\phi}_0^!)) \longrightarrow \hat{H}^{n+1}(\mathbb{Z}_2;\operatorname{coker}(\tilde{\phi}_1^!))$$

which appear in a transfer map of generalized Rothenberg exact sequences

$$\begin{array}{ccccccccc}
\cdots \longrightarrow & L_n^h(\rho) & \longrightarrow & L_n^{\ker\tilde{\phi}_0^!}(\rho) & \longrightarrow & \hat{H}^n(\mathbb{Z}_2;\ker\tilde{\phi}_0^!) & \longrightarrow & L_{n-1}^h(\rho) & \longrightarrow \cdots \\
& \downarrow{\scriptstyle \phi_L^!} & & \downarrow{\scriptstyle \phi_L^!} & & \downarrow{\scriptstyle \phi_H^!} & & \downarrow{\scriptstyle \phi_L^!} & \\
\cdots \longrightarrow & L_{n+1}^{\mathrm{im}\tilde{\phi}_1^!}(\pi) & \longrightarrow & L_{n+1}^h(\pi) & \longrightarrow & \hat{H}^{n+1}(\mathbb{Z}_2;\operatorname{coker}\tilde{\phi}_1^!) & \longrightarrow & L_n^{\mathrm{im}\tilde{\phi}_1^!}(\pi) & \longrightarrow \cdots .
\end{array}$$

In particular, for the trivial S^1-bundle $E = B \times S^1$, $\pi = \rho\times\mathbb{Z}$, $t = z$, $\tilde{\phi}_m^! = 0$ $(m = 0,1)$ and the exact sequence

$$0 \longrightarrow \mathrm{Wh}(\rho\times\mathbb{Z}) \xrightarrow{\tilde{j}} \mathrm{Wh}(\phi^!) \xrightarrow{\tilde{\partial}} \tilde{K}_0(\mathbb{Z}[\rho]) \longrightarrow 0$$

is split by the map

$$\tilde{\Sigma} : \tilde{K}_0(\mathbb{Z}[\rho]) \rightarrowtail \mathrm{Wh}(\phi^!) ;$$

$$[\mathrm{im}(X)] \longmapsto (X, 0, \begin{pmatrix} -X \\ X-z \end{pmatrix}, (z^{-1}(1-X)-1 \;\; -z^{-1}(1-X))) ,$$

which is related to the duality involutions $*$ by

$$\tilde{\Sigma}* - *\tilde{\Sigma} = \tilde{j}h'* : \tilde{K}_0(\mathbb{Z}[\rho]) \longrightarrow \mathrm{Wh}(\phi^!)$$

with

$$h' : \tilde{K}_0(\mathbb{Z}[\rho]) \rightarrowtail \mathrm{Wh}(\rho\times\mathbb{Z}) ; \quad [\mathrm{im}(X)] \longmapsto \tau(-zX+1-X) .$$

The transfer map in this case consists of split injections

$$\begin{array}{ccccccccc}
\cdots \longrightarrow & L_n^h(\rho) & \longrightarrow & L_n^p(\rho) & \longrightarrow & \hat{H}^n(\mathbb{Z}_2;\tilde{K}_0(\mathbb{Z}[\rho])) & \longrightarrow & L_{n-1}^h(\rho) & \longrightarrow \cdots \\
& \downarrow{\scriptstyle \phi_L^!} & & \downarrow{\scriptstyle \phi_L^!} & & \downarrow{\scriptstyle \phi_H^! = \hat{h}'} & & \downarrow{\scriptstyle \phi_L^!} & \\
\cdots \longrightarrow & L_{n+1}^s(\rho\times\mathbb{Z}) & \longrightarrow & L_{n+1}^h(\rho\times\mathbb{Z}) & \longrightarrow & \hat{H}^{n+1}(\mathbb{Z}_2;\mathrm{Wh}(\rho\times\mathbb{Z})) & \longrightarrow & L_n^h(\rho\times\mathbb{Z}) & \longrightarrow \cdots ,
\end{array}$$

although not the standard such injections - see Ranicki [7] for a further discussion.

BIBLIOGRAPHY

[0] H.Bass — Algebraic K-theory, Benjamin (1968)

[1] H.J.Munkholm — Transfer on algebraic K-theory and Whitehead torsion for PL fibrations, J. Pure and App. Alg. 20, 195 - 225 (1981)

[2] ——— and E.K.Pedersen — On the Wall finiteness obstruction for the total space of certain fibrations, Trans. A.M.S. 261, 529 - 545 (1980)

[3] ——— Whitehead transfers for S^1-bundles, an algebraic description, Comm. Math. Helv. 56, 404 - 430 (1981)

[4] ——— Transfers in algebraic K- and L-theory induced by S^1-bundles, these proceedings

[5] E.K.Pedersen and A.A.Ranicki — Projective surgery theory, Topology 19, 234 - 254 (1980)

[6] A.A.Ranicki — Exact sequences in the algebraic theory of surgery, Mathematical Notes 26, Princeton (1981)

[7] ——— Algebraic and geometric splittings of the K- and L-groups of polynomial extensions, preprint

[8] ——— Splitting theorems in the algebraic theory of surgery, in preparation

[9] R.G.Swan — K-theory of finite groups and orders, Springer Lecture Notes 149 (1970)

[10] C.T.C.Wall — Finiteness conditions for CW complexes, Ann. of Maths. 81, 56 - 69 (1965)

INSTITUTE FOR ADVANCED STUDY, PRINCETON (H.J.M. & A.A.R.)
ODENSE UNIVERSITET, DENMARK (H.J.M.)
PRINCETON UNIVERSITY (A.A.R.)

ABCDEFGHIJ–AMS–898765432